The Second World Ocean Assessment

WORLD OCEAN ASSESSMENT II

Volume I

United Nations

Cover photo: Yung-Sen Wu
United Nations World Oceans Day Photo Competition

United Nations publication
Sales no.: E.21.V.5
ISBN: 978-92-1-130422-0
eISBN: 978-92-1-604006-2

Printed at the United Nations, New York

Foreword

by the Secretary-General

The past year has presented unprecedented challenges. The coronavirus disease (COVID-19) pandemic has disrupted lives and livelihoods and exposed our societies' fragility. Sadly, the pandemic is not the only crisis that humanity faces. Climate change and biodiversity loss continue unabated, threatening sustainable development and our viability as a species. These challenges are particularly evident when we look at the state of our planet's life support system, the ocean.

In 2015, the first *World Ocean Assessment* warned that many areas of the ocean had been seriously degraded, the greatest threat to the ocean being the failure to deal with the many pressures caused by human activities. The message in the second *World Ocean Assessment* is that the situation has not improved, with the many benefits that the ocean provides at risk. The *Assessment* advises that, to ensure sustainability, we must work together to improve integrated ocean management, including through joint research, capacity development and the sharing of data, information and technology.

The ocean plays a crucial role in the achievement of the Sustainable Development Goals and the livelihoods of billions of people. We urgently need to change how we interact with it. The forthcoming United Nations Decade of Ocean Science for Sustainable Development and the United Nations Decade on Ecosystem Restoration provide opportunities for us to understand more and to reverse the damage that has already been done. The information in the second *Assessment* can assist in this process, as well as inform relevant intergovernmental conferences scheduled for 2021.

I urge leaders and all stakeholders to heed the warnings in the *Assessment* as we work to conserve and sustainably manage our planet's marine environment. Let us foster not only a green but also a blue recovery from the COVID-19 pandemic.

ANTÓNIO GUTERRES

Summary

In its resolutions 57/141 and 58/240, the General Assembly decided to establish a regular process under the United Nations for global reporting and assessment of the state of the marine environment, including socioeconomic aspects, both current and foreseeable, building on existing regional assessments. In its resolution 71/257, the Assembly recalled that the scope of the first cycle of the Regular Process focused on establishing a baseline and decided that the scope of the second cycle would extend to evaluating trends and identifying gaps. The programme of work for the period 2017–2020 of the second cycle of the Regular Process includes the preparation by the Group of Experts of the Regular Process for Global Reporting and Assessment of the State of the Marine Environment, including Socioeconomic Aspects, of the second *World Ocean Assessment*, building on the baselines established by the First Global Integrated Marine Assessment (first *World Ocean Assessment*). In its resolution 72/73, the Assembly decided that the Group of Experts should proceed on the basis of a single comprehensive assessment. The present document was prepared by the Group of Experts in accordance with those decisions.

Preface

The goal for the General Assembly in creating the Regular Process for the Global Reporting and Assessment of the State of the Marine Environment, including Socioeconomic Aspects, was to ensure a comprehensive overview of the ocean and the relationships between the ocean and humans, covering all environmental, social and economic aspects. Such an overview would serve as a background to the many decisions that must be taken in that field at the international, national and local levels in pursuit of sustainable development. The first *World Ocean Assessment* was completed in 2015 and represents a major step towards that goal.

Inevitably, with such an ambitious goal, not only were some aspects not fully covered in the first output of the Regular Process, but also, as time passed, the assessment that was made up to 2015 needed to be updated. The General Assembly therefore provided for further global integrated marine assessments to record developments from the baseline provided by the first Assessment and, where possible, to show trends. In 2016, it decided that a second comprehensive assessment should be prepared by the end of 2020.

The present volume contains the second *World Ocean Assessment*. It provides more information on aspects of the ocean and its relationships with humans, including separate assessments of the abyssal plains and marine hydrates, and brings together in specific chapters matters that were addressed in different sections of the first Assessment, such as the state of fish species and marine infrastructure.

As with the first Assessment, the production of the present Assessment has been a major task, relying essentially on voluntary efforts of hundreds of experts in many fields, with support from the regular budget of the United Nations. As before, it has been a privilege for the Group of Experts of the Regular Process for Global Reporting and Assessment of the State of the Marine Environment, including Socioeconomic Aspects, to organize, contribute to and finalize the Assessment. Crucial support has again been provided by the Secretariat, including the Division for Ocean Affairs and the Law of the Sea, several international organizations and a number of States Members of the United Nations, as detailed in chapter 2. The Group of Experts is grateful to all those people and institutions but, under the terms of reference and working methods endorsed by the General Assembly, is ultimately responsible for the final text.

The bulk of the text was written before the outbreak of the coronavirus disease (COVID-19) pandemic. Some mention of the effects of that pandemic has been included (for example, in the sections of chapter 8A dealing with fisheries, shipping and tourism), but the full implications of the pandemic on human interactions with the ocean are still being worked out and will need to be explored fully in the third cycle of the Regular Process. Nevertheless, the ocean and the services that it provides will have an important role in the recovery from the pandemic. It is hoped that the information in the present Assessment will help with that process.

As with the first Assessment, the present document contains no policy analysis or recommendations, in line with the guidance endorsed by the General Assembly. It is therefore for national Governments and competent international authorities to decide what action should be taken in the light of the assessments under the Regular Process.

RENISON RUWA and ALAN SIMCOCK
Joint Coordinators of
the Group of Experts of the Regular Process

JÖRN SCHMIDT
Member of the Group of Experts of the
Regular Process, assisting the Joint Coordinators

Contents

Volume I

	Page
Foreword by the Secretary-General	iii
Summary	v
Preface	vii
Part one: Summary	1
Chapter 1: Overall summary	3
Keynote points	5
1. Introduction	5
2. Drivers	6
3. Cleaning up the ocean	7
4. Protecting marine ecosystems	10
5. Understanding of the ocean for sustainable management	13
6. Promoting safety from the ocean	15
7. Sustainable food from the ocean	16
8. Sustainable economic use of the ocean	19
9. Effective implementation of international law as reflected in the United Nations Convention on the Law of the Sea	21
Part two: Introduction	37
Chapter 2: Approach to the assessment	39
Keynote points	41
1. Purpose of the second *World Ocean Assessment*	41
2. Primary audience and framework of the second *World Ocean Assessment*	42
3. Preparation of the second *World Ocean Assessment*	43
4. Terminology	44
5. Acknowledgements	45
References	45
Chapter 3: Scientific understanding of the ocean	47
Keynote points	49
1. Introduction	49
2. Description of changes in data, technology and models since the first *World Ocean Assessment* and their consequences for overall understanding, including socioeconomic consequences	50

	Page
3. Key region-specific changes and consequences	51
4. Outlook for scientific understanding of the ocean	56
5. Key remaining knowledge gaps	56
6. Key remaining capacity-building gaps	57
References	58
Part three: Drivers of changes in the marine environment	63
Chapter 4: Drivers	65
Keynote points	67
1. Introduction	67
2. Drivers of change in the marine environment	69
3. Key region-specific issues or aspects associated with drivers	73
4. Outlook	74
5. Key remaining knowledge and capacity-building gaps	76
References	77
Part four: Current state of the marine environment and its trends	81
Chapter 5: Trends in the physical and chemical state of the ocean	83
Keynote points	85
1. Introduction	85
2. Physical and chemical state of the ocean	87
3. Knowledge gaps	100
4. Summary	101
References	103
Chapter 6: Trends in the biodiversity of the main taxa of marine biota	111
Introduction	113
Chapter 6A: Plankton (phytoplankton, zooplankton, microbes and viruses)	115
Keynote points	117
1. Introduction	117
2. Summary of chapter 6 of the first *World Ocean Assessment*	118
3. Regions targeted in the present *World Ocean Assessment*	119
4. Estimating plankton diversity	120
5. Microbial plankton	121
6. Metazoan zooplankton	124
7. Documented trends	125
8. Outlook	128
References	130

	Page
Chapter 6B: Marine invertebrates	141
Keynote points	143
1. Introduction	143
2. Summary of the situation recorded in the first *World Ocean Assessment*	143
3. Description of environmental changes (2010–2020)	144
4. International and governmental responses	151
5. Achievement of relevant Sustainable Development Goals and contribution to Aichi Biodiversity Target 11	153
6. Key remaining knowledge gaps and capacity-building gaps	153
References	154
Addendum by the Group of Experts of the Regular Process for Global Reporting and Assessment of the State of the Marine Environment, including Socioeconomic Aspects	158
References	159
Chapter 6C: Fishes	161
Keynote points	163
1. Introduction	163
2. Documented change in the state of fish biodiversity	165
3. Consequences of biodiversity change on human communities, economies and well-being	168
4. Key region-specific changes and consequences	169
5. Outlook	171
References	172
Chapter 6D: Marine mammals	177
Keynote points	179
1. Introduction	179
2. Cetaceans	181
3. Pinnipeds	184
4. Sirenians	186
5. Otters and polar bear	186
6. Consequences of changes on human communities, economies and well-being	187
7. Outlook	188
8. Key remaining knowledge gaps	189
9. Key remaining capacity-building gaps	189
References	190

	Page
Chapter 6E: Marine reptiles	195
Keynote points	197
1. Introduction	197
2. Conservation status of marine reptiles	197
3. Regional trends	199
4. Threats	201
5. Economic and social consequences of the changes to marine reptile populations	203
6. Key knowledge and capacity-building gaps	204
References	205
Chapter 6F: Seabirds	211
Keynote points	213
1. Introduction	213
2. Description of environmental changes between 2010 and 2020	214
3. Consequences of changes in seabird populations on human communities, economies and well-being	217
4. Outlook	218
5. Key remaining knowledge gaps	219
6. Key remaining capacity-building gaps	220
References	220
Chapter 6G: Marine plants and macroalgae	225
Keynote points	227
1. Introduction	227
2. Mangroves	227
3. Salt marsh plants	229
4. Seagrasses	230
5. Macroalgae	232
6. Consequences of changes on human communities, economies and well-being	240
7. Key remaining knowledge and capacity-building gaps	241
8. Outlook	241
References	242
Chapter 7: Trends in the state of biodiversity in marine habitats	251
Introduction	253
Chapter 7A: Intertidal zone	255
Keynote points	257
1. Introduction	257
2. Description of the environmental changes between 2010 and 2020	260

	Page
3. Economic and social consequences	261
4. Key region-specific changes and consequences	261
5. Outlook	262
6. Key remaining knowledge gaps	263
7. Key remaining capacity-building gaps	263
References	264
Chapter 7B: Biogenic reefs and sandy, muddy and rocky shore substrates	267
Keynote points	269
1. Introduction	269
2. Documented change in state of biogenic reefs and sandy, muddy and rocky shore substrates	272
3. Consequences of the changes on human communities, economies and well-being	275
4. Key region-specific changes and consequences	277
5. Outlook	279
6. Key remaining knowledge and capacity-building gaps	280
References	281
Chapter 7C: Atoll and island lagoons	289
Keynote points	291
1. Introduction	291
2. Documented changes in state of atolls and island lagoons	292
3. Consequences of the changes on human communities, economies and well-being	295
4. Key region-specific changes and consequences	296
5. Outlook	296
6. Key remaining knowledge gaps	297
7. Key remaining capacity-building gaps	298
References	299
Chapter 7D: Tropical and subtropical coral reefs	305
Keynote points	307
1. Introduction	307
2. Description of environmental changes between 2010 and 2020	308
3. Description of economic and social consequences and/or other economic or social changes	309
4. Key region-specific changes and consequences	310
5. Outlook	312
6. Key remaining knowledge gaps	313
7. Key remaining capacity-building gaps	313
References	314

	Page
Chapter 7E: Cold water corals	321
Keynote points	323
1. Introduction and summary of the first *World Ocean Assessment*	323
2. Description of environmental changes between 2010 and 2020	324
3. Economic and social consequences	329
4. Key region-specific changes and consequences	330
5. Outlook	330
6. Key remaining knowledge gaps	331
7. Key remaining capacity-building gaps	332
References	333
Chapter 7F: Estuaries and deltas	339
Keynote points	341
1. Introduction	341
2. Documented changes in the state of estuaries and deltas	342
3. Consequences of the changes for human communities, economies and well-being	344
4. Key region-specific changes and consequences	345
5. Outlook	346
6. Key remaining knowledge and capacity-building gaps	347
References	348
Chapter 7G: Seagrass meadows	353
Keynote points	355
1. Introduction	355
2. Socioeconomic consequences	356
3. Region-specific changes	357
4. Outlook	358
5. Key remaining knowledge gaps	358
6. Key remaining capacity-building gaps	359
References	362
Chapter 7H: Mangroves	365
Keynote points	367
1. Introduction	367
2. Documented change in state of mangroves between 2010 and 2020	368
3. Consequences of the changes for human communities, economies and well-being	370
4. Key region-specific changes and consequences	372
5. Outlook	373
6. Key remaining knowledge and capacity-building gaps	373
References	374

	Page
Chapter 7I: Salt marshes	381
Keynote points	383
1. Introduction	383
2. Description of the environmental changes between 2010 and 2020	385
3. Consequences of the changes for human communities, economies and well-being	386
4. Key region-specific changes and consequences	386
5. Outlook	387
6. Key remaining knowledge gaps	388
7. Key remaining capacity-building gaps	389
References	389
Chapter 7J: Continental slopes and submarine canyons	395
Keynote points	397
1. Introduction	397
2. Developments in understanding of slopes and canyons	399
3. Ecosystem services and benefits on slopes and in canyons	404
4. Human impacts	405
5. Key remaining knowledge gaps	406
6. Key remaining capacity-building gaps	407
References	408
Chapter 7K: High-latitude ice	421
Keynote points	423
1. Introduction	423
2. Description of the environmental changes between 2010 and 2020	424
3. Economic and social consequences	428
4. Outlook	430
5. Key remaining knowledge and capacity-building gaps	431
References	431
Chapter 7L: Seamounts and pinnacles	437
Keynote points	439
1. Introduction	439
2. Description of changes in knowledge between 2010 and 2020	440
3. Description of economic and social changes	441
4. Key region-specific research in recent years	442
5. Outlook	444
6. Key remaining knowledge gaps	444
7. Key remaining capacity-building gaps	445
References	446

	Page
Chapter 7M: Abyssal plains	453
Keynote points	455
1. Introduction	455
2. Shifting baselines and documenting status and change in abyssal biodiversity	456
3. Major natural and anthropogenic pressures	464
4. Consequences of the changes on human communities, economies and well-being	465
5. Outlook	468
6. Key remaining knowledge gaps	468
References	469
Chapter 7N: Open ocean	477
Keynote points	479
1. Introduction	479
2. Environmental changes in the open ocean since 2010	481
3. Consequences of the changes for human communities, economies and well-being	484
4. Key region-specific changes and consequences	486
5. Outlook	487
6. Key remaining knowledge gaps	488
7. Key remaining capacity-building gaps	488
References	488
Chapter 7O: Ridges, plateaux and trenches	495
Keynote points	497
1. Introduction and summary of the first *World Ocean Assessment*	497
2. Description of the environmental changes between 2010 and 2020	499
3. Description of economic and social changes between 2010 and 2020	502
4. Key region-specific changes and consequences	505
5. Outlook	506
6. Key remaining knowledge gaps	507
7. Key remaining capacity-building gaps	507
References	508
Chapter 7P: Hydrothermal vents and cold seeps	513
Keynote points	515
1. Introduction	515
2. Environmental changes since the first *World Ocean Assessment*	518
3. Economic and social consequences	519
4. Key region-specific changes and consequences	521

	Page
5. Outlook	523
6. Key remaining knowledge gaps	523
7. Key remaining capacity-building gaps	524
References	524
Chapter 7Q: Sargasso Sea	531
Keynote points	533
1. Introduction	533
2. Change of state	534
3. Institutional arrangements	537
4. Consequences of changes	538
5. Outlook	539
References	540

Volume II

Chapter 8: Trends in the state of human society in relation to the ocean	1
Chapter 8A: Coastal communities and maritime industries	3
Keynote points	5
1. Introduction	5
2. Coastal communities	6
3. Capture fisheries, shellfish harvesting and aquaculture	9
4. Shipping	10
5. Seabed mining	14
6. Offshore hydrocarbons	15
7. Tourism and recreation	15
8. Marine genetic resources	20
9. Marine renewable energy	21
10. Desalinization	21
11. Salt production	22
12. Key knowledge and capacity-building gaps	23
13. Outlook	24
References	24
Chapter 8B: Human health as affected by the ocean	31
Keynote points	33
1. Introduction	33
2. General aspects of the relationship between human health and the ocean	33
3. Health of coastal communities relative to inland communities	39
4. Effects of exposure to contaminated seawater	39

Page

5. Problems for human health posed by food from the sea 42
6. Key remaining knowledge and capacity-building gaps 44
7. Outlook 45
References 45

Part five: Trends in pressures on the marine environment 53

Chapter 9: Pressures from changes in climate and atmosphere 55

Keynote points 57
1. Introduction 57
2. Climate pressures: extreme climate events and pressures from changes in ocean physical and chemical properties 58
3. Capacity-building: Global Ocean Acidification Observing Network and Global Ocean Oxygen Network 67
4. Summary 68
References 69

Chapter 10: Changes in nutrient inputs to the marine environment 77

Keynote points 79
1. Introduction 79
2. Situation reported in the first *World Ocean Assessment* 81
3. Global-scale patterns and trends 82
4. Patterns and trends within regions 85
5. Outlook 91
References 92

Chapter 11: Changes in liquid and atmospheric inputs to the marine environment from land (including through groundwater), ships and offshore installations 101

Keynote points 103
1. Introduction 104
2. Situation recorded in the first *World Ocean Assessment* 104
3. Persistent organic pollutants, including run-off from the use of agricultural pesticides 105
4. Metals 112
5. Radioactive substances 122
6. Pharmaceuticals and personal care products 127
7. Atmospheric pollutants (nitrogen oxides, sulfur oxides) 131
8. Hydrocarbons from terrestrial sources, ships and offshore installations, including arrangements for response to spills and discharges 132
9. Other substances used on, and discharged from, offshore installations 134
10. Relationship to the Sustainable Development Goals 135
11. Key remaining knowledge gaps 136

	Page
12. Key remaining capacity-building gaps	138
References	139
Chapter 12: Changes in inputs and distribution of solid waste, other than dredged material, in the marine environment	151
Keynote points	153
1. Activities resulting in marine debris, including plastics, abandoned fishing gear, microparticles and nanoparticles, and estimates of sources from land, ships and offshore installations	153
2. Dumping at sea, including garbage from ships and sewage sludge	171
References	177
Chapter 13: Changes in erosion and sedimentation	185
Keynote points	187
1. Introduction	187
2. Changes in state of coastal erosion and sedimentation	188
3. Consequences of the changes for human communities, economies and well-being	192
4. Key region-specific changes and consequences	193
5. Outlook	195
6. Key remaining knowledge and capacity-building gaps	195
References	196
Chapter 14: Changes in coastal and marine infrastructure	201
Keynote points	203
1. Introduction	203
2. Documented changes in the state of marine and coastal infrastructures	204
3. Consequences of changes for human communities, economies and well-being	207
4. Key region-specific changes and consequences	207
5. Outlook	210
6. Key remaining knowledge and capacity-building gaps	211
References	211
Chapter 15: Changes in capture fisheries and harvesting of wild marine invertebrates	215
Keynote points	217
1. Introduction	217
2. Catch-landing disparities, Sustainable Development Goals and small-scale fisheries	220
3. Invertebrate landings	224
4. Levels of by-catch and side effects	225
5. Post-harvest fish losses	225

Page

6. Potential for fisheries enhancement 225
7. Marine protein and oils in agriculture and aquaculture 225
8. Illegal, unreported or unregulated fishing 226
9. Outlook 227
10. Key knowledge gaps 228
11. Key capacity-building gaps 228
References 228

Chapter 16: Changes in aquaculture 235
Keynote points 237
1. Current status and major improvements 237
2. Aquaculture and the environment 240
3. Aquaculture and society 241
4. Key remaining knowledge gaps 241
5. Key remaining capacity-building gaps 242
6. Outlook 243
References 244

Chapter 17: Changes in seaweed harvesting and use 247
Keynote points 249
1. Introduction 249
2. Documented changes in the state of seaweed production and uses (2012–2017) 250
3. Consequences of changes in seaweed harvesting and use for communities, economies and well-being 253
4. Key region-specific changes and consequences 253
5. Outlook 254
6. Key remaining knowledge and capacity-building gaps 254
References 255

Chapter 18: Changes in seabed mining 257
Keynote points 259
1. Introduction 259
2. Changes in scale and significance of sea floor mining 262
3. Environmental aspects 270
4. Economic and social impacts 273
5. Capacity-building needs 276
References 277

Chapter 19: Changes in hydrocarbon exploration and extraction 281
Keynote points 283
1. Introduction 283

	Page
2. Offshore hydrocarbon exploration, production and decommissioning	285
3. Economic, social, and environmental aspects of offshore hydrocarbon exploration, production and decommissioning	288
4. Key knowledge and capacity-building gaps	290
5. Role of the offshore hydrocarbon industry in facilitating the marine renewable energy industry	291
6. Conclusion	292
References	293
Chapter 20: Trends in inputs of anthropogenic noise into the marine environment	297
Keynote points	299
1. Introduction	299
2. Description of the environmental status	300
3. Description of economic and social consequences and other economic or social changes	308
4. Key region-specific changes and consequences	308
5. Outlook	310
6. Key remaining knowledge gaps	312
7. Key remaining capacity-building gaps	313
References	313
Chapter 21: Developments in renewable energy sources	321
Keynote points	323
1. Introduction	323
2. State of marine renewable energy at the global level	324
3. Potential environmental impacts of marine renewable energy development	329
4. Socioeconomic benefits and impacts from marine renewable energy deployment	332
5. Key remaining knowledge and capacity-building gaps	333
6. Anticipated future trends	335
References	336
Chapter 22: Invasive species	343
Keynote points	345
1. Introduction	345
2. Documented baseline and changes in non-indigenous species	347
3. Consequences for human communities, economies and well-being	348
4. Key region-specific baselines, changes and consequences	350
5. Outlook	354
6. Other	356
References	356

Page

Chapter 23: Developments in the exploration for and use of marine genetic resources . 363
Keynote points 365
1. Introduction. 365
2. Trends between 2010 and 2020 366
3. Economic and social consequences and changes 370
4. Key region-specific developments in knowledge and their consequences 371
5. Capacity-building gaps 371
6. Methodological challenges and future trends 373
7. Marine genetic resources and the Sustainable Development Goals 374
References. 376

Chapter 24: Marine hydrates – a potentially emerging issue 381
Keynote points 383
1. Introduction. 383
2. What are marine hydrates? 383
3. Potential risks from marine methane hydrates 386
4. Marine hydrates as a source of energy 388
5. Key knowledge and capacity-building gaps 390
6. Outlook 390
References. 390

Chapter 25: Cumulative effects. 395
Keynote points 397
1. Introduction. 397
2. Cumulative effects assessments. 398
3. Regional applications of cumulative effects assessments on the marine environment: distribution and approaches 402
4. Outlook 406
References. 413

Part six: Trends in management approaches to the marine environment 421

Chapter 26: Developments in marine spatial planning. 423
Keynote points 425
1. Introduction. 425
2. Types of marine spatial planning 426
3. Marine spatial planning: a step-by-step approach toward ecosystem-based management. 427
4. Tools for marine spatial planning. 428
5. Progress in implementing marine spatial planning 430
References. 436

Page

Chapter 27: Developments in management approaches . . . 441

Keynote points . . . 443

1. Introduction . . . 443
2. Management approaches . . . 444
3. Advances in ocean management approaches . . . 448
4. Management tools to support mitigation of and adaptation to climate change, including building resilience . . . 458
5. Key region-specific issues . . . 460
6. Capacity-building . . . 461
7. Gaps and future perspectives . . . 462
8. Outlook . . . 463

References . . . 465

Chapter 28: Developments in the understanding of overall benefits from the ocean to humans . . . 471

Keynote points . . . 473

1. Introduction . . . 473
2. Benefits and their distribution . . . 477
3. Disbenefits to humans . . . 478
4. Threats to ocean ecosystem services . . . 479
5. Safeguarding ocean benefits through regional and international cooperation and improved implementation of international law as reflected in the United Nations Convention on the Law of the Sea . . . 480

References . . . 483

Annexes . . . 487

Annex I: Original members of the writing teams approved by the Bureau . . . 489

Annex II: Peer reviewers nominated for each chapter . . . 497

Part one
Summary

Chapter 1
Overall summary

Contributors: Maria João Bebianno, Hilconida Calumpong, Sanae Chiba, Karen Evans, Carlos Garcia-Soto, Osman Keh Kamara, Enrique Marschoff, Essam Yassin Mohammed, Henn Ojaveer, Chul Park, Ylenia Randrianarisoa, Renison Ruwa, Jörn Schmidt, Alan Simcock, Anastasia Strati, Joshua T. Tuhumwire, Ca Thanh Vu, Juying Wang and Tymon Zielinski (Group of Experts of the Regular Process for Global Reporting and Assessment of the State of the Marine Environment, including Socioeconomic Aspects).

Keynote points

- Understanding of the ocean continues to improve. Innovations in sensors and autonomous observation platforms have substantially increased observations of the ocean. Regional observation programmes have expanded, with better coordination and integration.
- Some responses for mitigating or reducing pressures and their associated impacts on the ocean have improved since the first *World Ocean Assessment*.[1] They include the expansion and implementation of management frameworks for conserving the marine environment, including the establishment of marine protected areas and, in some regions, improved management of pollution and fisheries. However, many pressures from human activities continue to degrade the ocean, including important habitats, such as mangroves and coral reefs. Pressures include those associated with climate change; unsustainable fishing, including illegal, unreported and unregulated fishing; the introduction of invasive species; atmospheric pollution causing acidification and eutrophication; excessive inputs of nutrients and hazardous substances, including plastics, microplastics and nanoplastics; increasing amounts of anthropogenic noise; and ill-managed coastal development and extraction of natural resources.
- There continues to be a lack of quantification of the impacts of pressures and their cumulative effects. A general failure to achieve the integrated management of human uses of coasts and the ocean is increasing risks to the benefits that people draw from the ocean, including in terms of food safety and security, material provision, human health and well-being, coastal safety and the maintenance of key ecosystem services.
- Improving the management of human uses of the ocean to ensure sustainability will require improved coordination and cooperation to provide capacity-building in regions where it is lacking, innovations in marine technology, the integration of multidisciplinary observation systems, the implementation of integrated management and planning and improved access to, and exchange of, ocean knowledge and technologies.
- The coronavirus disease (COVID-19) pandemic is having a major effect on many human activities carried out in the ocean. The full implications of the pandemic on human interactions with the ocean are still to be fully assessed.

1. Introduction

The ocean covers more than 70 per cent of the surface of the planet and forms 95 per cent of the biosphere. Changes in the ocean drive weather systems that influence both land and marine ecosystems. The ocean and its ecosystems also provide significant benefits to the global community, including climate regulation, coastal protection, food, employment, recreation and cultural well-being. Those benefits depend, to a great extent, on the maintenance of ocean processes, marine biological diversity and related ecosystem services.

1 United Nations, *The First Global Integrated Marine Assessment: World Ocean Assessment I* (Cambridge, Cambridge University Press, 2017).

Concerned by the declining state of the ocean, States Members of the United Nations, through the General Assembly, established the Regular Process for Global Reporting and Assessment of the State of the Marine Environment, including Socioeconomic Aspects. The aim of the Regular Process is to provide an evaluation of the state of the global ocean, the services that it provides and the human activities that influence its state. The first *World Ocean Assessment* was completed in 2015. It concluded that many parts of the ocean had been seriously degraded and that, if the problems that it described were not addressed, they would produce a destructive cycle of degradation in which the ocean could no longer provide many of the benefits on which humans rely. As part of the work identified for the second cycle of the Regular Process, three process-specific technical abstracts were produced, summarizing the content of the first *World Ocean Assessment* in relation to climate change, biodiversity in areas beyond national jurisdiction and Sustainable Development Goal 14, on life below water (see General Assembly resolution 70/1).

The second *World Ocean Assessment* provides an update to the first Assessment, taking into account developments and changes known to have occurred since 2015, and complements it by describing further human interactions with the ocean. Most of the text of the second Assessment was written before the outbreak of the COVID-19 pandemic, and it will take time for the full implications of the pandemic to become apparent. Where appropriate, the second Assessment provides an evaluation of how the developments and changes since the first *World Ocean Assessment* contribute to the achievement of relevant Sustainable Development Goals. Developments and changes relevant to the societal goals of the United Nations Decade of Ocean Science for Sustainable Development (see resolution 72/73) are also indicated.

2. Drivers

In the second *World Ocean Assessment*, drivers are characterized as social, demographic and economic developments in societies, including changes in lifestyles and associated consumption and production patterns that apply pressures to the ocean (chap. 4). Relationships between drivers and pressures (and their impacts) are complex and dynamic, with interlinkages leading to cumulative interactions. The drivers identified in chapter 4 are:

(a) **Population growth and demographic changes.** The world's population continues to grow, although the rate of growth has slowed from the rates observed in the late 1960s, with rates of international migration also increasing. The extent to which an increasing global population places pressure on the marine environment varies, depending on a range of factors, including where and how people live, their consumption patterns and technologies used to produce energy, food and materials, provide transport and manage waste;

(b) **Economic activity.** Economies continue to grow globally, although at a slower pace than reported in the first *World Ocean Assessment*, as a result of weaker manufacturing and trade. As the global population has grown, demand for goods and services has increased, with associated increases in energy consumption and resource use. Many countries have developed, or are developing, strategies for growing ocean-based economies (the blue economy). However, an important constraint on the growth of ocean economies is the current declining health of the ocean and the *pressures being placed on it;*

(c) **Technological advances.** Advances in technology continue to increase efficiency, expand markets and enhance economic growth. Innovations have enabled outcomes for the marine environment that are both positive (such as increasing efficiencies in energy generation) and negative (such as overcapacity in fisheries);

(d) **Changing governance structures and geopolitical instability.** At both the international and national levels, improved methods of cooperation and implementation of effective policies across some regions have contributed to reducing some pressures on the ocean. However, in regions where there is conflict over access to resources and maritime boundaries, policies and agreements focused on sustainability can be undermined;

(e) **Climate change.** Anthropogenic greenhouse gas emissions have continued to rise, causing further long-term climate changes, with widespread effects throughout the ocean that will persist for centuries and affect the ocean. The impacts of climate change have been recognized by the Conference of the Parties to the United Nations Framework Convention on Climate Change in its decision 1/CP.21, by which it adopted the Paris Agreement,[2] aimed at strengthening the global response to threats from climate change.

The global influence of the five drivers is not uniformly distributed. Human populations are not evenly dispersed, and population growth varies among countries and regions. Geographical disparities in economic growth have been increasing since the 1980s. Associated differences in technological advances mean that some countries can extract resources from previously inaccessible areas, with the probability of increased pressures in those regions. Many regions, in particular those with least developed countries, still lack access to technologies that can assist in using marine resources sustainably.[3] Regional disputes and geopolitical instabilities may impede the implementation of global and regional treaties and agreements, thereby affecting economic growth, the transfer of technologies and the implementation of frameworks for managing ocean use. The effects of climate change are also not uniform, with some regions, including the Arctic Ocean, warming at higher rates than the global average (chap. 5).

3. Cleaning up the ocean

The lack of appropriate wastewater treatment and the release of pollutants from the manufacturing industry, agriculture, tourism, fisheries and shipping continue to put pressure on the ocean, with a negative impact on food security, food safety and marine biodiversity. Marine litter, ranging from nanomaterials to macromaterials, is a further problem, given that, in addition to the damage caused by its presence, it can also carry pollutants and non-indigenous species over long distances (chaps. 10–12).

2 See FCCC/CP/2015/10/Add.1, decision 1/CP.21, annex.

3 Unless otherwise indicated, "sustainable" and "sustainability" are used with reference to all aspects – environmental, social and economic.

3.1. Linkages with the Sustainable Development Goals and the United Nations Decade of Ocean Science for Sustainable Development

Sustainable Development Goal target 14.1:
By 2025, prevent and significantly reduce marine pollution of all kinds, in particular from land-based activities, including marine debris and nutrient pollution

Decade of Ocean Science outcome:
A clean ocean where sources of pollution are identified and reduced or removed

Concentrations of some pollutants (such as persistent organic pollutants and metals) in some regions are declining, but information on concentrations is not spatially uniform. Knowledge gaps remain with regard to not only recognized but also emerging pollutants. In several regions, capacity gaps remain in applying consistent, coherent policies and related enforcement to prevent and control inputs of pollutants into the ocean (chaps. 10–12 and 20).

The particular ways in which progress towards other Sustainable Development Goals will assist in the achievement of target 14.1 are set out in table 1, and the particular ways in which the achievement of that target will assist with progress towards other Goals are set out in table 2.

3.2. Nutrient pollution

Anthropogenic inputs of nitrogen and phosphorus into coastal ecosystems from direct discharges, land run-off, rivers and the atmosphere have generally continued to rise, even though better control of their release is reducing inputs into some bodies of water. Owing to excessive inputs of such nutrients, eutrophication is an increasing problem, and the number of hypoxic zones (sometimes called "dead zones") has increased from more than 400 globally in 2008 to approximately 700 in 2019. The ecosystems most affected include the northern part of the Gulf of Mexico, the Baltic Sea, the North Sea, the Bay of Bengal, the South China Sea and the East China Sea. It is estimated that coastal anthropogenic nitrogen inputs will double during the first half of the twenty-first century. In addition, deoxygenation is projected to worsen through increases in ocean temperatures and changes in stratification and ocean currents driven by climate change (chap. 5), in particular in coastal regions of Africa, South America, South and South-East Asia and Oceania (chap. 10).

3.3. Hazardous substances

Industrial development and the intensity of agriculture have continued to increase, resulting in both ongoing and new inputs of hazardous substances into the ocean. New types of input include pharmaceuticals, personal care products and nanomaterials that cannot be removed by wastewater treatment in many parts of the world. The detection of pharmaceuticals and personal care products is increasing across the ocean, including in the Arctic Ocean and the Southern Ocean. A number of such products have been observed to cause harm to plants and animals, but the scale of the impact on marine organisms is unknown, largely because they are generally not monitored (chap. 11).

Although the Stockholm Convention on Persistent Organic Pollutants[4] has generally had a positive effect on global concentrations, persistent organic pollutants continue to be detected in marine areas and in marine species far from their sources of production and use. Even low concentrations have been shown to reduce reproductive success in marine species, including Arctic seals. In most ocean regions, information on trends is lacking (chap. 11).

4 United Nations, *Treaty Series*, vol. 2256, No. 40214.

The Minamata Convention on Mercury[5] has generally reduced global mercury concentrations, with evidence, in most regions, that mercury concentrations in the ocean are levelling off. However, a slight increase in concentrations of some metals in higher trophic organisms has been reported. To better assess metal concentration trends, expanded coastal time-series analyses are needed globally, including of levels of metal nanomaterials in the ocean (chap. 11).

Concentrations of most radioactive substances continue to decrease through the decay of historical inputs. There have been no major nuclear accidents since 2011, and discharges from nuclear reprocessing plants in Europe continue to decrease substantially. Smaller amounts of radionuclides continue to be released by nuclear power reactors in 30 countries (chap. 11).

Globally, the number of shipping accidents has continued to decrease: an annual average of 88 ships of more than 100 gross tonnage were lost between 2014 and 2018, compared with 120 in the preceding five years. Progress is being made in reducing air pollution from ships. The number of oil spills has remained low: an annual average of 6 spills of more than seven tons from oil tankers occurred between 2010 and 2018, compared with an annual average of 18 spills in the previous decade. Offshore oil and gas installations also release hydrocarbons into the marine environment, but the long-term impacts of such releases remain unknown (chaps. 11 and 19).

3.4. Solid waste

Inputs of solid waste into the ocean (including marine litter) from unintentional releases and the intentional dumping of waste are largely unquantified around the world. Plastics represent up to 80 per cent of marine litter, with annual inputs into the ocean from rivers estimated at 1.15–2.41 million tons. The presence of plastics has been recorded in more than 1,400 marine species. Less is known about the effects of microplastics (pieces of less than 5 mm) and nanoplastics (pieces of less than 100 nm), although nanoplastics have been observed to enter the cells of organisms. Those two groups of plastics are derived from both the breakdown of macroplastics and deliberate manufacture (e.g., as ingredients in personal care products). The dumping of sewage sludge and organic and inorganic waste remains limited, with the dumping of sewage sludge continuing to decline as a result of the implementation of the Convention on the Prevention of Marine Pollution by Dumping of Wastes and Other Matter of 1972 (the London Convention)[6] and the 1996 Protocol thereto,[7] and many regional conventions. However, insufficient reporting under those agreements remain, resulting in uncertainties in the extent of the dumping of waste. Munitions dumped at sea continue to present low risks to the marine ecosystem and (when caught in nets) to fishers. Recent research, however, suggests that the release of compounds from munitions might have sublethal genetic and metabolic effects in marine organisms (chap. 12).

3.5. Noise

Anthropogenic noise affecting the oceans comes from many sources (e.g., vessels, oil and gas exploration and extraction, industrial activities and sonar) and varies across space and time. The regions most affected are those characterized by heavy industrial use, such as the Gulf of Mexico, the North Sea and the Atlantic Ocean. Unlike many other sources of marine pollution, noise does not persist once the sound source has been removed from the

5 UNEP(DTIE)/Hg/CONF/4, annex II. The Convention entered into force on 16 August 2017.
6 United Nations, *Treaty Series*, vol. 1046, No. 15749.
7 The London Protocol entered into force on 24 March 2006.

environment. Understanding the impacts of anthropogenic noise on marine biodiversity has increased over the past two decades, with a range of direct and indirect impacts observed across a number of taxa, from zooplankton to marine mammals. Understanding of those impacts has improved in parallel with increasing recognition of the need to monitor noise entering the marine environment and to identify and reduce its impacts. While some efforts are being made to reduce noise created by a variety of sources, increasing use of the ocean is likely to offset those efforts (chap. 20).

3.6. Key knowledge and capacity-building gaps

Methods for standardizing the monitoring of pollutants, including noise, and data sets are needed urgently, so that both spatial and temporal differences in pollutants can be evaluated and priorities established. Capacity-building is needed to reduce the input of pollutants into the ocean, in particular through the introduction of cleaner production, quieter technologies and cheaper and readily deployable wastewater-processing technologies. To reduce the duplication of efforts, the creation of a general database on hazardous substances and a baseline of ambient noise would be desirable to support risk assessment and modelling. As the extent of transboundary marine pollution is poorly understood in many parts of the world, in particular with regard to airborne pollutants, more accurate data on their emissions and transport are needed. Lastly, it is necessary to gain a much better understanding of the effects of pollutants, including anthropogenic noise, on the marine environment (chaps. 10–12 and 20).

4. Protecting marine ecosystems

The main threats to marine ecosystems come from human activities, such as fishing, aquaculture, shipping, sand and mineral extraction, oil and gas exploitation, the building of renewable energy infrastructure, coastal infrastructure development and pollution, including the release of greenhouse gases.

4.1. Linkages with the Sustainable Development Goals and the United Nations Decade of Ocean Science for Sustainable Development

Sustainable Development Goal target 14.2:
By 2020, sustainably manage and protect marine and coastal ecosystems to avoid significant adverse impacts, including by strengthening their resilience, and take action for their restoration in order to achieve healthy and productive oceans

Sustainable Development Goal target 14.5:
By 2020, conserve at least 10 per cent of coastal and marine areas, consistent with national and international law and based on the best available scientific information

Decade of Ocean Science outcome:
A healthy and resilient ocean where marine ecosystems are understood, protected, restored and managed

Many marine species and habitats continue to be adversely affected by increasing anthropogenic pressures (chaps. 6A–G and 7A–Q; see also sect. 5 below). Understanding of the distribution and status of species and habitats and how they are being affected by anthropogenic pressures is improving. In 2020, marine protected areas covered 18 per cent of the ocean within national jurisdictions, representing approximately 8 per cent of the entire ocean, while about 1 per cent of marine areas beyond national jurisdiction had been protected (chap. 27).

The particular ways in which progress towards other Sustainable Development Goals will assist in the achievement of targets 14.2 and 14.5 are set out in table 1, and the particular ways in which the achievement of those targets will assist with progress towards other Goals are set out in table 2.

The protection of marine ecosystems is embedded in various international agreements, such as the United Nations Convention on the Law of the Sea[8] and the Convention on Biological Diversity,[9] as well as in regional conventions and national legislation. Notwithstanding the objectives of such agreements and conventions, the status of many marine species and habitats continues to decline globally, thereby putting the functioning of ecosystems at risk. In addition, climate change is resulting in ocean warming, acidification, changes in circulation, dissolved oxygen concentrations and water cycle amplification. As a result, the transfer of nutrients associated with primary productivity from surface waters to the deep sea is declining. Globally, about 2,000 marine species have been introduced outside their natural range as a result of human activities (chaps. 5, 6A–G, 7A–Q and 22).

Many management frameworks for protecting marine ecosystems have a sectoral focus and can therefore have differing objectives for the protection of the marine environment across sectors. Management tools can be area-based (such as marine protected areas and fishery closures) or non-area-based (such as global emission controls, catch and effort controls and technical restrictions). Management approaches are increasingly moving away from being focused on sectoral use towards including diverse links between ecological and social, economic and cultural aspects. The ecosystem approach integrates environmental, social and economic aspects at the global, regional, national or local level. Cultural information is becoming an integral part of management frameworks, both in the context of community-based management and for safeguarding the cultural dimension of the marine environment. Such information can be diverse and intangible, such as traditional marine resource use, sea routes, ancient navigational skills, maritime identities, legends, rituals, beliefs and practices, aesthetic and inspirational qualities, cultural heritage and places of spiritual, sacred and religious importance (chap. 27).

In some areas, in particular in South-East Asia, "blue infrastructure development", as well as such approaches as nature-based solutions, are being introduced in an attempt to harmonize coastal development and protection with habitat and ecological protection (chaps. 8A, 13 and 14).

4.2. Coastal ecosystems

Notwithstanding increases in marine protected areas and the expansion of Ramsar Sites,[10] mangroves (except in the Red Sea) and seagrass meadows (in particular in South-East Asia) continue to decline, with 19 per cent of mangroves and 21 per cent of seagrass species identified as near-threatened. The combined effects of ocean warming and human activities are increasingly affecting tropical and subtropical coral reefs and kelp forests globally. In recent years, coral reefs have undergone mass bleaching on an annual basis, while kelp forests have been affected by marine heatwaves (chap. 9), resulting in rapid losses (chaps. 6G, 7D and 7H).

Overall, about 6 per cent of known fish species and nearly 30 per cent of elasmobranch species are listed as Near Threatened or

8 United Nations, *Treaty Series*, vol. 1833, No. 31363.
9 Ibid., vol. 1760, No. 30619.
10 See Convention on Wetlands of International Importance especially as Waterfowl Habitat (United Nations, *Treaty Series*, vol. 996, No. 14583).

Vulnerable. Globally, the status of marine mammals varies, with 75 per cent of species in some groups (sirenians, freshwater dolphins, polar bears and otters) being classified as Vulnerable, Endangered or Critically Endangered. Many large whale species are now recovering from past harvesting as a result of prohibitions on and the regulation of commercial catches and national recovery plans. The conservation status of marine reptiles has varied greatly: protection in certain regions has increased some populations, while those in other areas are declining because of continuing or increasing threats. The global conservation status of seabirds has worsened, with over 30 per cent of species now listed as Vulnerable, Endangered or Critically Endangered (chaps. 6C–F).

4.3. Open ocean and deep-sea ecosystems[11]

The open ocean continues to be affected by ocean warming, acidification, deoxygenation and marine pollution. Nutrient inputs derived from the Amazon River and brought up by upwelling off the coast of West Africa appear to have fuelled a massive seaweed bloom of floating sargassum: the 20-million-ton bloom began to develop in 2011 in the equatorial Atlantic Ocean and, by 2018, had extended 8,850 km across that area (chaps. 7N, 10 and 12).

Understanding of the distribution of cold-water corals has increased, and they are known to occur along continental margins, mid-ocean ridges and seamounts worldwide. They and other deep-sea features (seamounts, pinnacles, ridges, trenches, hydrothermal vents and cold seeps) remain under threat from fishing, offshore oil drilling, deep-sea mining and pollution, including plastic waste, and, to a lesser extent, climate change. Some efforts to curb deep-water bottom trawling and establish marine protected areas where cold-water corals occur have partially restored some damaged cold-water coral communities. However, such habitats can take decades or even centuries to recover, making it difficult to identify trends of improvement (chaps. 7E, 7L, 7O and 7P).

4.4. Key knowledge and capacity-building gaps

Since 2015, on average, one new species of fish has been described per week, highlighting how much remains to be discovered. Although knowledge of ecosystem composition and functioning has improved since the issuance of the first Assessment, gaps remain, in particular with regard to deep-sea ecosystems and open-ocean planktonic and benthic species. Gaps also remain in understanding the biology and ecology of coastal species, in particular in the territorial waters of developing countries. There is no well-organized structure to study the approximately 2,000 non-indigenous species that have spread to new areas as a result of human activities and their impacts on natural ecosystems. The conservation status of less than 1 per cent of macroalgal species has been assessed (chaps. 6A–C, 6G, 7N and 22).

While the ecosystem approach has been widely acknowledged as an effective framework for managing human impacts, further research and capacity-building are needed to realize its full potential across the world's oceans. In many regions, there is a lack of information needed to establish links between ecological causes and effects in order to balance them against socioeconomic priorities, in decision-making. Enhanced collaboration in monitoring will help in sharing capacity across sectors and institutions and provide more efficient monitoring, data and information. Increased capacity in understanding management approaches and implementing them will support Governments and other stakeholders in understanding options for the management and governance of marine areas (chap. 27).

---

11 See chap. 2, sect. 4, for a definition of the terms "open ocean" and "deep sea".

5. Understanding of the ocean for sustainable management

The sustainable use of the ocean cannot be achieved before acquiring a deep understanding of ocean processes and its functioning, as well as coherent knowledge of the impacts of human activities on the ocean (chaps. 8A and 27).

5.1. Linkages with the Sustainable Development Goals and the United Nations Decade of Ocean Science for Sustainable Development

Sustainable Development Goal target 14.3:
Minimize and address the impacts of ocean acidification, including through enhanced scientific cooperation at all levels

Sustainable Development Goal target 14.a:
Increase scientific knowledge, develop research capacity and transfer marine technology, taking into account the Intergovernmental Oceanographic Commission Criteria and Guidelines on the Transfer of Marine Technology, in order to improve ocean health and to enhance the contribution of marine biodiversity to the development of developing countries, in particular small island developing States and least developed countries

Decade of Ocean Science outcomes:

- **A predicted ocean where society understands and can respond to changing ocean conditions**
- **An accessible ocean with open and equitable access to data, information and technology and innovation**
- **An inspiring and engaging ocean where society understands and values the ocean in relation to human well-being and sustainable development**

The input of carbon dioxide into the ocean is continuing, albeit in an irregular manner, resulting in acidification of the ocean. Compounded with other pressures, it has a negative impact on a wide range of organisms, in particular those that form calcium carbonate shells, with the potential to alter biodiversity and ecosystem structure. Ocean acidification, combined with rising temperatures, sea level rise, deoxygenation and increasing extreme climate events, further threatens the goods and services provided by coastal ecosystems (chaps. 5 and 9).

Scientific understanding of the ocean, its functioning and the impacts on it grows ever faster. However, in many parts of the ocean, knowledge and capacity-building gaps remain, in particular in areas beyond national jurisdiction. Quantification of the cumulative effects of pressures on the ocean is nascent, as is the quantification of comprehensive and standardized indicators of ocean health. The capacity to enable people to have access to and use scientific understanding remains a requirement for applying integrated approaches to the management of human impacts on the ocean (chaps. 3, 25 and 27).

The particular ways in which progress towards other Sustainable Development Goals will assist in the achievement of targets 14.3 and 14.a are set out in table 1, and the particular ways in which the achievement of those targets will assist with progress towards other Goals are set out in table 2.

5.2. Global scientific understanding

Innovations in technology and engineering related to sensors and autonomous observation platforms have allowed for ocean data collection at finer temporal and spatial resolutions and expanded those observations into remote areas. Cost-effective and user-friendly sensors, along with mobile applications, the enhanced participation of citizens and the deployment of sensors on non-scientific ships, are also facilitating the expanded collection of ocean observations. Such developments have increased understanding of physical and biogeochemical systems in the ocean and how the ocean is changing in response to climate change, as well as enhanced ocean modelling capabilities on the global and regional scales (chaps. 3 and 5).

The promotion of networking and the coordination of regional observation programmes have contributed to the further development of global ocean observations within an integrated system. The standardization and harmonization of observation methods are also being pursued through international initiatives. Platforms to share best practices in ocean observation, data-sharing and community dialogues have also been established, with the aim of improving the effective use of ocean data for the benefit of society (chap. 3).

5.3. Sustainable management

Over the past two decades, many frameworks for assessing interactions between human activities and natural events ("cumulative effects") have been developed using different approaches and terminologies and applied on differing scales. Along with other assessments of the environment, they include environmental impact assessments and strategic environmental assessments and are useful tools for informing marine spatial planning and resource management (chaps. 25–27).

Both marine spatial planning and management frameworks comprise a spectrum of processes but have unified objectives of identifying users of the marine environment, planning the activities of those users and effecting some form of regulation of that use to ensure sustainability. In general, marine spatial planning has been most effectively developed with the involvement of all relevant authorities and stakeholders and has included economic, environmental and social perspectives. Social perspectives and social and cultural values are increasingly recognized in management frameworks, but reconciling a multiplicity of heterogenous values is a challenge. Addressing multiple values is best done by engaging with affected communities, hence the need to recognize community-based management that is sensitive to the cultural dimensions of the sea within ecosystem approaches to management. Increased understanding of the rights, tenures and traditional and indigenous customary uses of inshore marine environments has catalysed recognition of the strengths of community-based management. Culture is potentially powerful, as both a factor to be managed and monitored and the foundation upon which management-incorporating ecosystem approaches may be developed in the context of sustainable development (chaps. 26 and 27).

5.4. Key knowledge and capacity-building gaps

Globally, disparities remain in knowledge to support ecosystem-based management. Most research and information available (based on the number of publications) relates to the North Atlantic Ocean, the North Pacific Ocean and the Arctic Ocean. Disparities in infrastructure and professional capacities limit ocean research, resulting in regional and national disparities in scientific understanding. To better monitor significant changes in physical and biogeochemical environments and their impacts on ecosystems and society, further integration of multidisciplinary observation systems and improved models are needed. Innovation in funding strategies is also required to sustain such systems (chap. 3).

Most assessments of cumulative effects tend to be focused on existing and past activities in the marine environment. Similarly, much marine spatial planning has been carried out in areas where activities are ongoing, and many management frameworks are applied to existing activities with regard to resource extraction and use, making them retrospective in nature. Assessments that allow for "foresighting" are needed to inform planning of future activities and support management that is adaptive to future conditions and sustains ecosystems and human well-being. Developing such approaches is not straightforward and will require substantial effort. Increased capacity in transboundary cooperation, the strengthening of science-policy capacity, greater coordination between social and natural sciences and between science and civil society, including industry, and the recognition of traditional knowledge, culture and social history are needed to support holistic management (chaps. 25–27).

6. Promoting safety from the ocean

A wide range of events in and on the ocean threaten those who live near or work on the ocean or rely on it for food. Examples of such events are tsunamis, storm surges, rogue waves, cyclones, hurricanes and typhoons, coastal flooding, erosion, marine heatwaves and harmful algal blooms. The ocean plays an important role in driving hydrological variability, such as droughts and pluvials over land, on intraseasonal to interannual (and longer) timescales (chap. 9). Such events, together with various effects of hazardous substances and excessive nutrients, have the potential to threaten food security and hamper sustainable economic development.

6.1. Linkages with the Sustainable Development Goals and the United Nations Decade of Ocean Science for Sustainable Development

Sustainable Development Goal target 14.1:
By 2025, prevent and significantly reduce marine pollution of all kinds, in particular from land-based activities, including marine debris and nutrient pollution

Sustainable Development Goal target 14.3:
Minimize and address the impacts of ocean acidification, including through enhanced scientific cooperation at all levels

Decade of Ocean Science outcome:
A safe ocean where life and livelihoods are protected from ocean-related hazards

Marine heatwaves and tropical cyclones, hurricanes and typhoons are increasing in frequency and severity as a result of climate change, but such increases can be reduced by climate change mitigation efforts. As indicated above, the ocean also drives hydrological variability over land. The construction of dams and reservoirs is, in some areas, reducing sediment supply to the coast by more than 50 per cent, leading to the erosion of deltas and adjacent coasts. As a result of nutrient pollution, harmful algal blooms are becoming more frequent. The number of pollutants in the ocean continues to increase, and therefore the mixtures to which biotas are exposed and that are integrated into food systems are becoming more complex (chaps. 9–11 and 13).

The particular ways in which progress towards other Sustainable Development Goals will assist in the achievement of targets 14.1 and 14.3 are set out in table 1, and the particular ways in which the achievement of those targets will assist with progress towards other Goals are set out in table 2.

6.2. Hazards from the ocean

In addition to continuing threats such as tsunamis, climate change is increasingly affecting areas and their associated communities not previously exposed to rising sea levels. Such rises can also exacerbate coastal erosion. Precipitation, winds and extreme sea level events associated with tropical cyclones have increased in recent decades, as has the annual global proportion of category 4 or 5 tropical cyclones. There are increasing risks to locations that had historically not been exposed to storms, owing to unprecedented storm trajectories. The management of risks from changing storm trajectories and storm intensity proves challenging because of the difficulties of early warning and the reluctance of affected populations to respond (chaps. 9 and 13).

Over the past two decades, marine heatwaves have had negative impacts on marine organisms and ecosystems in all ocean basins. Such events are projected to increase in frequency, duration, spatial extent and intensity under future global warming, thus pushing some marine organisms, fisheries and ecosystems beyond the limits of their resilience, with cascading impacts on economies and societies. Coastal erosion, driven by, for example, decreased fluvial sediment supply to the coast owing to changed river management, coastal sand mining and longshore impoundment by coastal structures, is increasingly causing problems. Changes in the coastal profile following the destruction of mangroves, salt marshes and barrier islands add to such problems. Inputs of nitrogen and phosphorus to coastal ecosystems through river run-off and atmospheric deposition have increased owing to the use of synthetic fertilizers, the combustion of fossil fuels and the direct input of municipal waste. That is leading to an increase in harmful algal blooms, including toxic algal events, which, inter alia, can lead to shellfish and fish becoming poisonous, thus causing paralysis and other illnesses in humans (chaps. 9, 10 and 13).

6.3. Key knowledge and capacity-building gaps

Improved understanding of the ocean and its interrelation with the atmosphere is essential to improving human safety in extreme weather events. Similarly, better understanding of the scale, progress and distribution of pollution and of coastal dynamics is needed. The need to strengthen and harmonize warning systems for reducing the risks associated with ocean hazards is identified in the Sendai Framework for Disaster Risk Reduction 2015–2030.[12] Progress is needed on forecasting systems for hazards, emergency planning and warnings should be expanded and preparation frameworks should be implemented to ensure a rapid response for affected communities. Integrated systems that allow for forecasting, detection and response to multiple hazards are required (chaps. 9–14).

7. Sustainable food from the ocean

Animal protein from the seas provides about 17 per cent of all animal protein consumed by humans and supports about 12 per cent of human livelihoods. It is largely derived from wild fisheries, although the contribution of aquaculture to food security is growing rapidly and has greater potential for growth than capture fisheries. Fishing practices place multiple stressors on the marine environment in many regions, and the expansion of aquaculture brings new or increased pressures on marine ecosystems, in particular in coastal areas (chaps. 15–17).

[12] General Assembly resolution 69/283, annex II.

7.1. Linkages with the Sustainable Development Goals and the United Nations Decade of Ocean Science for Sustainable Development

Sustainable Development Goal target 14.4:
By 2020, effectively regulate harvesting and end overfishing, illegal, unreported and unregulated fishing and destructive fishing practices and implement science-based management plans, in order to restore fish stocks in the shortest time feasible, at least to levels that can produce maximum sustainable yield as determined by their biological characteristics

Sustainable Development Goal target 14.6:
By 2020, prohibit certain forms of fisheries subsidies which contribute to overcapacity and overfishing, eliminate subsidies that contribute to illegal, unreported and unregulated fishing and refrain from introducing new such subsidies, recognizing that appropriate and effective special and differential treatment for developing and least developed countries should be an integral part of the World Trade Organization fisheries subsidies negotiation[13]

Sustainable Development Goal target 14.7:
By 2030, increase the economic benefits to small island developing States and least developed countries from the sustainable use of marine resources, including through sustainable management of fisheries, aquaculture and tourism

Sustainable Development Goal target 14.b:
Provide access for small-scale artisanal fishers to marine resources and markets

Decade of Ocean Science outcome:
A productive ocean supporting sustainable food supply and a sustainable ocean economy

The particular ways in which progress towards other Sustainable Development Goals will assist in the achievement of targets 14.4, 14.6, 14.7 and 14.b are set out in table 1, and the particular ways in which the achievement of those targets will assist with progress towards other Goals are set out in table 2.

7.2. Marine capture fisheries

Estimated global landings of marine capture fisheries increased by 3 per cent to 80.6 million tons, valued at $127 billion (at 2017 prices), between 2012 and 2017. About 33 per cent of the world's fish stocks, especially at higher trophic levels, are classified as being fished at biologically unsustainable levels, with close to 60 per cent maximally sustainably fished.[14] The sustainability of many of the world's capture fisheries continues to be hampered by overexploitation, overcapacity, ineffective management, harmful subsidies, by-catch, in particular of threatened, endangered and protected species, and illegal, unreported and unregulated fishing, with ongoing habitat degradation and loss of gear creating further pressures on the marine environment. Overfishing is estimated to have led to an annual loss of $88.9 billion in net benefits. Fish markets continue to exhibit fast-paced globalization, thus increasing the vulnerability of small-scale fisheries to the depletion of locally important stocks. Negotiations under the auspices of the World Trade Organization on reducing harmful fishery subsidies have continued, although no firm agreement has yet been reached. Less than 40 per cent of States have signed the Agreement on Port State Measures to Prevent, Deter and Eliminate Illegal, Unreported and Unregulated Fishing[15] of 2009. The application of information technology to help to expand the opportunities of small-scale fisheries in areas such as safety, the sharing of local knowledge, capacity-building and governance have been outlined by the Food and Agriculture Organization of the United Nations in its Voluntary

13 Taking into account ongoing World Trade Organization negotiations, the Doha Development Agenda and the Hong Kong ministerial mandate.

14 "Maximally sustainably fished" is used here in the sense explained in chapter 15.

15 Food and Agriculture Organization of the United Nations, document C 2009/REP and Corr.1–3, appendix E.

Guidelines for Securing Sustainable Small-Scale Fisheries in the Context of Food Security and Poverty Eradication, and the growing use of human rights approaches is providing opportunities for the empowerment of such fisheries (chap. 15).

Promisingly, scientific stock assessments and management have been shown to lead to more sustainable outcomes across a number of regions. New approaches to identifying illegal, unreported and unregulated fishing are now being applied in some regions. Recent research has shown that, with appropriate governance, the median time required to rebuild overfished stocks could be less than 10 years, and, if reforms were to be implemented, 98 per cent of overfished stocks could be considered healthy by the middle of the twenty-first century.

The impacts of climate change are expected to include increases in the intensity of natural hazards and their frequency, thus affecting the local distribution and abundance of fish populations. Fishery-dependent developing States may be affected most severely and, because of expected changes in species distributions and consequent expected increases in transboundary migrations of stocks, future international governance may need to account for such redistributions (chap. 15).

7.3. Aquaculture

Aquaculture continues to grow faster than other major food production sectors, although its growth has slowed over the past decade. The sector was valued at $249.6 billion in 2017. It supports the livelihoods of 540 million people, 19 per cent of whom were women in 2014. The importance of that form of food production lies in its high content of proteins and essential micronutrients and fatty acids. The reliance of aquaculture on fish meal decreased from 4.20 million tons in 2005 to 3.35 million tons in 2015. Aquaculture sustainability is more likely to be closely linked with the sustained supply of terrestrial animal and plant proteins, oils and carbohydrate sources for aquafeeds. Diseases continue to pose a challenge to global aquaculture and are among the primary deterrents to aquaculture development for many species. In general, the environmental performance of aquaculture has improved significantly over the past decade. Challenges to be met in expanding aquaculture production include reducing impacts on valuable coastal ecosystems such as mangroves, the sustainable provision of external feed, the management of fish diseases and the effects of escaped fish on native species (chap. 16).

7.4. Seaweed production

Seaweed for direct human consumption amount to 80 per cent of total seaweed harvesting. Since 2012, global harvesting of seaweed has risen at a rate of about 2.6 per cent a year, mostly from aquaculture, to 32 million tons in 2017, with an estimated value of $12 billion. In addition to being used as food, seaweed is used increasingly in industrial applications, such as cosmetics, pharmaceuticals and nutraceuticals, and as feed for livestock. Macroalga cultivation amounts to 96 per cent of total aquaculture production. Benefits from production include the provision of high-quality food and the creation of new jobs and increased incomes for coastal inhabitants. In addition, such production supports carbon sequestration and oxygen production and reduces eutrophication (chap. 17).

7.5. Key knowledge and capacity-building gaps

There is limited understanding of the extent to which changing conditions could contribute to shifts in marine ecosystem structures and functioning and the subsequent impacts on marine productivity. There have been improvements in approaches to assessing fisheries and accounting for their contributions in data-poor environments, but further work is needed to fill capacity-building gaps for coastal fisheries in developing regions. The science of fish stock propagation is still in its early stages, but shows some potential for increasing fishery yield beyond what is achievable through the exploitation of wild stocks alone. However, understanding of ecological consequences is lacking. Capacity-building gaps in the management of fisheries include those associated with identifying impacts on target species and incorporating the effects on other species into management frameworks. Ongoing capacity-building gaps in developing countries also hinder their ability to take part in regional and international negotiations for reaching consensus on management practices for sustaining healthy fish stocks.

To boost sustainable aquaculture development, improved extension services are needed. The training of extension services providers needs to incorporate information delivery methods, as well as practical farming techniques, to help them to better assist farmers in improving production practices. Information technology and media, farmers' associations, development agencies, private sector suppliers and others will need to come together to enhance sectoral training. The establishment of offshore aquaculture and mariculture will need to be supported by sufficient marine services to ensure the sustainability and safety of operations. Many knowledge gaps remain with regard to the large-scale production of seaweed and the likely impacts of climate change. Some efforts to address the knowledge and capacity-building gaps are under way. The biology of many seaweed species is still unknown, even for those species currently harvested or farmed (chaps. 15–17).

8. Sustainable economic use of the ocean

The ocean supports a wide range of economic activities, including maritime transport as part of world trade, tourism and recreation, extraction of natural resources such as hydrocarbons and other minerals, provision of renewable energy, and the use of marine genetic resources.

8.1. Linkages with the Sustainable Development Goals and the United Nations Decade of Ocean Science for Sustainable Development

Sustainable Development Goal target 14.2: **By 2020, sustainably manage and protect marine and coastal ecosystems to avoid significant adverse impacts, including by strengthening their resilience, and take action for their restoration in order to achieve healthy and productive oceans**

Sustainable Development Goal target 14.7: **By 2030, increase the economic benefits to small island developing States and least developed countries from the sustainable use of marine resources, including through sustainable management of fisheries, aquaculture and tourism**

Sustainable Development Goal target 14.c: **Enhance the conservation and sustainable use of oceans and their resources by implementing international law as reflected in the United Nations Convention on the Law of the Sea, which provides the legal framework for the conservation and sustainable use of oceans and their resources, as recalled in paragraph 158 of "The future we want"**

Economic use of the ocean has increased globally. Many countries are developing or have developed strategies for increasing such maritime activities as renewable ocean energy, aquaculture, marine biotechnology, coastal tourism and seabed mining (growth sectors of the blue economy – a term that can include environmentally sustainable shipping and fisheries). The distribution around the world of the economic benefits drawn from the ocean, however, is still very uneven (chaps. 4, 8A, 18 and 28).

The particular ways in which progress towards other Sustainable Development Goals will assist with the achievement of targets 14.2, 14.7 and 14.c, among others, are set out in table 1, and the particular ways in which the achievement of those targets will assist with progress towards other Goals are set out in table 2.

8.2. Seabed mining

Seabed mining for sand and gravel within national jurisdiction has increased to supplement diminished land-based sources. The scale of extraction can have significant effects on the local marine environment and cause coastal erosion. The scale of other major mining activities (such as for diamonds, phosphate, iron ore and tin) remains more or less stable. Deep seabed mining in areas beyond national jurisdiction is closer to becoming a commercial reality; however, exploiting many mineral resources requires advanced technology and is thus largely limited to those able to access such technology (chap. 18).

8.3. Extraction of offshore hydrocarbons

The offshore oil and gas sector is expanding at the global level into deep and ultradeep waters. Over the next decade, growth is likely to be focused in such areas as the eastern Mediterranean Sea and areas off the coast of Guyana and the west coast of Africa. Mature areas such as the North Sea and the Gulf of Mexico are seeing the exhaustion of some resources and the resulting increased decommissioning of offshore installations, although some may be used for producing renewable marine energy. Extraction techniques continue to evolve to reduce their impact on the marine environment (chap. 19).

8.4. Maritime transport

The increase in tonnage of cargo carried by international shipping has mirrored the growth in world trade, following the recovery of the world's economy after 2012. Such growth, however, has occurred against a weak competitive background. A large proportion of the world's tonnage continues to be associated with a relatively small number of registries, and ownership and control of shipping remain concentrated in the hands of firms in a relatively small number of countries. This concentration has significant implications for future port development, as it may result in fewer and larger main ports serving as distribution hubs for intercontinental trade. There was a slight decline in the total number of attempted and actual cases of piracy and armed robbery against ships between 2015 and 2019 (chap. 8A).

8.5. Tourism and recreation

International travel and associated tourism are economically important in many parts of the world, in particular in the "sun, sea and sand" type of tourism, which is concentrated in coastal marine regions. In all touristic areas, the major impact on the marine environment comes from coastal development, including the proportion of land covered by constructions, such as hotels, restaurants, retail shops and transport infrastructure, including airports and train terminals, and the need for "armoured" coastal defences, street lighting and sewerage. Snorkelling, diving and wildlife viewing continue to be significant elements in coastal tourism (chap. 8A).

8.6. Marine genetic resources

Marine genetic resources continue to be the focus of an expanding range of commercial and non-commercial applications. Rapidly shrinking costs of gene sequencing and synthesis, as well as rapid advances in metabolic engineering and synthetic biology, have reduced dependency on the acquisition of physical samples from the ocean. Sponges and algae continue to attract significant interest for the bioactive properties of their natural compounds (chap. 23).

8.7. Marine renewable energy

The marine renewable energy sector (offshore wind energy, tidal and ocean current energy, wave energy, ocean thermal energy and osmotic power and marine biomass energy) is evolving and developing at different rates. Of those power sources, offshore wind technology is mature and technically advanced. Although in 2018 it represented only 1 per cent of total renewable energy sources, it is growing rapidly: between 2017 and 2018, it accounted for 4 per cent of all growth in renewable energy. From 2017 to 2018, it grew by 59 per cent in Asia and by 17 per cent in Europe. In the next decade, Asia and the United States of America could be major drivers of offshore wind power development and installation. Tidal energy converters have reached the commercial stage, while other marine renewable energy technologies are currently under development. Among emerging marine renewable energy sources, offshore solar energy is the most promising, as components of the relevant technology are well developed (chap. 21).

8.8. Key knowledge and capacity-building gaps

All maritime industries are highly dependent on technology to operate safely and without damaging the marine environment. With regard to marine genetic resources, capacity-building remains an issue, as most work in this field is carried out in a small number of countries. There is a need to build skills in many countries to plan and develop their blue economy sustainably and to manage the related human activities (chaps. 8A, 14, 18, 19, 21, 23, 25 and 27).

9. Effective implementation of international law as reflected in the United Nations Convention on the Law of the Sea

Effective implementation of international law as reflected in the United Nations Convention on the Law of the Sea (which sets out the legal framework within which all activities in the oceans and seas must be carried out), is essential for the conservation and sustainable use of the ocean and its resources and for safeguarding the many ecosystem services that the ocean provides, both for current and future generations.

9.1. Linkages with the Sustainable Development Goals and the United Nations Decade of Ocean Science for Sustainable Development

Sustainable Development Goal target 14.c: **Enhance the conservation and sustainable use of oceans and their resources by implementing international law as reflected in the United Nations Convention on the Law of the Sea, which provides the legal framework for the conservation and sustainable use of oceans and their resources, as recalled in paragraph 158 of "The future we want"**

Steps have already been taken at all levels to strengthen the implementation of international law as reflected in the United Nations Convention on the Law of the Sea, including by increasing the level of participation of States in the numerous global and regional treaties that supplement its provisions. Examples at the global level include international conventions such as the London Convention and the London Protocol, the International Convention for the Prevention of Pollution from Ships, 1973, as modified by the Protocol of 1978 relating thereto and by the Protocol of 1997[16] (including its annex VI on the reduction in sulfur emissions from ships, which entered into force in 2020), and the Agreement on Port State Measures to Prevent, Deter and Eliminate Illegal, Unreported and Unregulated Fishing of FAO, which entered into force in 2016 (chaps. 8A, 11, 12, 15 and 28).

There are still major challenges to ensuring participation in international instruments, addressing resource and capacity constraints, strengthening intersectoral cooperation, ensuring coordination and information-sharing at all levels and developing new instruments to address emerging challenges in a timely fashion (chap. 28).

The particular ways in which progress towards other Sustainable Development Goals will assist in the achievement of target 14.c are set out in table 1, and the particular ways in which achievement of that target will assist with progress towards other Goals are set out in table 2.

9.2. Implementation of international law as reflected in the United Nations Convention on the Law of the Sea

The integration of environmental, social and economic dimensions is at the core of the United Nations Convention on the Law of the Sea. The Convention establishes a delicate balance between the need for economic and social development through the use of the ocean and its resources and the need to conserve and manage those resources in a sustainable manner and to protect and preserve the marine environment. The integrated approach to ocean management as reflected in the Convention is essential for promoting sustainable development, as sectoral and fragmented approaches lack coherence and may lead to solutions that are of limited benefit to the conservation and sustainable use of the ocean and its resources.

The Convention is, in many fields, supplemented by more specific, sectoral instruments. In addition to its two implementing agreements,[17] there are numerous global and regional legal instruments covering many aspects of ocean use. Effective conservation and sustainable use of the ocean and its resources will only be achieved through the full and effective implementation of this body of international law. Actions and efforts should focus primarily on implementation gaps or any regulatory gaps, especially in areas beyond national jurisdiction.

16 See www.imo.org/en/About/Conventions/Pages/International-Convention-for-the-Prevention-of-Pollution-from-Ships-(MARPOL).aspx.

17 Agreement relating to the implementation of Part XI of the United Nations Convention on the Law of the Sea of 10 December 1982; and Agreement for the Implementation of the Provisions of the United Nations Convention on the Law of the Sea of 10 December 1982 relating to the Conservation and Management of Straddling Fish Stocks and Highly Migratory Fish Stocks.

9.3. Implementation and regulatory gaps

Resource capacity, including financial capacity, remains a significant constraint for the protection and preservation of the marine environment and marine scientific research, while technological constraints are often an impediment to the effective implementation of a State's obligations. Gaps also exist with regard to the material scope (e.g., no comprehensive rules on plastics and microplastics) or geographical scope of application of relevant instruments (e.g., geographical coverage by the regional fisheries management organizations and arrangements) (chaps. 27 and 28). Many small island developing States and least developed countries lack access to the detailed knowledge and skilled human resources needed for ocean management, and resources for managing the large marine areas under their jurisdiction are often limited. Filling these gaps will ensure that economic benefits can be maximized in an environmentally sustainable manner. Specific challenges exist in the enforcement of management measures in areas beyond national jurisdiction, owing to regulatory gaps and a lack of cross-sectoral coordination. These issues are currently being discussed at the United Nations in the context of the intergovernmental negotiations on the development of an international legally binding instrument under the United Nations Convention on the Law of the Sea on the conservation and sustainable use of marine biological diversity of areas beyond national jurisdiction (chaps. 27 and 28).

Table 1
Contribution made by other Sustainable Development Goals to achieving Goal 14

Targets under Sustainable Development Goal 14	Sustainable Development Goals contributing to the achievement of Goal 14	Mechanism
Cleaning up the ocean		
Target 14.1: By 2025, prevent and significantly reduce marine pollution of all kinds, in particular from land-based activities, including marine debris and nutrient pollution	Goal 6: Ensure availability and sustainable management of water and sanitation for all	Improved wastewater management
	Goal 7: Ensure access to affordable, reliable, sustainable and modern energy for all	Improved sources and efficiencies in energy and associated reduction in emissions
	Goal 11: Make cities and human settlements inclusive, safe, resilient and sustainable	Sustainable urbanization and reduction in the environmental impact of cities
	Goal 12: Ensure sustainable consumption and production patterns	Environmentally sound management of chemicals and all wastes, including by reducing waste generation
	Goal 17: Strengthen the means of implementation and revitalize the Global Partnership for Sustainable Development	Improved access to science, technology and innovation, enhanced knowledge-sharing and transfer of technology, and capacity-building

Targets under Sustainable Development Goal 14	Sustainable Development Goals contributing to the achievement of Goal 14	Mechanism
Protecting marine ecosystems		
Target 14.2: By 2020, sustainably manage and protect marine and coastal ecosystems to avoid significant adverse impacts, including by strengthening their resilience, and take action for their restoration in order to achieve healthy and productive oceans Target 14.5: By 2020, conserve at least 10 per cent of coastal and marine areas, consistent with national and international law and based on the best available scientific information	Goal 6: Ensure availability and sustainable management of water and sanitation for all	Improved wastewater management and protection and restoration of wetlands
	Goal 7: Ensure access to affordable, reliable, sustainable and modern energy for all	Improved sources and efficiencies in energy and associated reduction in emissions
	Goal 9: Build resilient infrastructure, promote inclusive and sustainable industrialization and foster innovation	Use of clean technologies and associated reduction in emissions
	Goal 11: Make cities and human settlements inclusive, safe, resilient and sustainable	Sustainable urbanization and reduction in the environmental impact of cities
	Goal 12: Ensure sustainable consumption and production patterns	Sustainable management and use of natural resources and reduction in waste along supply chains
	Goal 13: Take urgent action to combat climate change and its impacts[a]	Implementation of climate change mitigation, adaptation and impact reduction measures
	Goal 15: Protect, restore and promote sustainable use of terrestrial ecosystems, sustainably manage forests, combat desertification, and halt and reverse land degradation and halt biodiversity loss	Reduction in the degradation of natural habitats and loss of biodiversity, and prevention of the extinction of species
	Goal 17: Strengthen the means of implementation and revitalize the Global Partnership for Sustainable Development	Improved access to science, technology and innovation, enhanced knowledge-sharing and transfer of technology, and capacity-building
Understanding of the ocean for sustainable management		
Target 14.3: Minimize and address the impacts of ocean acidification, including through enhanced scientific cooperation at all levels	Goal 9: Build resilient infrastructure, promote inclusive and sustainable industrialization and foster innovation	Enhancement of scientific research, upgrade of the technological capabilities of industrial sectors in all countries, in particular developing countries, and encouragement of innovation

Targets under Sustainable Development Goal 14	Sustainable Development Goals contributing to the achievement of Goal 14	Mechanism
Target 14.a: Increase scientific knowledge, develop research capacity and transfer marine technology, taking into account the Intergovernmental Oceanographic Commission Criteria and Guidelines on the Transfer of Marine Technology, in order to improve ocean health and to enhance the contribution of marine biodiversity to the development of developing countries, in particular small island developing States and least developed countries	Goal 13: Take urgent action to combat climate change and its impacts[a]	Implementation of climate change mitigation, adaptation and impact reduction measures
	Goal 17: Strengthen the means of implementation and revitalize the Global Partnership for Sustainable Development	Improved access to science, technology and innovation, enhanced knowledge-sharing and transfer of technology, and capacity-building
Promoting safety from the ocean		
Target 14.1: By 2025, prevent and significantly reduce marine pollution of all kinds, in particular from land-based activities, including marine debris and nutrient pollution	Goal 1: End poverty in all its forms everywhere	Reduction in exposure and vulnerability to climate-induced extreme events and building of resilience to environmental shocks and disasters
	Goal 2: End hunger, achieve food security and improved nutrition and promote sustainable agriculture	Strengthening of capacity to adapt to climate change, extreme weather and other disasters
	Goal 6: Ensure availability and sustainable management of water and sanitation for all	Reduction in pollution, improved wastewater management and protection and restoration of water-related ecosystems
	Goal 11: Make cities and human settlements inclusive, safe, resilient and sustainable	Reduction in the number of people affected by disasters, strengthening of national and regional development planning and implementation of integrated policies and plans for mitigation and adaptation to climate change, resilience to disasters and the development and implementation of holistic disaster risk management
	Goal 12: Ensure sustainable consumption and production patterns	Environmentally sound management of chemicals and all waste
	Goal 13: Take urgent action to combat climate change and its impacts[a]	Strengthening of resilience and adaptive capacity to climate-related and other natural disasters and support for impact reduction and early warning

Targets under Sustainable Development Goal 14	Sustainable Development Goals contributing to the achievement of Goal 14	Mechanism
	Goal 15: Protect, restore and promote sustainable use of terrestrial ecosystems, sustainably manage forests, combat desertification, and halt and reverse land degradation and halt biodiversity loss	Conservation, restoration and sustainable use of terrestrial and inland freshwater ecosystems, and reduction in the degradation of habitats
Sustainable food from the ocean		
Target 14.4: By 2020, effectively regulate harvesting and end overfishing, illegal, unreported and unregulated fishing and destructive fishing practices and implement science-based management plans, in order to restore fish stocks in the shortest time feasible, at least to levels that can produce maximum sustainable yield as determined by their biological characteristics Target 14.6: By 2020, prohibit certain forms of fisheries subsidies which contribute to overcapacity and overfishing, eliminate subsidies that contribute to illegal, unreported and unregulated fishing and refrain from introducing new such subsidies, recognizing that appropriate and effective special and differential treatment for developing and least developed countries should be an integral part of the World Trade Organization fisheries subsidies negotiation[b] Target 14.7: By 2030, increase the economic benefits to small island developing States and least developed countries from the sustainable use of marine resources, including through sustainable management of fisheries, aquaculture and tourism Target 14.b: Provide access for small-scale artisanal fishers to marine resources and markets	Goal 2: End hunger, achieve food security and improved nutrition and promote sustainable agriculture	Increase in agricultural productivity (including aquaculture and mariculture), ensuring sustainable food production and maintaining ecosystems and the genetic diversity of wild species
	Goal 8: Promote sustained, inclusive and sustainable economic growth, full and productive employment and decent work for all	Improved resource efficiency in consumption and production
	Goal 9: Build resilient infrastructure, promote inclusive and sustainable industrialization and foster innovation	Enhancement of scientific research and technological development, research and innovation in developing countries
	Goal 12: Ensure sustainable consumption and production patterns	Sustainable management and efficient use of natural resources, reduction in food losses along production and supply chains, including post-harvest losses, strengthening of scientific and technological capacity to move towards more sustainable patterns of consumption and production, implementation of methods to ensure that tourism remains sustainable, creates jobs and promotes local products, and phasing out of harmful subsidies, where they exist, to reflect their environmental impacts
	Goal 13: Take urgent action to combat climate change and its impacts[a]	Implementation of climate change mitigation, adaptation and impact reduction measures
	Goal 17: Strengthen the means of implementation and revitalize the Global Partnership for Sustainable Development	Improved access to science, technology and innovation, enhanced knowledge-sharing and transfer of technology, and capacity-building

Targets under Sustainable Development Goal 14	Sustainable Development Goals contributing to the achievement of Goal 14	Mechanism
Sustainable economic use of the ocean		
Target 14.2: By 2020, sustainably manage and protect marine and coastal ecosystems to avoid significant adverse impacts, including by strengthening their resilience, and take action for their restoration in order to achieve healthy and productive oceans Target 14.7: By 2030, increase the economic benefits to small island developing States and least developed countries from the sustainable use of marine resources, including through sustainable management of fisheries, aquaculture and tourism Target 14.c: Enhance the conservation and sustainable use of oceans and their resources by implementing international law as reflected in the United Nations Convention on the Law of the Sea, which provides the legal framework for the conservation and sustainable use of oceans and their resources, as recalled in paragraph 158 of "The future we want"	Goal 6: Ensure availability and sustainable management of water and sanitation for all	Improved wastewater management and protection and restoration of wetlands
	Goal 7: Ensure access to affordable, reliable, sustainable and modern energy for all	Improved sources and efficiencies in energy and associated reduction in emissions
	Goal 11: Make cities and human settlements inclusive, safe, resilient and sustainable	Sustainable urbanization and reduction in the environmental impact of cities
	Goal 12: Ensure sustainable consumption and production patterns	Sustainable management and use of natural resources
	Goal 13: Take urgent action to combat climate change and its impacts[a]	Implementation of climate change mitigation, adaptation and impact reduction measures
	Goal 15: Protect, restore and promote sustainable use of terrestrial ecosystems, sustainably manage forests, combat desertification, and halt and reverse land degradation and halt biodiversity loss	Reduction in the degradation of natural habitats and loss of biodiversity, and prevention of the extinction of species
	Goal 16: Promote peaceful and inclusive societies for sustainable development, provide access to justice for all and build effective, accountable and inclusive institutions at all levels	Promotion of the rule of law at the national and international levels
	Goal 17: Strengthen the means of implementation and revitalize the Global Partnership for Sustainable Development	Improved access to science, technology and innovation, enhanced knowledge-sharing and transfer of technology, and capacity-building
Effective implementation of international law as reflected in the United Nations Convention on the Law of the Sea		
Target 14.c: Enhance the conservation and sustainable use of oceans and their resources by implementing international law as reflected in the United Nations Convention on the Law of the Sea, which provides the legal framework for the conservation and sustainable use of oceans and their resources, as recalled in paragraph 158 of "The future we want"	Goal 2: End hunger, achieve food security and improved nutrition and promote sustainable agriculture	Ensuring of sustainable food production systems, maintenance of ecosystems and strengthening of capacity to adapt to climate change, extreme weather, drought, flooding and other disasters
	Goal 3: Ensure healthy lives and promote well-being for all at all ages	Reduction in hazardous chemicals, pollution and contamination

Targets under Sustainable Development Goal 14	Sustainable Development Goals contributing to the achievement of Goal 14	Mechanism
	Goal 6: Ensure availability and sustainable management of water and sanitation for all	Reduction in pollution, improved wastewater management and protection and restoration of water-related ecosystems
	Goal 11: Make cities and human settlements inclusive, safe, resilient and sustainable	Protection and safeguarding of cultural and natural heritage
	Goal 12: Ensure sustainable consumption and production patterns	Environmentally sound management of chemicals and all wastes throughout their life cycle, within agreed international frameworks
	Goal 13: Take urgent action to combat climate change and its impacts[a]	Integration of climate change measures into national policies, strategies and planning
	Goal 17: Strengthen the means of implementation and revitalize the Global Partnership for Sustainable Development	Enhancement of policy coherence for sustainable development

a Acknowledging that the United Nations Framework Convention on Climate Change is the primary international, intergovernmental forum for negotiating the global response to climate change.

b Taking into account ongoing World Trade Organization negotiations, the Doha Development Agenda and the Hong Kong ministerial mandate.

Table 2
Contribution made by Sustainable Development Goal 14 to achieving other Goals

Targets under Sustainable Development Goal 14	Sustainable Development Goals contributed to by the achievement of Goal 14	Mechanism
Target 14.1: By 2025, prevent and significantly reduce marine pollution of all kinds, in particular from land-based activities, including marine debris and nutrient pollution	Goal 3: Ensure healthy lives and promote well-being for all at all ages	Reduction in hazardous chemicals, pollution and contamination
	Goal 6: Ensure availability and sustainable management of water and sanitation for all	Reduction in pollution and the release of hazardous chemicals and materials and wastewater
	Goal 11: Make cities and human settlements inclusive, safe, resilient and sustainable	Sustainable urbanization and reduction in the environmental impact of cities
	Goal 12: Ensure sustainable consumption and production patterns	Environmentally sound management of chemicals and all wastes, including by reducing waste generation
	Goal 17: Strengthen the means of implementation and revitalize the Global Partnership for Sustainable Development	Improved access to science, technology and innovation, enhanced knowledge-sharing and transfer of technology, and capacity-building
Target 14.2: By 2020, sustainably manage and protect marine and coastal ecosystems to avoid significant adverse impacts, including by strengthening their resilience, and take action for their restoration in order to achieve healthy and productive oceans	Goal 1: End poverty in all its forms everywhere	Reduction in exposure and vulnerability to climate-induced extreme events and building of resilience to environmental shocks and disasters
	Goal 2: End hunger, achieve food security and improved nutrition and promote sustainable agriculture	Increase in agricultural productivity (including aquaculture and mariculture), ensuring sustainable food production and maintaining ecosystems and the genetic diversity of wild species
	Goal 8: Promote sustained, inclusive and sustainable economic growth, full and productive employment and decent work for all	Provision of opportunities for sustained economic growth and sustainable tourism
	Goal 11: Make cities and human settlements inclusive, safe, resilient and sustainable	Preservation of and support for those ecosystems that afford protection from disasters to coastal communities
	Goal 13: Take urgent action to combat climate change and its impacts[a]	Contribution to resilience to climate-related hazards

Targets under Sustainable Development Goal 14	Sustainable Development Goals contributed to by the achievement of Goal 14	Mechanism
Target 14.3: Minimize and address the impacts of ocean acidification, including through enhanced scientific cooperation at all levels	Goal 1: End poverty in all its forms everywhere	Reduction in exposure and building of resilience to environmental shocks and disasters
	Goal 2: End hunger, achieve food security and improved nutrition and promote sustainable agriculture	Ensuring of sustainable food production systems, maintenance of ecosystems, strengthening of capacity to adapt to climate change and enhancement of cooperation in research and technological development
	Goal 12: Ensure sustainable consumption and production patterns	Support for developing countries in strengthening their scientific and technological capacity
	Goal 13: Take urgent action to combat climate change and its impacts[a]	Implementation of climate change mitigation, adaptation and impact reduction measures
	Goal 17: Strengthen the means of implementation and revitalize the Global Partnership for Sustainable Development	Improved access to science, technology and innovation, enhanced knowledge-sharing and transfer of technology, and capacity-building
Target 14.4: By 2020, effectively regulate harvesting and end overfishing, illegal, unreported and unregulated fishing and destructive fishing practices and implement science-based management plans, in order to restore fish stocks in the shortest time feasible, at least to levels that can produce maximum sustainable yield as determined by their biological characteristics	Goal 2: End hunger, achieve food security and improved nutrition and promote sustainable agriculture	Increase in agricultural productivity (including aquaculture and mariculture), ensuring sustainable food production and maintaining ecosystems and the genetic diversity of wild species
	Goal 8: Promote sustained, inclusive and sustainable economic growth, full and productive employment and decent work for all	Support for productive activities
	Goal 12: Ensure sustainable consumption and production patterns	Achievement of sustainable management and efficient use of natural resources, reduction in food losses along production and supply chains, including post-harvest losses, strengthening of scientific and technological capacity to move towards more sustainable patterns of consumption and production, and phasing out of harmful subsidies
	Goal 17: Strengthen the means of implementation and revitalize the Global Partnership for Sustainable Development	Enhancement of partnerships for sustainable development

Targets under Sustainable Development Goal 14	Sustainable Development Goals contributed to by the achievement of Goal 14	Mechanism
Target 14.5: By 2020, conserve at least 10 per cent of coastal and marine areas, consistent with national and international law and based on the best available scientific information	Goal 2: End hunger, achieve food security and improved nutrition and promote sustainable agriculture	Maintenance of ecosystems, strengthening of capacity to adapt to climate change, and enhancement of cooperation in research and technological development
	Goal 11: Make cities and human settlements inclusive, safe, resilient and sustainable	Preservation of and support for those ecosystems that afford protection from disasters to coastal communities
	Goal 15: Protect, restore and promote sustainable use of terrestrial ecosystems, sustainably manage forests, combat desertification, and halt and reverse land degradation and halt biodiversity loss	Reduction in the degradation of natural habitats and loss of biodiversity, and prevention of the extinction of species
	Goal 17: Strengthen the means of implementation and revitalize the Global Partnership for Sustainable Development	Improved access to science, technology and innovation, enhanced knowledge-sharing and transfer of technology, and capacity-building
Target 14.6: By 2020, prohibit certain forms of fisheries subsidies which contribute to overcapacity and overfishing, eliminate subsidies that contribute to illegal, unreported and unregulated fishing and refrain from introducing new such subsidies, recognizing that appropriate and effective special and differential treatment for developing and least developed countries should be an integral part of the World Trade Organization fisheries subsidies negotiation[b]	Goal 8: Promote sustained, inclusive and sustainable economic growth, full and productive employment and decent work for all	Support for productive activities
	Goal 12: Ensure sustainable consumption and production patterns	Achievement of sustainable management and efficient use of natural resources, reduction in food losses along production and supply chains, including post-harvest losses, strengthening of scientific and technological capacity to move towards more sustainable patterns of consumption and production, and phasing out of harmful subsidies
	Goal 17: Strengthen the means of implementation and revitalize the Global Partnership for Sustainable Development	Enhancement of partnerships for sustainable development
Target 14.7: By 2030, increase the economic benefits to small island developing States and least developed countries from the sustainable use of marine resources, including through sustainable management of fisheries, aquaculture and tourism	Goal 1: End poverty in all its forms everywhere	Reduction in exposure and building of resilience to environmental shocks and disasters
	Goal 2: End hunger, achieve food security and improved nutrition and promote sustainable agriculture	Increase in agricultural productivity (including aquaculture and mariculture), ensuring sustainable food production and maintaining ecosystems and the genetic diversity of wild species

Targets under Sustainable Development Goal 14	Sustainable Development Goals contributed to by the achievement of Goal 14	Mechanism
	Goal 8: Promote sustained, inclusive and sustainable economic growth, full and productive employment and decent work for all	Provision of opportunities for sustained economic growth and sustainable tourism
	Goal 9: Build resilient infrastructure, promote inclusive and sustainable industrialization and foster innovation	Enhancement of scientific research, upgrade of the technological capabilities of industrial sectors in all countries, in particular developing countries, and encouragement of innovation
	Goal 12: Ensure sustainable consumption and production patterns	Achievement of sustainable management and efficient use of natural resources, and strengthening of scientific and technological capacity
	Goal 13: Take urgent action to combat climate change and its impacts[a]	Implementation of climate change mitigation, adaptation and impact reduction measures
	Goal 17: Strengthen the means of implementation and revitalize the Global Partnership for Sustainable Development	Improved access to science, technology and innovation, enhanced knowledge-sharing and transfer of technology, and capacity-building
Target 14.a: Increase scientific knowledge, develop research capacity and transfer marine technology, taking into account the Intergovernmental Oceanographic Commission Criteria and Guidelines on the Transfer of Marine Technology, in order to improve ocean health and to enhance the contribution of marine biodiversity to the development of developing countries, in particular small island developing States and least developed countries	Goal 9: Build resilient infrastructure, promote inclusive and sustainable industrialization and foster innovation	Enhancement of scientific research, upgrade of the technological capabilities of industrial sectors in all countries, in particular developing countries, and encouragement of innovation
	Goal 12: Ensure sustainable consumption and production patterns	Achievement of sustainable management and efficient use of natural resources, and strengthening of scientific and technological capacity
	Goal 17: Strengthen the means of implementation and revitalize the Global Partnership for Sustainable Development	Improved access to science, technology and innovation, enhanced knowledge-sharing and transfer of technology, and capacity-building
Target 14.b: Provide access for small-scale artisanal fishers to marine resources and markets	Goal 2: End hunger, achieve food security and improved nutrition and promote sustainable agriculture	Increase in agricultural productivity (including aquaculture and mariculture), ensuring sustainable food production and maintaining ecosystems and the genetic diversity of wild species

Targets under Sustainable Development Goal 14	Sustainable Development Goals contributed to by the achievement of Goal 14	Mechanism
	Goal 8: Promote sustained, inclusive and sustainable economic growth, full and productive employment and decent work for all	Improved resource efficiency in consumption and production
	Goal 9: Build resilient infrastructure, promote inclusive and sustainable industrialization and foster innovation	Enhancement of scientific research and technological development, research and innovation in developing countries
	Goal 12: Ensure sustainable consumption and production patterns	Sustainable management and efficient use of natural resources, and implementation of tools for monitoring sustainable development impacts for sustainable tourism that creates jobs and promotes local culture and products
	Goal 17: Strengthen the means of implementation and revitalize the Global Partnership for Sustainable Development	Improved access to science, technology and innovation, enhanced knowledge-sharing and transfer of technology, and capacity-building
Target 14.c: Enhance the conservation and sustainable use of oceans and their resources by implementing international law as reflected in the United Nations Convention on the Law of the Sea, which provides the legal framework for the conservation and sustainable use of oceans and their resources, as recalled in paragraph 158 of "The future we want"	Goal 2: End hunger, achieve food security and improved nutrition and promote sustainable agriculture	Ensuring of sustainable food production systems, maintenance of ecosystems and strengthening of capacity to adapt to climate change, extreme weather, drought, flooding and other disasters
	Goal 3: Ensure healthy lives and promote well-being for all at all ages	Reduction in hazardous chemicals, pollution and contamination
	Goal 6: Ensure availability and sustainable management of water and sanitation for all	Reduction in pollution, improved wastewater management and protection and restoration of water-related ecosystems
	Goal 7: Ensure access to affordable, reliable, sustainable and modern energy for all	Improved sources and efficiencies in energy and associated reduction in emissions
	Goal 11: Make cities and human settlements inclusive, safe, resilient and sustainable	Sustainable urbanization and reduction in the environmental impact of cities, and protection and safeguarding of cultural and natural heritage
	Goal 12: Ensure sustainable consumption and production patterns	Sustainable management and use of natural resources, environmentally sound management of chemicals and all wastes throughout their life cycle, within agreed international frameworks

Targets under Sustainable Development Goal 14	Sustainable Development Goals contributed to by the achievement of Goal 14	Mechanism
	Goal 13: Take urgent action to combat climate change and its impacts[a]	Implementation of climate change mitigation, adaptation and impact reduction measures, and integration of climate change measures into national policies, strategies and planning
	Goal 15: Protect, restore and promote sustainable use of terrestrial ecosystems, sustainably manage forests, combat desertification, and halt and reverse land degradation and halt biodiversity loss	Reduction in the degradation of natural habitats and loss of biodiversity, and prevention of the extinction of species
	Goal 16: Promote peaceful and inclusive societies for sustainable development, provide access to justice for all and build effective, accountable and inclusive institutions at all levels	Promotion of the rule of law at the national and international levels
	Goal 17: Strengthen the means of implementation and revitalize the Global Partnership for Sustainable Development	Enhancement of policy coherence for sustainable development

a Acknowledging that the United Nations Framework Convention on Climate Change is the primary international, intergovernmental forum for negotiating the global response to climate change.

b Taking into account ongoing World Trade Organization negotiations, the Doha Development Agenda and the Hong Kong ministerial mandate.

Landscape of subgoals under Sustainable Development Goal 14 and relevant chapters

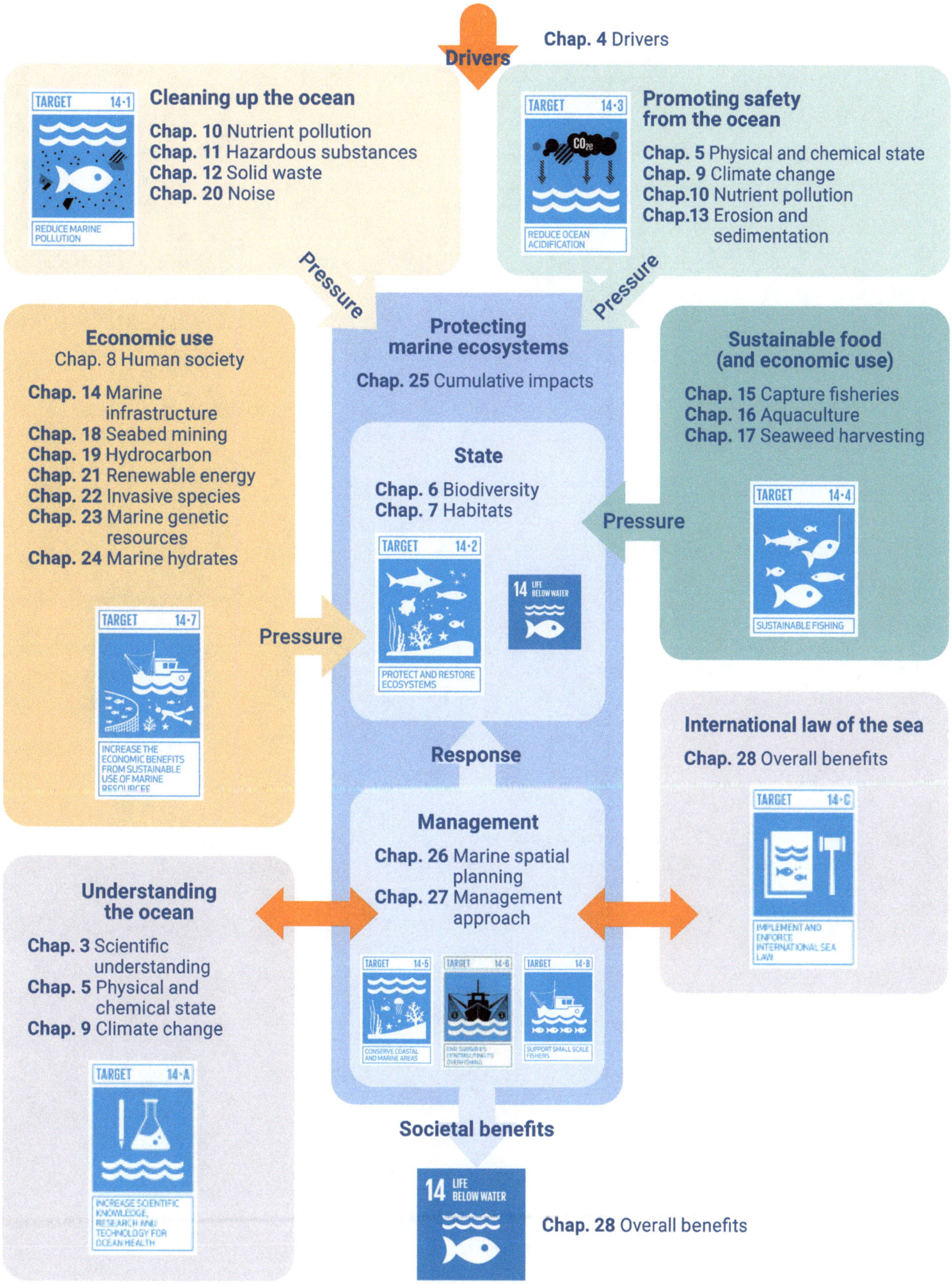

Part two
Introduction

Chapter 2
Approach to the assessment

Contributors: Maria João Bebianno, Hilconida Calumpong, Sanae Chiba, Karen Evans, Carlos Garcia-Soto, Osman Keh Kamara, Enrique Marschoff, Essam Yassin Mohammed, Henn Ojaveer, Chul Park, Ylenia Randrianarisoa, Renison Ruwa, Jörn Schmidt, Alan Simcock, Anastasia Strati, Joshua T. Tuhumwire, Ca Thanh Vu, Juying Wang and Tymon Zielinski (Group of Experts of the Regular Process for Global Reporting and Assessment of the State of the Marine Environment, including Socioeconomic Aspects).

Keynote points

- The purpose of the second *World Ocean Assessment* is to update the first *World Ocean Assessment* by providing an understanding of changes that have occurred in the global ocean since 2010 and associated trends.
- The Assessment also provides an overview of understanding of some aspects not fully covered in the first Assessment, such as inputs of anthropogenic noise, marine hydrates, cumulative effects, marine spatial planning and management approaches.
- The Assessment follows a modified approach to the drivers-pressures-state-impact-response framework, supported through a series of workshops aimed at identifying region-specific information and input for the Assessment, a peer-review process and a process of review by States.

1. Purpose of the second *World Ocean Assessment*

The purpose of the second *World Ocean Assessment* is derived from the principles guiding the Regular Process for Global Reporting and Assessment of the State of the Marine Environment, including Socioeconomic Aspects, and its objective and scope as set out in relevant decisions of the General Assembly, the Ad Hoc Working Group of the Whole of the General Assembly on the Regular Process and its Bureau. The overall objective is set out in the recommendations of the Working Group on the proposed framework of the Regular Process (A/64/347, annex):

> The regular process under the United Nations would be recognized as the global mechanism for reviewing the state of the marine environment, including socioeconomic aspects, on a continual and systematic basis by providing regular assessments at the global and supraregional levels and an integrated view of environmental, economic and social aspects. Such assessments would support informed decision-making and thus contribute to managing in a sustainable manner human activities that affect the oceans and seas, in accordance with international law, including the United Nations Convention on the Law of the Sea and other applicable international instruments and initiatives.

The recommendations of the Working Group were endorsed by the General Assembly in its resolution 64/71, and the principles guiding the Regular Process and its objective and scope were reaffirmed by the Assembly in its resolution 71/257.

With regard to the principles guiding the Regular Process, the Working Group states in its recommendations that:

The regular process would be guided by international law, including the United Nations Convention on the Law of the Sea and other applicable international instruments and initiatives, and would include reference to the following principles:

(a) Viewing the oceans as part of the whole Earth system;

(b) Regular evaluation by Member States of assessment products and the regular process itself to support adaptive management;

(c) Use of sound science and the promotion of scientific excellence;

(d) Regular analysis to ensure that emerging issues, significant changes and gaps in knowledge are detected at an early stage;

(e) Continual improvement in scientific and assessment capacity, including the promotion and development of capacity-building activities and transfer of technology;

(f) Effective links with policymakers and other users;

(g) Inclusiveness with respect to communication and engagement with all stakeholders through appropriate means for their participation, including appropriate representation and regional balance at all levels;

(h) Recognition and utilization of traditional and indigenous knowledge and principles;

(i) Transparency and accountability for the regular process and its products;

(j) Exchange of information at all levels;

(k) Effective links with, and building on, existing assessment processes, in particular at the regional and national levels;

(l) Adherence to equitable geographical representation in all activities of the regular process.

During the first cycle, the scope of the Regular Process and of the first *World Ocean Assessment* was to establish a baseline assessment of all aspects of the ocean – environmental, social and economic. In its resolution 72/73, the General Assembly decided that the scope of the second cycle would extend to evaluating trends and identifying gaps.

The present Assessment, which is the first follow-up to the first Assessment, is aimed at providing a global overview of trends since 2010 in all aspects of the ocean. In addition, it contains reports on some aspects of the ocean that were not fully covered in the first Assessment, including inputs of anthropogenic noise, marine hydrates, cumulative effects, marine spatial planning and management approaches.

2. Primary audience and framework of the second *World Ocean Assessment*

The Regular Process is primarily accountable to the General Assembly (see A/65/358). Given the objective of the Regular Process to "support informed decision-making and thus contribute to managing in a sustainable manner human activities that affect the ocean and seas", the main intended audience of the present Assessment are people in all sectors who will be making decisions that will affect the marine environment. They need to be able to gain an overview of the entire marine environment, as well as focus on the aspects most relevant to their field.

The present Assessment follows the drivers-pressures-state-impacts-response framework, as discussed in chapter 3 of the first Assessment, with some modifications. That modified approach resulted from the discussions in the first round of regional workshops (described below) on the structure of the Assessment. Consequently, in the present Assessment:

(a) Relevant drivers of change in the ocean are set out (Part 3);

(b) Trends in the current state of the main components of the marine environment, such as groups of species, types of habitats and human society, including maritime industries, are described (Part 4);

(c) Pressures and their impacts on the ocean, including relevant socioeconomic components, are identified (Part 5);

(d) Developments in the management measures adopted in response to those pressures and impacts are described (Part 6).

In the rest of the present Part, an overview is given of current scientific understanding of the ocean, in order to provide the background to the Assessment.

3. Preparation of the second *World Ocean Assessment*

In accordance with the terms of reference and working methods of the Group of Experts of the Regular Process for Global Reporting and Assessment of the State of the Marine Environment, including Socioeconomic Aspects, for the second cycle of the Regular Process, and the guidance for contributors prepared by the Group of Experts, the present Assessment was prepared by the Group and by writing teams drawn from the pool of experts.

The Group of Experts was constituted to oversee the work required to undertake the present Assessment. The Group comprises experts nominated by each of the United Nations regional groups of Member States. Those who wrote the individual chapters of the Assessment and reviewed drafts of those chapters were drawn from both the Group of Experts and a pool of experts. The pool comprises experts who served in the pool of experts during the first cycle of the Regular Process and additional experts nominated by States specifically for the second cycle.

The process of preparing the present Assessment began with the Group of Experts setting out the structure. An initial structure, based on that of the first Assessment, consisted of a summary followed by four parts focused on components of the ocean: the ocean and its circulation; the food web; coastal and shelf seas; and the open ocean. The proposed structure was discussed in an initial round of five regional workshops held in 2017, at which the outcomes of the first Assessment were outlined, recent regional assessments were reviewed and regional priorities were identified for incorporation into the present Assessment. On the basis of inputs from those attending the workshops, the proposed structure of the Assessment was revised by the Group so that it addressed two main issues raised: that it more explicitly follow the internationally recognized drivers-pressures-state-impacts-response framework (Smeets and Weterings, 1999), and that it include specific coverage of management issues. The revised structure was embodied in the outline for the present Assessment, which was considered, amended and approved by the Ad Hoc Working Group of the Whole at its 10th meeting.

In support of the development of the chapters, a second round of regional workshops was held in 2018, attended by members of the Group of Experts and experts (including members of the pool of experts and proposed members of writing teams) nominated by States, especially those in the regions concerned. The workshops were focused on developing specific chapters of the Assessment, with participants highlighting regional inputs, regional capacity-building needs and other issues.

The Group of Experts ensured that there were appropriate lists of members of the writing teams, and submitted them to the Bureau of the Ad Hoc Working Group of the Whole for approval. A range of methods was adopted to identify possible members of the writing teams: several experts with relevant expertise were already in the pool of experts and agreed to participate in the drafting when approached by the Group; some members took part in a regional workshop and were later nominated to the pool of experts; and some were added following a request for specific expertise by the Group or through self-nomination. The writing teams primarily conducted their work through teleconferences and correspondence.

The guidance for contributors was developed by the Group of Experts, which noted, inter alia, the need to strive for a global overview, how to describe risks, how to handle uncertainty, and ethics in authoring and evaluating material for the Regular Process (UNGA, 2017b; UNGA, 2018). Lead and co-lead members of the Group for each chapter provided guidance on acceptable types of information and balance within the chapter. Members of the writing team for each chapter were expected to consider the overall balance of the draft chapter and ensure that, as far as possible, each chapter was based on the best available data and information and that the conclusions made were sound and well supported.

The draft of each chapter, once completed and considered fit for peer review, was sent for review by at least two peer reviewers drawn from the pool of experts. Peer reviewers acted in a totally independent capacity and were not involved in drafting the chapter that they reviewed. They were requested to evaluate each chapter from the point of view of overall balance and to consider whether the best available data and information had been used and whether the conclusions were sound and well supported.

Following the peer review and the subsequent revision of each chapter by the writing teams, the chapters were compiled and edited to produce an integrated document to be submitted for review by States. Following that review, the chapters were further revised by the writing teams, and a final draft assessment was compiled. That was submitted by the Group of Experts to the Ad Hoc Working Group of the Whole in order for the Working Group to authorize submission to the General Assembly.

4. Terminology

There is an important distinction to be made between the terminology used in the scientific description of the ocean and the legal terminology used to describe States' rights and obligations in the ocean. With the exception of some aspects of the continental shelf beyond 200 nautical miles, the limits of the maritime zones established by the United Nations Convention on the Law of the Sea are not based on geomorphic criteria.

In the present Assessment, unless stated otherwise, the term "continental shelf" refers to the geomorphic continental shelf, and not to the continental shelf as defined in the Convention (see in particular chapters 7J, 7M and 7N). The geomorphic continental shelf is usually defined in terms of the submarine extension of a continent or island as far as the point at which there is a marked discontinuity in the slope, and the continental slope begins its fall down to the continental rise or the abyssal plain (Hobbs, 2003).

Similarly, the term "open ocean" refers to the water column of deep-water areas that are beyond (i.e., seawards of) the geomorphic continental shelf. It covers the whole of the water column (pelagic zone) in the areas beyond the geomorphic continental shelf.

The term "deep sea" refers to the sea floor of deep-water areas that are beyond (i.e., seawards of) the geomorphic continental shelf. It is the benthic zone that lies in deep water (generally, water with a depth greater than 200 m).

Furthermore, the term "areas beyond national jurisdiction" refers to the high seas and the Area (i.e., the seabed and ocean floor and subsoil thereof, beyond the limits of national jurisdiction) as defined in the Convention.

5. Acknowledgements

Very importantly, resources have been provided from the regular budget of the United Nations for the present Assessment. That has made a significant difference in carrying out the work.

None of the members of the Group of Experts or the pool of experts have received any remuneration for their work.

The Regular Process is grateful to Australia, Brazil, Ecuador, Estonia, Germany, Ghana, Greece, Indonesia, Japan, Malta, New Zealand, Palau, Poland, Portugal, Qatar, the Republic of Korea, Thailand, Ukraine, the United Kingdom of Great Britain and Northern Ireland, the United Republic of Tanzania and the United States of America for their support for the present Assessment. Support was also provided by the Intergovernmental Oceanographic Commission, the Permanent Commission for the South Pacific and an anonymous private donor. The Regular Process is also grateful to Estonia, Ireland, the Netherlands, New Zealand and the Republic of Korea for their voluntary contributions to the trust fund for the Regular Process during the second cycle.

References

Hobbs, Carl, III (2003). "Continental Shelf". In *Encyclopedia of Geomorphology*, ed. Andrew Goudie, Routledge, London and New York.

Smeets, E., and R. Weterings (1999). Environmental indicators: typology and overview. Technical report No. 25/1999. European Environment Agency, Copenhagen.

United Nations, Ad Hoc Working Group of the Whole of the General Assembly on the Regular Process for Global Reporting and Assessment of the State of the Marine Environment, including Socioeconomic Aspects. Report on the work of the Ad Hoc Working Group of the Whole to recommend a course of action to the General Assembly on the regular process for global reporting and assessment of the state of the marine environment, including socioeconomic aspects (A/64/347, annex, paras. 7 and 9).

United Nations General Assembly (UNGA) (2010). Resolution 64/71, para. 177. See also resolution 72/73, para. 302.

________ (2016). Resolution 71/257 (Oceans and the Law of the Sea), para. 299.

________ (2017a). Resolution 72/73 (Oceans and the Law of the Sea), paras. 304 and 330.

________ (2017b). Guidance for Contributors: Part I (A/72/494, annex IV).

________ (2018). Guidance for Contributors: Part II (A/73/74, annex II).

Chapter 3
Scientific understanding of the ocean

Contributors: Bing Qiao (convener), Carlos Francisco Andrade, Paulo Antunes Horta, Nene Bi Trace Boniface, Sanae Chiba (co-lead member), Mohammad Zahedur Rahman Chowdhury, Antonio Di Natale, Karen Evans, Carlos Garcia-Soto (co-lead member), Enrique Marschoff, Colin Moffat, Jocelyne Mpemba Kazadi, Henn Ojaveer, Renison Ruwa (lead member), Jörn Schmidt (co-lead member), Hoinsoude Segniagbeto, Sekou Tidiane Bangoura, Kedong Yin, Chang-Ik Zhang and Tymon Zielinski (co-lead member).

Keynote points

- Innovations in technology and engineering regarding sensors and autonomous observation platforms have substantially increased observations of the ocean and allowed for those observations to be collected at finer temporal and spatial resolutions.
- The networking and coordination of regional observation programmes has been promoted and has enabled better the coordination and integration of efforts and the standardization and harmonization of observation methods.
- Global disparities in understanding and knowledge gaps at the continental regional level remain, in particular across Africa, Oceania and South America.
- Most observation networks do not incorporate the economic, social and cultural aspects of the ocean and, as a consequence, there is a lack of focused, publicly accessible observations of such aspects in standardized formats at the regional and global levels; such observations may be provided through work on supplemental national accounts.

1. Introduction

In the present chapter, changes related to the scientific basis for understanding the marine environment are described. Evidence-based science is considered to be the basis for understanding all aspects of the world. The natural sciences have been particularly important for the discovery and advancement of understanding of the environment, while social sciences and the humanities are important for understanding values placed on the marine environment and human behaviour in both using and valuing the ocean. Those disciplines, when combined, have been essential to understanding the challenges faced by humanity, from individuals to communities and societies, in achieving sustainable use of the marine environment that preserves those values and ensures that the marine environment is conserved. Interdisciplinary and transdisciplinary approaches are increasingly encouraged in marine sciences, and new funding schemes that support such approaches have been implemented by several international funding bodies, such as BiodivERsA,[1] JPI Oceans[2] and the Belmont Forum,[3] and national agencies, as well as through scientific diplomacy efforts and initiatives.[4,5]

The present chapter provides an overview of the advances in science that underly understanding of the ocean, as well as the changes in scientific capacity since the first *World Ocean Assessment* (United Nations, 2017c). It summarizes new developments in science and progress in relation to scientific capacity and builds upon chapter 3 of the first Assessment, on scientific understanding of ecosystem services (United Nations, 2017a), and chapter 30 of the first Assessment, on marine scientific research (United Nations, 2017b). However, it does not provide an update on the concept of ecosystem services or details of the new concept of nature's contributions to people as outlined in the recent report of the Intergovernmental Science-Policy Platform on Biodiversity and Ecosystem Services (IPBES)

1 See www.biodiversa.org.
2 See http://jpi-oceans.eu.
3 See www.belmontforum.org.
4 See https://allatlanticocean.org/main.
5 See https://meetings.pices.int.

(Pascual and others, 2017), as they are included in chapter 28 of the present Assessment.

The present chapter also covers more general developments since the first Assessment in relation to specific disciplines and how they have changed understanding of the ocean (see sect. 2). It summarizes key region-specific changes (see sect. 3), outlines changes that can be expected in the coming years (see sect. 4) and provides an overview of existing knowledge gaps (see sect. 5) and capacity-building gaps (see sect. 6).

2. Description of changes in data, technology and models since the first *World Ocean Assessment* and their consequences for overall understanding, including socioeconomic consequences

Following Valdés and others (International Oceanographic Commission (UNESCO-IOC), 2017a), changes and growth in scientific understanding are identified for eight global categories of marine scientific research disciplines, namely: (a) marine ecosystem functions and processes; (b) ocean and climate; (c) ocean crust and marine geohazards; (d) blue growth; (e) ocean health; (f) human health and well-being; (g) ocean technology and engineering; and (h) ocean observations and marine data. Innovations in technology and engineering related to sensors (e.g., Wang and others, 2019) and autonomous observation platforms (Zolich and others, 2019) have allowed for data collection at finer temporal and spatial resolutions and expanded those observations into remote areas (Camus and others, 2019). Cost-effective and user-friendly sensors, along with mobile applications, the enhanced participation of citizens (e.g., Simoniello and others, 2019) and the deployment of sensors on non-scientific ships are also facilitating the expanded collection of ocean observations (Jiang and others, 2019). That has increased understanding of physical and biogeochemical systems in the ocean (e.g., Moore and others, 2019) and benefited further development of capacity in the early warning and prediction of hazards (Luther and others, 2017). Data sets and methods for the accurate assessment of anthropogenic CO_2 emissions and their redistribution in the atmosphere, oceans and terrestrial biosphere have been developed (Le Quéré and others, 2018).

Advances in computing technology and in statistical approaches to analysing large data sets, such as through machine learning and artificial intelligence, have resulted in advances in remote sensing and the utility of ocean data sets, notably in the monitoring and surveillance of fisheries (Toonen and Bush, 2020) and bioinvasion management (Koerich and others, 2020). Advances in genomic approaches to ocean observation, such as through eDNA methods (Ruppert and others, 2019), are advancing understanding of the distribution and composition of species (Canonico and others, 2019) in the ocean and providing greater insights into food webs, trophic linkages and the connectivity of species throughout regions. New frameworks and tools that identify and assess the cumulative effects of multiple pressures on marine ecosystems (Stelzenmüller and others, 2018; see also chap. 25) and allow for the exploration of management options for the sustainable development of human society have been developed (Halpern and others, 2017; Audzijonyte and others, 2019). Projects such as the Seabed 2030 project[6] of

[6] See https://seabed2030.gebco.net.

the Nippon Foundation and the General Bathymetric Chart of the Oceans have been initiated, with ambitious goals to map 100 per cent of the ocean floor by 2030.

To further develop global ocean observations within an integrated system and ensure that ocean data are comparable, the networking and coordination of regional observation programmes has been promoted (Moltmann and others, 2019). Observation methods are being standardized and harmonized through international initiatives such as the Global Climate Observing System essential climate variables initiative (Bojinski and others, 2014) and the Global Ocean Observing System essential ocean variables initiative (Miloslavich and others, 2018). Findable, accessible, interoperable and reusable data services and principles have been proposed for the ocean (Tanhua and others, 2019a), and platforms for sharing best practices in ocean observing, data sharing and community dialogue have also been established (Pearlman and others, 2019), with the aim of improving the effective use of ocean data for the benefit of society.

3. Key region-specific changes and consequences

3.1. Arctic Ocean

The Arctic Council, including the Arctic Monitoring and Assessment Programme and the Conservation of Arctic Flora and Fauna Circumpolar Biodiversity Monitoring Program, regularly publishes reports on the state of the terrestrial, freshwater and marine environment of the Arctic. Recent reports on the state of Arctic biodiversity (Conservation of Arctic Flora and Fauna (CAFF), 2017), ocean acidification (Arctic Monitoring and Assessment Programme (AMAP), 2018) and climate change effects (AMAP, 2019) have provided new information on rapid changes in the Arctic marine environment, including increasing river discharges associated with low ice coverage that have resulted in an increase in carbon and nutrients and, thus, primary production in coastal regions. Such changes in production, as well as in the timing and intensity of marine algae blooms, are having profound impacts on the whole food web. Warming of the Arctic has also resulted in the introduction of 20 species, and changes in the distributional range of 59 others have been confirmed in the Chukchi Sea and the Beaufort Sea in the past 15 years. According to observations, ocean acidification is severely affecting the Arctic food web, including commercial species such as cod (AMAP, 2019). Despite significant changes in the Arctic Ocean, several regions and ecosystem components continue to be understudied and are lacking long-term monitoring (CAFF, 2017).

3.2. North Atlantic Ocean, Baltic Sea, Black Sea, Mediterranean Sea and North Sea

The joint Baltic Sea research and development programme, BONUS,[7] has made significant progress in improving understanding of the Baltic Sea. Some major trend reversals, such as the return of top predators, the recovery of certain fish stocks and the reduced input of nutrients and harmful substances in the Baltic, have been noted recently (Reusch and others, 2018). A spatially explicit end-to-end Atlantis ecosystem model was recently developed for the Baltic Sea with the aim of evaluating the effects of anthropogenic pressures on the

7 See www.bonusportal.org.

marine ecosystem (Bossier and others, 2018). In the Second Holistic Assessment of the Ecosystem Health of the Baltic Sea, the Helsinki Commission showed that, although there were limited signs of improvement in the state of the Baltic Sea, the goals and ecological objectives of the Baltic Sea Action Plan had not yet been reached. The results of economic and social analyses were also included for themes for which information was available at the subregional level (Helsinki Commission, 2018).

The Commission for the Protection of the Marine Environment of the North-East Atlantic[8] publishes updates from time to time on the status of the marine environment. According to its assessment (Commission for the Protection of the Marine Environment of the North-East Atlantic, 2017), marine protected areas had expanded, and there had been a decrease in contaminants and radioactive discharge, in particular from oil and gas installations. However, eutrophication was still an issue, and an increase in marine litter, especially plastics, was noted. Although the population of some marine mammals, such as the harbour seal (*Phoca vitulina*) and the grey seal (*Halichoerus grypus*), was increasing, the population of others, including the harbour porpoise (*Phocoena phocoena*) and the bottlenose dolphin (*Tursiops truncatus*), was declining. More than a quarter of the marine bird species assessed were declining, and benthic habitats continued to be affected by bottom trawling.

Under the Convention for the Protection of the Marine Environment and the Coastal Region of the Mediterranean,[9] several well-elaborated action plans are being developed that target priority issues for the Mediterranean. They include pollution, the conservation of habitats and species, climate change, integrated coastal zone management and the sustainable use of resources.

Large numbers of local hypoxic "dead zone" vortices in the eastern part of the tropical North Atlantic have recently been discovered. North of 12° north, the vortices bring low-salinity seawater from the upwelling area of the eastern boundary of the North Atlantic to the high seas, while south of 12° north, the eddies appear to be generated in the open ocean (Schütte and others, 2016a). Increased chlorophyll concentrations associated with enhanced oxygen consumption within the eddy cores result in an increase in total oxygen consumption in the open eastern tropical North Atlantic Ocean. That is thought to contribute to the formation of the shallow oxygen minimum zone in the region (Schütte and others, 2016b).

3.3. South Atlantic Ocean and wider Caribbean

Significant progress has been made in the observation, understanding and prediction of multiple coupled climate change effects in the tropical Atlantic, such as continental rainfall, hurricane activity, marine biological productivity, heatwaves, atmospheric circulation with the equatorial Pacific, correlation with and impact on social phenomena, and freshwater input from the Amazon (Foltz and others, 2019; Rodrigues and others, 2019). The Prediction and Research Moored Array in the Tropical Atlantic[10] has been transitioned to the next generation of mooring to expand and enhance its capability for ocean and climate research and forecasting. More in situ observations have been obtained through repeated hydrographic and volunteer ship surveys. There has been a long-term relaxation of upwelling in the coastal regions of Senegal, resulting in diatom blooms. That is expected to result in anoxia and nitrogen loss in the region (Machu and others, 2019). Greater understanding of the cause, movement and ecological impacts

8 See www.ospar.org.
9 See www.unep.org/unepmap.
10 See http://pirata.ccst.inpe.br/en/home.

of sargassum blooms in the Caribbean Sea is required (Wang and Hu, 2017).

Progress has been made on the Coral Reef Early Warning System, in particular through new partnerships, including between the Caribbean Community Climate Change Centre and the National Oceanic and Atmospheric Administration. Under that agreement, the Atlantic Ocean and Meteorological Laboratory, which is partially funded by the Coral Reef Conservation Program, provides consultation and information systems support, including programming of the data-gathering buoy and transmission of the data to the Laboratory.

3.4. Indian Ocean, Arabian Sea, Bay of Bengal, Red Sea, Gulf of Aden and Persian Gulf

Advances in understanding of the Indian Ocean and its ecosystems since the first Assessment are largely due to the second International Indian Ocean Expedition, which has been operational since 2015 and was extended in 2020 for another five years (Hood and others, 2015; Hood and others, 2019). The multinational collaborative effort has observed that the subsurface depletion of oxygen is expanding along the western boundary of the Arabian Sea and has led to a dramatic shift in ecosystems in both the Arabian Sea and the Bay of Bengal (Gomes and others, 2014; Bristow and others, 2017). The expedition has also discovered new submarine canyons and provided enhanced understanding of the benthic habitats of the abyssal nodule field of the central Indian Ocean basin, the western continental margin of the Arabian Sea and the western regions of the Bay of Bengal (Hood and others, 2019). Massive changes in the biogeochemistry and ecosystems of the Persian Gulf resulting from human activities and the first reported measurements of primary production, nitrogen uptake and phytoplankton diversity across biogeochemical provinces in the central oligotrophic Indian Ocean have also been noted as part of the expedition (Hood and others, 2015).

A review of the Indian Ocean Observing System has led to the redesign of the Research Moored Array for African-Asian-Australian Monsoon Analysis and Prediction moorings to include new sites in the Arabian Sea and eight additional sites planned for just outside the exclusive economic zone of India. The moorings provide oceanographic and meteorological data in near real time and are directly available to climate and weather prediction centres for climate modelling and weather forecasting (Hermes and others, 2019). Growing numbers of Argo floats with biogeochemical sensors deployed under the system provide insights into key processes associated with plankton blooms and oxygen minimum zones (Hermes and others, 2019).

The Indonesian Throughflow, the leakage of western tropical Pacific water into the south-eastern tropical Indian Ocean through the Indonesian seas, is an important pathway for the transfer of climate signals and their anomalies in the global ocean (Fan and others, 2018; Feng and others, 2017; Iwatani and others, 2018; Lee and others, 2019; Maher and others, 2018; Zhou and others, 2016). A large amount of uncertainty still remains in measuring and modelling the physical and biogeochemical variability within the Indonesian seas.

3.5. North Pacific Ocean

The North Pacific components of the Integrated Ocean Observing System have expanded their capacity in coastal monitoring and started to include social science disciplines. That has led to a better understanding of the mechanism and ecological impacts of the Alaskan heatwave in 2014–2016 (Yang and others, 2019).

Since the first Assessment, the North Pacific Marine Science Organization[11] has enhanced

[11] See https://meetings.pices.int.

its role in coordinating regional observation networks in the North Pacific and serves as the platform for sharing knowledge among scientists and the bridge between science and policymakers. In the period since the first Assessment, it has released two special publications: one on ocean acidification and deoxygenation in the North Pacific Ocean (Christian and Ono, 2019); and another on the effects of marine debris caused by the 2011 tsunami in Japan (Clarke Murray and others, 2019). It has also furthered understanding of climate and ecosystem predictability, drivers of algal and jellyfish blooms, marine ecosystems and the services that they provide, human well-being and top predators (Watanuki and others, 2016; Makino and Perry, 2017; Trainer, 2017; Uye and Brodeur, 2017; Zhang and others, 2015; Jang and Curchitser, 2018). The organization periodically produces a North Pacific ecosystem status report aimed at reviewing and summarizing the status and trends of marine ecosystems in the North Pacific, in which it considers the factors that are causing or are expected to cause change in the near future. The third report, which will contain details of the trends of physical, chemical and biological properties of the North Pacific Ocean throughout the 2010s, is currently being prepared.

Intensive expansion of marine research capability and capacity, including remote sensing and in situ platforms and land-based infrastructure, by China (Chen and Lei, 2019), has enhanced monitoring capacity in waters off South-East Asia. The system has supported progress in regional cooperation in sustainable development and marine and climate research.

3.6. South Pacific Ocean

New understanding of the effects of climate change and ocean warming has helped to identify major hotspots within the South Pacific Ocean, including south-east of Australia, west of the Galapagos Islands, eastern Micronesia and the Drake Passage, where regions are warming at rates above the global average.[12] At the same time, descriptions and understanding of marine heatwaves and their impacts on marine ecosystems have progressed (Oliver and others, 2018; Fordyce and others, 2019). Assessments of coral atolls across the region have revealed no widespread signs of physical destabilization in the face of sea level rise, with land area remaining stable (Duvat, 2018). Observing systems in the region are now collecting time series of a variety of ocean observations, including the physical and chemical environment, biological productivity and marine animals for which trends and changes are being reported.[13]

New regional partnerships among the members of the Permanent Commission for the South Pacific (Chile, Colombia, Ecuador and Peru) have been developed with the aim of monitoring and forecasting oceanographic and climatic variability.[14] In the recent report on the Tropical Pacific Observing System,[15] recommendations were made for a redesigned moored array[16] that could improve observations in the tropical Pacific Ocean.

Every five years, the Government of Australia produces a report on the state of the Australian environment, the most recent of which was issued in 2016 (Clark and Johnston, 2016; Evans and others, 2016; Evans and others, 2018). It was concluded in the marine and coastal thematic reports that the overall state of the Australian coastal and marine environments could be regarded as good. However, the historical

12 See www.marinehotspots.org.
13 See www.imosoceanreport.org.au.
14 See www.met.igp.gob.pe/elnino/enfen.
15 See http://tpos2020.org.
16 See www.pmel.noaa.gov/gtmba/mission.

impacts of a number of pressures, such as commercial and recreational fishing, and ongoing pressures caused by activities that are currently inadequately managed, such as climate change and marine debris, have led to a deterioration in those environments and are continuing to have a negative effect on them. As a result, the outlook for the coastal and marine environment was regarded as mixed and largely depended on the escalating trajectory of climate-related pressures and the ongoing expansion of coastal and marine development.

New Zealand also regularly produces reports on the state of its marine environment; two have been released since the first Assessment, in 2016 and 2019.[17] The most recent report highlighted ongoing issues, including the fact that many species and habitats are under threat, pollution inputs are increasing, as is sediment accumulation of the marine environment, and boat activity and shipping are increasing, resulting in the spread of non-native species and pollution, increased coastal development and unprecedented change in the marine environment associated with climate change. It is worth noting that the report highlighted that the cumulative effect of such pressures was the most urgent problem faced by the ocean.

3.7. Southern Ocean

The Southern Ocean Observing System, a joint initiative of the Scientific Committee on Antarctic Research and the Scientific Committee on Oceanic Research established in 2011, facilitates the collection of essential physical, chemical and biological oceanographic observations in the Southern Ocean. Regional networks of observational activities operating under the framework of the Southern Ocean Observing System facilitate information exchange, technology transfer, the standardization of measurements and data sharing.[18] Tools developed by the system include an open-access interactive web-based platform, which allows users to explore circumpolar data sets and facilitate the exchange of scientific information. A database of upcoming expeditions to the Southern Ocean enables users to discover which expeditions, such as voyages, flights or traverses, are planned to help to facilitate the coordination of field activities (Newman and others, 2019). The system has supported progress in the number of observations collected since the first Assessment, in particular with regard to monitoring increases in ocean temperature (Roemmich and others, 2015), increases in westerly winds over the Antarctic circumpolar current (Gent, 2016) and freshening of the ocean, most notably close to the continent (Schmidtko and others, 2014). Deployment of biochemical sensors has increased measurements of chlorophyll *a*, nitrate, oxygen, light, optical properties and pH throughout the Southern Ocean (Newman and others, 2019). Ice-capable bio-Argo floats are now collecting information on biogeochemical cycles during ice-covered periods (Briggs and others, 2017), and gliders are adding to the collection of ocean observations (Newman and others, 2019). As ecosystems are changing, variable effects on marine predators have been observed; some populations of the Adélie penguin (*Pygoscelis adeliae*) and the chinstrap penguin (*Pygoscelis antarcticus*) have declined, while some populations of the gentoo penguin (*Pygoscelis papua*) have increased (Trivelpiece and others, 2011; Hinke and others, 2017; see also chap. 7K). Long-term monitoring of marine species, including penguins and seals, continues to be undertaken through the Convention on the Conservation of Antarctic Marine Living Resources as part of the management of krill fisheries and is increasing understanding of their foraging behaviour and demographics (Newman and others, 2019).

17 See www. mfe.govt.nz.

18 See http://soos.aq/activities/cwg/soflux.

4. Outlook for scientific understanding of the ocean

Further scientific research will help in assessing the achievement of targets under Sustainable Development Goal 14, especially during the United Nations Decade of Ocean Science for Sustainable Development.[19] As part of the Decade, innovative approaches to science, involving many disciplines and many sectors of society, are recognized as being necessary to achieve the 2030 Agenda for Sustainable Development. With regard to ocean and coastal observation in general, the OceanObs'19 Conference[20] has put forward a series of recommendations that are focused, inter alia, on sustaining ocean observations; connecting with users and stakeholders; identifying the benefits to society of observations; further developing indicators for the ocean; and fostering trans-disciplinary approaches to research. Work has begun on road maps to further the development of a global ocean observing system that includes and integrates abiotic and biotic observations and goes beyond traditional observation technologies (Speich and others, 2019). Together with advances in computing technology and analytical methods, the outputs of eDNA studies will help in the analysis of biodiversity observations, with a resulting improvement in the input of information into ecosystem models and their use in ecosystem-based management.

The International Convention for the Control and Management of Ships' Ballast Water and Sediments entered into force in 2017.[21] It is aimed at preventing the spread of harmful aquatic organisms from one region to another by establishing standards and procedures for the management and control of ships' ballast water and sediments. Further scientific work is needed to generate the required evidence and knowledge, including on the basis of observations and technology development, to assist managers and stakeholders, including government authorities, in implementing the Convention.

5. Key remaining knowledge gaps

The near-future scientific challenges are related to topics such as understanding and anticipating El Niño Southern Oscillation events and marine ecosystem tipping points, quantifying the cumulative effects of multiple pressures on marine environments, developing adaptive management approaches and making them more operational, and encouraging broader consideration and integration of local, traditional and indigenous knowledge in marine ecosystem assessment and management.

Global disparities in understanding and knowledge gaps at the continental regional level remain. The bulk of the research and the information readily available (based on the number of publications) relates to the North Atlantic Ocean, the North Pacific Ocean and the Arctic Ocean. For other areas, in particular Africa, Oceania and South America (UNESCO-IOC, 2017b), there is less information available.

Timely dissemination of collected measurements is very important for the effective usage

19 See General Assembly resolution 72/73; see also www.oceandecade.org.
20 See www.oceanobs19.net/sessions.
21 International Maritime Organization, document BWM/CONF/36, annex; see also www.imo.org/en/About/Conventions/Pages/International-Convention-for-the-Control-and-Management-of-Ships'-Ballast-Water-and-Sediments-(BWM).aspx.

of data in today's connected ocean prediction and monitoring systems. That aspect of making data available and the software for quality control are essential to making the best use of ocean observations.

Currently, most global observation networks do not incorporate economic, social and cultural aspects of the ocean and, as a consequence, focused, sustained and publicly accessible observations of those aspects of marine systems in standardized formats at the regional and global levels are lacking (Evans and others, 2019). The compilation of economic, social and cultural information in useable formats for inclusion in an assessment framework for synthesis on a global scale requires considerable effort, often beyond the ability of those individuals or groups of individuals involved in contributing to present Assessment. It is an area in which an extension of current observation frameworks to incorporate sustained and standardized monitoring of economic, social and cultural aspects of the ocean would significantly improve assessments undertaken in the framework of the Assessment (Evans and others, 2019). IPBES has made clear the need to increase capacity not only to monitor biodiversity but also to understand its functions and the effect that human activities, including climate change, have on biodiversity (IPBES, 2019). One of the aims of the variables being developed under the Global Ocean Observing System is to expand observations of pressures placed on marine ecosystems by human activities to include ocean noise and marine debris, including plastics. The outputs of the Assessment could assist in guiding the process for identifying such variables and, in so doing, could provide a mutual pathway for further improvements to the observations that contribute to future assessments.

6. Key remaining capacity-building gaps

Advances in global understanding of scientific knowledge depend on uniformity in efforts to engage in research globally across the continental regions. The uniformity of research efforts globally further depends on how advanced infrastructure, specialized scientific human capacity and technology are distributed and shared through partnerships. Many natural science disciplines, such as physical, chemical and biological oceanography and marine geology, require research vessels or other specialized equipment and upgraded modern technology and the support of land-based laboratories equipped with modern equipment to support research surveys in the entire depth range of the global ocean. Further support needs to be provided through the use of satellites for remote-sensing studies of the ocean. Innovation for cost-effective in situ observation tools and methods is also needed.

Currently, the level of scientific understanding is regionally skewed because of disparities in the capacities of regional infrastructure and in specialized professional human capacity. Such disparities, therefore, affect possibilities for engaging in competitive ocean research and, in turn, lead to the observed disparities in scientific understanding of oceans at the regional level.

For improving the forecast capacities for the El Niño Southern Oscillation and other ocean-climatic variations, ocean observing systems need to be strengthened and partnerships with regional countries promoted in order to enhance local capacities.[22] In order to monitor

[22] See http://soos.aq/activities/cwg/soflux.

the significant changes in physical and biogeochemical environments and their impacts on ecosystems and society, further integration of multidisciplinary observations and a reduction in the uncertainty of prediction models are needed. Innovation in funding strategies is also required to sustain integrated observing systems.

The ocean science community has proposed action plans for the coming decade (Speich and others, 2019), which include efforts to increase the efficiency of the ocean information value chain (Tanhua and others, 2019b). To maximize the value of ocean data for societal use, the interface of each service, scientific observation, data assembly and management and policy should be smoothly streamlined. For example, the integration of observing systems and findable, accessible, interoperable and reusable data principals must be implemented in a harmonized manner. The aim of the present Assessment is to enable scientific knowledge to be conveyed as information that is usable and understandable for non-academic users and thus can serve as an important link in the ocean data value chain.

Local, traditional and indigenous knowledge needs to be further integrated, and concepts relating to facilitating collaboration to provide opportunities for recognizing synergies and for sharing and exchanging information (Wright and others, 2019) need to become best practices.

References

Arctic Monitoring and Assessment Programme (AMAP) (2018). *AMAP Assessment 2018: Arctic Ocean Acidification*. Tromsø, Norway: AMAP.

________ (2019). *AMAP Climate Change Update 2019: An Update to Key Findings of Snow, Water, Ice and Permafrost in the Arctic (SWIPA) 2017*. Oslo, Norway: AMAP, p. 12.

Audzijonyte, Asta, and others (2019). Atlantis: a spatially explicit end-to-end marine ecosystem model with dynamically integrated physics, ecology and socio-economic modules. *Methods in Ecology and Evolution*, vol. 10, No. 10, pp. 1814–1819. https://doi.org/10.1111/2041-210X.13272.

Bojinski, Stephan, and others (2014). The concept of essential climate variables in support of climate research, applications, and policy. *Bulletin of the American Meteorological Society*, vol. 95, No. 9, pp. 1431–1443.

Bossier, Sieme, and others (2018). The Baltic Sea Atlantis: An integrated end-to-end modelling framework evaluating ecosystem-wide effects of human-induced pressures. *PloS One*, vol. 13, No. 7.

Briggs, Ellen M., and others (2017). Physical and biological drivers of biogeochemical tracers within the seasonal sea ice zone of the Southern Ocean from profiling floats. *Journal of Geophysical Research: Oceans*, 123(2), pp. 746–758. https://doi.org/10.1002/2017JC012846.

Bristow, L.A., and others (2017). N2 production rates limited by nitrite availability in the Bay of Bengal oxygen minimum zone. *Nature Geoscience*, vol. 10, No. 1, pp. 24–29. https://doi.org/10.1038/ngeo2847.

Camus, Lionel, and others (2019). Autonomous surface and underwater vehicles reveal new discoveries in the arctic ocean. In *OCEANS 2019-Marseille*, pp. 1–8. IEEE.

Canonico, Gabrielle, and others (2019). Global observational needs and resources for marine biodiversity. *Frontiers in Marine Science*, vol. 6, art. 367. https://doi.org/10.3389/fmars.2019.00367.

Chen, Lianzeng, and Bo Lei (2019). Marine science and technology development over the past 70 years in China. *Haiyang Xuebao*, 41(10): 3–22. https://doi.org/10.3969/j.issn.0253-4193.2019.10.002.

Christian, James R., and Tsuneo Ono, eds. (2019). *Ocean Acidification and Deoxygenation in the North Pacific Ocean*. PICES Special Publication 5. North Pacific Marine Science Organization (PICES).

Clark G.F., and E.L. Johnston (2016). Coasts: coasts. In *Australia State of the Environment 2016*. Canberra: Australian Government Department of the Environment and Energy. https://soe.environment.gov.au/theme/coasts.

Clarke Murray, Cathryn, and others, eds. (2019). *The Effects of Marine Debris Caused by the Great Japan Tsunami of 2011*. PICES Special Publication 6. North Pacific Marine Science Organization (PICES).

Commission for the Protection of the Marine Environment of the North-East Atlantic (2017). *Intermediate Assessment 2017*. https://oap.ospar.org/en/ospar-assessments/intermediate-assessment-2017.

Conservation of Arctic Flora and Fauna (CAFF) (2017). State of the Arctic Marine Biodiversity. www.arcticbiodiversity.is/marine.

Delory, E., and J. Pearlman, eds. (2018), *Challenges and Innovations in Ocean: In Situ Sensors*, 1st edition. ISBN: 9780128098868.

Duvat, Virginie K.E. (2018). A global assessment of atoll island planform changes over the past decades. *WIREs Climate Change*, vol. 10, No. 1, p. e557. https://doi.org/10.1002/wcc.557.

Dziak, R.P., and others (2017): Ambient sound at Challenger Deep, Mariana Trench. Oceanography, 30(2), 186–197, https://doi.org/10.5670/oceanog.2017.240.

Evans, Karen, and others (2016). Marine environment: marine environment. In *Australia State of the Environment 2016*. Canberra: Australian Government Department of the Environment and Energy. Canberra.

Evans, Karen, and others (2019). The global integrated world ocean assessment: linking observations to science and policy across multiple scales. *Frontiers in Marine Science*, vol. 6, art. 298. https://doi.org/10.3389/fmars.2019.00298.

Evans, Karen, and others (2018). Enhancing the robustness of a national assessment of the marine environment. *Marine Policy*, vol. 98, pp. 133–145. https://doi.org/10.1016/j.marpol.2018.08.011.

Fan, W., and others (2018) Variability of the Indonesian Throughflow in the Makassar Strait over the Last 30ka. *Scientific Reports*, 8(1):5678. https://doi.org/10.1038/s41598-018-24055-1.

Feng, M., and others (2017). Contribution of the deep ocean to the centennial changes of the Indonesian Throughflow. Geophysical Research Letters, 44(6): 2859–2867. https://doi.org/10.1002/2017GL072577.

Fernandez C., and others (2019). Temporal and spatial variability of biological nitrogen fixation off the upwelling system of central Chile (35–38.5°S), *Journal of Geophysical Research Oceans*, , vol. 120, pp.3330–3349. https://doi.org/10.1002/2014JC010410.

Foltz, G.R., and others (2019). The tropical Atlantic observing system. *Frontiers in Marine Science*, vol. 6, art. 206. https://doi.org/10.3389/fmars.2019.00206.

Fordyce, Alexander J., and others (2019). Marine Heatwave Hotspots in Coral Reef Environments: Physical Drivers, Ecophysiological Outcomes, and Impact Upon Structural Complexity. *Frontiers in Marine Science*, vol. 6, art. 498. https://doi.org/10.3389/fmars.2019.00498.

Gent, Peter R. (2016). Effects of Southern Hemisphere wind changes on the meridional overturning circulation in ocean models. *Annual Review of Marine Science*, vol. 8, No. 1, pp. 79–94. https://doi.org/10.1146/annurev-marine-122414-033929.

Gomes, Helga do Rosário, and others (2014). Massive outbreaks of noctiluca scintillans blooms in the Arabian Sea due to spread of hypoxia. *Nature Communications*, vol. 5, No. 1, p. 4862. https://doi.org/10.1038/ncomms5862.

Halpern, Benjamin S., and others (2017). Drivers and implications of change in global ocean health over the past five years. *PLOS ONE*, vol. 12, No. 7, pp. 1–23. https://doi.org/10.1371/journal.pone.0178267.

Helsinki Commission (2018). State of the Baltic Sea: second HELCOM holistic assessment 2011-2016. In *Baltic Sea Environment Proceedings 155*. Helsinki, Finland.

Hermes, J.C., and others (2019). A sustained ocean observing system in the Indian Ocean for climate related scientific knowledge and societal needs. *Frontiers in Marine Science*, vol. 6, art. 355. https://doi.org/10.3389/fmars.2019.00355.

Hinke, Jefferson T., and others (2017). Variable vital rates and the risk of population declines in Adélie penguins from the Antarctic Peninsula region. *Ecosphere*, vol. 8, No. 1, p. e01666. https://doi.org/10.1002/ecs2.1666.

Hood, Raleigh R., and others (2015). *Science Plan of the Second International Indian Ocean Expedition (IIOE-2): A Basin-Wide Research Program*. Newark, Delaware: Scientific Committee on Oceanic Research.

Hood, Raleigh R., and others (2019). The second International Indian Ocean Expedition (IIOE-2): Motivating new exploration in a poorly understood ocean basin (volume 2). *Deep Sea Research Part II: Topical Studies in Oceanography*, vol. 166, pp. 3–5. https://doi.org/10.1016/j.dsr2.2019.07.016.

Huang, Zhi, and Xiao Hua Wang (2019). Mapping the spatial and temporal variability of the upwelling systems of the Australian south-eastern coast using 14-year of MODIS data. *Remote Sensing of Environment*, vol. 227, pp. 90–109.

Intergovernmental Oceanographic Commission (UNESCO-IOC) (2017a). *Global Ocean Science Report: The Current Status of Ocean Science around the World*. ed. Luis Valdés. Paris: UNESCO Publishing.

_________ (2017b). Research productivity and science impact. In *Global Ocean Science Report: The Current Status of Ocean Science around the World*, ed. Luis Valdés. Paris: UNESCO Publishing.

Intergovernmental Science-Policy Platform on Biodiversity and Ecosystem Services (IPBES) (2019). *Summary for Policymakers of the Global Assessment Report on Biodiversity and Ecosystem Services of the Intergovernmental Science-Policy Platform on Biodiversity and Ecosystem Services, IPBES/7/10/Add.1*.

Jang, Chan Joo, and Enrique Curchitser, eds. (2018). Report of working group 29 on regional climate modeling. *PICES Scientific Report*, No. 54, pp. 1–177.

Iwatani, Hokuto, and others (2018). Intermediate-water dynamics and ocean ventilation effects on the Indonesian Throughflow during the past 15,000 years: Ostracod evidence. *Geology*. https://doi.org/10.1130/G40177.1.

Jiang, Zong-Pei, and others (2019). Enhancing the observing capacity for the surface ocean by the use of Volunteer Observing Ship. *Acta Oceanologica Sinica*, vol. 38, No. 7, pp. 114–120. https://doi.org/10.1007/s13131-019-1463-3.

Koerich, Gabrielle, and others (2020). How experimental physiology and ecological niche modelling can inform the management of marine bioinvasions? *Science of The Total Environment* 700: 134692. https://doi.org/10.1016/j.scitotenv.2019.134692.

Lee, T., and others (2019). Maritime Continent water cycle regulates low-latitude chokepoint of global ocean circulation. *Nature Communications*. https://doi.org/10.1038/s41467-019-10109-z.

Le Quéré, C., and others (2018). Global carbon budget 2018. *Earth System Science Data*, vol. 10, No. 4, pp. 2141–2194. https://doi.org/10.5194/essd-10-2141-2018.

Luther, Jochen, and others (2017). World Meteorological Organization: Concerted International Efforts for Advancing Multi-hazard Early Warning Systems. In *Advancing Culture of Living with Landslides*, eds. Kyoji Sassa, Matjaž Mikoš, and Yueping Yin, pp. 129–41. Cham, Switzerland: Springer International Publishing.

Machu, E., and others (2019). First evidence of anoxia and nitrogen loss in the southern Canary upwelling system. *Geophysical Research Letters*, vol. 46, No. 5, pp. 2619–2127. https://doi.org/10.1029/2018GL079622.

Maher, N., and others (2018). Role of Pacific trade winds in driving ocean temperatures during the recent slowdown and projections under a wind trend reversal. *Climate Dynamics*, 51(1-2):321-336. https://doi.org/10.1007/s00382-017-3923-3.

Makino, Mitsutaku, and R. Ian Perry, eds. (2017). Marine Ecosystems and Human Well-being: The PICES-Japan MAFF MarWeB Project. *PICES Scientific Report*, No. 52, pp. 1–234.

Miloslavich, Patricia, and others (2018). Essential ocean variables for global sustained observations of biodiversity and ecosystem changes. *Global Change Biology*, vol. 24, No. 6, pp. 2416–2133. https://doi.org/10.1111/gcb.14108.

Molina, Verónica, and Laura Farías (2009). Aerobic ammonium oxidation in the oxycline and oxygen minimum zone of the eastern tropical South Pacific off northern Chile (~20°S). *Deep Sea Research Part II: Topical Studies in Oceanography*, vol. 56, pp. 1032–1041. https://doi.org/10.1016/jdsr2.2008.09.006.

Moltmann, Tim, and others (2019). A Global Ocean Observing System (GOOS), delivered through enhanced collaboration across regions, communities, and new technologies. *Frontiers in Marine Science*, vol. 6, art. 291. https://doi.org/10.3389/fmars.2019.00291.

Moore, Andrew M., and others (2019). Synthesis of ocean observations using data assimilation for operational, real-time and reanalysis systems: a more complete picture of the state of the ocean. *Frontiers in Marine Science*, vol. 6, art. 90. https://doi.org/10.3389/fmars.2019.00090.

Newman, Louise, and others (2019). Delivering sustained, coordinated, and integrated observations of the Southern Ocean for global impact. *Frontiers in Marine Science*, vol. 6, art. 433. https://doi.org/10.3389/fmars.2019.00433.

Oliver, Eric C.J., and others (2018). Marine heatwaves off eastern Tasmania: trends, interannual variability, and predictability. *Progress in Oceanography*, vol. 161, pp. 116–130.

Pascual, Unai, and others (2017). Valuing nature's contributions to people: the IPBES approach. *Current Opinion in Environmental Sustainability*, vol. 26–27, pp. 7–16. https://doi.org/10.1016/j.cosust.2016.12.006.

Pearlman, Jay, and others (2019). Evolving and sustaining ocean best practices and standards for the next decade. *Frontiers in Marine Science*, vol. 6, art. 277. https://doi.org/10.3389/fmars.2019.00277.

Reusch, Thorsten B.H., and others (2018). The Baltic Sea as a time machine for the future coastal ocean. *Science Advances*, vol. 4, No. 5. https://doi.org/10.1126/sciadv.aar8195.

Rignot, Eric, and others (2002). Rapid bottom melting widespread near Antarctic Ice Sheet grounding lines. Science (New York), 296(5575): 2020–3. https://doi.org/10.1126/science.1070942.

Rodrigues, R.R., and others (2019). Common cause for severe droughts in South America and marine heatwaves in the South Atlantic. *Nature Geoscience*, 12(8), 620–626. https://doi.org/10.1038/s41561-019-0393-8.

Roemmich, Dean, and others (2015). Unabated planetary warming and its ocean structure since 2006. *Nature Climate Change*, vol. 5, No. 3, pp. 240–45. https://doi.org/10.1038/nclimate2513.

Ruppert, Krista M., and others (2019). Past, present, and future perspectives of environmental DNA (eDNA) metabarcoding: A systematic review in methods, monitoring, and applications of global eDNA. *Global Ecology and Conservation*, vol. 17, p. e00547. https://doi.org/10.1016/j.gecco.2019.e00547.

Schmidtko, Sunke, and others (2014). Multidecadal warming of Antarctic waters. *Science*, vol. 346, No. 6214, pp. 1227–1231. https://doi.org/10.1126/science.1256117.

Schütte, Florian, and others (2016a). Occurrence and characteristics of mesoscale eddies in the tropical northeastern Atlantic Ocean. *Ocean Science*, 12(3), pp. 663–685. https://doi.org/10.5194/os-12-663-2016.

Schütte, Florian, and others (2016b). Characterization of "dead-zone" eddies in the tropical northeast Atlantic Ocean. *Biogeosciences (BG)*, 13, pp. 5865–5881.

Simoniello, Christina, and others (2019). Citizen-science for the future: advisory case studies from around the globe. *Frontiers in Marine Science*, vol. 6, art. 225. https://doi.org/10.3389/fmars.2019.00225.

Speich, Sabrina, and others (2019). Editorial: OceanObs'19: an ocean of opportunity. *Frontiers in Marine Science*, vol. 6, art. 570. https://doi.org/10.3389/fmars.2019.00570.

Stelzenmüller, Vanessa, and others (2018). A risk-based approach to cumulative effect assessments for marine management. *Science of The Total Environment*, vol. 612, pp. 1132–1140. https://doi.org/10.1016/j.scitotenv.2017.08.289.

Tanhua, Toste, and others (2019a). Ocean fair data services. *Frontiers in Marine Science*, vol. 6, art. 440.

Tanhua, Toste, and others (2019b). What we have learned from the framework for ocean observing: evolution of the global ocean observing system? *Frontiers in Marine Science*, vol. 6, art. 471. https://doi.org/10.3389/fmars.2019.00471.

Toonen, Hilde M., and Simon R. Bush (2020). The digital frontiers of fisheries governance: fish attraction devices, drones and satellites. *Journal of Environmental Policy & Planning*, vol. 22, No. 1, pp. 125–137. https://doi.org/10.1080/1523908X.2018.1461084.

Trainer, Vera L., ed. (2017). Conditions Promoting Extreme Pseudo-nitzschia Events in the Eastern Pacific but not the Western Pacific. *PICES Scientific Report*, No. 53, pp. 1–52.

Trivelpiece, Wayne Z., and others (2011). Variability in krill biomass links harvesting and climate warming to penguin population changes in Antarctica. *Proceedings of the National Academy of Sciences*, vol. 108, No. 18, pp. 7625–7628. https://doi.org/10.1073/pnas.1016560108.

United Nations (2017a). Chapter 3: Scientific understanding of ecosystem services. In *The First Global Integrated Marine Assessment: World Ocean Assessment I*. Cambridge: Cambridge University Press.

________ (2017b). Chapter 30: Marine scientific research. In *The First Global Integrated Marine Assessment: World Ocean Assessment I*. Cambridge: Cambridge University Press.

________ (2017c). *The First Global Integrated Marine Assessment: World Ocean Assessment I*. Cambridge: Cambridge University Press.

Uye, Shin-ichi, and Richard D. Brodeur, eds. (2017). Report of working group 26 on jellyfish blooms around the North Pacific rim: causes and consequences. *PICES Scientific Report*, No. 51, pp. 1–222.

Wang, M.Q., and C.M. Hu (2017). Predicting sargassum blooms in the Caribbean Sea from MODIS observations. *Geophysical Research Letters* 44: 3265–3273. https://doi.org/10.1002/2017GL072932.

Wang, Zhaohui Aleck, and others (2019). Advancing observation of ocean biogeochemistry, biology, and ecosystems with cost-effective in situ sensing technologies. *Frontiers in Marine Science*, vol. 6, art. 519. https://doi.org/10.3389/fmars.2019.00519.

Watanuki, Yutaka, and others, eds. (2016). Spatial ecology of marine top predators in the North Pacific: tools for integrating across datasets and identifying high use areas. *PICES Scientific Report*, No. 50, pp. 1–55.

Wright, A.L., and others (2019). Using two-eyed seeing in research with indigenous people: an integrative review. *International Journal of Qualitative Methods*. https://doi.org/10.1177/1609406919869695.

Yang, Qiong, and others (2019). How "The Blob" affected groundfish distributions in the Gulf of Alaska. *Fisheries Oceanography*, vol. 28, No. 4, pp. 434–453. https://doi.org/10.1111/fog.12422.

Zhang, Chang Ik, and others (2015). An extended ecosystem-based fisheries assessment. In *Proceedings of the Twelfth International Conference on the Mediterranean Coastal Environment*, MEDCOAST 15, 6–10 October 2015, Varna, Bulgaria, E. Ozhan (ed.), vol. 1467–1490.

Zhou, L., and others (2016). A Central Indian Ocean Mode and Heavy Precipitation during Indian Summer Monsoon. *Journal of Climate*, 30(6):2055-2067. https://doi.org/10.1175/JCLI-D-16-0347.1.

Zolich, Artur, and others (2019). Survey on communication and networks for autonomous marine systems. *Journal of Intelligent & Robotic Systems*, vol. 95, No. 3, pp. 789–813. https://doi.org/10.1007/s10846-018-0833-5.

Part three

Drivers of changes in the marine environment

Chapter 4
Drivers

Contributors: Chang-Ik Zhang (convener), Karen Evans (co-lead member), Andrew F. Johnson, Osman Keh Kamara (co-lead member), Renison Ruwa (lead member), Jörn Schmidt (co-lead member) and Thomas W. Therriault.

Keynote points

- Drivers that have the greatest influence on the marine environment and its sustainability are: (a) population growth and demographic changes; (b) economic activity; (c) technological advances; (d) changing governance structures and geopolitical instability; and (e) climate change.
- Relationships between drivers and pressures (and their impacts) are complex and dynamic, with interlinkages between drivers leading to cumulative interactions and effects of pressures.
- Drivers vary regionally as a result of global variability in population distribution and demographics, the degree of economic development, technological capacity and the uneven effects of climate change; therefore human activities and pressures vary globally. The most notable differences are between temperate and tropical regions, and developed and least developed regions.
- Integrated modelling frameworks, within which scenarios can be explored – including changes to people and economies, governance structures and the effects of climate change on maritime industries and the environment that are multisectoral and therefore provide whole-of-system approaches – allow for the identification of sustainable ocean use.

1. Introduction

The drivers-pressures-state-impacts-response conceptual framework (Smeets and Weterings, 1999) is a widely used approach to assess causes and consequences of ecosystem change and the actions that might be implemented in response to that change. Since its development, it has been further refined, and many derivatives have been formulated to address limitations and apply the framework to specific environments (e.g., Patricio and others, 2016). Although there are many variants, the underlying framework helps to characterize the effect of human activities on the environment and can be used to inform decision-making and policymaking (Maxim and others, 2009). The framework has been used to structure the second Assessment, and a detailed description is included in chapter 2.

The present chapter is focused on drivers of change in the marine environment, their development since the first Assessment (United Nations, 2017a) and projected changes for the future. Drivers of change in the marine environment were not specifically detailed in the first Assessment, although they were considered in some chapters.

There is no universally agreed set of drivers that has been defined for the marine environment. Different programmes and assessment processes have defined drivers in varying ways and, in some cases, drivers and pressures, whether natural or anthropogenic in nature, are used interchangeably. In the Millennium Ecosystem Assessment, drivers are defined as any natural or human-induced factor that directly or indirectly results in a change in an ecosystem (Millennium Ecosystem Assessment, 2003). The Intergovernmental Science-Policy Platform on Biodiversity and Ecosystem Services takes a similar approach in its global assessment, identifying drivers as direct human influences on nature and factors behind human choices that affect nature (Balvanera and others, 2019). The European Environment Agency considers only human-induced factors as drivers (European Environment Agency,

2005), while the Intergovernmental Panel on Climate Change defines drivers within the context of global emissions as those elements that directly or indirectly contribute to greenhouse gas emissions (Blanco and others, 2014).

In the context of the present Assessment, drivers have been characterized according to social, demographic and economic developments in societies, including corresponding changes in lifestyles and associated overall consumption and production patterns (European Environment Agency, 2019), that are applying pressures on the marine environment, as detailed in Part 5. Pressures are the immediate factors that lead to changes in the state of the marine environment and occur in addition to changes resulting from natural processes (United Nations Environment Programme, 2019). The drivers that have the greatest influence on the marine environment and its sustainability are:

(a) Population growth and demographic changes;

(b) Economic activity;

(c) Technological advances;

(d) Changing governance structures and geopolitical instability;

(e) Climate change.[1]

Increases in the global population, together with global economic growth and technological change, have led to changes in lifestyle and thus an increase in the demand for resources, including food, energy and natural resources such as rare earth elements, sand and metals. Population growth and the associated demand are causing increases in greenhouse gases emissions, the production of waste, including plastic, the use of chemicals in agricultural production, energy production and the extraction of resources.

The relationships between drivers and pressures, and their impacts, are complex and dynamic, with interlinkages between drivers. For example, technological advances can influence economic growth, and changing governance regimes can influence access to and use of technologies. With increasing affluence and access to technologies, efficiencies can be achieved in resource extraction, leading to greater pressures being placed on the ocean (see also sect. 2).

The Sustainable Development Goals[2] were developed to translate human aspirations for a sustainable and equitable future into specific development goals, while recognizing explicitly adverse ecological threats and the strategies required to mitigate them (United Nations, 2017b). While the marine environment is directly addressed in Goal 14 on life below water, the Goals are interlinked, with progress made in one influencing the others. Accordingly, realization of the sustainable use of the marine environment will depend on successfully addressing all of the Goals (International Council for Science, 2017).

---

1 Strictly speaking, the driver is the increase in greenhouse gases that is causing a changed climate. However, the term "climate change" is widely used to describe human activity that directly or indirectly alters the composition of the global atmosphere.

2 See General Assembly resolution 70/1.

2. Drivers of change in the marine environment

2.1. Population growth and demographic changes

Although the global human population increased from 7 billion in 2011 to 7.7 billion in 2019,[3] the growth rate has been steadily decreasing, from 2.1 per cent in 1968 to 1.08 per cent in 2019. Projections of global population growth suggest an uneven but continuing increase, at a lower rate of change, to a mean population size of 9.7 billion by 2050. The decrease in the growth rate is associated with declining numbers of births and, combined with lower mortality rates and increasing longevity associated with increasing health standards, means that the average age of the global population is rising (Baxter and others, 2017).

The number of international migrants has increased from 2.8 per cent of the global population in 2000 to 3.5 per cent in 2019 (United Nations, Department of Economic and Social Affairs, Population Division (UNDESA), 2019a). Most migrations have occurred between countries within the same region, with the exception of North America and Oceania, where, 97.5 per cent and 87.9 per cent, respectively, of international migrants were born in another region (UNDESA, 2019a).

More than 600 million people live in coastal regions that are less than 10 m above sea level, and nearly 2.5 billion people live within 100 km of the coast (UNDESA, 2019b). Those regions are experiencing higher rates of population growth and urbanization than inland regions (Neumann and others, 2015). Such growth has resulted in many economic benefits to coastal regions, including improved transportation and increased trade, tourism and food production, as well as social, recreational and cultural benefits (Clark and Johnston, 2017). However, as the population in those regions grows, it is placing ever more pressure on coastal ecosystems. The extent to which an increasing global population places pressures on the marine environment varies and depends on a variety of factors, including where and how people live, the amount that is consumed and the technologies used to produce energy, food and materials, provide transport and manage the waste produced. The implications of changes to the global population on coastal regions, the use of marine resources and the generation of waste are described in detail in chapter 8 and Part 5.

2.2. Economic activity

Economic growth, measured as gross domestic product (GDP) per capita, has steadily increased globally,[4] although it has slowed as a result of declining trade volume. Growth in the first half of 2019 stood at 1 per cent, the weakest level since 2012 (IMF, 2019). Economic growth, when averaged across the global population (with vast geographical variability in economic growth, as noted in sect. 3), has resulted in the average annual income of an individual increasing from $3,300 in 1950 to $14,574 in 2016. The slowdown in growth is largely associated with weak manufacturing and trade. In contrast, service industries such as tourism have grown (IMF, 2019).

As the global population has grown and the demand for goods and services has increased, there has been an associated increase in energy consumption and resource use. Understanding the relationship between increasing economic activity and the use of natural resources is essential to identifying future sustainability and limiting impacts associated with extraction, production, consumption and waste generation (Jackson, 2017).

---

[3] See https://population.un.org/wpp/Graphs/DemographicProfiles/Line/900.

[4] See https://ourworldindata.org/economic-growth.

Total energy demand, as measured in million tons of oil equivalent (Mtoe), grew from 13,267 Mtoe in 2014 to 13,978 Mtoe in 2018.[5] At the same time, primary energy intensity, an indicator of how much energy is used by the global economy, slowed from 1.7 per cent in 2017 to 1.2 per cent in 2019 (IEA, 2019a). The slowing of efficiencies (i.e., the amount of GDP generated for the amount of energy used) is the result of a number of short-term factors, such as growth in fossil fuel-based electricity generation, and longer-term structural changes, such as a slowing transition towards less energy-intensive industries. At the same time, investment targeting energy efficiencies has been stable since 2014. Improvements in technical efficiency reduced energy-related carbon emissions by 3.5 gigatons of CO_2 between 2015 and 2018 (IEA, 2019a). In addition, renewable energy production has grown as many countries shift to energy strategies that rely on it as part of efforts to reduce greenhouse gas emissions. Ocean energy production is part of many strategic developments and grew from 1 TWh (Terawatt hour) in 2014 to 1.2 TWh in 2018 (IEA, 2019b). Changes in energy production, including marine renewable energy and the pressures generated on the marine environment, are detailed in chapters 19 and 21.

Economic activity associated with the extraction of marine resources also continues to grow as the global population expands. Marine and freshwater food production was a key protein provider and source of income for approximately 59.6 million people globally in 2016, an increase from 56.6 million in 2014. Although marine capture fisheries remain stable at around 80 million tons, mariculture is steadily increasing, from 26.8 million tons in 2014 to 28.7 million tons in 2016 (Food and Agriculture Organization of the United Nations, 2018). The implications for increasing marine food production demands, including overfishing, bycatch of endangered species and habitat loss or degradation by fishing and aquaculture, are described in detail in chapters 15–17.

Many countries are developing or have developed strategies for the potential growth of maritime activities, such as ocean energy, aquaculture, marine biotechnology, coastal tourism and seabed mining (i.e., growth of the blue economy). However, an important constraint on the growth of ocean economies is the current declining health of the ocean and the pressures already being placed on it (Organization for Economic Cooperation and Development, 2016), many of which are detailed in Part 5.

2.3. Technological advances

As maritime activities have expanded and demands on resources have increased, technological advances have been key to increasing efficiencies, expanding markets and enhancing economic growth associated with activities. Such innovations have led to both positive and negative outcomes for the marine environment. Some advances in fishing technologies have led to an overall increase in capacity and, in many regions in Asia, Europe and North America, to overcapacity (Eigaard and others, 2014). Increased efficiencies generated through the use of technologies (also known as "technological creep"), for example, allowing for more efficient and accurate targeting of catches, have also resulted in effort gains within fisheries, thus contributing to overfishing of stocks (Finkbeiner and others, 2017). Conversely, advances in remote sensing, camera technologies, field deployment of genetic approaches to species identification and the use of artificial intelligence and machine learning approaches are now contributing to better monitoring of illegal, unregulated and unreported catches (Detsis and others, 2012), improving the reporting of catches (Ruiz and others, 2014), allowing for the traceability of

5 See https://yearbook.enerdata.net/total-energy/world-consumption-statistics.html.

products (Lewis and Boyle, 2017) and reducing wastage along supply chains (Hafliðason and others, 2012). Such technologies are also assisting in improved monitoring of the movements of fishing fleets, thus ensuring more effective management of protected areas (Rowlands and others, 2019).

Technological advances, including digitalization, are modernizing energy efficiency by reducing energy use, shifting demand from peak to off-peak periods, increasing connectivity and providing flexible loads (which account for increasing shares of intermittent energy generation in the renewable sector), with positive outcomes in terms of greenhouse gas emissions (IEA, 2019a). Improvements in vehicle engines to burn fossil fuels more efficiently and innovations in solar and wind energy to produce clean energy are also helping to reduce greenhouse gas emissions.

2.4. Changing governance structures and geopolitical instability

Many international treaties and agreements, including the United Nations Convention on the Law of the Sea,[6] the Convention on the Prevention of Marine Pollution by Dumping of Wastes and Other Matter of 1972, [7] the Agreement for the Implementation of the Provisions of the United Nations Convention on the Law of the Sea of 10 December 1982 relating to the Conservation and Management of Straddling Fish Stocks and Highly Migratory Fish Stocks,[8] the Convention on Biological Diversity[9] and the 2030 Agenda for Sustainable Development,[10] are aimed at reducing pressures on the marine environment and improving conservation outcomes. Targets set in association with international agreements, such as the Aichi Biodiversity Targets[11] and the Sustainable Development Goals,[12] have led to an increase in the establishment of marine protected areas and an associated increase in the protection of the marine environment. Regional fisheries management organizations provide for the coordination of efforts aimed at managing shared fishery resources (Haas and others, 2020), and in some regions they have provided for the implementation of effective stock rebuilding frameworks following overfishing (Hillary and others, 2016).

Supporting policies implemented nationally have also improved the management of marine activities in some areas (Evans and others, 2017). However, global inequities, including those associated with wealth, gender, geography, rights and access to resources, can have implications on the effectiveness of policies designed to manage the marine environment (Balvanera and others, 2019). Furthermore, the consolidation and concentration of company ownership has resulted in a few corporations or financiers often controlling large shares of the flows in any market (e.g., Bailey and others, 2018). Corporations have increased potential to negotiate directly with Governments, which could hamper progress towards sustainable outcomes for the marine environment. Where there is conflict over access to resources and property rights, policies and agreements focused on sustainability can be undermined by such conflicts (Suárez-de Vivero and Rodríguez Mateos, 2017). In addition, instability in Governments can result in the slow or ineffectual development of policies and management frameworks, resulting in ongoing or increasing overexploitation of resources.

6 United Nations, *Treaty Series*, vol. 1833, No. 31363.
7 Ibid., vol. 1046, No. 15749.
8 Ibid., vol. 2167, No. 37924.
9 Ibid., vol. 1760, No. 30619.
10 See General Assembly resolution 70/1.
11 See United Nations Environment Programme, document UNEP/CBD/COP/10/27, annex, decision X/2.
12 See General Assembly resolution 70/1.

2.5. Climate change

Climate has always been a major influence on the marine environment, with high natural variability from year to year and longer-term variability associated with climate phenomena at the regional and global levels. However, there is strong evidence that the climate is changing at a rate unprecedented in the geological record. The Intergovernmental Panel on Climate Change, in its special report on the ocean and cryosphere in a changing climate (Intergovernmental Panel on Climate Change (IPCC), 2019), summarizes historic and recent patterns in the global climate and provides projections of changes under different greenhouse gas emissions scenarios.

Greenhouse gas emissions have continued to rise over the period since the first Assessment, with global CO_2 emissions increasing from 30.4 gigatons in 2010 to 33.3 gigatons in 2019.[13] The growth in emissions has resulted in widespread reduction of the cryosphere (frozen-water parts of the planet), continued increases in ocean temperature, decreases in ocean pH and oxygen, shifts in currents and increases in extreme events such as heatwaves (IPCC, 2019). Those changes are described in detail in chapter 5, and the pressures that they are generating, including socioeconomic impacts, are described in detail in chapter 9.

Following on from the United Nations Framework Convention on Climate Change (which entered into force in 1994) and the Kyoto Protocol (which entered into force in 2005), the Conference of the Parties to the Convention, at its twenty-first session, adopted the Paris Agreement in December 2015.[14] The Agreement is aimed at strengthening the global response to the threat of climate change by holding the global average temperature well below 2°C above pre-industrial levels and to pursue efforts to limit the temperature increase to 1.5°C above pre-industrial levels. It is recognized in the Agreement that climate change represents an urgent and potentially irreversible threat to human societies and the planet and thus requires the widest possible cooperation of all countries. It is also recognized that deep reductions in global emissions will be required in order to achieve the ultimate objective of the Convention.

In its report on global warming of 1.5°C (IPCC, 2018), the Intergovernmental Panel on Climate Change outlined the mitigation pathways compatible with a 1.5°C warming of the global climate, likely impacts associated with such warming and what would be needed in response to such a change. It highlighted that warming from anthropogenic emissions would persist for centuries to millenniums and would continue to cause further long-term changes in the climate system, including the ocean.

Interactions between climate change and other drivers include influencing the distribution of global populations as people shift from increasingly uninhabitable areas, economic impacts, including those associated with food production (e.g., aquaculture and fisheries), and an ever greater need for technological innovations and solutions to reduce greenhouse gases, including further reliance on marine renewable energy.

---

13 See www.iea.org/articles/global-co2-emissions-in-2019.

14 See FCCC/CP/2015/10/Add.1, decision 1/CP.21, annex.

3. Key region-specific issues or aspects associated with drivers

Geographical variability in the distribution of populations, economic development, access to technological advances, capacity in implementing governance and management frameworks and effects and responses to climate change result in considerable variability in the influence of each of the drivers described in section 2 across ocean regions.

3.1. Population growth and demographic changes

Fertility rates in high-income regions are lower than those in middle-income and low-income regions (Baxter and others, 2017). Varying fertility rates present challenges for those countries where fertility rates and population growth are both high (UNDESA, 2019c), as well as for those countries where fertility rates are low and the ageing component of the population is growing (see also sect. 4). Sub-Saharan Africa, Central and Southern Asia and Eastern and South-Eastern Asia are all regions with high population growth. The average rate of population growth in the least developed countries[15] was 2.3 per cent over the period 2015–2020, more than double the global rate. That presents challenges for those countries in achieving sustainable development and the conservation of coastal and marine areas, and is further compounded by their vulnerability to climate change, climate variability and sea level rise (UNDESA, 2019c).

3.2. Economic growth

Geographic disparities in economic growth have been increasing since the 1980s, reflecting economic gains in some regions and stagnation in others. While most countries experienced positive growth between 1950 and 2016, others, such as the Central African Republic and the Democratic Republic of the Congo, experienced negative growth, largely as a result of political instability (Karnane and Quinn, 2019). Notably, disparities within countries in employment and productivity have also been growing, with large differences in the extent of disparities across developed economies (IMF, 2019). Climate change may further exacerbate such disparities, in particular where there is geographical variability in the distribution of susceptible industries such as agriculture (including fisheries and aquaculture). In general, economic activity is affected by non-linear increases in temperature, which may bring benefits to economic activities in very cold regions (e.g., opening up the Arctic Ocean to shipping routes and greater trade potential) but, beyond a certain optimum temperature, there are negative impacts for economic output and labour potential (IMF, 2019).

3.3. Technological advances

Areas beyond national jurisdiction have become increasingly accessible owing to technological advancements that facilitate the exploration and exploitation of deep-sea resources, including biodiversity, minerals, oil and gas. Ensuring the sustainable development of those regions will require international cooperation in order to manage them effectively. Negotiations on a legally binding international instrument under the United Nations Convention on the Law of the Sea on the conservation and sustainable use of the marine biological diversity of areas beyond national jurisdiction[16] are focused on ensuring the sustainable development and

---

[15] The Group of Least Developed Countries comprises 47 countries: 32 in sub-Saharan Africa; 2 in Northern Africa and Western Asia; 4 in Central and Southern Asia; 4 in Eastern and South-Eastern Asia; 1 in Latin America and the Caribbean; and 4 in Oceania. For further information, see http://unohrlls.org/about-ldcs.

[16] See General Assembly resolution 72/249.

conservation of those areas (see also chap. 28). The International Seabed Authority has a dual mandate of promoting the development of deep-sea minerals, while ensuring that such development is not harmful to the environment. In areas beyond national jurisdiction, appropriate planning will be required to minimize impacts on the marine environment. The uptake of technological advances for accessing and utilizing marine resources, sustainably developing marine industries and effectively managing those uses is not globally even. Many regions, in particular those where the least developed countries are located, still lack access to technologies that can assist with the sustainable use of marine resources.

3.4. Changing governance structures and geopolitical instability

There has been an increase in nationalism and protectionism over the past decade, contributing to changing trade agreements and, more recently, the implementation of tariffs on goods between specific countries. The Democracy Index[17] fell from 5.55 in 2014 to 5.44 in 2019, largely driven by the regional deterioration of conditions in Latin America and sub-Saharan Africa. When indices for individual countries are calculated, stark regional differences are evident. Countries in Scandinavia, the far north of North America and the South-West Pacific had the highest indices, while those in sub-Saharan Africa, the Middle East and parts of Asia had the lowest. Those differences have an impact on the implementation of global and regional treaties and agreements, thereby affecting economic growth, the transfer of technologies and the implementation of frameworks for managing ocean use, including the development of national ocean-related policies. That, therefore, affects the sustainability of human activities and the protection of marine ecosystems in those areas.

3.5. Climate change

Climate change effects are not uniform across the global ocean. A number of regions are warming at higher rates than the global average and are identified as marine hotspots (Hobday and Pecl, 2014). A number of those hotspots are located where human dependence on marine resources is greatest, such as South-East Asia and Western Africa, with substantive implications for food security compared with other regions. The Arctic is another region where the ocean is warming at 2 to 3 times above the global average (IPCC, 2018). Similarly, decreases in the pH and carbonate ion concentrations of the ocean, associated with ocean acidification, and other effects of climate change, such as deoxygenation, stratification and sea level rise, are regionally variable, with highly variable impacts on the marine environment. Regional differences in such changes are described in detail in chapter 5, and the pressures that they are generating, including socioeconomic impacts, are described in detail in chapter 9.

4. Outlook

Within coastal regions, projections under shared socioeconomic pathways estimate a 71 per cent increase in the global human population across the period 2000–2050 to over 1 billion, as a result of overall global population growth as well as migration into those areas (Merkens and others, 2016). Under the same scenarios, populations in low-density to medium-density areas (<1,000 persons km^2) are projected to decline, while those in higher-density areas are projected to increase (Jones and O'Neill, 2016), with an expansion of

17 See www.eiu.com/topic/democracy-index.

urban footprints in high-density areas and an increasing strain on associated infrastructure. How and where global populations live and their associated impacts on the environment will be influenced by climate change in many ways. As areas become increasingly unliveable as a result of declining precipitation, increasing temperatures, sea level rise and the loss of ecosystem goods and services, people will redistribute themselves to more liveable regions, increasing urban footprints in those regions.

As the global population ages and overall growth slows, the size of the labour force is expected to decline, with an impact on the global economy. It is estimated that the global population considered to have the highest proportion of contributors to global economies, namely the population aged 20–64, will grow less than half as fast over the period 2015–2040 compared with the previous 25 years, while the population over the age of 65 will grow five times faster than the working-age population (Baxter and others, 2017). How global economies respond to the influence of changing population growth and demographics will depend on public policy, such as introducing policies that reduce barriers to the employment of women, and their ability to use advances in technologies to maintain productivity. Those changes in population growth and the distribution and densities of the population, as well as changing economies, will influence the marine environment in ways yet to be determined.

Economic activity in the ocean is expanding rapidly, with projections that, by 2030, under a business-as-usual scenario, the ocean economy could more than double to a value of more than $3 trillion, with approximately 40 million full-time jobs (Organization for Economic Co-operation and Development, 2016). Technological advances and innovations will be critical to identifying sustainable pathways that allow for the development of global economies, including the ocean economy, while addressing many of the challenges facing the ocean at present.

In the context of such rapid change, regulation and governance will struggle to keep up. The integration of emerging ocean industries into existing, fragmented regulatory frameworks will restrict the ability to address pressures generated by industries in an effective and timely way. More effective integrated ocean management will be required to ensure a sustainable future for the ocean in the light of the drivers of change detailed in the present chapter and also in chapter 27.

If greenhouse gas emissions continue to be released at the current rate, it is estimated that the surface temperature will warm by 1.5°C sometime between 2030 and 2052 (IPCC, 2018). Many changes to marine ecosystems as a result of climate change have already been observed, and future climate-related change and associated risks will depend on whether (and when) net zero greenhouse gas emissions are achieved and the associated rate, peak and duration of surface warming (IPCC, 2018). Even if net zero global anthropogenic CO_2 emissions are achieved, sustained warming will persist for centuries to millenniums and will continue to cause further long-term changes in the climate system and, by association, in the ocean, including sea level rise and ocean acidification (IPCC, 2018). The upscaling and acceleration of mitigating and adaptive approaches will be required to reduce future climate-related risks to food security, maritime industries and coastal communities associated with changes to the marine environment.

The COVID-19 pandemic is currently sweeping across the world, causing major disruption to national economies and people. In many regions, because of mitigating efforts to reduce the spread of the virus, pressures immediately affecting the ocean, such as fishing, tourism activities, pollution and greenhouse gas emissions, have been temporarily reduced.[18] With

[18] See www.carbonbrief.org/analysis-coronavirus-has-temporarily-reduced-chinas-co2-emissions-by-a-quarter.

restrictions being placed on the movement of people and on business operations, along with the closing of borders, disruption to supply chains and declining markets have affected a number of marine industries, notably fisheries.[19] The likely impacts that reduced pressures might have on longer-term change by drivers such as climate change, however, are expected to be minimal, and it is currently unclear what benefits might be afforded to marine ecosystems. Disruptions to global supply chains have highlighted the need in many countries to strengthen local supply chains and, in particular, to explore e-commerce options for supporting supply chains in general.

5. Key remaining knowledge and capacity-building gaps

All five drivers detailed in the present chapter interact with each other in varying ways. Understanding of such interactions varies, and, in particular, understanding of the mechanisms by which interactions between drivers influence the marine environment, although recognized as essential for developing holistic approaches to ocean management, is an emerging area of research. Integrated management that takes into account social, economic, ecosystem and cultural values and needs – a whole-of-system approach – allows for the identification of sustainable pathways that support national economies and human well-being.

Modelling frameworks need to be developed, within which scenarios that include changes to population, exploration of governance structures and environmental and economic effects resulting from climate change, can be explored. The initial development of integrated socioecological models that incorporate the marine environment and fisheries in shared socioeconomic pathways is now being implemented to explore future structuring of ocean fisheries (Maury and others, 2017; Bograd and others, 2019). Alternative approaches to integrated models are also being used to explore future states of the marine ecosystem and fisheries (Tittensor and others, 2018). There is a need to advance such efforts, not only to expand modelling approaches to explore the effects of multiple drivers and their cumulative effects on marine ecosystems, but also to develop tools that provide an interface between modelling approaches and decision-making frameworks and allow for the planning and implementation of sustainable approaches to the use of the ocean.

The ability to measure and, therefore, understand the key components that contribute to the drivers of change outlined in the present chapter, namely social, demographic and economic developments in societies, including corresponding changes in lifestyles and associated overall consumption and production patterns, is not equal across the planet. There is a need for capacity development, in particular in the least developed countries, in collecting observations that provide for the understanding of key drivers that affect the marine environment, their interactions and the outcomes of change in each for the marine environment. Similarly, the development of capacity to record changes caused by pressures associated with drivers of change and, thereby, understand impacts on the marine environment is also needed (Evans and others, 2019). Finally, the capacity to effectively plan, assess and manage ocean activities within frameworks that recognize key drivers of change and their interactions is necessary, in particular in regions where there is currently little capacity to implement such frameworks.

---

19 See www.ices.dk/news-and-events/news-archive/news/Pages/wgsocialCOVID.aspx.

References

Bailey, Megan, and others (2018). The role of corporate social responsibility in creating a Seussian world of seafood sustainability. *Fish and Fisheries*, vol. 19, No. 5, pp. 782–790.

Balvanera, Patricia, and others (2019). Chapter 2: Status and trends; indirect and direct drivers of change. In *IPBES Global Assessment on Biodiversity and Ecosystem Services*, ed. IPBES. Bonn: IPBES Secretariat.

Baxter, David, and others (2017). Population aging and the global economy: weakening demographic tailwinds reduce economic growth. In *Berkeley Forum on Aging and the Global Economy*. Issue brief No. 1.

Blanco, Gabriel, and others (2014). Chapter 5: Drivers, trends and mitigation. In *Climate Change 2014: Mitigation of Climate Change. IPCC Working Group III Contribution to AR5*. Cambridge University Press.

Bograd, Steven J., and others (2019). Developing a Social-Ecological-Environmental System Framework to Address Climate Change Impacts in the North Pacific. *Frontiers in Marine Science*, vol. 6, art. 333.

Clark, G.F., and E.L. Johnston (2017). *Australia State of the Environment 2016: Coasts, Independent Report to the Australian Government Minister for Environment and Energy*. Canberra: Australian Government Department of the Environment and Energy.

Detsis, Emmanouil, and others (2012). Project catch: a space-based solution to combat illegal, unreported and unregulated fishing: Part I: vessel monitoring system. *Acta Astronautica*, vol. 80, pp. 114–123.

Eigaard, Ole Ritzau, and others (2014). Technological development and fisheries management. *Reviews in Fisheries Science & Aquaculture*, vol. 22, No. 2, pp. 156–174. https://doi.org/10.1080/23308249.2014.899557.

European Environment Agency (2005). *Sustainable Use and Management of Natural Resources*. EEA Report, 9/2005. Copenhagen: European Environment Agency.

________ (2019). *The European Environment — State and Outlook 2020, Knowledge for Transition to a Sustainable Europe*. EEA Report, 9/2005. Copenhagen: European Environment Agency.

Evans, Karen, and others (2019). The global integrated world ocean assessment: linking observations to science and policy across multiple scales. *Frontiers in Marine Science*, vol. 6, art. 298.

Evans, Karen, and others (2017). *Australia State of the Environment 2016: Marine Environment, Independent Report to the Australian Government Minister for the Environment and Energy*. Canberra: Australian Government Department of the Environment and Energy.

Food and Agriculture Organization of the United Nations (2018). *The State of World Fisheries and Aquaculture 2018-Meeting the Sustainable Development Goals*. Rome.

Finkbeiner, Elena M., and others (2017). Reconstructing overfishing: moving beyond malthus for effective and equitable solutions. *Fish and Fisheries*, vol. 18, No. 6, pp. 1180–1191.

Haas, Bianca, and others (2020). Factors influencing the performance of regional fisheries management organizations. *Marine Policy*, vol. 113.

Hafliðason, Tómas, and others (2012). Criteria for temperature alerts in cod supply chains. *International Journal of Physical Distribution & Logistics Management*, vol. 42, No. 2, pp. 355–371.

Hillary, Richard M., and others (2016). A scientific alternative to moratoria for rebuilding depleted international tuna stocks. *Fish and Fisheries*, vol. 17, No. 2, pp. 469–482. https://doi.org/10.1111/faf.12121.

Hobday, Alistair J., and Gretta T. Pecl (2014). Identification of global marine hotspots: sentinels for change and vanguards for adaptation action. *Reviews in Fish Biology and Fisheries*, vol. 24, No. 2, pp. 415–425.

Intergovernmental Panel on Climate Change (IPCC) (2018). *Global Warming of 1.5° C. An IPCC Special Report on the Impacts of Global Warming of 1.5° C above Pre-Industrial Levels and Related Global Greenhouse Gas Emission Pathways, in the Context of Strengthening the Global Response to the Threat of Climate Change, Sustainable Development, and Efforts to Eradicate Poverty*. eds. Valérie Masson-Delmotte and others. Intergovernmental Panel on Climate Change.

________ (2019). Summary for policymakers. In *IPCC Special Report on the Ocean and Cryosphere in a Changing Climate*, eds. Hans-Otto Pörtner and others, Intergovernmental Panel on Climate Change.

International Council for Science (2017). *A Guide to SDG Interactions: From Science to Implementation*. eds. D.J. Griggs and others. Paris: International Council for Science, Paris.

International Energy Agency (IEA) (2019a). *Energy Efficiency 2019*. Paris: International Energy Agency.

________ (2019b). *Tracking Power*. Paris: International Energy Agency. www.iea.org/reports/tracking-power-2019.

International Monetary Fund (IMF) (2019). *World Economic Outlook: Global Manufacturing Downturn, Rising Trade Barriers*. Washington, D.C.: International Monetary Fund.

Jackson, W.J. (2017). *Australia State of the Environment 2016: Drivers, Independent Report to the Australian Government Minister for the Environment and Energy*. Canberra: Australian Government Department of the Environment and Energy.

Jones, Bryan, and Brian C. O'Neill (2016). Spatially explicit global population scenarios consistent with the shared socioeconomic pathways. *Environmental Research Letters*, vol. 11, No. 8, p. 084003.

Karnane, Pooja, and Michael A. Quinn (2019). Political instability, ethnic fractionalization and economic growth. *International Economics and Economic Policy*, vol. 16, No. 2, pp. 435–461. https://doi.org/10.1007/s10368-017-0393-3.

Lewis, Sara G., and Mariah Boyle (2017). The expanding role of traceability in seafood: tools and key initiatives. *Journal of Food Science*, vol. 82, No. S1, pp. A13–21. https://doi.org/10.1111/1750-3841.13743.

Maury, Olivier, and others (2017). From shared socio-economic pathways (SSPs) to oceanic system pathways (OSPs): Building policy-relevant scenarios for global oceanic ecosystems and fisheries. *Global Environmental Change*, vol. 45, pp. 203–216.

Maxim, Laura, and others (2009). An analysis of risks for biodiversity under the DPSIR framework. *Ecological Economics*, vol. 69, No. 1, pp. 12–23.

Merkens, Jan-Ludolf, and others (2016). Gridded population projections for the coastal zone under the shared socioeconomic pathways. *Global and Planetary Change*, vol. 145, pp. 57–66.

Millennium Ecosystem Assessment (2003). *Millennium Ecosystem Assessment: Ecosystems and Human Well-Being: A Framework for Assessment*. Washington, D.C.: Island Press.

Neumann, Barbara, and others (2015). Future coastal population growth and exposure to sea-level rise and coastal flooding – a global assessment. *PloS One*, vol. 10, No. 3, p. e0118571.

Organization for Economic Cooperation and Development (2016). *The Ocean Economy in 2030*. https://doi.org/10.1787/9789264251724-en.

Patrício, Joana, and others (2016). DPSIR – two decades of trying to develop a unifying framework for marine environmental management? *Frontiers in Marine Science*, vol. 3, art. 177. https://doi.org/10.3389/fmars.2016.00177.

Rowlands, Gwilym, and others (2019). Satellite surveillance of fishing vessel activity in the Ascension Island Exclusive Economic Zone and Marine Protected Area. *Marine Policy*, vol. 101, pp. 39–50.

Ruiz, J., and others (2014). Electronic monitoring trials on in the tropical tuna purse-seine fishery. *ICES Journal of Marine Science*, vol. 72, No. 4, pp. 1201–1213. https://doi.org/10.1093/icesjms/fsu224.

Smeets, Edith, and Rob Weterings (1999). *Environmental Indicators: Typology and Overview*. Copenhagen: European Environment Agency.

Suárez-de Vivero, Juan L., and Juan C. Rodríguez Mateos (2017). Forecasting geopolitical risks: Oceans as source of instability. *Marine Policy*, vol. 75, pp. 19–28.

Tittensor, Derek P., and others (2018). A protocol for the intercomparison of marine fishery and ecosystem models: Fish-MIP v1. 0. *Geoscientific Model Development*, vol. 11, No. 4, pp. 1421–1442.

United Nations (2017a). *The First Global Integrated Marine Assessment: World Ocean Assessment I*. Cambridge: Cambridge University Press.

________ (2017b). *The Conservation and Sustainable Use of Marine Biological Diversity of Areas beyond National Jurisdiction. A Technical Abstract of the First Global Integrated Marine Assessment*. New York: United Nations.

United Nations, Department of Economic and Social Affairs, Population Division (UNDESA) (2019a). *International Migrant Stock 2019*. United Nations. www.un.org/en/development/desa/population/migration/data/estimates2/estimates19.asp.

________(2019b). *Percentage of Total Population Living in Coastal Areas*. New York: United Nations. https://sedac.ciesin.columbia.edu/es/papers/Coastal_Zone_Pop_Method.pdf.

________ (2019c). *World Population Prospects 2019: Highlights (ST/ESA/SER.A/423)*. New York: United Nations.

United Nations Environment Programme, ed. (2019). *Global Environment Outlook – GEO-6: Healthy Planet, Healthy People*. Cambridge University Press. https://doi.org/10.1017/9781108627146.

Part four
Current state of the marine environment and its trends

Chapter 5
Trends in the physical and chemical state of the ocean

Contributors: Carlos Garcia-Soto (convener and lead member), Levke Caesar, Anny Cazenave, Lijing Cheng, Alicia Cheripka, Paul Durack, Karen Evans (co-lead member), David Halpern, Libby Jewett, Sung Yong Kim, Guancheng Li, Ignatius Rigor, Sunke Schmidtko, Juying Wang (co-lead member) and Tymon Zielinski (co-lead member).

Keynote points

- Thermal expansion from a warming ocean and land ice melt are the main causes of the accelerating global rise in the mean sea level.
- Global warming is also affecting many circulation systems. The Atlantic meridional overturning circulation has already weakened and will most likely continue to do so in the future. The impacts of ocean circulation changes include a regional rise in sea levels, changes in the nutrient distribution and carbon uptake of the ocean and feedbacks with the atmosphere, such as altering the distribution of precipitation.
- More than 90 per cent of the heat from global warming is stored in the global ocean. Oceans have exhibited robust warming since the 1950s from the surface to a depth of 2,000 m. The proportion of ocean heat content has more than doubled since the 1990s compared with long-term trends. Ocean warming can be seen in most of the global ocean, with a few regions exhibiting long-term cooling.
- The ocean shows a marked pattern of salinity changes in multidecadal observations, with surface and subsurface patterns providing clear evidence of a water cycle amplification over the ocean. That is manifested in enhanced salinities in the near-surface, high-salinity subtropical regions and freshening in the low-salinity regions such as the West Pacific Warm Pool and the poles.
- An increase in atmospheric CO_2 levels, and a subsequent increase in carbon in the oceans, has changed the chemistry of the oceans to include changes to pH and aragonite saturation. A more carbon-enriched marine environment, especially when coupled with other environmental stressors, has been demonstrated through field studies and experiments to have negative impacts on a wide range of organisms, in particular those that form calcium carbonate shells, and alter biodiversity and ecosystem structure.
- Decades of oxygen observations allow for robust trend analyses. Long-term measurements have shown decreases in dissolved oxygen concentrations for most ocean regions and the expansion of oxygen-depleted zones. A temperature-driven solubility decrease is responsible for most near-surface oxygen loss, though oxygen decrease is not limited to the upper ocean and is present throughout the water column in many areas.
- Total sea ice extent has been declining rapidly in the Arctic, but trends are insignificant in the Antarctic. In the Arctic, the summer trends are most striking in the Pacific sector of the Arctic Ocean, while, in the Antarctic, the summer trends show increases in the Weddell Sea and decreases in the West Antarctic sector of the Southern Ocean. Variations in sea ice extent result from changes in wind and ocean currents.

1. Introduction

In the present chapter, the current physical and chemical state of the ocean and its trends are analysed using seven key climate change indicators:

- **Sea level.** Sea level integrates changes occurring in the Earth's climate system in response to unforced climate variability, as well as natural and anthropogenic influences. It is therefore a leading indicator of global climate change and variability.

- **Ocean circulation.** Ocean circulation plays a central role in regulating the Earth's climate and influences marine life by transporting heat, carbon, oxygen and nutrients. The main drivers of ocean circulation are surface winds and density gradients (determined by ocean temperature and salinity), and any changes in those drivers can induce changes to ocean circulation.
- **Sea temperature and ocean heat content.** The rapid warming of the global ocean over the past few decades has affected the weather, climate, ecosystems, human society and economies (Intergovernmental Panel on Climate Change, 2019). More heat in the ocean is manifested in many ways, including an increasing interior ocean temperature (Cheng and others, 2019b), a rising sea level caused by thermal expansion, melting ice sheets, an intensified hydrological cycle, changing atmospheric and oceanic circulations and stronger tropical cyclones with heavier rainfall (Trenberth and others, 2018).
- **Salinity.** With the advent of improved observational salinity products, more attention has been paid to ocean salinity in Intergovernmental Panel on Climate Change assessment reports (fourth report, Bindoff and others, 2007; and fifth report, Rhein and others, 2013) and in the first *World Ocean Assessment* (United Nations, 2017). Changes to ocean salinity are important given that the global ocean covers 71 per cent of the Earth's surface and contains 97 per cent of the Earth's free water (Durack, 2015). Any global water changes will be expressed in the changing patterns of ocean salinity, a water cycle marker of the largest reservoir of the climate system.
- **Ocean acidification.** Rising concentrations of CO_2 in the atmosphere also have a direct effect on the chemistry of the ocean through the absorption of CO_2. The ocean absorbed roughly 30 per cent of all CO_2 emissions in the period from 1870 to 2015 (Le Quéré and others, 2016; Gruber and others, 2019), and the increased CO_2 level in the water lowers its pH through the formation of carbonic acid.
- **Dissolved oxygen.** Variations in oceanic oxygen have a profound impact on marine life, from nutrient cycling to pelagic fish habitat boundaries (e.g., Worm and others, 2005; Diaz and Rosenberg, 2008; Stramma and others, 2012; Levin, 2018) and can influence climate change through emissions of nitrous oxide, a potent greenhouse gas (e.g., Voss and others, 2013).
- **Sea ice.** Sea ice in the polar regions covers about 15 per cent of the global ocean and affects the global climate system through its influence on global heat balance and global thermohaline circulation. In addition, sea ice has a high albedo, reflecting more sunlight than the liquid ocean, and its melt releases fresh water, which slows the global ocean conveyor belt (the constantly moving system of deep-ocean circulation driven by temperature and salinity).

The present chapter, using those indicators, contains details of the impacts of climate change on the physical and chemical state of the ocean and its evolution and spatial patterns. It is to be read in conjunction with chapter 9, in which extreme climate events (marine heatwaves, extreme El Niño events and tropical cyclones) are analysed and the pressures of some of the physical and chemical changes on marine ecosystems and human populations are described in more detail. Some additional aspects are covered in the section on high-latitude ice in chapter 7 on trends in the state of biodiversity in marine habitats.

2. Physical and chemical state of the ocean

2.1. Sea level

Since the early 1990s, sea level has been routinely monitored at the global and regional levels by a series of high-precision altimetry missions (TOPEX/POSEIDON, Jason-1, Jason-2, Jason-3, Envisat, SARAL/AltiKa, Sentinel-3A and Sentinel-3B).

The most recently updated global mean sea level curve based on satellite altimetry is shown in figure I (update of Legeais and others, 2018). Since 1993, the global mean sea level has been rising at a mean rate of 3.1 ± 0.3 mm per year, with a clear superimposed acceleration of approximately 0.1 mm per year^2 (Chen and others, 2017; Dieng and others, 2017; Yi and others, 2017; Nerem and others, 2018; World Climate Research Programme Global Sea Level Budget Group, 2018).[1] Satellite altimetry has also revealed strong regional variability in the rates of sea level change, with regional rates up to 2–3 times more than the global mean in some regions over the altimetry era (see figure II).

Figure I
Global mean sea level evolution from multi-mission satellite altimetry

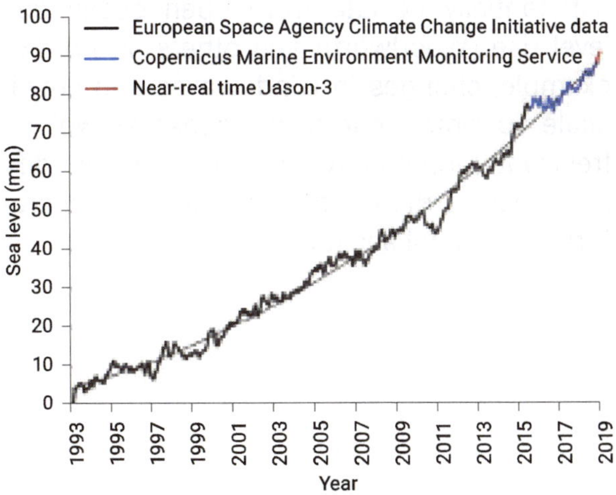

Source: Legeais and others, 2018 (updated).

There are now various observing systems that make it possible to quantify the different contributions of global and regional sea level changes. The Argo system of autonomous profiling floats[2] measures sea water temperature and salinity to a depth of 2,000 m with almost global coverage. The Gravity Recovery and Climate Experiment, a space gravimetry mission, allows for monitoring of ocean mass changes owing to glacier and ice sheet mass loss, as well as land water storage change. It also measures individual water mass changes of glaciers, ice sheets and terrestrial water bodies. Other techniques, such as interferometric synthetic aperture radar and radar and laser altimetry, are also used to estimate ice sheet mass balances.

The study of the sea level budget is important as it provides constraints on missing or poorly known contributions, such as the deep ocean, that are undersampled by current observing systems. Global mean sea level corrected for ocean mass change helps independently to estimate changes in total ocean heat content over time, from which the Earth's energy imbalance can be deduced. Figure III presents annual averages since 2005 of the global mean sea level and sum of ocean thermal expansion and ocean mass increase owing to land ice melt and terrestrial water storage change (World Climate Research Programme Global Sea Level Budget Group, 2018). The figure shows that annual residuals remain below the 2 mm level. In terms of trends, the sea level budget since 2005 is close to 0.3 mm per year, similar to the mean sea level rise uncertainty. Other studies (Dieng and others, 2017; Nerem and others, 2018) also show closure of the sea level budget over the whole altimetry era (since 1993).

1 See General Assembly resolution 70/1.
2 See www.argo.net.

Figure II
Regional trend patterns in sea level from satellite altimetry (January 1993 to October 2019)

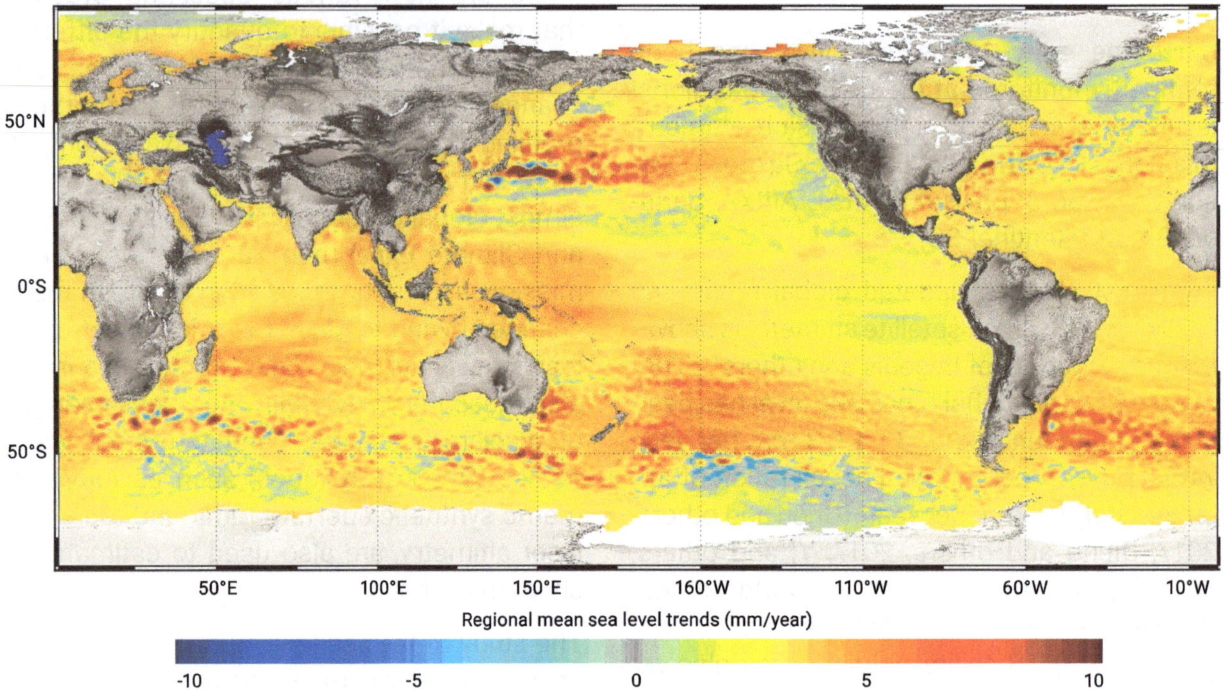

Source: Copernicus Marine Environment Monitoring Service.

Figure III
Yearly global mean sea level budget since 2005

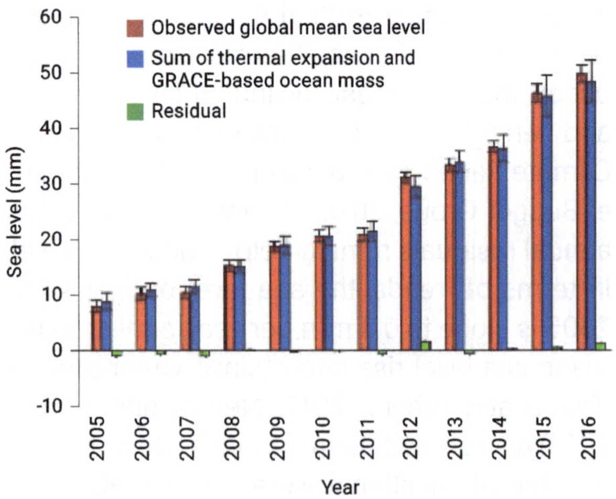

Source: World Climate Research Programme Global Mean Sea Level Budget Group, 2018.
Abbreviation: GRACE, Gravity Recovery and Climate Experiment.

At the local level, in particular in coastal areas, additional small-scale processes are added to the global mean and regional sea level components and can make coastal sea level substantially deviate from open ocean sea level rise (Woodworth and others, 2019). For example, changes in wind, waves and small-scale currents close to the coast, as well as freshwater input in river estuaries, can modify the density structure of sea waters, and therefore the coastal sea level.

2.2. Ocean circulation

The observed changes in the ocean circulation system occur globally and are derived from a variety of data sources. Changes in sea level height, measured with high-precision satellite altimetry since 1993, seem to indicate a widening and strengthening of the subtropical gyres in the North Pacific (Qiu and Chen, 2012) and South Pacific (Cai, 2006; Hill and others, 2008). The data, furthermore, show a poleward movement of many ocean currents, including the Antarctic circumpolar current and the subtropical gyres in the southern hemisphere (Gille, 2008), as well as western boundary currents in all ocean basins (Wu and others, 2012).

The most severe changes, however, are observed in the Atlantic Ocean. One of the major ocean current systems, the Atlantic meridional overturning circulation, has long been predicted to slow down in response to global warming (Intergovernmental Panel on Climate Change, 2013). As the current system transports heat from the southern hemisphere and the tropics into the North Atlantic, its evolution can be deduced from that of sea surface temperature. The observed cooling in the subpolar North Atlantic since the end of the nineteenth century has already been linked to a slowing Atlantic meridional overturning circulation (Dima and Lohmann, 2010; Latif and others, 2006; Rahmstorf and others, 2015). Furthermore, different and largely independent proxy indicators of the evolution of the circulation published in recent years indicate that it is at its weakest for several hundreds of years (see figure IV) and has been weakening during the past century (see figure V; Caesar and others, 2018). Such weakening can also be seen in the direct measurements of the RAPID research programme[3] (Smeed and others, 2018) over the past decade.

Information about the circulations and their changes can be inferred from direct measurements, proxies or model simulations. The main uncertainties regarding the trends in ocean circulation arise from the short timespans of direct, continuous measurements, the incompleteness when representing a circulation through proxies and the inherent uncertainties of the models. It is therefore important that the existing research programmes of observations, such as the Global Drifter Program (Dohan, 2010) and the Argo programme (Freeland and others, 2010), are sustained. That also includes the main projects for observing the Atlantic meridional overturning circulation, namely the RAPID array (Smeed and others, 2014) measuring the strength of the circulation since 2004 at roughly 26° north, the Overturning in the Subpolar North Atlantic Program[4] (Lozier and others, 2017) measuring the overturning that has been feeding the circulation since 2014 and the Observatoire de la variabilité interannuelle et décennale en Atlantique Nord[5] line measuring ocean parameters along a line between Greenland and Portugal (Mercier and others, 2015).

---

3 The RAPID programme is aimed at determining the variability of the Atlantic meridional overturning circulation and its link to climate. An array deployed in 2004 continuously observes the strength of the circulation at about 26° north.

4 This is an international program designed to provide a continuous record of the fluxes of heat, mass and fresh water in the subpolar North Atlantic.

5 The project documents the variability of the circulation and water mass properties in the northern North Atlantic.

Figure IV
Trend of the strength of the Atlantic meridional overturning circulation in observations since 1400 from various proxies

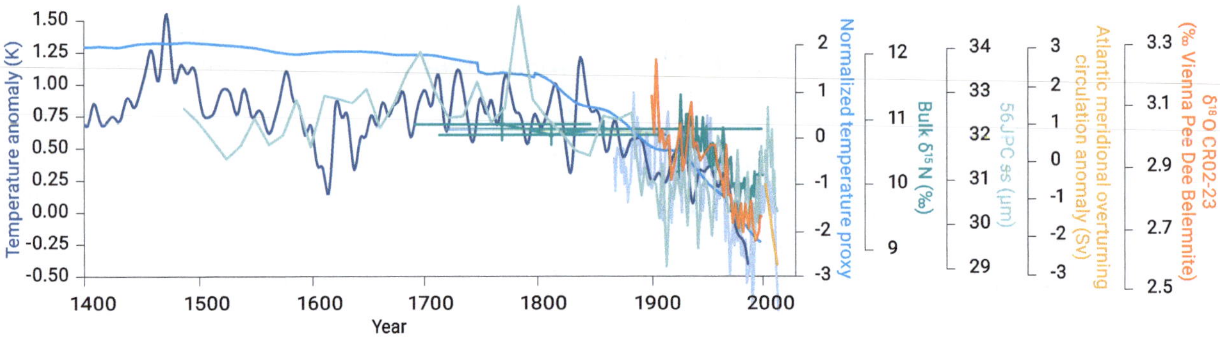

The figure shows the long-term evolution of the sea surface and land temperatures in the North Atlantic region (light blue (Caesar and others, 2018), navy (Rahmstorf and others, 2015), blue (Thornalley and others, 2018)), data from deep-sea cores (dark green (Sherwood and others, 2011), light green (Thornalley and others, 2018), orange (Thibodeau and others, 2018)) and the linear trend of in situ circulation monitoring by the RAPID project (tan (Smeed and others, 2018)).

Figure V
Trend of the strength of the Atlantic meridional overturning circulation in observations

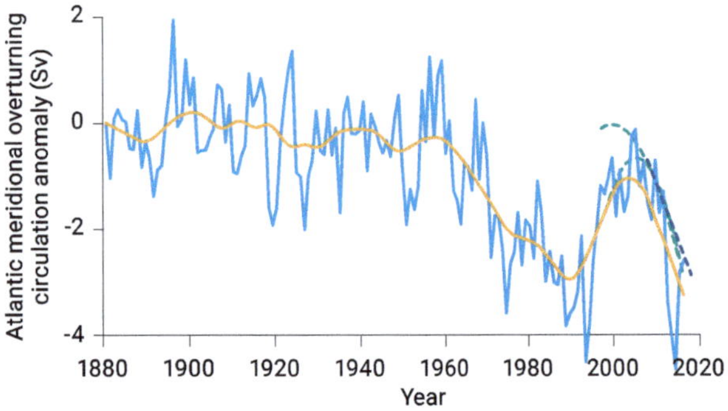

The figure shows the long-term (20-year locally weighted scatterplot smoothing filtering; blue, navy and green lines are annual values) sea surface temperature proxy (tan), the quadratic trend of an ocean reanalysis product (Meteorological Office Global Seasonal Forecasting System, version 5; Jackson and others, 2016), a reconstruction from satellite altimetry and cable measurements (Frackja-Williams, 2015) and the linear trend of in situ circulation monitoring by the RAPID project.
Source: Caesar and others, 2018.

The impacts of the changes in the ocean circulation system vary. The Atlantic meridional overturning circulation is crucial for meridional heat transport and therefore strongly influences the climate in the North Atlantic region. Its slowdown can reduce ocean carbon uptake (Zickfeld and others, 2008) and will enhance sea level rise along the east coast of the United States of America (Goddard and others, 2015). The stronger North Pacific subtropical gyre, however, leads to regional sea level rise in the western tropical North Pacific Ocean (Timmermann and others, 2010). These are the dynamic responses of sea level height to changes in ocean circulation. The poleward displacement of the western boundary currents leads to warming in regions previously unaffected by those warm and strong currents. The consequent thermal expansion will cause a rise in sea level in adjacent coastal areas, such as in the Southern Ocean and the Indian Ocean (Alory and others, 2007; Gille, 2008). Other possible impacts that need further investigation include changes in marine ecosystems and primary production, given that currents transport nutrients, and effects on weather systems, such as the occurrence of heatwaves, droughts or flooding, because ocean circulation has a considerable impact on atmospheric circulation, and with that precipitation, patterns (Duchez and others, 2016).

2.3. Sea temperature and ocean heat content

Sea surface temperature

The global sea surface temperature analyses assessed here are derived from four published data sets (see figure VI). All data sets reveal an increase in global mean sea surface temperature since the early twentieth century. The globally averaged sea surface temperature data as calculated by a linear trend over the period 1900–2018 show an incontrovertible warming of 0.60°C ± 0.07°C (centennial in situ observation-based estimates of sea surface temperature, version 1, COBE1) (Ishii and others, 2005), 0.62°C ± 0.11°C (centennial in situ observation-based estimates of sea surface temperature, version 2, COBE2) (Hirahara and others, 2014), 0.56°C ± 0.07°C (Hadley Centre sea ice and sea surface temperature data set, HadISST) (Rayner and others, 2003), 0.72°C ± 0.10°C (extended reconstructed sea surface temperature, ERSST) (Huang and others, 2017) per century (c^{-1}), with a 90 per cent confidence interval provided. Considering all data sets, the mean sea surface temperature rate is 0.62°C ± 0.12°C c^{-1} over the same period. Differences between the data sets are mainly due to how each methodology treats areas with little or no data, and how each analysis accounts for changes in measurement methods. Among all data sets, the 10 warmest years on record have all occurred since 1997, with the 5 warmest years occurring since 2014. The recent decade (2009–2018) shows a much higher rate of warming than the long-term trend: 2.41°C ± 1.79°C (COBE1), 2.97°C ± 1.81°C (COBE2), 2.05°C ± 1.85°C (HadISST) and 2.81°C ± 1.98°C (ERSST) c^{-1}. The mean rate is 2.56°C ± 0.68°C c^{-1} in the period 2009–2018. In addition to the in situ observations, satellite-based data gave consistent changes in sea surface temperature in the period from 1981 to 2016 (Good and others, 2020; see also figure VI).

Most ocean areas around the globe are warming (see figure VI.B). The broad warming over the global ocean surface is direct evidence of human influence on the climate system (Bindoff and others, 2013). A few regions, such as the subpolar North Atlantic Ocean, have experienced cooling over the past century (often named the "cold blob" or the "North Atlantic warming hole"). A number of studies suggest that the "cold blob" indicates a weakening Atlantic meridional overturning circulation, possibly in response to increased CO_2 concentrations in the atmosphere (Caesar and others, 2018). On other hand, lower warming rates have characterized the Equatorial Pacific and Eastern Tropical Pacific. In the South-East Pacific, from central Peru to northern Chile, a multidecadal surface cooling trend was detected until the late 2000s (Gutiérrez and others, 2016, and references therein), probably associated with coastal upwelling enhancement or remotely driven circulation changes (Dewitte and others, 2012).

Ocean heat content

Climate change from human activities is mainly due to interference with the natural flows of energy through the climate system, creating an energy imbalance caused by increased heat-trapping (greenhouse) gases (Hansen and others, 2011; Trenberth and others, 2018) in the atmosphere. More than 90 per cent of the energy imbalance accumulates in the ocean (Rhein and others, 2013). The heat imbalance is manifested by the increase in ocean heat content. Locally, ocean heat content (OHC) can be estimated by integrating sea temperature (T) from ocean depth z_1 to z_2:

$$OHC = c_p \int_{z_1}^{z_2} \rho T dz$$

Where ρ is the density of the seawater and C_p is the specific heat capacity of the seawater.

Figure VI.A
Global average surface temperature anomalies (°C, annual mean)

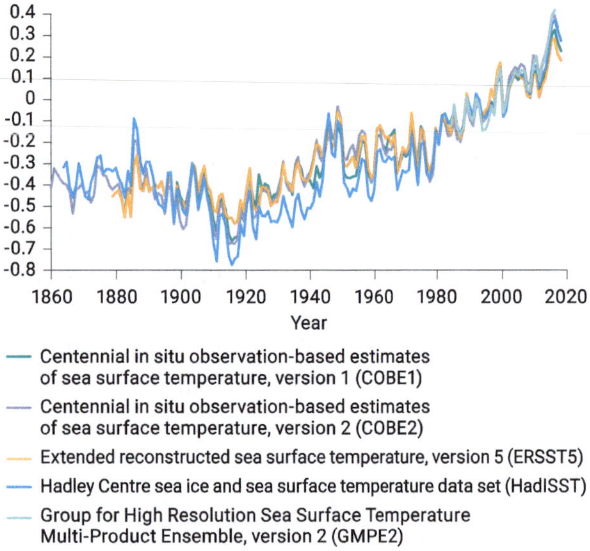

Note: In situ estimates are shown from the COBE1, COBE2, ERSST5, HadISST and GMPE2 data sets

Figure VI.B
Spatial pattern of the long-term sea surface temperature trend (°C per century) from 1854 to 2018 for ERSST data

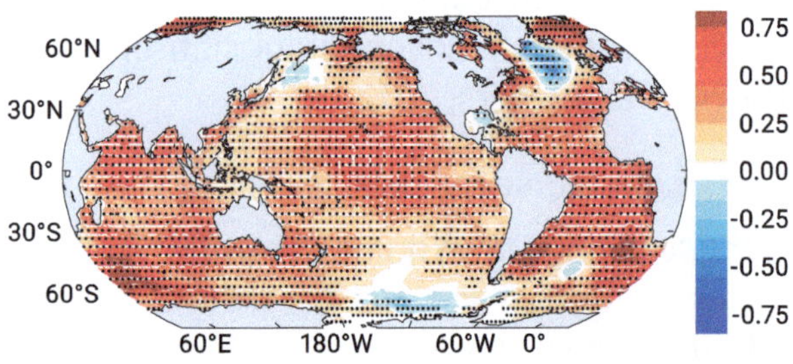

Note: All data use a common 1981–2010 baseline. Black dot signs indicate grid boxes where trends are significant (i.e., a trend of 0 lies outside the 90 per cent confidence interval).

The Earth's energy imbalance and ocean heat content are the fundamental metrics for global warming (Hansen and others, 2011; Trenberth and others, 2018; Von Schuckmann and others, 2016; Cheng and others, 2018). The ocean heat content record is much less affected by internal variability in the climate system than the more commonly used sea surface temperature records, so it is better suited to detecting and attributing human influences (Cheng and others, 2018) than other measures.

Since the fifth assessment report of the Intergovernmental Panel on Climate Change (Rhein and others, 2013), substantial progress has been made in improving long-term ocean heat content records, and a number of sources of uncertainty in prior measurements and analyses have been identified and are better accounted for (Abraham and others, 2013; Boyer and others, 2016; Cheng and others, 2016, 2017a; Ishii and others, 2017). At the same time, efforts have been made to improve how spatial or temporal gaps are accounted for in historical ocean temperature measurements. For example, a new spatial interpolation method was proposed (Cheng and others, 2017a), and a correction to an existing estimate was made available (Ishii and others, 2017). It is becoming clearer that many traditional gap-filling strategies introduced a conservative bias towards low-magnitude changes. Those with less bias include Cheng and others (2017a), Domingues and others (2008) and Ishii and others (2017).

The three recent ocean heat content estimates based on observations show highly consistent ocean warming since the late 1950s (see figure VII). They suggest a linear rate of 0.36 ± 0.06 Wm^{-2} (Ishii and others (2017) and 0.33 ± 0.10 Wm^{-2} (Cheng and others, 2017a) (averaged over the Earth's surface) in the period 1955–2018, with the mean rate of 0.34 ± 0.08 Wm^{-2} among all data sets. The new estimates are collectively higher than previous estimates (Rhein and others, 2013) and more consistent with each other (Cheng and others, 2019a). The rate of ocean warming for the upper 2,000 m has increased in the decades after the 1990s, with linear trends of 0.58 ± 0.06 Wm^{-2} (Cheng and others, 2017a), 0.61 ± 0.08 Wm^{-2} (Ishii and others, 2017) and 0.66 ± 0.02 Wm^{-2} (Domingues and others, 2008; Levitus and others, 2012) between 1999 and 2018. The mean rate is 0.62 ± 0.05 Wm^{-2}. In the recent decade (2009–2018), the rate of ocean heat content increase is: 0.56 ± 0.06 Wm^{-2} (Cheng and others,

2017a), 0.66 ± 0.09 Wm^{-2} (Ishii and others, 2017) and 0.66 ± 0.03 Wm^{-2} (Domingues and others, 2008; Levitus and others, 2012). The mean rate is 0.65 ± 0.07 Wm^{-2}. For ocean heat content, the past 10 years are the 10 warmest on record (Cheng and others, 2019a), as the heat content is less affected by natural variability.

Increases in ocean heat content are observed practically throughout the global ocean, to a depth of 2,000 m (see figure VII). Some intriguing patterns emerge for long-term content change in the period 1960–2018: stronger warming in the Southern Ocean (approximately 70° south to approximately 40° south) and Atlantic Ocean (approximately 40° south to approximately 50° north) than other regions and weaker warming throughout the Pacific Ocean and Indian Ocean (approximately 30° south to approximately 60° north) (see figure VII). The long-term warming of the Southern Ocean has been identified and attributed primarily to greenhouse gases (Cheng and others, 2017a; Swart and others, 2018), driven predominantly by air-sea flux changes associated with upper-ocean overturning circulation and mixing (Swart and others, 2018). Southern Ocean warming has important consequences owing to its influence on the southern hemisphere ice reservoir. Near-surface Southern Ocean heat content is key in limiting the seasonal development of sea ice, and warming can therefore feed back into the global climate by limiting the Earth's albedo. In addition, ocean warming accelerates the melting of Antarctic ice shelves, threatening the stability of the Antarctic ice sheet, with global implications in terms of sea level rise (Sallée and others, 2018).

Over the period 1998–2013, a slowdown in the increase of sea surface temperature and global surface temperature led to numerous assertions about a "climate hiatus" (Hartmann, 2013). The updated record until 2018 (see figure V) shows that the linear trend of sea surface temperature for the period 1998–2018 is 1.25°C ± 0.52°C c^{-1}, which is greater than the linear trend during the reference period (1982–1997) (1.00°C ± 0.46°C c^{-1}). That effectively indicates the end of the slowdown in surface temperature increase with the appearance of the extreme 2015/16 El Niño event (Hu and Fedorov, 2017). In addition, it is clear that the rate of ocean heat content increase has risen since the late 1990s (see figure VII). The unabated increase in the rate of sea surface temperature and ocean heat content refute the concept of a slowdown of human-induced global warming.

Figure VII.A
Observational ocean heat content changes

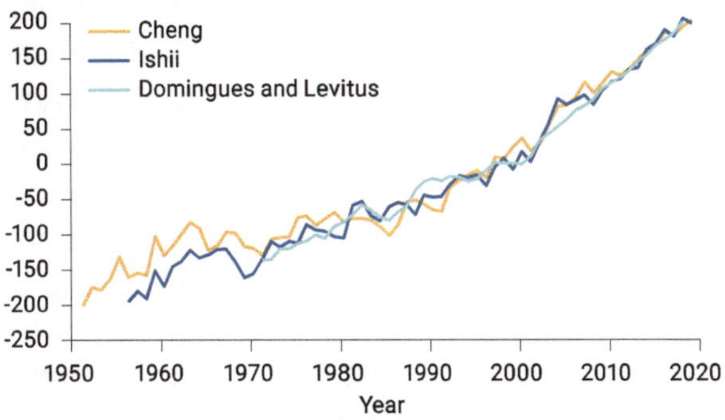

Note: Annual mean for the upper 2,000 m, in zettajoules (10^{21} joules) (Cheng and others, 2017a; Domingues and others, 2008; Levitus and others, 2012; Ishii and others, 2017). The estimate of Domingues (0–700 m) is combined with that of Levitus (700–2,000 m) to produce a 0–2,000 m time series, following the fifth assessment report of the Intergovernmental Panel on Climate Change (Rhein and others, 2013).

Figure VII.B
Spatial pattern of long-term ocean heat content trend (Wm^{-2}), 1955–2018

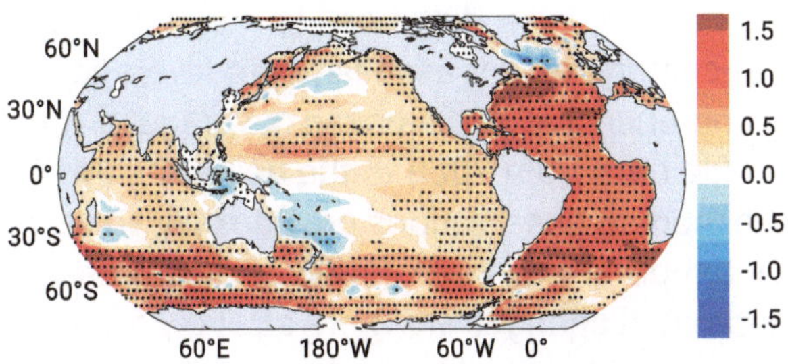

Note: All data use a common 1981–2010 baseline. Black dot signs indicate grid boxes where trends are significant (i.e., a trend of 0 lies outside the 90 per cent confidence interval).
Source: Cheng and others, 2017a.

2.4. Salinity

The studies described in the fourth and fifth assessment reports of the Intergovernmental Panel on Climate Change documented spatial patterns in near-surface and subsurface salinity that represent long-term change (Bindoff and others, 2007; Rhein and others, 2013). In the first *World Ocean Assessment* (United Nations, 2017), the marked long-term multidecadal changes to global ocean salinity were documented throughout the historical period.

The studies noted above provide clear evidence that the near-surface, high-salinity subtropical ocean regions and the entire Atlantic basin have become more saline, and low-salinity regions, such as the West Pacific Warm Pool, and high latitude regions have become fresher when comparing the earlier historical data (from about the 1950s) with present-day salinities (e.g., Boyer and others, 2005; Hosoda and others, 2009; Durack and Wijffels, 2010; Helm and others, 2010; Skliris and others, 2014). The pattern of changes reflects an amplification of climatological mean salinity and has been linked through model simulations (e.g., Durack and others, 2012, 2013; Terray and others, 2012; Vinogradova and Ponte, 2013; Durack, 2015; Levang and Schmitt, 2015; Zika and others, 2015) to indicate a coincident amplification of the atmospheric water cycle (e.g., Held and Soden, 2006).

While long-term historical assessments of change are complicated by the sparse observing network extending back to the mid-twentieth century, recent assessments leverage the comprehensive global ocean coverage of Argo profile data from 2008 to the near-present. As the modern observations provide only 10 years of temporal coverage (2008 to the present), estimated changes are more strongly affected by unforced variability modes, which influence ocean salinity regionally more than long-term estimates, but their spatial and temporal coverage allows for more accurate estimates of change. The latest Argo-only analyses have shown for the first time that nearly all salinity anomalies in 2017 in the Atlantic between 0 and 1,500 m are positive (> 0.05 Practical Salinity Scale-78), mirroring the long-term trends noted above, with the Pacific showing a general freshening, similar to long-term trends.

Since the first Assessment, salinity retrievals from the Soil Moisture and Ocean Salinity Aquarius and Soil Moisture Active Passive satellites (e.g., Berger and others, 2002; Lagerloef and others, 2008; Tang and others, 2017) have become more prominent. While satellite salinity data are only available since 2010 and work is ongoing to intercompare and homogenize data products across satellite platforms, they are beginning to provide key insights into ocean salinity variability owing to precipitation events (e.g., Boutin and others, 2013, 2014; Drushka and others, 2016). In addition, the comparative high temporal and spatial coverage of satellite salinity, when contrasted with the in situ platforms (e.g., Argo), for the first time provides insights into water cycle interactions with the terrestrial and oceanic water cycles, such as the Amazon outlet plume (Grodsky and others, 2014).

Considering all available analyses, it is extremely likely that near-surface and subsurface salinity changes have occurred across the globe since the 1950s. A salinity pattern amplification is apparent, with fresh regions becoming fresher and salty regions becoming saltier, and is supported by all available observational studies that have considered salinity change since the advent of instrumental records. For example, high-latitude oceans have shown significant rates of freshening. More modern assessments are currently too short to confirm consistent changes over the past decade. However, the most recent analyses suggest that consistent patterns are beginning to emerge for the Atlantic and, to a lesser degree, the upper Pacific Ocean basins.

2.5. Ocean acidification

Global surface ocean pH has declined on average by approximately 0.1 since the Industrial Revolution (Caldeira and others, 2003), an increase in acidity of about 30 per cent. Ocean pH is projected to decline, approximately, by an additional 0.2–0.3 over the next century (Caldeira and others, 2003; Feely and others, 2009) unless global carbon emissions are significantly curtailed. Those changes can be observed in extended ocean time series (see figure VIII), and the rate of change is likely to be unparalleled in at least the past 66 million years (Hönisch and others, 2012; Zeebe and others, 2016). Carbonate chemistry varies according to large-scale oceanic features, including depth, distance from continents owing to land influence, upwelling regime, freshwater and nutrient input and latitude (Jewett and Romanou, 2017). Owing to that variability, as determined by the various characteristics, only longer-term, observational time series can detect the predicted long-term increase in acidity at individual sites on account of rising atmospheric CO_2 levels. The time of emergence of the signal varies from 8 to 15 years for open ocean sites and from 16 to 41 years for coastal sites (Sutton and others, 2019), making it necessary to commit to long-term observational records, especially in the coastal zone where most commercially and culturally important marine resources reside.

It has now been documented that ocean acidification is making it harder for some marine organisms, such as corals, oysters and pteropods (Hoegh-Guldberg and others, 2017; Lemasson and others, 2017; Bednarsek and others, 2016; Feely and others, 2004; Orr and others, 2005), to form calcium carbonate shells and skeletons. In some cases, ocean acidification has also been shown to lower fitness in some species such as coccolithophores, crabs and sea urchins (Campbell and others, 2016; Dodd and others, 2015; Riebesell and others, 2017; Munday and others, 2009). Although individual species, when tested, are vulnerable to ocean acidification in laboratory settings, how that is going to translate into changes in actual ecosystems and species populations remains unclear and mostly undocumented (McElhany, 2017). Research efforts over the past decade have begun to build understanding of how marine species, ecosystems and biogeochemical cycles may be influenced by ocean acidification alone and in concert with other stressors, including eutrophication, warming and hypoxia (Baumann, 2019; Murray, 2019). The interaction of ocean acidification in coastal zones with coastal processes, such as upwelling of undersaturated water and land-based nutrient influxes, has become a high priority area of research (Borgesa and Gypensb, 2010; Feely and others, 2008). Natural variability in carbonate chemistry, such as coastal upwelling and seasonal fluctuations in primary productivity, is compounded by anthropogenic changes to create particularly extreme ocean acidification conditions in some regions of the global ocean (Feely and others, 2008; Cross and others, 2014). Intensive national and international efforts focused on carbonate chemistry monitoring, biological observations and biogeochemical or ecological forecast modelling over the past decade have shed light on the status and impacts of ocean acidification from the local to the global level. Gaps in the current understanding of ocean chemistry are being addressed through global monitoring capacity-building efforts, such as the Global Ocean Acidification Observing Network, increased biological impact studies and biogeochemical ecosystem modelling.

Figure VIII
Trends in surface (< 50 m) ocean carbonate chemistry calculated from observations obtained at the Hawaii Ocean Time-series Program in the North Pacific from 1988 to 2018

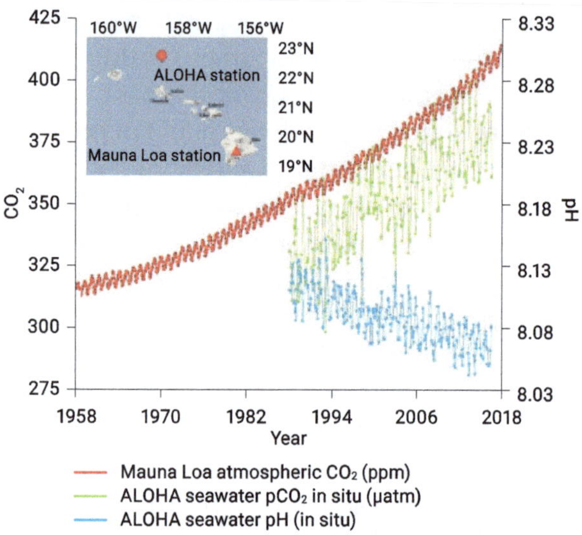

The figure shows the linked increase in atmospheric CO_2 concentrations (red), seawater pCO_2 concentrations (green) and a corresponding decline in seawater pH (blue, secondary y-axis). Ocean chemistry data were obtained from the Hawaii Ocean Time-series Data Organization and Graphical System
Source: National Oceanic and Atmospheric Administration Pacific Marine Environmental Laboratory Carbon Program.

2.6. Dissolved oxygen

Since chemical analysis methods have essentially not changed (Carpenter, 1965; Wilcock and others, 1981; Knapp and others, 1991), long-term oceanic oxygen trends can be estimated fairly robustly where there is sufficient data coverage. Dissolved oxygen samples are analysed by Winkler titration, which was established in 1903 and has since been used to calibrate all means of oceanic dissolved oxygen measurements. That allows a robust analysis of long-term trends in all areas with sufficient data coverage. Modern Winkler titration is computer-aided, providing analysis with higher accuracy, though a bias of historic measurements could not be shown (Schmidtko and others, 2017). The postulated possible bias of 0.5 per cent reagent changes (Knapp and others, 1991) was tested on a global oxygen data set and found to be very unlikely, since the mapped pattern of oxygen change for a deliberately introduced bias does not match any observed pattern (Schmidtko and others, 2017).

In the open ocean, most regional long-term series data show a small long-term decrease despite temporal variations on many timescales (e.g., Keeling and others, 2010). Increasing oxygen levels are found only in very limited time series (Keeling and others, 2010). Coastal changes have mostly been fuelled by riverine run-off of fertilizers, but in some cases may have been affected by larger-scale oxygen changes. They can lead to an increased occurrence of dead zones, with consequences for the regional ecology and economy (Diaz and Rosenberg, 2008).

Globally, the ocean has been losing oxygen in recent decades. Both methods, comparing decadal oxygen data snapshots and local regression analyses (Schmidtko and others, 2017; Ito and others, 2017), show large-scale oxygen declines (see figures IX.A and IX.B). Despite various methods, the derived rates agree within the same water layers and given uncertainties. Deoxygenation rates vary with depth and region, resembling the manifold processes modifying the oxygen content, with isolated regions showing an increase in oxygen. The overall oxygen budget has decreased by 2 per cent in the past five decades, a loss of 4.8 ± 2.1 petamoles since 1960 (Schmidtko and others, 2017). In the upper water column, temperature-driven solubility decrease is dominating (see figure IX.C). For the period 1970–2010, the oxygen concentration in the upper 1,000 m has decreased by 0.046 ± 0.047 $\mu mol\ l^{-1}\ yr^{-1}$, including a solubility change of 0.025 $\mu mol\ l^{-1}\ yr^{-1}$ (Schmidtko and others, 2017). Analysing shallower layers increases the solubility-related change significantly (see figure IX.C), in accordance with the heat gain in the upper water column (see figure IX.C, upper section). However, for the full ocean column, solubility-driven changes from 1970 to 2010 are small, -0.006 μmol l^{-1} yr^{-1} compared with the overall oxygen loss 0.063 ± 0.031 μmol l^{-1} yr^{-1}. Nevertheless, temperature cannot be ruled out as the key source of such changes, through mechanisms other than solubility change. The mechanisms include stratification increase, circulation changes and thermal impacts on biogeochemical cycles (e.g., Keeling and others, 2002; Bianchi and others, 2013; Stendardo and Gruber, 2012).

Figure IX.A
Mean dissolved water column oxygen concentration

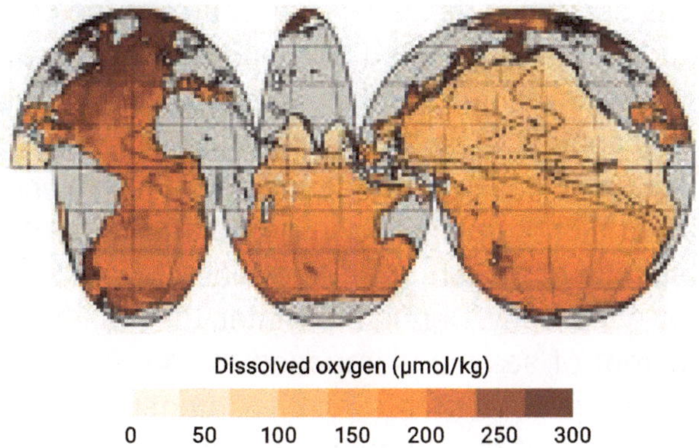

Figure IX.B
Dissolved oxygen changes in per cent per decade

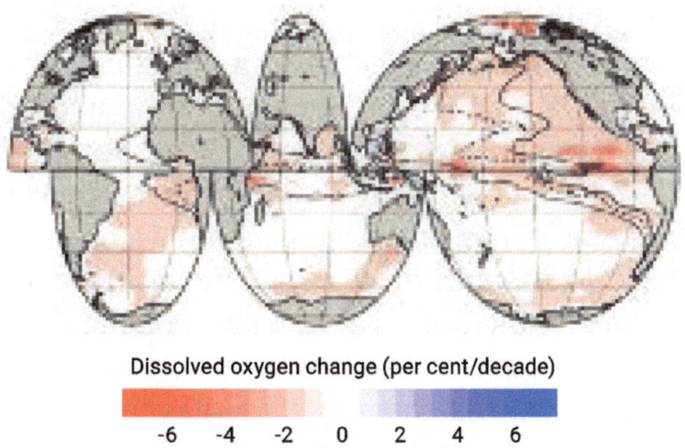

Note: Solid, dotted and dashed lines indicate the presence of low oxygen (40, 80 and 120 umol l^{-1}) at some depth within the water column.

Figure IX.C
Vertical distribution of oxygen loss per decade of oxygen change

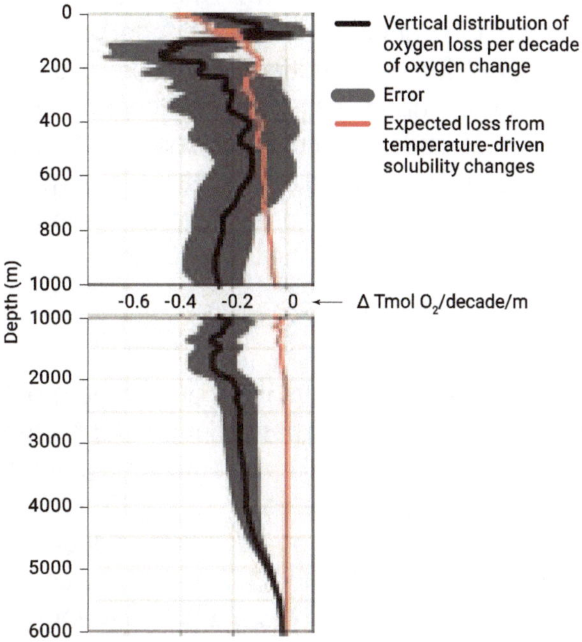

Figure IX.D
Water column cumulative oxygen loss owing to solubility change as a percentage of observed deoxygenation

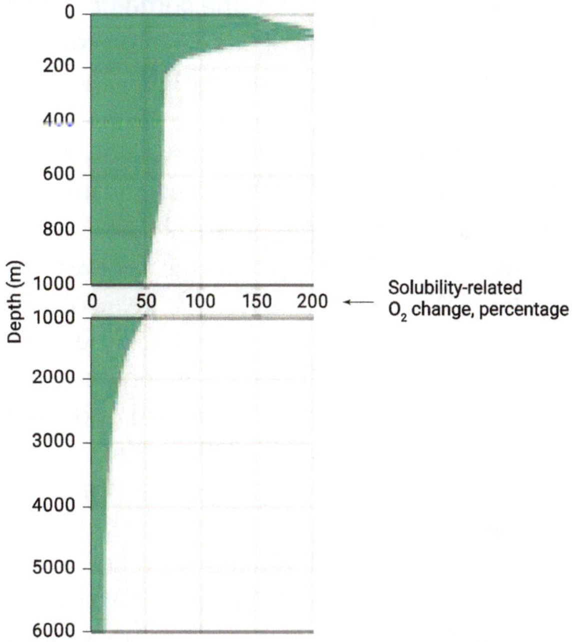

Note: Solubility changes above 100 per cent are due to processes that increase upper-ocean oxygen content and counteract warming.
Source: Schmidtko and others, 2017.

The area of oxygen minimum zones has typically been expanding in recent decades, although there is significant regional variability (Diaz and Rosenberg, 2008). Oxygen minimum zones have potential impacts on climate change because they emit large quantities of nitrous oxide, a potent greenhouse gas, owing to denitrification processes under anoxic conditions (e.g., Codispoti, 2010; Santoro and others, 2011). In particular, oxygen minimum zones have increased in the Pacific Ocean and the Indian Ocean.

2.7. Sea ice

Sea ice in the Arctic has been one of the most iconic indicators of climate change. During the boreal winter, the areal extent of Arctic sea ice reaches a maximum area of 15.4 x 10^6 km^2 in March and, during the boreal summer, declines to 6.4 x 10^6 km^2 in September. Arctic sea ice areal extent is declining by -2.7 ± 0.4 per cent per decade during the winter (March 1979–2019), and -12.8 ± 2.3 per cent per decade during the summer (September 1979–2018) (see figure X; Feterrer and others, 2017). While the decreasing trends during the winter are more evenly distributed around the pole, the summer trends are almost twice as high in the Pacific sector of the Arctic Ocean (upper right of maps, figure X). In that area, the changes in wind related to the Arctic Oscillation have been increasingly blowing the ice away from coastal areas and into the North Atlantic (Rigor and others, 2002), leaving in its wake a much younger and thinner ice pack (Rigor and Wallace, 2004). The thickness of Arctic sea ice has decreased by at least 40 per cent (Rothrock and others, 1999, comparing submarine observations from 1958 to 1976 and from 1993 to 1997)), and Kwok (2018) shows that those changes persist today. The observed trends in sea ice extent (area) and thickness together indicate that the volume of Arctic sea ice has decreased by over 75 per cent since 1979. That estimate is coincident with many modelling studies, such as the Pan-Arctic Ice Ocean Modeling and Assimilation System (Zhang and Rothrock, 2003; Schweiger and others, 2011), which estimates that the average volume of Arctic sea ice of 11.5 x 10^3 km^3 in September has decreased between 1979 and 2017 by -2.8 × 10^3 km^3 per decade, with the record minimum in total ice volume set in 2010.

In Antarctica, sea ice advances to its maximum extent of 19–20 x 10^6 km^2 in September (austral winter) and decreases to a minimum of 3.1 x 10^6 km^2 in February (austral summer). The trends in Antarctic sea ice extent are 0.6 ± 0.6 per cent per decade during the summer (February 1979–2019) and 1.1 ± 3.7 per cent per decade during the winter (September 1979–2018). Net Antarctic sea ice extent showed a statistically significant increase from 1979 to 2015. From 2016 onwards, it has been consistently below average and has set new record low values. Given that the sudden variability in Antarctic sea ice cover is largely attributed to changes in the ocean mixed layer, it is highly relevant to expand the explanation. The net overall changes in sea ice cover have varied greatly between regions. That dichotomy between Arctic and Antarctic sea ice has been attributed to limits imposed by geography. During the winter, the maximum extent of sea ice is imposed by the Antarctic circumpolar currents and the underlying bathymetry of the Southern Ocean (Nghiem and others, 2016) and, during the summer, the sea ice can only retreat to the edge of the Antarctic continent. However, figure X (C and D) shows that, regionally, the trends are more pronounced. During the summer, sea ice extent is increasing in the Weddell Sea but decreasing in the Bellingshausen Sea and the Amundsen Sea (West Antarctica), where the ice sheet is more vulnerable to ocean processes. The regional trends in sea ice extent have been related to changes in wind (and ocean currents) related to the Southern Annular Mode and the El Niño Southern Oscillation (Parkinson, 2019; and references therein). The 40-year record reveals gradual Antarctic sea ice increases followed by decreases at rates far exceeding those seen in the Arctic.

Since sea ice floats on the ocean, the contribution of melting sea ice to sea level rise is negligible. However, sea ice acts as a shield, keeping insolation from warming the ocean, and acts as a buttress for land ice, which terminates over the ocean, keeping warm waters and waves from the ocean from eroding the ice sheet. The loss of sea ice has made many ice sheets more vulnerable and increased the rate of sea level rise owing to the melt of the terrestrial ice sheets (e.g., Stewart and others, 2019).

Figure X
Arctic and Antarctic sea ice concentration trends (per cent per decade)

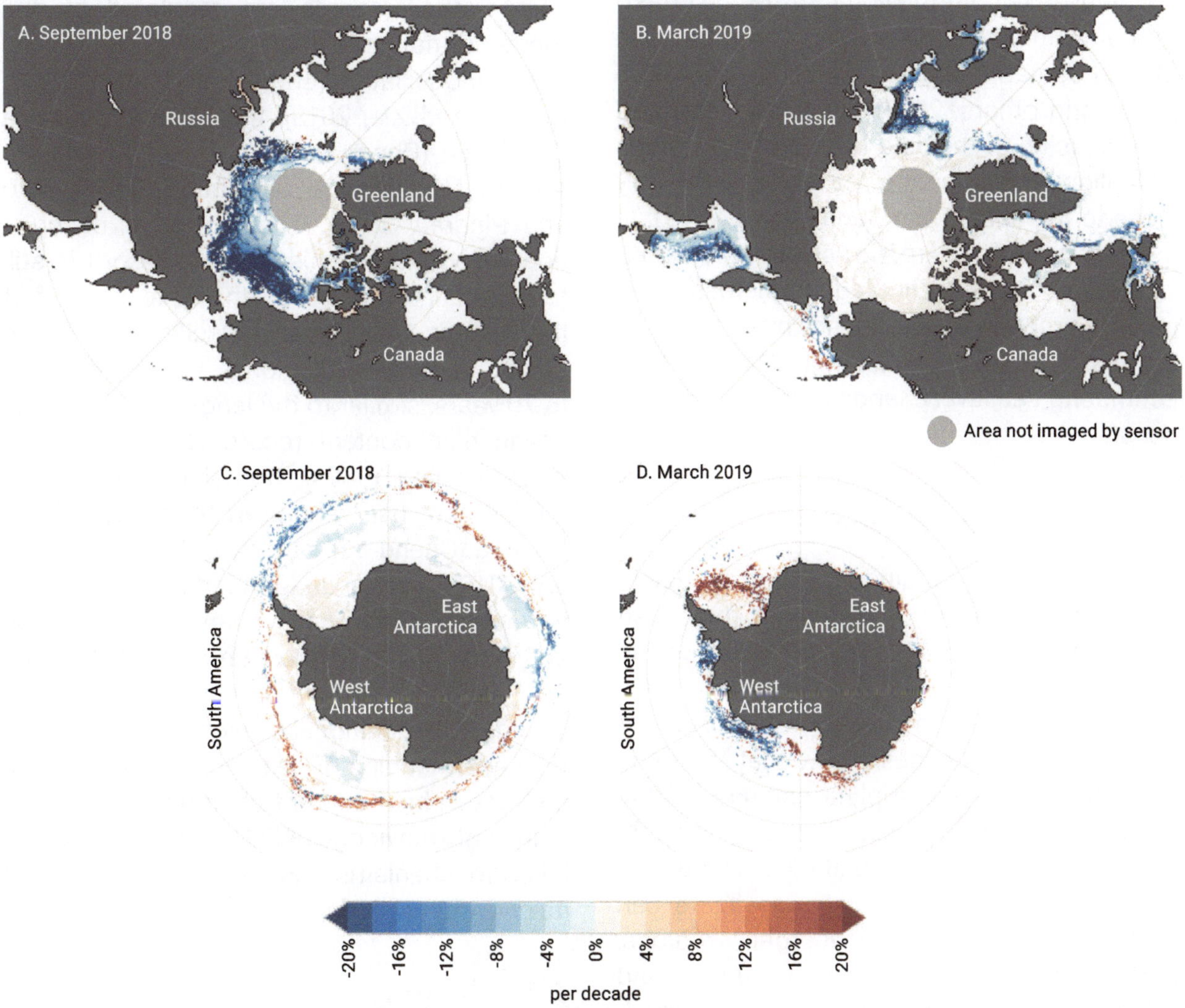

Trends for the Arctic are shown in the top row, and those for the Antarctic are shown in the bottom row, for September 1979–2018 in the left column, and for March 1979–2019 in the right column.

Source: National Snow and Ice Data Center, University of Colorado Boulder; Fetterer and others, 2017.

3. Knowledge gaps

3.1. Sea level

Unlike global mean and regional sea level measured by satellite altimetry missions, coastal sea level changes remain poorly known. Coastal zones are indeed highly undersampled by tide gauges and currently unsurveyed (within 10 kilometres of the coast) by conventional altimetry missions because of land contamination on the radar signal (Cipollini and others, 2018). However, dedicated reprocessing of the data from those missions now allow for estimating sea level change very close to the coast (Marti and others, 2019). In the near future, systematic use of new synthetic aperture radar technology implemented in recent European Space Agency missions (e.g., CryoSat-2 and Sentinel-3) will also allow for estimating sea level changes very close to the coast.

3.2. Ocean circulation

Some limitations remain for the current ocean observation network, in particular for coastal regions, marginal seas and deep ocean regions below 2,000 m. It is important to establish a deep ocean system in the future to monitor ocean changes below that depth, in order to provide a complete estimate of the Earth's energy imbalance (Johnson and others, 2015). Currently, boundary currents are not fully represented by Argo, as floats can swiftly pass through the energetic regions, such as the western boundary current and Atlantic circumpolar current regions, which could induce an inverse cascade of kinetic energy and affect large-scale low-frequency variability (Wang and others, 2017). Achieving adequate sampling will require an observing system design based on a mixture of observing technologies adapted to the different operating environments. There is a need to develop and maintain multiple platform observations for cross-validation and calibration purposes (Meyssignac and others, 2019), including the validation of climate models.

3.3. Sea surface temperature and ocean heat content

Temperature records are regulated by the modes of natural climate, such as the Pacific Decadal Oscillation (England and others, 2014; Kosaka and Xie, 2013), the El Niño-Southern Oscillation (Cheng and others, 2018) and the Atlantic Multidecadal Oscillation (Garcia-Soto and Pingree, 2012). The caveat of observation-based analyses is that the record is still too short: in other words, the typical period of the Atlantic Multidecadal Oscillation and the Pacific Decadal Oscillation is approximately 30 to 70 years, similar to the length of the reliable ocean heat content record (about 60 years since the late 1950s). Combined analyses of models and observations are the proposed way forward (Cheng and others, 2018; Liu and others, 2016) to better understand the sea surface temperature and ocean heat content change and variability on different timescales. The lack of global long-term surface energy flux observations is an additional challenge that prevents full understanding of sea surface temperature and ocean heat content changes. There is insufficient knowledge of El Niño Southern Oscillation mechanisms and feedbacks, as well as its diversity related to global warming.

3.4. Salinity

While the observed salinity changes appear robustly in all observation-based analyses to date, knowledge gaps exist in the definite source of those changes, in particular in near-coastal regions, which are linked to terrestrial and cryospheric water reservoirs. Many observational and model studies have conclusively linked open ocean changes to surface-forced water cycle change, with coincident enhancement of evaporation and precipitation patterns as the primary driver of change. Continued changes will have significant impacts on marine ecosystems, including effects on the life cycle timing, fitness and survivability of ecologically and economically important species.

3.5. Ocean acidification

More research is needed to better inform models and improve predictions of the Earth system response to ocean acidification, its impacts on marine populations and communities and the capacity of organisms to acclimatize or adapt to the changes in ocean acidification-induced ocean chemistry. There remains a strong need for more extensive monitoring in coastal regions, and high-quality, low-cost sensors to do such monitoring, increased access to satellite data and research into the long-term trends in ocean chemistry beyond the observational record (paleo-ocean acidification). A good example is the extension of the Argo programme to include biogeochemical parameters, including pH.[6]

3.6. Sea ice

Maintaining the in situ observing networks in the polar regions is a challenge owing to the harsh environment and access being typically limited to the spring and summer seasons. Retrievals of geophysical parameters by satellite are improving, but in situ observations are required to validate such retrievals. In particular, in situ measurements of snow on sea ice, and the thickness of sea ice, are invaluable to advancing understanding of physical processes in the polar regions. Such measurements are rare in the Arctic, and even rarer in the Antarctic.

4. Summary

Ocean warming and land ice melt are the main causes of present-day accelerating global mean sea level rise. The global mean sea level has been rising since 1993 (the altimetry era) at a mean rate of 3.1 ± 0.3 mm per year, with a clear superimposed acceleration of approximately 0.1 mm per year.[7] Satellite altimetry has also revealed strong regional variability in the rates of sea level change, with regional rates up to two or three times greater than the global mean in some regions. Owing to global warming, many circulation systems also experience changes.

Changes in sea level height, measured with high-precision satellite altimetry, hint at the widening and strengthening of the subtropical gyres in the North and South Pacific. The studies, furthermore, show a poleward movement of many ocean currents, including the Antarctic circumpolar current and the subtropical gyres in the southern hemisphere, as well as western boundary currents in all ocean basins. One of the major ocean current systems, the Atlantic meridional overturning circulation, has already weakened, and it is very likely that it will continue to do so in the future. Impacts that follow such changes include regional sea level rise, changes in nutrient distribution and carbon uptake and feedbacks with the atmosphere.

The globally averaged ocean surface temperature data show a warming of 0.62±0.12 °C per century over the period 1900–2018. In the recent decade (2009–2018), the rate of ocean surface warming is 2.56±0.68 °C c^{-1}. The warming happens in most ocean regions,

6 See https://biogeochemical-argo.org.
7 See General Assembly resolution 70/1.

with some areas, such as in the North Atlantic, showing long-term cooling. Since 1955, the upper 2,000 m of the ocean has also exhibited signs of robust warming, as evidenced by the increase in ocean heat content.

The spatial patterns of multidecadal salinity changes provide convincing evidence of global-scale water cycle change in the global ocean coincident with warming over the period. The resolved changes are replicated in all observed analyses of long-term salinity changes, and more recently have been reproduced in forced climate model simulations. Those changes are manifested in enhanced salinities in the near-surface, high-salinity subtropical regions and corresponding freshening in the low-salinity regions such as the West Pacific Warm Pool and the poles. Similar changes are also seen in the ocean subsurface, with similar patterns of freshening low-salinity waters and enhanced high-salinity waters represented in each of the ocean basins, Atlantic, Pacific and Indian, and across the Southern Ocean.

Global surface ocean pH has declined on average by approximately 0.1 since the Industrial Revolution, an increase in acidity of about 30 per cent. Ocean pH is projected to decline by approximately an additional 0.3 over the next century unless global carbon emissions are significantly curtailed. The changes can be observed in extended ocean time series, and the rate of change is likely to be unparalleled in at least the past 66 million years. The time of emergence of the signal varies from 8 to 15 years for open ocean sites and 16 to 41 years for coastal sites, making it necessary to commit to long-term observational records, especially in the coastal zone, where most commercially and culturally important marine resources reside.

Oceanic oxygen levels have declined in recent decades, with strong regional variations. While the overall oxygen content has decreased by about 2 per cent in five decades, oxygen in coastal areas or near oxygen minimum zones shows larger variations. Coastal changes are mostly fuelled by riverine run-off, and the open ocean changes are likely related to a combination of changes in ocean circulation and biogeochemical cycles. Temperature-driven solubility decrease is responsible for most near-surface oxygen loss, while other processes have to be accountable for deep-ocean oxygen loss. A further decrease in oxygen in and near oxygen minimum zones can lead to climate feedback through consequent greenhouse gas emissions.

Sea ice covers 15 per cent of the global ocean and affects global heat balance and global thermohaline circulation. Total sea ice extent has been declining rapidly in the Arctic, but trends are insignificant in the Antarctic. Arctic sea ice extent is declining by -2.7 ± 0.4 per cent c^{-1} during the winter, and -2.8 ± 2.3 per cent c^{-1} during the summer. In contrast, trends in total Antarctic sea ice extent are insignificant, 0.6 ± 0.6 per cent c^{-1} during the summer and 1.1 ± 3.7 per cent c^{-1} during the winter. Regionally, the spatial distribution of the trends is dramatic. In the Arctic, the summer trends are most striking in the Pacific sector of the Arctic Ocean, while, in the Antarctic, the summer trends show increases in the Weddell Sea and decreases in the West Antarctic sector of the Southern Ocean. The spatial distribution of the changes in sea ice is attributed to changes in wind and ocean currents related to the Arctic Oscillation in the northern hemisphere and the Southern Annular Mode and El Niño in the southern hemisphere.

References

Sea level

Chen, Xianyao, and others (2017). The increasing rate of global mean sea-level rise during 1993–2014. *Nature Climate Change*, vol. 7, No. 7, p. 492.

Cipollini, Paolo, and others (2018). Satellite altimetry in coastal regions. In *Satellite Altimetry over Oceans and Land Surfaces*, eds. Detlef Stammer and Anny Cazenave, pp. 343–373. CRC Press.

Dieng, H.B., and others (2017). New estimate of the current rate of sea level rise from a sea level budget approach. *Geophysical Research Letters*, vol. 44, No. 8, pp. 3744–3751.

Legeais, Jean-François, and others (2018). An improved and homogeneous altimeter sea level record from the ESA Climate Change Initiative. *Earth System Science Data*, vol. 10, pp. 281–301.

Marti, Florence, and others (2019). Altimetry-based sea level trends along the coasts of Western Africa. *Advances in Space Research*.

Nerem, Robert S., and others (2018). Climate-change–driven accelerated sea-level rise detected in the altimeter era. *Proceedings of the National Academy of Sciences*, vol. 115, No. 9, pp. 2022–2025.

Woodworth, Philip L., and others (2019). Forcing factors affecting sea level changes at the coast. *Surveys in Geophysics*, pp. 1–47.

World Climate Research Programme Global Sea Level Budget Group (2018). Global sea-level budget 1993–present. *Earth System Science Data*, vol. 10, No. 3, pp. 1551–1590. https://doi.org/10.5194/essd-10-1551-2018.

Yi, Shuang, and others (2017). Acceleration in the global mean sea level rise: 2005–2015. *Geophysical Research Letters*, vol. 44, No. 23, p. 11905.

Ocean circulation

Alory, Gaël, and others (2007). Observed temperature trends in the Indian Ocean over 1960–1999 and associated mechanisms. *Geophysical Research Letters*, vol. 34, No. 2.

Caesar, Levke, and others (2018). Observed fingerprint of a weakening Atlantic Ocean overturning circulation. *Nature*, vol. 556, No. 7700, p. 191.

Cai, Wenju (2006). Antarctic ozone depletion causes an intensification of the Southern Ocean super-gyre circulation. *Geophysical Research Letters*, vol. 33, No. 3.

Dima, Mihai, and Gerrit Lohmann (2010). Evidence for two distinct modes of large-scale ocean circulation changes over the last century. *Journal of Climate*, vol. 23, No. 1, pp. 5–16.

Dohan, Kathleen, and others (2010). Measuring the global ocean surface circulation with satellite and in situ observations. *Proceedings of OceanObs*, vol. 9.

Duchez, Aurélie, and others (2016). Drivers of exceptionally cold North Atlantic Ocean temperatures and their link to the 2015 European heat wave. *Environmental Research Letters*, vol. 11, No. 7, p. 074004.

Frackja-Williams, Eleanor (2015). Estimating the Atlantic overturning at 26 N using satellite altimetry and cable measurements. *Geophysical Research Letters*, vol. 42, No. 9, pp. 3458–3464.

Freeland, Howard, and others (2010). Argo – a decade of progress. *Proceedings of OceanObs*, vol. 9, pp. 357–370.

Gille, Sarah T. (2008). Decadal-scale temperature trends in the Southern Hemisphere ocean. *Journal of Climate*, vol. 21, No. 18, pp. 4749–4765.

Goddard, Paul B., and others (2015). An extreme event of sea-level rise along the Northeast coast of North America in 2009–2010. *Nature Communications*, vol. 6, No. 6346.

Hill, K.L., and others (2008). Wind forced low frequency variability of the East Australia Current. *Geophysical Research Letters*, vol. 35, No. 8.

Intergovernmental Panel on Climate Change (2013). *Climate Change 2013: The Physical Science Basis. Contribution of Working Group I to the Fifth Assessment Report of IPCC the Intergovernmental Panel on Climate Change*. eds. Thomas F. Stocker and others Cambridge: Cambridge University Press.

Jackson, Laura C., and others (2016). Recent slowing of Atlantic overturning circulation as a recovery from earlier strengthening. *Nature Geoscience*, vol. 9, No. 7, p. 518.

Latif, Mojib, and others (2006). Is the thermohaline circulation changing? *Journal of Climate*, vol. 19, No. 18, pp. 4631–4637.

Lozier, M.S., and others (2017). Overturning in the Subpolar North Atlantic Program: A new international ocean observing system. *Bulletin of the American Meteorological Society*, vol. 98, No. 4, pp. 737–752.

Mercier, H., and others (2015). Variability of the meridional overturning circulation at the Greenland–Portugal OVIDE section from 1993 to 2010. *Progress in Oceanography*, vol. 132, pp. 250–261.

Qiu, Bo, and Shuiming Chen (2012). Multidecadal sea level and gyre circulation variability in the northwestern tropical Pacific Ocean. *Journal of Physical Oceanography*, vol. 42, No. 1, pp. 193–206.

Rahmstorf, Stefan, and others (2015). Exceptional twentieth-century slowdown in Atlantic Ocean overturning circulation. *Nature Climate Change*, vol. 5, No. 5, p. 475.

Sherwood, Owen, and others (2011). Nutrient regime shift in the western North Atlantic indicated by compound-specific delta15N of deep-sea gorgonian corals. *Proceedings of the National Academy of Sciences of the United States of America*, vol. 108, pp. 1011–1015. https://doi.org/10.1073/pnas.1004904108.

Smeed, D.A., and others (2014). Observed decline of the Atlantic meridional overturning circulation 2004–2012. *Ocean Science*, vol. 10, No. 1, pp. 29–38.

Smeed, D.A., and others (2018). The North Atlantic Ocean is in a state of reduced overturning. *Geophysical Research Letters*, vol. 45, No. 3, pp. 1527–1533.

Thibodeau, B., and others (2018). Last Century Warming Over the Canadian Atlantic Shelves Linked to Weak Atlantic Meridional Overturning Circulation. *Geophysical Research Letters*, vol. 45, pp. 12376–12385. https://doi.org/10.1029/2018gl080083.

Thornalley, David J.R., and others (2018). Anomalously weak Labrador Sea convection and Atlantic overturning during the past 150 years. *Nature*, vol. 556, No. 7700, p. 227.

Timmermann, Axel, and others (2010). Wind effects on past and future regional sea level trends in the southern Indo-Pacific. *Journal of Climate*, vol. 23, No. 16, pp. 4429–4437.

Wu, Lixin, and others (2012). Enhanced warming over the global subtropical western boundary currents. *Nature Climate Change*, vol. 2, No. 3, p. 161.

Zanna, L., and others (2019). Global reconstruction of historical ocean heat storage and transport. *Proceedings of the National Academy of Sciences of the United States of America*, vol. 116, p. 1126. https://doi.org/10.1073/pnas.1808838115.

Zickfeld, Kirsten, and others (2008). Carbon-cycle feedbacks of changes in the Atlantic meridional overturning circulation under future atmospheric CO_2. *Global Biogeochemical Cycles*, vol. 22, No. 3.

Sea temperature and ocean heat content

Abraham, John P., and others (2013). A review of global ocean temperature observations: Implications for ocean heat content estimates and climate change. *Reviews of Geophysics*, vol. 51, No. 3, pp. 450–483.

Bindoff, Nathaniel L., and others (2013). Detection and attribution of climate change: from global to regional.

Boyer, Tim, and others (2016). Sensitivity of global upper-ocean heat content estimates to mapping methods, XBT bias corrections, and baseline climatologies. *Journal of Climate*, vol. 29, No. 13, pp. 4817–4842.

Caesar, Levke, and others (2018). Observed fingerprint of a weakening Atlantic Ocean overturning circulation. *Nature*, vol. 556, No. 7700, p. 191.

Cheng, Lijing, and others (2016). XBT Science: Assessment of instrumental biases and errors. *Bulletin of the American Meteorological Society*, vol. 97, No. 6, pp. 924–933.

Cheng, Lijing, and others (2017a). Improved estimates of ocean heat content from 1960 to 2015. *Science Advances*, vol. 3, No. 3, p. e1601545.

Cheng, Lijing, and others (2017b). Taking the pulse of the planet. *Earth and Space Science News, Eos*, vol. 99, pp. 14–16.

Cheng, Lijing, and others (2018). Decadal Ocean Heat Redistribution Since the Late 1990s and Its Association with Key Climate Modes. *Climate*, vol. 6, No. 4, p. 91.

Cheng, Lijing, and others (2019a). 2018 Continues Record Global Ocean Warming. *Advances in Atmospheric Sciences*, vol. 36, No. 3, pp. 249–252.

Cheng, Lijing, and others (2019b). How fast are the oceans warming? *Science*, vol. 363, No. 6423, pp. 128–129.

Dewitte, B., and others. 2012. Change in El Niño flavours over 1958–2008: Implications for the long-term trend of the upwelling off Peru. *Deep-Sea Research II*, 77–80 (2012), pp. 143–156.

Domingues, Catia M., and others (2008). Improved estimates of upper-ocean warming and multi-decadal sea-level rise. *Nature*, vol. 453, No. 7198, p. 1090.

Durack, Paul J. (2015). Ocean salinity and the global water cycle. *Oceanography*, vol. 28, No. 1, pp. 20–31.

England, Matthew H., and others (2014). Recent intensification of wind-driven circulation in the Pacific and the ongoing warming hiatus. *Nature Climate Change*, vol. 4, No. 3, p. 222.

Garcia-Soto, Carlos, and Robin D. Pingree (2012). Atlantic Multidecadal Oscillation (AMO) and sea surface temperature in the Bay of Biscay and adjacent regions. *Journal of the Marine Biological Association of the United Kingdom*, vol. 92, No. 2, pp. 213–234.

Good, S.A. (2020): ESA Sea Surface Temperature Climate Change Initiative (SST_cci): GHRSST Multi-Product ensemble (GMPE), v2.0. Centre for Environmental Data Analysis.

Gutiérrez, D., and others. 2016. Productivity and Sustainable Management of the Humboldt Current Large Marine Ecosystem under Climate Change.

Hansen, James, and others (2011). Earth's energy imbalance and implications. *Atmospheric Chemistry and Physics*, vol. 11, No. 24, pp. 13421–13449.

Hartmann, Dennis L., and others (2013). Observations: atmosphere and surface. In *Climate Change 2013 the Physical Science Basis: Working Group I Contribution to the Fifth Assessment Report of the Intergovernmental Panel on Climate Change*, pp. 159–254. Cambridge University Press.

Hirahara, Shoji, and others (2014). Centennial-scale sea surface temperature analysis and its uncertainty. *Journal of Climate*, vol. 27, pp. 57–75.

Hu, Shineng, and Alexey V. Fedorov (2017). The extreme El Niño of 2015–2016 and the end of global warming hiatus. *Geophysical Research Letters*, vol. 44, No. 8, pp. 3816–3824.

Huang, Boyin, and others (2017). Extended reconstructed sea surface temperature, version 5 (ERSSTv5): upgrades, validations, and intercomparisons. *Journal of Climate*, vol. 30, No. 20, pp. 8179–8205.

Intergovernmental Panel on Climate Change (2019). Summary for policymakers. IPCC Special Report on the Ocean and Cryosphere in a Changing Climate, H-O. Pörtner and others, eds. (in press).

Ishii, Masayoshi, and others (2005). Objective analyses of sea-surface temperature and marine meteorological variables for the 20th century using ICOADS and the Kobe collection. *International Journal of Climatology*, vol. 25, No. 7, pp. 865–879.

Ishii, Masayoshi, and others (2017). Accuracy of global upper ocean heat content estimation expected from present observational data sets. *Sola*, vol. 13, pp. 163–167.

Johnson, Gregory C., and others (2015). Informing deep Argo array design using Argo and full depth hydrographic section data. *Journal of Atmospheric and Oceanic Technology*, vol. 32, No. 11, pp. 2187–2198.

Kosaka, Yu, and Shang-Ping Xie (2013). Recent global-warming hiatus tied to equatorial Pacific surface cooling. *Nature*, vol. 501, No. 7467, pp. 403.

Levitus, Sydney, and others (2012). World ocean heat content and thermosteric sea level change (0–2000 m), 1955–2010. *Geophysical Research Letters*, vol. 39, No. 10.

Liu, Wei, and others (2016). Tracking ocean heat uptake during the surface warming hiatus. *Nature Communications*, vol. 7, p. 10926.

Meyssignac, Benoit, and others (2019). Measuring global ocean heat content to estimate the earth energy imbalance. *Frontiers in Marine Science*, vol. 6, art. 432

Rayner, N.A.A., and others (2003). Global analyses of sea surface temperature, sea ice, and night marine air temperature since the late nineteenth century. *Journal of Geophysical Research: Atmospheres*, vol. 108, No. D14.

Rhein, M., and others (2013). Observations: ocean. In *Climate Change 2013 the Physical Science Basis: Working Group I Contribution to the Fifth Assessment Report of the Intergovernmental Panel on Climate Change*, pp. 159–254. Cambridge University Press.

Sallée, Jean-Baptiste (2018). Southern Ocean warming. *Oceanography*, vol. 31, No. 2, pp. 52–62.

Swart, Neil C., and others (2018). Recent Southern Ocean warming and freshening driven by greenhouse gas emissions and ozone depletion. *Nature Geoscience*, vol. 11, No. 11, p. 836.

Trenberth, Kevin E., and others (2018). Hurricane Harvey links to ocean heat content and climate change adaptation. *Earth's Future*, vol. 6, No. 5, pp. 730–744.

Von Schuckmann, K., and others (2016). An imperative to monitor Earth's energy imbalance. *Nature Climate Change*, vol. 6, No. 2, p. 138.

Wang, Gongjie, and others (2017). Consensuses and discrepancies of basin-scale ocean heat content changes in different ocean analyses. *Climate Dynamics*, vol. 50, Nos. 7–8, pp. 2471–2487.

Salinity

Berger, Michael, and others (2002). Measuring ocean salinity with ESA's SMOS Mission–advancing the science.

Bindoff, Nathaniel L., and others (2007). Observations: oceanic climate change and sea level.

Boutin, Jacqueline, and others (2013). Sea surface freshening inferred from SMOS and ARGO salinity: impact of rain. *Ocean Science*, vol. 9, No. 1.

Boutin, Jacqueline, and others (2014). Sea surface salinity under rain cells: SMOS satellite and in situ drifters observations. *Journal of Geophysical Research: Oceans*, vol. 119, No. 8, pp. 5533–5545.

Boyer, Timothy P., and others (2005). Linear trends in salinity for the World Ocean, 1955–1998. *Geophysical Research Letters*, vol. 32, No. 1.

Drushka, Kyla, and others (2016). Understanding the formation and evolution of rain-formed fresh lenses at the ocean surface. *Journal of Geophysical Research: Oceans*, vol. 121, No. 4, pp. 2673–2689.

Durack, Paul J. (2015). Ocean salinity and the global water cycle. *Oceanography*, vol. 28, No. 1, pp. 20–31.

Durack, Paul J., and Susan E. Wijffels (2010). Fifty-year trends in global ocean salinities and their relationship to broad-scale warming. *Journal of Climate*, vol. 23, No. 16, pp. 4342–4362.

Durack, Paul J., and others (2013). Chapter 28: Long-term Salinity Changes and Implications for the Global Water Cycle. In *Ocean Circulation and Climate*, eds. Gerold Siedler and others, vol. 103, pp. 727–57. International Geophysics. Academic Press. https://doi.org/10.1016/B978-0-12-391851-2.00028-3.

Durack, Paul J., and others (2012). Ocean salinities reveal strong global water cycle intensification during 1950 to 2000. *Science*, vol. 336, No. 6080, pp. 455–458.

Grodsky, Semyon A., and others (2014). Year-to-year salinity changes in the Amazon plume: Contrasting 2011 and 2012 Aquarius/SACD and SMOS satellite data. *Remote Sensing of Environment*, vol. 140, pp. 14–22.

Held, Isaac M., and Brian J. Soden (2006). Robust responses of the hydrological cycle to global warming. *Journal of Climate*, vol. 19, No. 21, pp. 5686–5699.

Helm, Kieran P., and others (2010). Changes in the global hydrological-cycle inferred from ocean salinity. *Geophysical Research Letters*, vol. 37, No. 18.

Hosoda, Shigeki, and others (2009). Global surface layer salinity change detected by Argo and its implication for hydrological cycle intensification. *Journal of Oceanography*, vol. 65, No. 4, pp. 579–596.

Lagerloef, Gary, and others (2008). The Aquarius/SAC-D mission: Designed to meet the salinity remote-sensing challenge. *Oceanography*, vol. 21, No. 1, pp. 68–81.

Levang, Samuel J., and Raymond W. Schmitt (2015). Centennial changes of the global water cycle in CMIP5 models. *Journal of Climate*, vol. 28, No. 16, pp. 6489–6502.

Rhein, M., and others (2013). Observations: ocean. In *Climate Change 2013 the Physical Science Basis: Working Group I Contribution to the Fifth Assessment Report of the Intergovernmental Panel on Climate Change*, pp. 159–254. Cambridge University Press.

Skliris, Nikolaos, and others (2014). Salinity changes in the World Ocean since 1950 in relation to changing surface freshwater fluxes. *Climate Dynamics*, vol. 43, Nos. 3–4, pp. 709–736.

Tang, Wenqing, and others (2017). Validating SMAP SSS with in situ measurements. *Remote Sensing of Environment*, vol. 200, pp. 326–340.

Terray, Laurent, and others (2012). Near-surface salinity as nature's rain gauge to detect human influence on the tropical water cycle. *Journal of Climate*, vol. 25, No. 3, pp. 958–977.

United Nations (2017). *The First Global Integrated Marine Assessment: World Ocean Assessment I*. Cambridge: Cambridge University Press.

Vinogradova, Nadya T., and Rui M. Ponte (2013). Clarifying the link between surface salinity and freshwater fluxes on monthly to interannual time scales. *Journal of Geophysical Research: Oceans*, vol. 118, No. 6, pp. 3190–3201.

Zika, Jan D., and others (2015). Maintenance and broadening of the ocean's salinity distribution by the water cycle. *Journal of Climate*, vol. 28, No. 24, pp. 9550–9560.

Ocean acidification

Baumann, Hannes (2019). Experimental assessments of marine species sensitivities to ocean acidification and co-stressors: how far have we come? *Canadian Journal of Zoology*, vol. 97, No. 5, pp. 399–408.

Bednaršek, Nina, and others (2016). Pteropods on the edge: Cumulative effects of ocean acidification, warming, and deoxygenation. *Progress in Oceanography*, vol. 145, pp. 1–24.

Borgesa, Alberto V., and Nathalie Gypensb (2010). Carbonate chemistry in the coastal zone responds more strongly to eutrophication than ocean acidification. *Limnology and Oceanography*, vol. 55, No. 1, pp. 346–353.

Breitburg, Denise L., and others (2015). And on top of all that... Coping with ocean acidification in the midst of many stressors. *Oceanography*, vol. 28, No. 2, pp. 48–61.

Caldeira, Ken, and Michael E. Wickett (2003). Oceanography: anthropogenic carbon and ocean pH. *Nature*, vol. 425, No. 6956, p. 365.

Campbell, Anna L., and others (2016). Ocean acidification changes the male fitness landscape. *Scientific Reports*, vol. 6, p. 31250.

Cross, Jessica N., and others (2014). Annual sea-air CO_2 fluxes in the Bering Sea: Insights from new autumn and winter observations of a seasonally ice-covered continental shelf. *Journal of Geophysical Research: Oceans*, vol. 119, No. 10, pp. 6693–6708.

Dodd, Luke F., and others (2015). Ocean acidification impairs crab foraging behaviour. *Proceedings of the Royal Society B: Biological Sciences*, vol. 282, No. 1810, p. 20150333.

Feely, Richard A., and others (2004). Impact of anthropogenic CO_2 on the $CaCO_3$ system in the oceans. *Science*, vol. 305, No. 5682, pp. 362–366.

Feely, Richard A., and others (2008). Evidence for upwelling of corrosive "acidified" water onto the continental shelf. *Science*, vol. 320, No. 5882, pp. 1490–1492.

Feely, Richard A., and others (2009). Ocean acidification: Present conditions and future changes in a high-CO_2 world. *Oceanography*, vol. 22, No. 4, pp. 36–47.

Gruber, Nicolas, and others (2019). The oceanic sink for anthropogenic CO_2 from 1994 to 2007. *Science*, vol. 363, No. 6432, pp. 1193–1199.

Hoegh-Guldberg, Ove, and others (2017). Coral reef ecosystems under climate change and ocean acidification. *Frontiers in Marine Science*, vol. 4, art. 158.

Hönisch, Bärbel, and others (2012). The geological record of ocean acidification. *Science*, vol. 335, No. 6072, pp. 1058–1063.

Jewett, L., and A. Romanou (2017). Ocean acidification and other ocean changes. *Climate Science Special Report: Fourth National Climate Assessment*, vol. 1, pp. 364–392.

Le Quéré, Corinne, and others (2016). Global carbon budget 2016.

Lemasson, Anaelle J., and others (2017). Linking the biological impacts of ocean acidification on oysters to changes in ecosystem services: a review. *Journal of Experimental Marine Biology and Ecology*, vol. 492, pp. 49–62.

McElhany, Paul (2017). CO_2 sensitivity experiments are not sufficient to show an effect of ocean acidification. *ICES Journal of Marine Science*, vol. 74, No. 4, pp. 926–928.

Munday, Philip L., and others (2009). Ocean acidification impairs olfactory discrimination and homing ability of a marine fish. *Proceedings of the National Academy of Sciences*, vol. 106, No. 6, pp. 1848–1852.

Murray, Christopher S. (2019). An Experimental Evaluation of the Sensitivity of Coastal Marine Fishes to Acidification, Hypoxia, and Warming.

Orr, James C., and others (2005). Anthropogenic ocean acidification over the twenty-first century and its impact on calcifying organisms. *Nature*, vol. 437, No. 7059, pp. 681–686.

Riebesell, Ulf, and others (2017). Competitive fitness of a predominant pelagic calcifier impaired by ocean acidification. *Nature Geoscience*, vol. 10, No. 1, p. 19.

Sutton, Adrienne J., and others (2019). Autonomous seawater pCO_2 and pH time series from 40 surface buoys and the emergence of anthropogenic trends. *Earth System Science Data*, p. 421.

Zeebe, Richard E., and others (2016). Anthropogenic carbon release rate unprecedented during the past 66 million years. *Nature Geoscience*, vol. 9, No. 4, pp. 325–329.

Dissolved oxygen

Bianchi, Daniele, and others (2013). Intensification of open-ocean oxygen depletion by vertically migrating animals. *Nature Geoscience*, vol. 6, No. 7, pp. 545–548.

Carpenter, James H. (1965). The accuracy of the Winkler method for dissolved oxygen analysis. *Limnology and Oceanography*, vol. 10, No. 1, pp. 135–140.

Codispoti, Louis A. (2010). Interesting times for marine N_2O. *Science*, vol. 327, No. 5971, pp. 1339–1340.

Diaz, Robert J., and Rutger Rosenberg (2008). Spreading dead zones and consequences for marine ecosystems. *Science*, vol. 321, No. 5891, pp. 926–929.

Ito, Takamitsu, and others (2017). Upper ocean O_2 trends: 1958–2015. *Geophysical Research Letters*, vol. 44, No. 9, pp. 4214–4223.

Keeling, Ralph F., and Hernan E. Garcia (2002). The change in oceanic O_2 inventory associated with recent global warming. *Proceedings of the National Academy of Sciences*, vol. 99, No. 12, pp. 7848–7853.

Keeling, Ralph F., and others (2010). Ocean deoxygenation in a warming world. *Annual Review of Marine Science*, vol. 2, pp. 199–229.

Knapp, George P., and others (1991). Iodine losses during Winkler titrations. *Deep Sea Research Part A. Oceanographic Research Papers*, vol. 38, No. 1, pp. 121–128.

Levin, L.A. (2018). Manifestation, Drivers, and Emergence of Open Ocean Deoxygenation. *Annual Review of Marine Science*, vol. 10, pp. 229–260, https://doi.org/10.1146/annurev-marine-121916-063359.

Santoro, Alyson E., and others (2011). Isotopic signature of N_2O produced by marine ammonia-oxidizing archaea. *Science*, vol. 333, No. 6047, pp. 1282–1285.

Schmidtko, Sunke, and others (2017). Decline in global oceanic oxygen content during the past five decades. *Nature*, vol. 542, No. 7641, pp. 335–339. https://doi.org/10.1038/nature21399.

Stendardo, I., and N. Gruber (2012). Oxygen trends over five decades in the North Atlantic. *Journal of Geophysical Research: Oceans*, vol. 117, No. C11.

Stramma, Lothar, and others (2012). Expansion of oxygen minimum zones may reduce available habitat for tropical pelagic fishes. *Nature Climate Change*, vol. 2, No. 1, pp. 33–37.

Voss, Maren, and others (2013). The marine nitrogen cycle: Recent discoveries, uncertainties. *Philosophical Transactions of the Royal Society B: Biological Sciences*, vol. 368.

Wilcock, R.J., and others (1981). An interlaboratory study of dissolved oxygen in water. *Water Research*, vol. 15, No. 3, pp. 321–325.

Worm, Boris, and others (2005). Global patterns of predator diversity in the open oceans. *Science*, vol. 309, No. 5739, pp. 1365–1369.

Sea ice

Fetterer, F., and others (2017). *Sea Ice Index, Version 3*. Boulder, Colorado, United States of America: NSIDC: National Snow and Ice Data Center. https://doi.org/10.7265/N5K072F8.

Kwok, Ron (2018). Arctic sea ice thickness, volume, and multiyear ice coverage: losses and coupled variability (1958–2018). *Environmental Research Letters*, vol. 13, No. 10, p. 105005.

Massom, R.A., and others (2018). Antarctic Ice shelf disintegration triggered by sea ice loss and ocean swell. *Nature*, vol. 558, pp. 383–389, https://doi.org/10.1038/s41586-018-0212-1.

Meehl, G.A., and others (2019). Sustained ocean changes contributed to sudden Antarctic sea ice retreat in late 2016. *Nature Communications*, vol. 10(1), p. 14. https://doi.org/10.1038/s41467-018-07865-9.

Nghiem, S.V., and others (2016). Geophysical constraints on the Antarctic sea ice cover. *Remote Sensing of Environment*, vol. 181, pp. 281–292.

Parkinson, Claire L. (2019). A 40-y record reveals gradual Antarctic sea ice increases followed by decreases at rates far exceeding the rates seen in the Arctic. *Proceedings of the National Academy of Sciences*, vol. 116, No. 29, pp. 14414–14423.

Reid, P., and others (2019): Sea ice extent, concentration, and seasonality. In *State of the Climate in 2018. Bulletin of the American Meteorological Society*, vol. 100 (9), pp. S178–S181.

Rigor, Ignatius G., and John M. Wallace (2004). Variations in the age of Arctic sea-ice and summer sea-ice extent. *Geophysical Research Letters*, vol. 31, No. 9.

Rigor, Ignatius G., and others (2002). Response of sea ice to the Arctic Oscillation. *Journal of Climate*, vol. 15, No. 18, pp. 2648–2668.

Rothrock, Drew A., and others (1999). Thinning of the Arctic sea-ice cover. *Geophysical Research Letters*, vol. 26, No. 23, pp. 3469–3472.

Schweiger, Axel, and others (2011). Uncertainty in modeled Arctic sea ice volume. *Journal of Geophysical Research: Oceans*, vol. 116, No. C8.

Stewart, Craig L., and others (2019). Basal melting of Ross Ice Shelf from solar heat absorption in an ice-front polynya. *Nature Geoscience*, vol. 12, No. 6, pp. 435–440.

Zhang, Jinlun, and D.A. Rothrock (2003). Modeling global sea ice with a thickness and enthalpy distribution model in generalized curvilinear coordinates. *Monthly Weather Review*, vol. 131, No. 5, pp. 845–861.

Chapter 6
Trends in the biodiversity of the main taxa of marine biota

Chapter lead member: Chul Park.

Introduction

In the first *World Ocean Assessment*, published in 2017, biological diversity was considered from three viewpoints: geographical region, taxonomic group and habitats identified as of concern. In the second *World Ocean Assessment*, biological diversity is considered by taxonomic group (chap. 6) and by habitats (chap. 7) for all regions for which data are available. For the taxonomic groups that were included in the first Assessment, the focus is on changes that have occurred since its publication, including new information. For the taxonomic groups that were not included in the first Assessment, the focus is on general information, in order to establish a baseline on their current state.

Subchapter 6A expands on the information on plankton contained in the first Assessment by describing the biodiversity of that group, providing in particular information on single-celled phytoplankton, bacteria, viruses and metazoan zooplankton. Information on benthic invertebrates, which were not discussed separately in the first Assessment, is provided in subchapter 6B. Among the pelagic invertebrates, planktonic forms are included in subchapter 6A. Data on pelagic invertebrates (cephalopods) remain a gap that will need to be filled in a future assessment, even though some information on those invertebrates has been provided in an addendum to subchapter 6B by the Group of Experts of the Regular Process for Global Reporting and Assessment of the State of the Marine Environment, including Socioeconomic Aspects. New and expanded information on fish diversity, in particular in relation to species that were not considered in the first Assessment, is included in subchapter 6C. Chapter 6 also contains information on recent changes in the biodiversity of marine mammals (subchap. 6D), marine reptiles (subchap. 6E), seabirds (subchap. 6F) and marine plants and macroalgae (subchap. 6G). That last subchapter incorporates trends in the status of kelp forests and algal beds. Marine plants are also described in subchapters 7G–I from the viewpoint of habitats.

Chapter 6A
Plankton (phytoplankton, zooplankton, microbes and viruses)

Contributors: Thomas Malone (convener), Maurizio Azzaro, Russell Hopcroft, Chul Park (chapter lead member), Kazuaki Tadokoro, Michael Thorndyke and Sinjae Yoo

Keynote points

- Single-celled microbes are the most abundant and diverse form of marine life. Food webs based on them sustain most ocean biodiversity.
- Marine phytoplankton account for about 50 per cent of the Earth's primary production, oxygen supply and N_2-fixation. Diatoms and picoplankton (< 2 μm) account for most marine primary production.
- Driven by upper-ocean warming, increases in the vertical separation of layers of water (stratification) and decreases in inorganic nutrient inputs to the part of the ocean where photosynthesis is possible (the euphotic zone) are likely to result in:
    - Decreases in phytoplankton productivity and cell size;
    - Increases in energy flow through microbial food webs relative to that through metazoan food webs (plankton > 20 μm);
    - Decreases in the export of biological production to the deep ocean. Such decreases would reduce the capacity of the ocean to absorb CO_2, accelerating global atmospheric warming;
    - Decrease in biological production of higher trophic level.
    - Climate-driven ocean acidification may reduce the abundance and distribution of calcareous plankton.
- Current global ocean observations do not specifically monitor plankton diversity. An international, integrated observing system of ocean life is needed as a component of the Global Earth Observing System of Systems.

1. Introduction

Marine plankton communities are comprised of viruses, prokaryotes (archaea and bacteria) and eukaryotes (protists and Metazoa). Prokaryotes and eukaryotes include both primary producers and heterotrophic consumers, and marine plankton represent the most phylogenetically diverse group of organisms on Earth (Colomban and others, 2015; United Nations, 2017a). The focus in the present subchapter is on plankton assemblages of the upper ocean (0 to -1,000 m) and climate-driven changes in plankton that are most likely to have an impact on ecosystem services.

Unicellular microbes account for most biomass, biodiversity and metabolic activity in the oceans (Gasol and others, 1997; Azam and Malfatti 2007; Salazar and Sunagawa, 2017; Bar-On and others, 2018) and play critical roles in the provision of marine ecosystem services (Palumbi and others, 2009; Liquete and others, 2013). In particular, phytoplankton account for some 50 per cent of the Earth's net primary production (NPP) that fuels marine food webs and for some 50 per cent of the Earth's oxygen supply (Field and others, 1998; Westberry and others, 2008); and planktonic food webs support most fisheries (Blanchard and others, 2012; Boyce and others, 2015), fuel the biological pump[1] (Honjo and others, 2014) and sustain biodiversity (Beaugrand and others, 2013; Vallina and others, 2014). Phytoplankton NPP and the flows of nutrients through planktonic food webs make significant contributions to

[1] Biologically mediated export of particulate organic matter and calcium carbonate to the deep ocean (below 1,000 m).

at least 14 Sustainable Development Goals (Wood and others, 2018), first of all Goal 14 (Conserve and sustainably use the oceans, seas and marine resources for sustainable development).[2]

The objectives of the present subchapter are to: (a) describe the current composition of plankton assemblages and past trends in their diversity and productivity at the global and regional levels; (b) summarize predicted, climate-driven trends in those plankton assemblages; and (c) identify gaps in current knowledge. The climate-driven changes in the upper-ocean environment targeted in the present section are ocean warming and ocean acidification.[3] This information is particularly relevant to chapters 5 (trends in the physical and chemical state of the ocean) and 10 (changes in nutrient inputs to the marine environment). Subjects addressed in the present subchapter that were not specifically covered in chapter 6 of the first *World Ocean Assessment* (United Nations, 2017b) include past and projected trends in the diversity of plankton.

2. Summary of chapter 6 of the first *World Ocean Assessment*

Regional and global patterns of NPP by phytoplankton and benthic macrophytes, nutrient cycling in the upper ocean and anthropogenic impacts on those processes were studied, leading to the following findings:

- With the exception of coastal waters subject to riverine inputs of nutrients and high-nutrient low-chlorophyll zones, the global pattern of phytoplankton NPP reflects the pattern of deep-water nutrient (nitrogen and phosphorus) inputs to the euphotic zone.[4]
- Phytoplankton NPP in subtropical gyres decreased from 1998 to 2006 as a result of climate-driven upper-ocean warming and associated decreases in nutrient supply, while NPP has increased in coastal ecosystems as a result of increases in land-based nutrient inputs. This has led to a global spread of hypoxia in the ocean, a decline in the spatial extent of seagrass beds and increases in the occurrence of toxic phytoplankton events.
- Phytoplankton species diversity tends to be lowest in polar and subpolar waters, where fast-growing species account for most NPP, and highest in tropical and subtropical waters, where small phytoplankton (< 10 μm) account for most NPP.
- As the upper ocean warms and becomes more stratified, it is likely that small phytoplankton species will account for an increasingly large fraction of NPP, resulting in decreases in fish stocks and organic carbon export to the deep sea.
- With increases in upper-ocean temperatures at high latitudes, the spatial ranges of copepod species in the North Atlantic have expanded to higher latitudes and seasonal peaks in abundance are occurring earlier in the year for temperate species.

2 See General Assembly resolution 70/1.

3 Ocean acidification refers to a reduction in the pH of the ocean over an extended period of time, caused primarily by the uptake of CO_2 from the atmosphere.

4 The euphotic zone is the upper layer of the ocean into which sufficient light penetrates for photosynthesis to occur.

3. Regions targeted in the present *World Ocean Assessment*

Mean sea surface chlorophyll *a* concentration (1997–2010) and six regions targeted in the present report

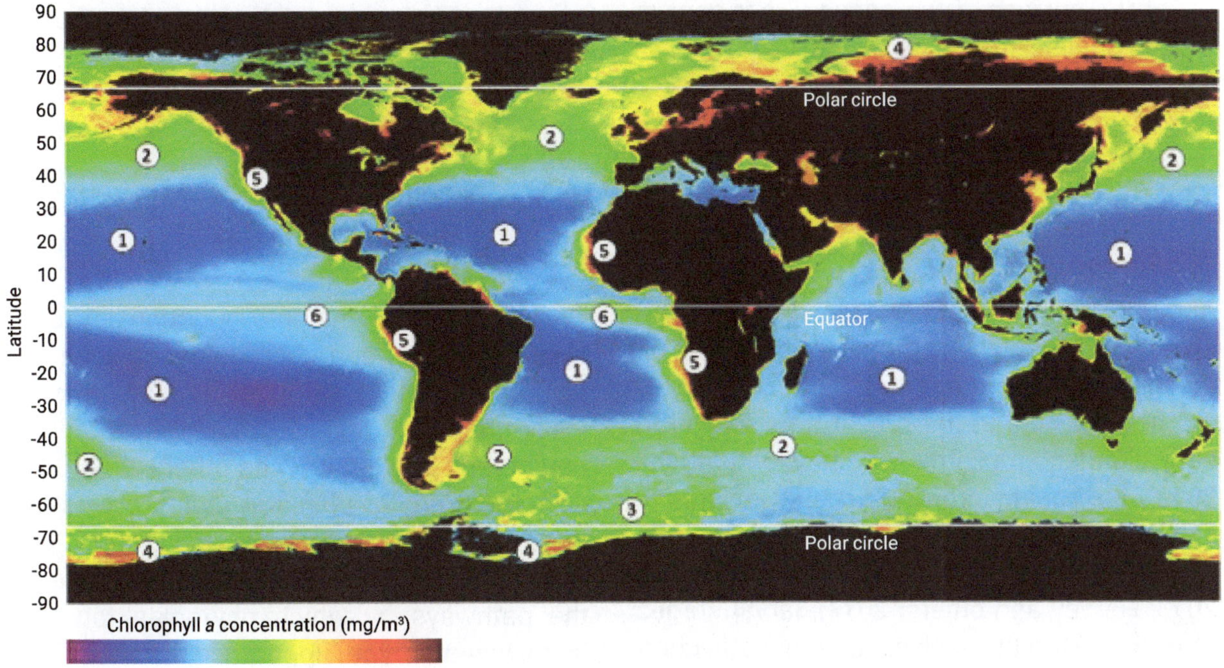

Source: adapted from Sundby and others (2016).

Note: 1, central gyres; 2, high-latitude spring bloom regions; 3, Antarctic circumpolar region; 4, polar sea ice region; 5, coastal upwelling region; and 6, equatorial upwelling region (blue: < 0.1 mg/m^{-3}; green: 0.1–1.0 mg/m^{-3}; yellow: 1–3 mg/m^{-3}; and red: > 3 mg/m^{-3}).

Phytoplankton NPP varies regionally (Behrenfeld and others, 2006; Ultz and others, 2010; United Nations, 2017c), and the global ocean has been subdivided into regions by the Intergovernmental Panel on Climate Change accordingly (Hoegh-Guldberg and Poloczanska, 2017). Of the seven regions defined by the Panel, those targeted in the present section represent high- and low-latitude systems, systems in which the primary input of nutrients is from deep water through vertical mixing or upwelling and systems that exhibit a broad range of trophic states[5] (that do not reflect land-based inputs of nutrients). The six targeted regions (as numbered in the figure above) are as follows:

1. The five subtropical central gyres, the largest biomes of the upper ocean (about 40 per cent of the surface of the ocean, 22 per cent of the annual NPP of the ocean);
2. High-latitude spring bloom regions (about 25 per cent of the surface of the ocean, 43 per cent of the annual NPP of the ocean);
3. Antarctic circumpolar region of the Southern Ocean (about 12 per cent of the surface of the ocean, about 9 per cent of the annual NPP of the ocean);

[5] From oligotrophic regions with low mean annual chlorophyll *a* concentrations (< 0.1 mg m^{-3}) to eutrophic regions with relatively high mean concentrations (1–30 mg m^{-3}).

4. Polar sea ice regions of the Arctic Ocean and Southern Ocean (about 4 per cent of the surface of the ocean, 1–2 per cent of annual NPP of the ocean);
5. Coastal upwelling regions (about 2 per cent of the surface of the ocean, 7 per cent of the annual NPP of the ocean);
6. Equatorial upwelling regions (about 8 per cent of the surface of the ocean, 9 per cent of the annual NPP of the ocean).

Collectively, those six regions encompass 90 per cent of the surface of the ocean and account for 90 per cent of the NPP of the ocean.

4. Estimating plankton diversity

4.1. Species diversity

Accurate estimates of plankton species diversity at the regional and global levels based on the microscopic examination of ocean samples are not possible at the presnt time, owing to severe undersampling[6] (Appeltans and others, 2012), the rapidly growing number of cryptic species[7] revealed by metagenomics (Delong, 2009; Goetze, 2010; Lindeque and others, 2013; Harvey and others, 2017), larval stages of zooplankton that lack clear diagnostic characteristics (Bucklin and others, 2016) and a lack of consensus among microbiologists on the definition of species (Amaral-Zettler and others, 2010). The problem of undersampling can only be addressed by increasing the time-space resolution of sampling. In this regard, the importance of expanding and sustaining support for the Global Alliance of Continuous Plankton Recorder Surveys (Batten and others, 2019) and for the development of an integrated ocean life observing system (Canonico and others, 2019) must be stressed.

4.2. Functional diversity

Grouping organisms into functional groups that share common characteristics (size and ecological roles) can be more useful ecologically than taxonomic groupings (Litchman and others, 2010; Mitra and others, 2016). The size spectrum of plankton spans over seven orders of magnitude (Boyce and others, 2015; Sommer and others, 2017) and is reflected in the pathways by which phytoplankton NPP is recycled in the upper ocean, channelled to fisheries or exported to the interior of the ocean through the biological pump (Ward and others, 2012; Acevedo-Trejos and others, 2018). Plankton have been categorized by size into picoplankton (0.2–2 μm), nanoplankton (2–20 μm), microplankton (20–200 μm), mesoplankton (200 μm to 20 mm), macroplankton (20–200 mm) and megaplankton (> 200 mm) (Sieburth and others, 1978; Sommer and others, 2017). At the level of the major ocean basins, phytoplankton biomass and the fraction of large phytoplankton generally increase as the availability of dissolved inorganic nutrients increase, a pattern that reflects the importance of nutrient supplies as a parameter of phytoplankton NPP and community composition (Mousing and others, 2018).

---

6 Measurements are too sparse in time and space to estimate plankton biodiversity accurately at the regional and global levels.

7 Genetically distinct species (based on operational taxonomic units) that do not exhibit clear morphological differences. Such units are used to estimate species richness on the basis of genetic differences (Caron and others, 2009).

From a functional perspective, marine food webs can be divided into two categories based on size (Fenchel, 1988; Pomeroy and others, 2007):

(a) Microbial food webs populated by picophytoplankton and nanophytoplankton, heterotrophic bacteria and protozoan consumers that are fuelled primarily by picophytoplankton NPP (including the release of dissolved organic matter by primary producers and consumers);

(b) Metazoan food webs populated by microphytoplankton and metazoan plankton that are fuelled primarily by microphytoplankton productivity and microbial food webs.

Microbial food webs account for most living biomass and nutrient recycling in the ocean (Del Giorgio and Duarte 2002; Sunagawa and others, 2015), while metazoan food webs support most fisheries and the biological pump (Legendre and Michaud, 1998; Sommer and others, 2002). Thus, changes in the balance between those two food webs are likely to have major impacts on the provision of ecosystem services (Müren and others, 2005; Worm and others, 2006; Sommer and others, 2016).

5. Microbial plankton

5.1. Phytoplankton

Major taxa in terms of their contribution to global NPP include prokaryotic cyanobacteria and eukaryotic diatoms, coccolithophores and chlorophytes (Not and others, 2007; Simon and others, 2009; Uitz and others, 2010; Flombaum and others, 2013).

5.1.1. Diversity and functional groups of phytoplankton

Five functional groups of phytoplankton have been recognized on the basis of their size and roles in pelagic food webs and nutrient cycles (Chisholm, 1992; Le Quéré and others, 2005; Marañón and others, 2012): photosynthetic picoplankton, silicifying microplankton, calcifying nanoplankton, nitrogen-fixing mesoplankton and dimethyl sulphide-producing nanoplankton. For the purposes of the present subchapter, toxic microplankton have been added to this list. Changes in the relative abundance, productivity and toxicity of those functional groups have major repercussions on their capacity to support ecosystem services.

5.1.1.1. Picoplankton

Picoplankton include two genera of cyanobacteria (*Prochlorococcus* and *Synechococcus*) and a diverse ensemble of picoeukaryotes from several phyla (Not and others, 2007; Kirkham and others, 2013). They are globally ubiquitous, account for some 50 per cent of the NPP of the ocean (Agusti and others, 2019) and fuel microbial food webs (Marañón and others, 2001, 2015). It is estimated that *Prochlorococcus* accounts for 17–39 per cent of picoplankton biomass globally, *Synechococcus* for 12–15 per cent and picoeukaryotes for 49–62 per cent (Buitenhuis and others, 2012). *Prochlorococcus* dominates the phytoplankton in warm (> 15 °C), nutrient-poor waters (Chisholm, 2017). *Synechococcus* has a broader, more uniform distribution and is more abundant than *Prochlorococcus* under cooler, nutrient-rich conditions (Follows and others, 2007; Flombaum and others, 2013). Picoeukaryotes tend to increase in abundance with increasing nutrient levels, often dominating the phytoplankton at high-latitudes (Li, 1994; Worden and Not, 2008; Kirkham and others, 2013). Those organisms exhibit extraordinary genomic diversity that underlies their broad geographical distribution (Vaulot, 2008; Kent and others, 2016).

5.1.1.2. Silicifying microplankton: diatoms

Diatoms dominate microphytoplankton in cold, turbulent, nutrient-rich waters (Malone, 1980; Rousseaux and Gregg, 2015). They account for 40–50 per cent of global marine NPP, fuel metazoan food webs and account for some 40 per cent of carbon export through the biological pump (Honjo and others, 2014; Tréguer and others, 2018). Diatoms are therefore important players in the global carbon cycle.

5.1.1.3. Calcifying nanoplankton[8]

Coccolithophores (dominated by *Emiliania huxleyi*) are globally ubiquitous, function as both a sink (photosynthesis) and a source (calcification) of CO_2 and are therefore important players in the global carbon cycle (Sarmiento and others, 2002; Balch and others, 2016). *E. huxleyi* forms the "great calcite belt" that surrounds Antarctica between the subantarctic and polar fronts (Balch and others, 2016; Nissen and others, 2018). There is evidence that *E. huxleyi* produces more biogenic $CaCO_3$ than any other organism on Earth (Iglesias-Rodríguez and others, 2002). Blooms tend to occur following seasonal diatom blooms (Brown and Yoder, 1994; Smith and others, 2017). *E. huxleyi* harbours a pan-genome of extensive genetic variability that underpins its cosmopolitan distribution and its capacity to bloom under a wide variety of environmental conditions (Read and others, 2013).

5.1.1.4. Nitrogen-fixing mesoplankton[9]

Planktonic cyanobacteria account for about half of the Earth's N_2-fixation (Karl and others, 2002; Landolfi and others, 2018) and are the largest source of fixed nitrogen in the global ocean (Galloway and others, 2004; Gruber, 2004). The group includes unicellular symbionts (diatom-diazotroph associations) and colonial genera (e.g., *Trichodesmium*) (Delmont and others, 2018; White and others, 2018). Most marine nitrogen fixation occurs in the subtropical gyres (Gruber, 2019) where *Trichodesmium* is most abundant at temperatures above 20°C (Breitbarth and others, 2007; Monteiro and others, 2010).

5.1.1.5. Dimethylsulphoniopropionate-producing nanoplankton[10]

Over 90 per cent of dimethyl sulphide emissions to the atmosphere come from dimethylsulphoniopropionate produced in the ocean, most of which by Prymnesiophyceae (e.g., *Phaeocystis* spp. and *Emiliania huxleyi*) and Dinophyceae (e.g., *Prorocentrum minimum*) during blooms (Keller and others, 1989; Bullock and others, 2017). *Phaeocystis* is a cosmopolitan genus with a life cycle that alternates between free-living nanoplankton (3–9 µm) and large (> 2 mm) gelatinous colonies (Schoemann and others, 2005). The latter develop during massive summer blooms in high-latitude spring bloom regions and during summer blooms in polar sea ice regions and in the Antarctic circumpolar region (Schoemann and others, 2005; Vogt and others, 2012). Blooms of *Prorocentrum minimum* occur in regions with relatively high anthropogenic nutrient inputs, and its global distribution is expected to expand, given that anthropogenic nutrient inputs are projected to more than double by 2050, unless inputs are controlled more effectively on a global scale (Glibert and others, 2008).

---

8 Calcifying plankton include taxa that create shells, skeletons or other structures from calcium carbonate. This is a taxonomically diverse group that includes phytoplankton such as coccolithophores, zooplankton such as pteropods and the larval stages of benthic bivalve molluscs and echinoderms.

9 Nitrogen fixation is not limited to mesozooplankton. There is evidence that there are non-cyanobacterial diazotrophs (bacteria and archaea) in the oceans (Benavides and others, 2018).

10 Dimethylsulphoniopropionate is the biogenic precursor of dimethyl sulphide, which represents a large source of sulfur going into the Earth's atmosphere, where it helps to drive the formation of clouds that block solar radiation from reaching the Earth's surface and reflect it back into space.

5.1.1.6. Toxin-producing microplankton

Among the 5,000 species of extant marine phytoplankton (Sournia and others, 1991), some 80 or so species have the capacity to produce potent toxins that find their way through fish and shellfish into humans (Hallegraeff and others, 2004). Most toxic species are dinoflagellates that cause paralytic shellfish poisoning (e.g., *Alexandrium* spp.), diarrhoeic shellfish poisoning (e.g., *Dinophysis* spp.), neurotoxic shellfish poisoning (e.g., *Karenia* spp.), azaspiracid shellfish poisoning (e.g., *Protoperidinium crassipes*) and ciguatera fish poisoning (e.g., *Gambierdiscus toxicus*). One diatom genus (*Pseudo-nitzschia* spp.) also causes amnesic shellfish poisoning (Lelong and others, 2012). Toxin-producing microplankton have a cosmopolitan distribution (Hallegraeff and others, 2004).

5.2. Protozoan consumers

Most heterotrophic protozoa fall into the nano- and micro-zooplankton size classes and are major consumers in microbial food webs and important links to metazoan food webs (Landry and Calbet, 2004; Mitra and others, 2016). Their diversity can be described in terms of three basic body plans that broadly determine their ecological roles: amoeboid, flagellated and ciliated forms (Fuhrman and Caron, 2016).

Amoeboid foraminifers are most abundant in high-latitude spring bloom regions and least abundant in subtropical gyres (Berger, 1969). They are major producers of marine calcareous shells deposited on the ocean floor (Schiebel and Hemleben, 2005). Radiolaria are common in the euphotic zone in tropical and subtropical oceanic regions globally and much less abundant in coastal upwelling, high-latitude spring bloom and polar regions (Caron and Swanberg, 1990).

Heterotrophic nanoflagellates are the most abundant protozoan consumers and control the abundance of bacterioplankton (Fenchel, 1982; Massana and Jürgens, 2003). While nanoflagellates are important grazers of picophytoplankton in oligotrophic habitats, heterotrophic microflagellates (e.g., dinoflagellates) can be important consumers of microphytoplankton, including bloom-forming diatoms (Sherr and Sherr, 2007; Calbet, 2008).

Microzooplankton (dinoflagellates and ciliates) have been estimated to graze over half of daily global phytoplankton NPP and exert significant top-down control on phytoplankton blooms in ecosystems from the Southern Ocean (Swalethorp and others, 2019) and the western Arctic Ocean (Sherr and others, 2009) to temperate coastal ecosystems (Pierce and Turner, 1992).

5.3. Heterotrophic bacteria and archaea

Bacterial assemblages are typically dominated by a small number of phylotypes[11] (Yooseph and others, 2010), the 20 most abundant of which fall into one of four groups (Amaral-Zettler and others, 2010; Luo and Moran, 2014): α-Proteobacteria (SAR11, Rhodobacteraceae), γ-Proteobacteria (SAR86), Bacteroidetes (Flavobacteriaceae) and Actinobacteria, most abundant of which are α-Proteobacteria (Lefort and Gasol, 2013; Giovannoni, 2017). Species richness tends to decrease toward the poles for both the animal and plant kingdoms (Wietz and others, 2010).

Four major groups of archaea (marine groups I–IV) are abundant in the ocean (Church and others, 2003; Danovaro and others, 2017). Marine group I archaea are among the most abundant and widely distributed, from polar to tropical waters (Karner and others, 2001;

[11] A group of genetically similar organisms that may be grouped at different taxonomic levels, such as species, family, class or phylum.

Santoro and others, 2019). Although bacteria tend to outnumber archaea, the latter make an important contribution to microbial biomass in deep waters (Danovaro and others, 2015).

5.4. Viruses

Viruses play important roles in marine food webs and nutrient recycling via their control of the abundance of microbial populations and the release of dissolved organic matter by cell lysis (Rohwer and Thurber, 2009; Sieradzki and others, 2019). Viruses, including free-living virions, are the most abundant biological entities in the oceans and are a major reservoir of genetic diversity (Suttle, 2007; Simmonds and others, 2017). The majority of viruses are bacteriophages (Coutinho and others, 2017), and their abundance is correlated with the abundance of bacteria from regional to global scales (Fuhrman and Caron, 2016). Metagenomic analyses indicate that there are thousands of different virions in a few litres, with the most abundant genotypes represented by a relatively small fraction of the entire assemblage (Breitbart and others, 2004; Angly and others, 2006). However, despite recent advancements in metagenomics such as those, it is clear that this is the "tip of the iceberg" in terms of viral biodiversity (Paez-Espino and others, 2019).

6. Metazoan zooplankton

6.1. Holoplankton[12]

Metazoan holoplankton have been described from 15 phyla (Bucklin and others, 2010; Wiebe and others, 2010). As a group, they exhibit diverse feeding types (Kiørboe, 2011), from filter-feeders (e.g., copepods, euphausiids and tunicates) to passive ambush predators (e.g., ctenophores and some pteropods) and active ambush predators (e.g., chaetognaths and some amphipods). Like other groups of animals, the diversity of holozooplankton tends to decrease toward the poles (Lindley and Batten, 2002; Burridge and others, 2017). Diversity also tends to be higher when biomass is low (e.g., subtropical gyres) and lower when biomass is high (e.g., coastal upwelling and high-latitude spring bloom regions) (United Nations, 2017a).

6.1.1. Crustaceans

About half of the known species of holoplankton are crustaceans (Verity and Smetacek, 1996; United Nations, 2017a). Copepods are by far the most abundant and are a key trophic link between phytoplankton and fisheries (e.g., Möllmann and others, 2003; Beaugrand, 2005). While copepod abundance is generally highest in regions where high NPP occurs seasonally, biodiversity is generally highest in warm water regions where NPP is relatively low (Rombouts and others, 2009; Valdés and others, 2017).

With nearly 100 documented species (Baker and others, 1990), euphausiids (krill) occur throughout the global ocean and, like copepods, are most abundant during periods of high phytoplankton productivity (Baker and others, 1990). They are especially abundant in the Southern Ocean, where they play a crucial role in the food web and are a target for fisheries (Mangel and Nicol, 2000; Boopendranath, 2013).

There are approximately 200 described species of planktonic ostracods (Angel and others, 2007) and approximately 300 species

[12] Species that live their entire life cycles as plankton.

of hyperiid amphipods[13] (Vinogradov, 1996; Boltovskoy and others, 2003). Ostracod species richness tends to be highest in the mesopelagic zone at low latitudes (< 50°N) and in the epipelagic zone at higher latitudes. The majority of hyperiids spend at least part of their life cycle living as commensals of salps, jellyfish, ctenophores or siphonophores (Madin and Harbison, 1977; Gasca and Haddock, 2004), and their species richness is highest in regions where gelatinous zooplankton are most abundant.

6.1.2. Gelatinous zooplankton

This diverse group includes cnidarians (jellyfish),[14] ctenophores (comb jellies), chaetognaths (arrow worms), tunicates (salps, doliolids and appendicularians) and molluscs (pteropods and heteropods) (Alldredge, 1984; Jennings and others, 2010). As a group, tunicates are well adapted to life in oligotrophic oceans, where their diversity and abundance are often greater than those of planktonic crustaceans (Alldredge and Madin, 1982; Madin and Harbison, 2001). Species richness is highest in jellyfish (more than 1,000 species) (Purcell and others, 2007; Pitt and others, 2018), followed by molluscs (250 species) (Jennings and others, 2010), ctenophores (200 species) (Harbison, 1985; Madin and Harbison, 2001), tunicates (145 species) (Deibel and Lowen, 2012) and chaetognaths (100 species) (Daponte and others, 2004).

6.2. Meroplankton

Meroplankton are larval stages of benthic and pelagic adults (e.g., shellfish and fish) and are, therefore, temporary members of the plankton. Their contribution to plankton diversity occurs episodically or seasonally, and their abundance relative to holoplankton decreases with increasing depth and increasing latitude (Silberberger and others, 2016; Costello and Chaudhary, 2017). The distribution, diversity and fecundity of adults that have a planktonic larval stage are inextricably linked to the abundance and diversity of their meroplanktonic larvae which, in turn, influence the distribution and diversity of their adult stage (Miron and others, 1995; Hughes and others, 2000).

7. Documented trends

7.1. Global

The study of satellite time series (1998–2015) of sea surface chlorophyll *a* has not yet revealed a long-term trend in NPP on a global scale (Gregg and others, 2017). However, microplankton diatom biomass has declined relative to picophytoplankton in most regions during the period under review (Rousseaux and Gregg, 2015; Gregg and others, 2017), a trend that appears to be related to upper-ocean warming, increases in vertical stratification[15], and decreases in nutrient supplies from the deep sea (Daufresne and others, 2009; Basu and Mackey, 2018).

A comparison of known toxic events in 1970 with those observed in 2017[16] suggests that

13 An order of amphipods that is exclusively marine.

14 Although jellyfish have a life cycle with a benthic polyp stage and a planktonic medusa stage, they are considered to be holoplankton because the sexually reproducing stage (medusa) is planktonic.

15 A water column becomes vertically stratified when a less dense body of water develops (owing to an increase in temperature, a decrease in salinity, or both) over deeper, denser water. This process limits mixing between the surface mixed layer and the deep ocean.

16 United States National Office for Harmful Algal Blooms, "Distribution of HABs throughout the World". Available at www.whoi.edu/website/redtide/regions/world-distribution.

the public health and economic impacts of toxic events have increased in frequency and have spread globally (Hallegraeff and others, 2004):

- Paralytic shellfish poisoning events caused by *Alexandrium tamarense* and *A. catenella* increased from 19 coastal sites (including 12 in North America and 4 in western Europe) to 118 coastal sites (including 26 in North America, 25 in western Europe, 36 in the western Pacific, 9 in Australia and New Zealand, 7 in South America, 7 in Africa and 4 in India).
- Diarrhoeic shellfish poisoning events caused by *Dinophysis* spp. increased from 15 coastal sites (including 13 in western Europe) to 71 coastal sites (8 in North America, 37 in western Europe, 9 in South America, 7 in Australia and New Zealand, 6 in Japan and 4 in India).
- Amnesic shellfish poisoning events caused by *Pseudo-nitzschia* spp. increased from 1 coastal site, in North America, to 31 coastal sites (including 12 in North America, 9 in western Europe and 9 in Australia and New Zealand).

While there is reason to suspect that the combined effects of increases in coastal eutrophication, sea surface temperature and vertical stratification may favour the growth of dinoflagellates, the underlying causes of those trends remain a matter of speculation (Wells and others, 2015).

Upper-ocean warming is influencing the biogeography and phenology of plankton species (Hays and others, 2005; Thackeray and others, 2010; Mackas and others, 2012). On average, seasonal spring peaks in biomass have advanced by 4.4 days per decade, with a standard error of 0.7 days, and the leading edges of species distributions have extended polewards by 72 km per decade (1920–2010), with a standard error of 0.35 km (Hoegh-Guldberg and others, 2014). While holoplankton show large shifts in both biogeography and phenology in response to upper-ocean warming, meroplankton show relatively small shifts in distribution but greater changes in phenology (Edwards and Richardson, 2004), changes that are likely to have feedback effects on the abundance of adult populations.

7.2. Polar sea ice regions

7.2.1. Southern Ocean

A significant interannual trend in NPP in the Southern Ocean as a whole has not been documented (Arrigo and others, 2008). However, opposing trends in NPP in the Ross Sea (increasing) and the West Antarctic Peninsula (decreasing) coincided with increases (Ross Sea) and decreases (West Antarctic Peninsula) in sea ice extent[17] (Montes-Hugo and others, 2009; Ducklow and others, 2013). The decrease in NPP was associated with a shift in the size spectrum of phytoplankton from microplankton-dominated (diatoms) assemblages to nanoplankton and picoeukaryotes as sea surface temperature increased (Moline and others, 2004; Montes-Hugo and others, 2009). Warming and the shift to smaller phytoplankton have also been associated with a range extension of *Emiliania huxleyi* from the Antarctic circumpolar region into the polar sea ice region (Cubillos and others, 2007).

Inter-annual variations in the extent of sea ice off the Antarctic Peninsula also appear to be reflected in the relative abundance of two dominant grazers: krill (*Euphausia superba*) and salps (*Salpa thompsoni*). Krill recruitment, which depends on the survival of larval krill during winter, is the population parameter most likely to be altered by climate change (Flores and others, 2012). *E. superba* has been found to be more abundant following winters with extensive sea ice cover, while salps have been

17 See Michon Scott and Kathryn Hansen, "Sea ice", NASA Earth Observatory, 16 September 2016.

more abundant following winters when the spatial extent of sea ice is relatively low (Loeb and others, 1997). Thus, while krill populations may have suffered from sea ice decline, salps appear to have benefited from warming surface waters during the twentieth century (Loeb and Santora 2012). The observed decrease in sea ice extent portends a long-term shift from a food web dominated by *E. superba* to one dominated by salps, with unknown cascading effects on the abundance of vertebrate predators (Henschke and others, 2016).

7.2.2. Arctic Ocean

The Arctic Ocean is in the process of transitioning to a warmer state (cf., Buchholz and others, 2010). Unlike the Antarctic, sea ice extent has decreased (1998–2015) in all sectors of the Arctic, owing to increases in sea surface temperature (Kahru and others, 2016), a trend that is associated with increases in NPP (Arrigo and van Dijken, 2011; Hill and others, 2017) and increases in the biomass of picoeukaryotes at the expense of microplankton diatoms as the vertical stratification of the water column increased (Li and others, 2009).

As in coastal waters of the West Antarctic Peninsula, krill are an important prey for a number of species, including smelt. From 1984 to 1992 and from 2007 to 2015, krill abundance increased in the south-western and central Barents Sea, despite high smalt predation, probably as a result of increasing temperatures, stronger advection of krill into the Barents Sea (Slagstad and others, 2011) and increases in phytoplankton NPP (Dalpadado and others, 2014). Warming has also influenced the relative abundance of krill species, with the boreal species *Meganyctiphanes norvegica* increasing and the cold water species *Thysanoeassa raschii* decreasing (Rasmussen, 2018).

7.3. North Atlantic high-latitude spring bloom region

In the seasonally nutrient-rich waters of the North Atlantic high-latitude spring bloom region, upper-ocean warming and an earlier set-up of the seasonal pycnocline[18] combine to increase the length of the growing season and the availability of sunlight. As a result, NPP has been increasing in recent decades (1979–2010) (Dalpadado and others, 2014; Raitsos and others, 2014), a trend that has been accompanied by increases in picoeukaryotes and coccolithophores relative to diatoms (Li and others, 2009), a reduction in the average size of phytoplankton and zooplankton and an increase in the biodiversity of plankton assemblages (Hoegh-Guldberg and Bruno, 2010; Edwards and others, 2013).

Poleward expansions in the range of plankton species in response to upper-ocean warming have been well documented (Poloczanska and others, 2013), especially in the North Atlantic: *Emiliana huxleyi* into the Barents Sea (Smyth and others, 2004); *Calanus helgolandicus* replacing *C. finmarchicus* in the North Sea (Edwards and others, 2013); and a poleward expansion of the ranges of calcifying species of plankton (foraminifers, coccolithophores and pteropods) (Beaugrand and others, 2013; Winter and others, 2014).

The phenologies[19] of phytoplankton and zooplankton species are also changing in response to upper-ocean warming (1958–2002). For example, during the period from 1958 to

[18] A pycnocline is a vertical zone over which an increase in density separates a surface layer of relatively low density from a deeper layer of relatively high density. A seasonal pycnocline begins to form in the North Atlantic high-latitude spring bloom region when solar heating begins to warm the surface layer during late winter-early spring, a process that increases the availability of solar energy for photosynthesis.

[19] Phenology refer to the timing of biological events in plant and animal lives (e.g., reproduction and migration) in relation to changes in season and climate.

2002, the seasonal abundance of the copepod *Calanus finmarchicus* in the North Atlantic began to peak earlier, so that, by the turn of the century, abundance was peaking some 10 days earlier in the year than previously, while its food (microplankton diatoms and dinoflagellates) peaked some 30 days earlier (Edwards and Richardson, 2004). Similarly, diatom blooms in the North Sea are occurring earlier in the year than the peak in abundance of their macrozooplankton grazers (Hays and others, 2005). Such uncoupling of trophic levels has also been documented in the Baltic Sea, where the duration of the growing season during the period from 1988 to 2017 increased at a rate of 4.5 days per year, resulting in an earlier spring bloom, a prolongation of the summer biomass minimum and a later and more prolonged autumn bloom (Wasmund and others, 2019).

7.4. Upwelling regions

Diatom production has been increasing (1996–2011) in eastern boundary upwelling systems (Kahru and others, 2012), while NPP increased in Pacific equatorial upwelling (Chavez and others, 2011), apparently owing to increases in upwelling (Tim and others, 2016). However, ocean acidification in coastal upwelling systems is proving to be corrosive to pteropod shells (*Limacina helicina*) (Bednaršek and others, 2014). As the habitat suitability for pteropods declines, metazoan food webs are likely to be affected (Bednaršek and others, 2012; Lischka and others, 2011), and ocean acidification is likely to have a similar impact in the Southern Ocean and the Arctic Ocean (Comeau and others, 2009; Negrete-García and others, 2019).

7.5. Subtropical gyres

Owing largely to declines in diatoms and chlorophytes (Gregg and others, 2017), a significant downward trend in chlorophyll *a* (1998–2013) has been documented in all gyres, except in the South Pacific (Signorini and others, 2015). Rates of decline were greatest in the northern hemisphere and lowest in the South Atlantic Ocean and the Indian Ocean, trends that correspond to expansions of the gyres in the Atlantic Ocean and the North Pacific Ocean (Polovina and others, 2008)

Downward trends in NPP were observed in all five gyres that coincided with upper-ocean warming and decreases in phytoplankton cell size (Polovina and Woodworth, 2012). The latter is consistent with observed increases in the relative abundance of *Prochlorococcus* and *Synechococcus* (Flombaum and others, 2013; Agusti and others, 2019), trends that most likely reflect both warming temperatures (Daufresne and others, 2009; Morán and others, 2010) and decreasing nutrient supplies as the euphotic zone becomes more isolated from nutrient-rich deep water (Marañón and others, 2015; Sommer and others, 2016).

8. Outlook

Climate change during the course of the twenty-first century is expected to continue to drive changes in the upper ocean that have an impact on the diversity and productivity of plankton assemblages on the regional to global scales. These changes include an expansion of the subtropical gyres (Polovina and others, 2011), ocean warming and acidification, decreases in salinity, increases in vertical stratification and decreases in inorganic nutrient supplies to the euphotic zone in the open ocean (Bopp and others, 2013). Predicted biological responses to those changes on a global scale include the following:

(a) NPP is likely to decrease and the relative abundance of picophytoplankton is likely to increase (Daufresne and others, 2009; Morán and others, 2010) at the expense of microplankton diatoms (Bopp and others, 2005; Moore and others, 2018);

(b) These trends are likely to propagate through food webs resulting in decreases in the ocean's carrying capacity for fisheries (Worm and others, 2006; Chust and others, 2014) and in its capacity to sequester carbon through the biological pump (Boyd, 2015);

(c) The expansion of the subtropical gyres may promote increases in N2 fixation (Boatman and others, 2017; Follett and others, 2018), a trend that could further perturb the global nitrogen cycle (Jiang and others, 2018);

(d) Plankton food webs in the polar oceans and coastal upwelling regions will be the most affected by ocean acidification, owing to the high solubility of CO_2 in cold waters (Bednaršek and others, 2014; Gardner and others, 2018).

Regional exceptions during the course of the twenty-first century are predicted to occur polewards of the subtropical gyres as a result of environmental changes in the euphotic zone, including increases in the availability of sunlight as the surface mixed layer shoals in nutrient-rich environments (promoting increases in NPP), increases in temperature and decreases in salinity (favouring the growth of small phytoplankton) (Tréguer and others, 2018). Notable examples include:

(a) An increase in NPP and a decrease in phytoplankton size in the Arctic Ocean (Mueter and others, 2009; Kahru and others, 2011; Dalpadado and others, 2014);

(b) Increases in NPP, export production and the abundance of diatoms during the first half of the century in the polar sea ice region of Antarctica (Bopp and others, 2001; Kaufman and others, 2017; Moore and others, 2018);

(c) Expansion of the range of *Emiliania huxleyi* into the polar oceans (Winter and others, 2014) and increases in the frequency of coccolithophore blooms in high-latitude spring bloom regions (Bopp and others, 2013; Rivero-Calle and others, 2015);

(d) An increase in NPP and a decrease in the relative abundance of diatoms in the North Atlantic high-latitude bloom region (Bopp and others, 2005, 2013; Sundby and others, 2016).

Projections of future trends in NPP in coastal upwelling regions are less certain, owing to uncertainty concerning how interactions between increases in upwelling-favourable winds (increases in upwelling, NPP and the relative abundance of diatoms) and upper-ocean warming (decreases in upwelling, NPP and the relative abundance of diatoms) will play out (Chavez and others, 2011; García-Reyes and others, 2015).

In this context, it must be emphasized that the present analysis of the impacts of climate change on plankton communities does not consider transgenerational adaptation to climate-driven changes in the upper-ocean environment (e.g., Schlüter and others, 2014; Thor and Dupont, 2015).

References

Acevedo-Trejos, Esteban, and others (2018). Phytoplankton size diversity and ecosystem function relationships across oceanic regions. *Proceedings of the Royal Society B: Biological Sciences*, vol. 285, No. 1879, pp. 20180621.

Agusti, Susana, and others (2019). Projected changes in photosynthetic picoplankton in a warmer subtropical ocean. *Frontiers in Marine Science*, vol. 5, art. 506.

Alldredge, A.L., and L.P. Madin (1982). Pelagic tunicates: unique herbivores in the marine plankton. *Bioscience*, vol. 32, No. 8, pp. 655–663.

Alldredge, Alice L. (1984). The quantitative significance of gelatinous zooplankton as pelagic consumers. In *Flows of Energy and Materials in Marine Ecosystems*, pp. 407–433. Boston, MA: Springer.

Amaral-Zettler, Linda, and others (2010). A global census of marine microbes. *Life in the World's Oceans: Diversity, Distribution and Abundance*, pp. 223–245.

Angel, Martin V., and others (2007). Changes in the composition of planktonic ostracod populations across a range of latitudes in the North-east Atlantic. *Progress in Oceanography*, vol. 73, No. 1, pp. 60–78.

Angly, Florent E., and others (2006). The marine viromes of four oceanic regions. *PLoS Biology*, vol. 4, No. 11, e368.

Appeltans, Ward, and others (2012). The magnitude of global marine species diversity. *Current Biology*, vol. 22, No. 23, pp. 2189–2202.

Arrigo, Kevin R., and Gert L. van Dijken (2011). Secular trends in Arctic Ocean net primary production. *Journal of Geophysical Research: Oceans*, vol. 116, No.C 9.

Arrigo, Kevin R., and others (2008). Primary production in the Southern Ocean, 1997–2006. *Journal of Geophysical Research: Oceans*, vol. 113, No.C 8.

Azam, Farooq, and Francesca Malfatti (2007). Microbial structuring of marine ecosystems. *Nature Reviews Microbiology*, vol. 5, pp. 782–791.

Baker, A. de C., and others (1990). A practical guide to the euphausiids of the world. *British Museum (Natural History)*, vol. 96.

Balch, William M., and others (2016). Factors regulating the Great Calcite Belt in the Southern Ocean and its biogeochemical significance. *Global Biogeochemical Cycles*, vol. 30, No. 8, pp. 1124–1144.

Bar-On, Yinon M., and others (2018). The biomass distribution on Earth. *Proceedings of the National Academy of Sciences*, vol. 115, No. 25, pp. 6506–6511.

Basu, Samarpita, and Katherine R. M. Mackey (2018). Phytoplankton as key mediators of the biological carbon pump: Their responses to a changing climate. *Sustainability*, vol. 10, No. 3.

Batten, Sonia D., and others (2019). A global plankton diversity monitoring program. *Frontiers in Marine Science*, vol. 6, art. 321.

Beaugrand, Grégory (2005). Monitoring pelagic ecosystems using plankton indicators. *ICES Journal of Marine Science*, vol. 62, No. 3, pp. 333–338.

Beaugrand, Gregory, and others (2013). Long-term responses of North Atlantic calcifying plankton to climate change. *Nature Climate Change*, vol. 3, pp. 263–267.

Bednaršek, N., and others (2012). Extensive dissolution of live pteropods in the Southern Ocean. *Nature Geoscience*, vol. 5, No. 12, pp. 881–885.

Bednaršek, N., and others (2014). *Limacina helicina* shell dissolution as an indicator of declining habitat suitability owing to ocean acidification in the California Current Ecosystem. *Proceedings of the Royal Society B: Biological Sciences*, vol. 281, No. 1785, 20140123.

Behrenfeld, M.J., and others. 2006. Climate-driven trends in contemporary ocean productivity. *Nature*, vol. 444, pp. 752–755. https://doi.org/10.1038/nature05317.

Benavides M., and others (2018). Deep into oceanic N_2 fixation. *Frontiers in Marine Science*, vol. 5, art. 108. https://doi.org/10.3389/fmars.2018.00108.

Berger, Wolfgang H. (1969). Ecologic patterns of living planktonic foraminifera. In *Deep Sea Research and Oceanographic Abstracts*, vol. 16: pp. 1–24. Elsevier.

Blanchard, Julia L., and others (2012). Potential consequences of climate change for primary production and fish production in large marine ecosystems. *Philosophical Transactions of the Royal Society B: Biological Sciences*, vol. 367, No. 1605, pp. 2979–2989.

Boatman, Tobias G., and others (2017). A key marine diazotroph in a changing ocean: the interacting effects of temperature, CO_2 and light on the growth of *Trichodesmium erythraeum* IMS101. *PLoS One*, vol. 12, No. 1, e0168796.

Boltovskoy, Demetrio, and others (2003). Marine zooplanktonic diversity: a view from the South Atlantic. *Oceanologica Acta*, vol. 25, No. 5, pp. 271–278.

Boopendranath, M.R. (2013). Antarctic krill–A keystone species of Antarctica. *Science India*, vol. 16, pp. 4–10.

Bopp, Laurent, and others (2001). Potential impact of climate change on marine export production. *Global Biogeochemical Cycles*, vol. 15, No. 1, pp. 81–99.

Bopp, Laurent, and others (2005). Response of diatoms distribution to global warming and potential implications: A global model study. *Geophysical Research Letters*, vol. 32, No. 19.

Bopp, Laurent, and others (2013). Multiple stressors of ocean ecosystems in the 21st century: projections with CMIP5 models. *Biogeosciences*, vol. 10, pp. 6225–6245.

Boyce, Daniel G., and others (2015). Spatial patterns and predictors of trophic control in marine ecosystems. *Ecology Letters*, vol. 18, No. 10, pp. 1001–1011.

Boyd, Philip W. (2015). Toward quantifying the response of the oceans' biological pump to climate change. *Frontiers in Marine Science*, vol. 2, art. 77.

Breitbart, Mya, and others (2004). Diversity and population structure of a near-shore marine-sediment viral community. *Proceedings of the Royal Society of London. Series B: Biological Sciences*, vol. 271, No. 1539, pp. 565–574.

Breitbarth, Eike, and others (2007). Physiological constraints on the global distribution of *Trichodesmium*? Effect of temperature on diazotrophy. *Biogeosciences*, vol. 4, No. 1, pp. 53–61.

Brown, Christopher W., and James A. Yoder (1994). Coccolithophorid blooms in the global ocean. *Journal of Geophysical Research: Oceans*, vol. 99, No.C 4, pp. 7467–7482.

Buchholz, Friedrich, and others (2010). Ten years after: krill as indicator of changes in the macro-zooplankton communities of two Arctic fjords. *Polar Biology*, vol. 33, No. 1, pp. 101–113.

Bucklin, Ann, and others (2010). A census of zooplankton of the global ocean. *Life in the World's Oceans: Diversity, Distribution, and Abundance, Edited by: McIntyre, A*247–265.

Bucklin, Ann, and others (2016). Metabarcoding of marine zooplankton: prospects, progress and pitfalls. *Journal of Plankton Research*, vol. 38, No. 3, pp. 393–400.

Buitenhuis, Erik Theodoor, and others (2012). Picophytoplankton biomass distribution in the global ocean. *Earth System Science Data*, vol. 4, No. 1, pp. 37–46.

Bullock, Hannah A., and others (2017). Evolution of dimethylsulfoniopropionate metabolism in marine phytoplankton and bacteria. *Frontiers in Microbiology*, vol. 8, art. 637.

Burridge, Alice K., and others (2017). Diversity and distribution of hyperiid amphipods along a latitudinal transect in the Atlantic Ocean. *Progress in Oceanography*, vol. 158, pp. 224–235.

Calbet, Albert (2008). The trophic roles of microzooplankton in marine systems. *ICES Journal of Marine Science*, vol. 65, No3, pp. 325–331.

Canonico, Gabrielle, and others (2019). Global observational needs and resources for marine biodiversity. *Frontiers in Marine Science*, vol. 6, art. 367.

Caron, David A. (2016). Mixotrophy stirs up our understanding of marine food webs. *Proceedings of the National Academy of Sciences*, vol. 113, No. 11, pp. 2806–2808.

Caron, David A., and N. R. Swanberg (1990). The ecology of planktonic sarcodines. *Reviews in Aquatic Sciences*, vol. 3, pp. 147–80.

Chavez, Francisco P., and others (2011). Marine primary production in relation to climate variability and change. *Annual Review of Marine Science*, vol. 3, pp. 227–260.

Chisholm, Sallie W. (1992). Phytoplankton Size. In *Primary Productivity and Biogeochemical Cycles in the Sea*, eds. Paul G. Falkowski, Avril D. Woodhead, and Katherine Vivirito, pp. 213–237. Boston, MA: Springer US. https://doi.org/10.1007/978-1-4899-0762-2_12.

Chisholm, Sallie W. (2017). *Prochlorococcus*. *Current Biology*, vol. 27, No. 11, pp. R447–R448.

Church, Matthew J., and others (2003). Abundance and distribution of planktonic Archaea and Bacteria in the waters west of the Antarctic Peninsula. *Limnology and Oceanography*, vol. 48, No. 5, pp. 1893–1902.

Chust, Guillem, and others (2014). Biomass changes and trophic amplification of plankton in a warmer ocean. *Global Change Biology*, vol. 20, No. 7, pp. 2124–2139.

Comeau, S., and others (2009). Impact of ocean acidification on a key Arctic pelagic mollusc (*Limacina helicina*). *Biogeosciences*, vol. 6, No. 9, pp. 1877–1882.

Costello, Mark J., and Chhaya Chaudhary (2017). Marine biodiversity, biogeography, deep-sea gradients, and conservation. *Current Biology*, vol. 27, No. 11, pp. R511–R527.

Coutinho, Felipe H., and others (2017). Marine viruses discovered via metagenomics shed light on viral strategies throughout the oceans. *Nature Communications*, vol. 8, art. 15955.

Cubillos, J.C., and others (2007). Calcification morphotypes of the coccolithophorid *Emiliania huxleyi* in the Southern Ocean: changes in 2001 to 2006 compared to historical data. *Marine Ecology Progress Series*, vol. 348, pp. 47–54.

Dalpadado, Padmini, and others (2014). Productivity in the Barents Sea-response to recent climate variability. *PloS One*, vol. 9, No. 5, e95273.

Danovaro, R., and others (2015). Towards a better quantitative assessment of the relevance of deep-sea viruses, Bacteria and Archaea in the functioning of the ocean seafloor. *Aquatic Microbial Ecology*, vol. 75, No. 1, pp. 81–90.

Danovaro, Roberto, and others (2017). Marine archaea and archaeal viruses under global change. *F1000Research*, vol. 6.

Daponte, M.C., and others (2004). Sagitta friderici Ritter-Záhony (Chaetognatha) from South Atlantic waters: abundance, population structure, and life cycle. *ICES Journal of Marine Science*, vol. 61, No. 4, pp. 680–686.

Daufresne, Martin, and others (2009). Global warming benefits the small in aquatic ecosystems. *Proceedings of the National Academy of Sciences*, vol. 106, No. 31, pp. 12788–12793.

Delong, E.F. (2009) The microbial ocean from genomes to biomes. *Nature*, vol. 459, pp. 200–206.

Deibel, Don, and Ben Lowen (2012). A review of the life cycles and life-history adaptations of pelagic tunicates to environmental conditions. *ICES Journal of Marine Science*, vol. 69, No. 3, pp. 358–369.

Del Giorgio, Paul A., and Carlos M. Duarte (2002). Respiration in the open ocean. *Nature*, vol. 420, pp. 379–384.

Delmont, Tom O., and others (2018). Nitrogen-fixing populations of Planctomycetes and Proteobacteria are abundant in surface ocean metagenomes. *Nature Microbiology*, vol. 3, No. 7, pp. 804–813.

De Vargas, Colomban, and others (2015). Eukaryotic plankton diversity in the sunlit ocean. *Science*, vol. 348, No. 6237.

Ducklow, Hugh W., and others (2013). West Antarctic Peninsula: an ice-dependent coastal marine ecosystem in transition. *Oceanography*, vol. 26, No. 3, pp. 190–203.

Edwards, Martin, and others (2013). Impacts of climate change on plankton. *MCCIP Science Review*, vol. 2013, pp. 98–112.

Edwards, Martin, and Anthony J. Richardson (2004). Impact of climate change on marine pelagic phenology and trophic mismatch. *Nature*, vol. 430, pp. 881–884.

Fenchel, T. (1982). Ecology of heterotrophic microflagellates. IV. Quantitative occurrence and importance as bacterial consumers. *Marine Ecology Progress Series*, vol. 9, pp. 35–42.

Fenchel, Tom (1988). Marine plankton food chains. *Annual Review of Ecology and Systematics*, vol. 19, No. 1, pp. 19–38.

Field, Christopher B., and others (1998). Primary production of the biosphere: integrating terrestrial and oceanic components. *Science*, vol. 281, No. 5374, pp. 237–240.

Flombaum, Pedro, and others (2013). Present and future global distributions of the marine Cyanobacteria *Prochlorococcus* and *Synechococcus*. *Proceedings of the National Academy of Sciences*, vol. 110, No. 24, pp. 9824–9829.

Flores, Hauke, and others (2012). Impact of climate change on Antarctic krill. *Marine Ecology Progress Series*, vol. 458, pp. 1–19.

Follett, Christopher L., and others (2018). Seasonal resource conditions favor a summertime increase in North Pacific diatom–diazotroph associations. *The ISME Journal*, vol. 12, pp. 1543–1557.

Follows, Michael J., and others (2007). Emergent biogeography of microbial communities in a model ocean. *Science*, vol. 315, No. 5820, pp. 1843–1846.

Fuhrman, Jed A., and David A. Caron (2016). Heterotrophic planktonic microbes: virus, bacteria, archaea, and protozoa. In *Manual of Environmental Microbiology, Fourth Edition*, pp. 4–2. American Society of Microbiology.

Galloway, James N., and others (2004). Nitrogen cycles: past, present, and future. *Biogeochemistry*, vol. 70, No. 2, pp. 153–226.

García-Reyes, Marisol, and others (2015). Under pressure: climate change, upwelling, and eastern boundary upwelling ecosystems. *Frontiers in Marine Science*, vol. 2, art. 109.

Gardner, Jessie, and others (2018). Southern Ocean pteropods at risk from ocean warming and acidification. *Marine Biology*, vol. 165, art. 8.

Gasca, Rebeca, and Steven H.D. Haddock (2004). Associations between gelatinous zooplankton and hyperiid amphipods (Crustacea: Peracarida) in the Gulf of California. *Hydrobiologia*, vol. 530, Nos. 1–3, pp. 529–535.

Gasol, Josep M., and others (1997). Biomass distribution in marine planktonic communities. *Limnology and Oceanography*, vol. 42, No. 6, pp. 1353–1363.

Giovannoni, Stephen J. (2017). SAR11 bacteria: the most abundant plankton in the oceans. *Annual Review of Marine Science*, vol. 9, pp. 231–255.

Glibert, Patricia M., and others (2008). *Prorocentrum* minimum tracks anthropogenic nitrogen and phosphorus inputs on a global basis: application of spatially explicit nutrient export models. *Harmful Algae*, vol. 8, No. 1, pp. 33–38.

Goetze, Erica (2010). Species discovery in marine planktonic invertebrates through global molecular screening. *Molecular Ecology*, vol. 19, No. 5, pp. 952–967.

Gregg, Watson W., and others (2017). Global trends in ocean phytoplankton: a new assessment using revised ocean colour data. *Remote Sensing Letters*, vol. 8, pp. 1102–1111.

Gruber, Nicolas (2004). The dynamics of the marine nitrogen cycle and its influence on atmospheric CO 2 variations. In *The Ocean Carbon Cycle and Climate*, pp. 97–148. Springer.

________ (2019). A diagnosis for marine nitrogen fixation. *Nature*, vol. 566, pp. 191–193.

Hallegraeff, Gustaaf M., and others (2004). *Manual on Harmful Marine Microalgae*. United Nations Educational, Scientific and Cultural Organization.

Harbison, G.R. (1985). On the classification and evolution of the Ctenophora. In *The Origins and Relationships of Lower Invertebrates*, pp. 78–100.

Harvey, Julio B.J., and others (2017). Comparison of morphological and next generation DNA sequencing methods for assessing zooplankton assemblages. *Journal of Experimental Marine Biology and Ecology*, vol. 487, pp. 113–126.

Hays, Graeme C., and others (2005). Climate change and marine plankton. *Trends in Ecology & Evolution*, vol. 20, No. 6, pp. 337–344.

Henschke, Natasha, and others (2016). Rethinking the role of salps in the ocean. *Trends in Ecology & Evolution*, vol. 31, No. 9, pp. 720–733.

Hill, Victoria, and others (2017). Decadal trends in phytoplankton production in the Pacific Arctic Region from 1950 to 2012. *Deep Sea Research Part II: Topical Studies in Oceanography*, vol. 152, pp. 82–94.

Hoegh-Guldberg, Ove, and John F. Bruno (2010). The impact of climate change on the world's marine ecosystems. *Science*, vol. 328, No. 5985, pp. 1523–1528.

Hoegh-Guldberg, Ove, and others (2014): The Ocean. In *Climate Change 2014: Impacts, Adaptation, and Vulnerability.* Part B: Regional Aspects. Contribution of Working Group II to the Fifth Assessment Report of the Intergovernmental Panel on Climate Change, Barros, V.R., and others (eds.) (Cambridge University Press, Cambridge, United Kingdom and New York, United States), pp. 1655–1731.

Hoegh-Guldberg, Ove, and Elvira S. Poloczanska (2017). The Effect of Climate Change across Ocean Regions. *Frontiers in Marine Science*, vol. 4, art. 361.

Honjo, Susumu, and others (2014). Understanding the role of the biological pump in the global carbon cycle: an imperative for ocean science. *Oceanography*, vol. 27, No. 3, pp. 10–16.

Hughes, T.P., and others (2000). Supply-side ecology works both ways: the link between benthic adults, fecundity, and larval recruits. *Ecology*, vol. 81, No. 8, pp. 2241–2249.

Iglesias-Rodríguez, M. Débora, and others (2002). Representing key phytoplankton functional groups in ocean carbon cycle models: Coccolithophorids. *Global Biogeochemical Cycles*, vol. 16, No. 4, pp. 47–1.

Jennings, Robert M., and others (2010). Species diversity of planktonic gastropods (Pteropoda and Heteropoda) from six ocean regions based on DNA barcode analysis. *Deep Sea Research Part II: Topical Studies in Oceanography*, vol. 57, Nos. 24–26, pp. 2199–2210.

Jiang, Hai-Bo, and others (2018). Ocean warming alleviates iron limitation of marine nitrogen fixation. *Nature Climate Change*, vol. 8, pp. 709–712.

Kahru, M., and others (2011). Are phytoplankton blooms occurring earlier in the Arctic? *Global Change Biology*, vol. 17, No. 4, pp. 1733–1739.

Kahru, M., and others (2012). Trends in the surface chlorophyll of the California Current: Merging data from multiple ocean color satellites. *Deep Sea Research Part II: Topical Studies in Oceanography*, vol. 77, pp. 89–98.

Kahru, M., and others (2016). Effects of sea ice cover on satellite-detected primary production in the Arctic Ocean. *Biology Letters*, vol. 12, No. 11, 20160223.

Karl, D.M., and others (2002). Dinitrogen fixation in the World's oceans. *Biogeochemistry*, vols.57–58, pp. 47–98. https://doi.org/10.1023/A:1015798105851.

Karner, Markus B., and others (2001). Archaeal dominance in the mesopelagic zone of the Pacific Ocean. *Nature*, vol. 409, pp. 507–510.

Kaufman, Daniel E., and others (2017). Climate change impacts on southern Ross Sea phytoplankton composition, productivity, and export. *Journal of Geophysical Research: Oceans*, vol. 122, No. 3, pp. 2339–2359.

Keller, Maureen D., and others (1989). Dimethyl sulfide production in marine phytoplankton: The Importance of Species Composition and Cell Size. *Biological Oceanography*, vol. 6, Nos. 5–6, pp. 75–382.

Kent, Alyssa G., and others (2016). Global biogeography of *Prochlorococcus* genome diversity in the surface ocean. *The ISME Journal*, vol. 10, pp. 1856–1865.

Kiørboe, Thomas (2011). How zooplankton feed: mechanisms, traits and trade-offs. *Biological Reviews*, vol. 86, No. 2, pp. 311–339.

Kirkham, Amy R., and others (2013). A global perspective on marine photosynthetic picoeukaryote community structure. *The ISME Journal*, vol. 7, pp. 922–936.

Landolfi, Angela, and others (2018). Global marine N2 fixation estimates: From observations to models. *Frontiers in Microbiology*, vol. 9, art. 2112.

Landry, Michael R., and Albert Calbet (2004). Microzooplankton production in the oceans. *ICES Journal of Marine Science*, vol. 61, No. 4, pp. 501–507.

Le Quéré, Corinne, and others (2005). Ecosystem dynamics based on plankton functional types for global ocean biogeochemistry models. *Global Change Biology*, vol. 11, No. 11, pp. 2016–2040.

Lefort, Thomas, and Josep M. Gasol (2013). Short-time scale coupling of picoplankton community structure and single-cell heterotrophic activity in winter in coastal NW Mediterranean Sea waters. *Journal of Plankton Research*, vol. 36, No. 1, pp. 243–258.

Legendre, Louis, and Josée Michaud (1998). Flux of biogenic carbon in oceans: size-dependent regulation by pelagic food webs. *Marine Ecology Progress Series*, vol. 164, pp. 1–11.

Lelong, A. (2012). *Pseudo-nitzschia* (Bacillariophyceae) species, domoic acid and amnesic shellfish poisoning: revisiting previous paradigms. *Phycologia*, vol. 51 (2), pp. 168–216.

Li, William K.W. (1994). Primary production of prochlorophytes, cyanobacteria, and eucaryotic ultraphytoplankton: measurements from flow cytometric sorting. *Limnology and Oceanography*, vol. 39, No. 1, pp. 169–175.

Li, William K.W., and others (2009). Smallest algae thrive as the Arctic Ocean freshens. *Science*, vol. 326, No. 5952, pp. 539–539.

Lindeque, Penelope K., and others (2013). Next generation sequencing reveals the hidden diversity of zooplankton assemblages. *PloS One*, vol. 8, No. 11, e81327.

Lindley, J.A., and S. D. Batten (2002). Long-term variability in the diversity of North Sea zooplankton. *Journal of the Marine Biological Association of the United Kingdom*, vol. 82, No. 1, pp. 31–40.

Liquete, Camino, and others (2013). Current status and future prospects for the assessment of marine and coastal ecosystem services: a systematic review. *PloS One*, vol. 8, No. 7, e67737.

Lischka, Silke, and others (2011). Impact of ocean acidification and elevated temperatures on early juveniles of the polar shelled pteropod *Limacina helicina*: mortality, shell degradation, and shell growth. *Biogeosciences (BG)*, vol. 8, pp. 919–932.

Litchman, Elena, and others (2010). Linking traits to species diversity and community structure in phytoplankton. In *Fifty Years after the '"Homage to Santa Rosalia"': Old and New Paradigms on Biodiversity in Aquatic Ecosystems*, pp. 15–28. Springer.

Loeb, Valerie, and others (1997). Effects of sea-ice extent and krill or salp dominance on the Antarctic food web. *Nature*, vol. 387, pp. 897–900.

Loeb, V.J., and J.A. Santora (2012). Population dynamics of *Salpa thompsoni* near the Antarctic Peninsula: growth rates and interannual variations in reproductive activity (1993–2009). *Progress in Oceanography*, vol. 96, No. 1, pp. 93–107.

Luo, Haiwei, and Mary Ann Moran (2014). Evolutionary ecology of the marine Roseobacter clade. *Microbiology and Molecular Biology Reviews*, vol. 78, No. 4, pp. 573–587.

Mackas, D.L., and others (2012). Changing zooplankton seasonality in a changing ocean: Comparing time series of zooplankton phenology. *Progress in Oceanography*, vol. 97, pp. 31–62.

Madin, L.P., and G.R. Harbison (1977). The associations of Amphipoda Hyperiidea with gelatinous zooplankton—I. Associations with Salpidae. *Deep Sea Research*, vol. 24, No. 5, pp. 449–463.

________ (2001). Gelatinous zooplankton. *1st Edition of Encyclopedia of Ocean Sciences*, vol. 2, pp. 1120–1130.

Malone, T. C. (1980). Algal size. *The Physiological Ecology of Phytoplankton.*

Mangel, Marc, and Stephen Nicol (2000). Krill and the unity of biology. *Canadian Journal of Fisheries and Aquatic Sciences*, vol. 57, (S3), pp. 1–5.

Marañón, Emilio, and others (2001). Patterns of phytoplankton size structure and productivity in contrasting open-ocean environments. *Marine Ecology Progress Series*, vol. 216, pp. 43–56.

Marañón, Emilio, and others (2012). Temperature, resources, and phytoplankton size structure in the ocean. *Limnology and Oceanography*, vol. 57, No. 5, pp. 1266–1278.

Marañón, Emilio, and others (2015). Resource supply alone explains the variability of marine phytoplankton size structure. *Limnology and Oceanography*, vol. 60, No. 5, pp. 1848–1854.

Massana, Ramon, and Klaus Jürgens (2003). Composition and population dynamics of planktonic bacteria and bacterivorous flagellates in seawater chemostat cultures. *Aquatic Microbial Ecology*, vol. 32, No. 1, pp. 11–22.

Miron, Gilles, and others (1995). Use of larval supply in benthic ecology: testing correlations between larval supply and larval settlement. *Marine Ecology Progress Series*, vol. 124, pp. 301–305.

Mitra, Aditee, and others (2016). Defining planktonic protist functional groups on mechanisms for energy and nutrient acquisition: incorporation of diverse mixotrophic strategies. *Protist*, vol. 167, No. 2, pp. 106–120.

Moline, Mark A., and others (2004). Alteration of the food web along the Antarctic Peninsula in response to a regional warming trend. *Global Change Biology*, vol. 10, No. 12, pp. 1973–1980.

Möllmann, Christian, and others (2003). The marine copepod, *Pseudocalanus elongatus*, as a mediator between climate variability and fisheries in the Central Baltic Sea. *Fisheries Oceanography*, vol. 12, Nos. 4–5, pp. 360–368.

Monteiro, Fanny Meline, and others (2010). Distribution of diverse nitrogen fixers in the global ocean. *Global Biogeochemical Cycles*, vol. 24, No. 3.

Montes-Hugo, Martin, and others (2009). Recent changes in phytoplankton communities associated with rapid regional climate change along the western Antarctic Peninsula. *Science*, vol. 323, No. 5920, pp. 1470–1473.

Moore, J. Keith, and others (2018). Sustained climate warming drives declining marine biological productivity. *Science*, vol. 359, No. 6380, pp. 1139–1143.

Morán, Xosé Anxelu, and others (2010). Increasing importance of small phytoplankton in a Warmer Ocean. *Global Change Biology*, vol. 16, No. 3, pp. 1137–1144. https://doi.org/10.1111/j.1365-2486.2009.01960.x.

Mousing, Erik Askov, Katherine Richardson, and Marianne Ellegaard (2018). Global patterns in phytoplankton biomass and community size structure in relation to macronutrients in the open ocean. *Limnology and Oceanography*, vol. 63, No. 3, pp. 1298–1312.

Mueter, Franz J., and others (2009). Ecosystem responses to recent oceanographic variability in high-latitude Northern Hemisphere ecosystems. *Progress in Oceanography*, vol. 81, Nos. 1–4, pp. 93–110.

Müren, U., and others (2005). Potential effects of elevated sea-water temperature on pelagic food webs. *Hydrobiologia*, vol. 545, No. 1, pp. 153–166.

Negrete-García, Gabriela, and others (2019). Sudden emergence of a shallow aragonite saturation horizon in the Southern Ocean. *Nature Climate Change*, vol. 9, pp. 313–317.

Nissen, Cara, and others (2018). Factors controlling coccolithophore biogeography in the Southern Ocean. *Biogeosciences*, vol. 15, No. 22, pp. 6997–7024.

Not, Fabrice, and others (2007). Diversity and ecology of eukaryotic marine phytoplankton. In *Advances in Botanical Research*, vol. 64: pp. 1–53. Elsevier.

Paez-Espino, David, and others (2019). IMG/VR v. 2.0: an integrated data management and analysis system for cultivated and environmental viral genomes. *Nucleic Acids Research*, vol. 47, No. D 1, pp. D678–D686.

Palumbi, Stephen R., and others (2009). Managing for ocean biodiversity to sustain marine ecosystem services. *Frontiers in Ecology and the Environment*, vol. 7, No. 4, pp. 204–211.

Pierce, Richard W., and Jefferson T. Turner (1992). Ecology of planktonic ciliates in marine food webs. *Reviews in Aquatic Sciences*, vol. 6, No. 2, pp. 139–181.

Pineda, Jesús, and others (2010). Causes of decoupling between larval supply and settlement and consequences for understanding recruitment and population connectivity. *Journal of Experimental Marine Biology and Ecology*, vol. 392, Nos. 1–2, pp. 9–21.

Pitt, Kylie Anne, and others (2018). Claims that anthropogenic stressors facilitate jellyfish blooms have been amplified beyond the available evidence: a systematic review. *Frontiers in Marine Science*, vol. 5, art. 451.

Poloczanska, Elvira S., and others (2013). Global imprint of climate change on marine life. *Nature Climate Change*, vol. 3, pp. 919–925.

Polovina, Jeffrey J., and others (2011). Projected expansion of the subtropical biome and contraction of the temperate and equatorial upwelling biomes in the North Pacific under global warming. *ICES Journal of Marine Science*, vol. 68, No. 6, pp. 986–995.

Polovina, Jeffrey J., and others (2008). Ocean's least productive waters are expanding. *Geophysical Research Letters*, vol. 35, No. 3.

Polovina, Jeffrey J., and Phoebe A. Woodworth (2012). Declines in phytoplankton cell size in the subtropical oceans estimated from satellite remotely-sensed temperature and chlorophyll, 1998–2007. *Deep Sea Research Part II: Topical Studies in Oceanography*, vol. 77, pp. 82–88.

Pomeroy, Lawrence R., and others (2007). The microbial loop. *Oceanography*, vol. 20, No. 2, pp. 28–33.

Purcell, Jennifer E., and others (2007). Anthropogenic causes of jellyfish blooms and their direct consequences for humans: a review. *Marine Ecology Progress Series*, vol. 350, pp. 153–174.

Raitsos, Dionysios E., and others (2014). From silk to satellite: half a century of ocean colour anomalies in the Northeast Atlantic. *Global Change Biology*, vol. 20, No. 7, pp. 2117–2123.

Rasmussen, Astrid Fuglseth (2018). Changes in the abundance, species composition and distribution of the Barents Sea euphausiids (krill): with focus on the expansion and reproduction of *Meganyctiphanes norvegica*. Master's Thesis, Norwegian University of Life Sciences, Aas.

Read, Betsy A., and others (2013). Pan genome of the phytoplankton *Emiliania* underpins its global distribution. *Nature*, vol. 499, pp. 209–213.

Rivero-Calle, Sara, and others (2015). Multidecadal increase in North Atlantic coccolithophores and the potential role of rising CO_2. *Science*, vol. 350, No. 6267, pp. 1533–1537.

Rohwer, Forest, and Rebecca Vega Thurber (2009). Viruses manipulate the marine environment. *Nature*, vol. 459, pp. 207–212.

Rombouts, Isabelle, and others (2009). Global latitudinal variations in marine copepod diversity and environmental factors. *Proceedings of the Royal Society B: Biological Sciences*, vol. 276, No. 1670, pp. 3053–3062.

Rousseaux, Cecile S., and Watson W. Gregg (2015). Recent decadal trends in global phytoplankton composition. *Global Biogeochemical Cycles*, vol. 29, No. 10, pp. 1674–1688.

Salazar, Guillem, and Shinichi Sunagawa (2017). Marine microbial diversity. *Current Biology*, vol. 27, No. 11, pp. R489–R494.

Santoro, Alyson E., and others (2019). Planktonic marine archaea. *Annual Review of Marine Science*, vol. 11, pp. 131–158.

Sarmiento, Jorge Louis, and others (2002). A new estimate of the CaCO3 to organic carbon export ratio. *Global Biogeochemical Cycles*, vol. 16, No. 4, pp. 54-1–54-12.

Schiebel, Ralf, and Christoph Hemleben (2005). Modern planktic foraminifera. *Paläontologische Zeitschrift*, vol. 79, No. 1, pp. 135–148.

Schlüter, Lothar, and others (2014). Adaptation of a globally important coccolithophore to ocean warming and acidification. *Nature Climate Change*, vol. 4, pp. 1024–1030.

Schoemann, Véronique, and others (2005). *Phaeocystis* blooms in the global ocean and their controlling mechanisms: a review. *Journal of Sea Research*, vol. 53, Nos. 1–2, pp. 43–66.

Sherr, Evelyn B., and Barry F. Sherr (2007). Heterotrophic dinoflagellates: a significant component of microzooplankton biomass and major grazers of diatoms in the sea. *Marine Ecology Progress Series*, vol. 352, pp. 187–197.

Sherr, Evelyn B., and others (2009). Microzooplankton grazing impact in the Western Arctic Ocean. *Deep Sea Research Part II: Topical Studies in Oceanography*, vol. 56, No. 17, pp. 1264–1273.

Sieburth, John McN., and others (1978). Pelagic ecosystem structure: Heterotrophic compartments of the plankton and their relationship to plankton size fractions. *Limnology and Oceanography*, vol. 23, No. 6, pp. 1256–1263.

Sieradzki, Ella T., and others (2019). Dynamic marine viral infections and major contribution to photosynthetic processes shown by spatiotemporal picoplankton metatranscriptomes. *Nature Communications*, vol. 10, art. 1169.

Signorini, Sergio R., and others (2015). Chlorophyll variability in the oligotrophic gyres: mechanisms, seasonality and trends. *Frontiers in Marine Science*, vol. 2, art. 1.

Silberberger, Marc J., and others (2016). Spatial and temporal structure of the meroplankton community in a sub-Arctic shelf system. *Marine Ecology Progress Series*, vol. 555, pp. 79–93.

Simmonds, Peter, and others (2017). Consensus statement: virus taxonomy in the age of metagenomics. *Nature Reviews Microbiology*, vol. 15, pp. 161–168.

Simon, Nathalie, and others (2009). Diversity and evolution of marine phytoplankton. *Comptes Rendus Biologies*, vol. 332, Nos. 2–3, pp. 159–170.

Slagstad, D., and others (2011). Evaluating primary and secondary production in an Arctic Ocean void of summer sea ice: an experimental simulation approach. *Progress in Oceanography*, vol. 90, Nos. 1–4, pp. 117–131.

Smith, Helen E.K., and others (2017). The influence of environmental variability on the biogeography of coccolithophores and diatoms in the Great Calcite Belt. *Biogeosciences*, vol. 14, pp. 4905–4925.

Smyth, T.J., and others (2004). Time series of coccolithophore activity in the Barents Sea, from twenty years of satellite imagery. *Geophysical Research Letters*, vol. 31, No. 11.

Sommer, Ulrich, and others (2002). Pelagic food web configurations at different levels of nutrient richness and their implications for the ratio fish production: primary production. In *Sustainable Increase of Marine Harvesting: Fundamental Mechanisms and New Concepts*, pp. 11–20. Springer.

Sommer, Ulrich, and others (2016). Benefits, costs and taxonomic distribution of marine phytoplankton body size. *Journal of Plankton Research*, vol. 39, No. 3, pp. 494–508.

Sommer, Ulrich, and others (2017). Do marine phytoplankton follow Bergmann's rule sensu lato? *Biological Reviews*, vol. 92, No. 2, pp. 1011–1026.

Sournia, Alain, and others (1991). Marine phytoplankton: how many species in the world ocean? *Journal of Plankton Research*, vol. 13, No. 5, pp. 1093–1099.

Sunagawa, Shinichi, and others (2015). Structure and function of the global ocean microbiome. *Science*, vol. 348, No. 6237, 1261359.

Sundby, Svein, and others (2016). The North Atlantic spring-bloom system—Where the changing climate meets the winter dark. *Frontiers in Marine Science*, vol. 3, art. 28.

Suttle, Curtis A. (2007). Marine viruses—major players in the global ecosystem. *Nature Reviews Microbiology*, vol. 5, pp. 801–812.

Swalethorp, Rasmus, and others (2019). Microzooplankton distribution in the Amundsen Sea Polynya (Antarctica) during an extensive *Phaeocystis antarctica* bloom. *Progress in Oceanography*, vol. 170, pp. 1–10.

Thackeray, Stephen J., and others (2010). Trophic level asynchrony in rates of phenological change for marine, freshwater and terrestrial environments. *Global Change Biology*, vol. 16, No. 12, pp. 3304–3313.

Thor, Peter, and Sam Dupont (2015). Transgenerational effects alleviate severe fecundity loss during ocean acidification in a ubiquitous planktonic copepod. *Global Change Biology*, vol. 21, No. 6, pp. 2261–2271.

Tim, N., and others. (2016). The importance of external climate forcing for the variability and trends of coastal upwelling in past and future climate. *Ocean Science*, vol. 12, pp. 807–823.

Tréguer, Paul, and others (2018). Influence of diatom diversity on the ocean biological carbon pump. *Nature Geoscience*, vol. 11, pp. 27–37.

Uitz, Julia, and others (2010). Phytoplankton class-specific primary production in the world's oceans: Seasonal and interannual variability from satellite observations. *Global Biogeochemical Cycles*, vol. 24, No. 3.

United Nations (2017a). Chapter 34: Global patterns in marine biodiversity. In *The First Global Integrated Marine Assessment: World Ocean Assessment I*. Cambridge: Cambridge University Press, pp. 501–524. https://doi.org/10.1017/9781108186148.037.

________(2017b). *The First Global Integrated Marine Assessment: World Ocean Assessment I*. Cambridge: Cambridge University Press.

________(2017c). Chapter 6: Primary production, cycling of nutrients, surface layer and plankton. In United Nations (ed.), *The First Global Integrated Marine Assessment: World Ocean Assessment I*. Cambridge: Cambridge University Press, pp. 119–148. https://doi.org/10.1017/9781108186148.009.

Valdés, Valentina, and others (2017). Scaling copepod grazing in a coastal upwelling system: the importance of community size structure for phytoplankton C flux. *Latin American Journal of Aquatic Research*, vol. 45, No. 1, pp. 41–54.

Vallina, Sergio M., and others (2014). Global relationship between phytoplankton diversity and productivity in the ocean. *Nature Communications*, vol. 5, art. 4299.

Vaulot, Daniel, and others (2008). The diversity of small eukaryotic phytoplankton (≤ 3 μm) in marine ecosystems. *FEMS Microbiology Reviews*, vol. 32, No. 5, pp. 795–820.

Verity, Peter G., and Victor Smetacek (1996). Organism life cycles, predation, and the structure of marine pelagic ecosystems. *Marine Ecology Progress Series*, vol. 130, pp. 277–293.

Vinogradov, Mikhail Evgen'evich, and others (1996). Hyperiid amphipods (Amphipoda, Hyperiidea) of the world oceans.

Vogt, Meike, and others (2012). Global marine plankton functional type biomass distributions: *Phaeocystis* spp. *Earth System Science Data*, vol. 4, No. 1, pp. 107–120.

Ward, Ben A., and others (2012). A size-structured food-web model for the global ocean. *Limnology and Oceanography*, vol. 57, No. 6, pp. 1877–1891.

Wasmund, Norbert, and others (2019). Extension of the growing season of phytoplankton in the western Baltic Sea in response to climate change. *Marine Ecology Progress Series*, vol. 622, pp. 1–16.

Wells, Mark L., and others (2015). Harmful algal blooms and climate change: Learning from the past and present to forecast the future. *Harmful Algae*, vol. 49, pp. 68–93.

Westberry, T., and others (2008). Carbon-based primary productivity modeling with vertically resolved photoacclimation. *Global Biogeochemical Cycles*, vol. 22, No. 2.

White, Angelicque E., and others (2018). Temporal Variability of *Trichodesmium* spp. and Diatom-Diazotroph Assemblages in the North Pacific Subtropical Gyre. *Frontiers in Marine Science*, vol. 5, art. 27.

Wiebe, Peter H., and others (2010). Deep-sea sampling on CMarZ cruises in the Atlantic Ocean–An introduction. *Deep Sea Research Part II: Topical Studies in Oceanography*, vol. 57, Nos. 24–26, pp. 2157–2166.

Wietz, Matthias, and others (2010). Latitudinal patterns in the abundance of major marine bacterioplankton groups. *Aquatic Microbial Ecology*, vol. 61, No. 2, pp. 179–189.

Winter, Amos, and others (2014). Poleward expansion of the coccolithophore *Emiliania huxleyi*. *Journal of Plankton Research*, vol. 36, No. 2, pp. 316–325.

Wood, Sylvia L.R., and others (2018). Distilling the role of ecosystem services in the Sustainable Development Goals. *Ecosystem Services*, vol. 29, pp. 70–82.

Worden, Alexandra Z., and Fabrice Not (2008). Ecology and diversity of picoeukaryotes. *Microbial Ecology of the Oceans*, vol. 2, pp. 159–205.

Worm, Boris, and others (2006). Impacts of biodiversity loss on ocean ecosystem services. *Science*, vol. 314, No. 5800, pp. 787–790.

Yooseph, Shibu, and others (2010). Genomic and functional adaptation in surface ocean planktonic prokaryotes. *Nature*, vol. 468, pp. 60–66.

Chapter 6B
Marine invertebrates

Contributors: Lis Lindal Jørgensen (convener), Christos Arvanitidis, Silvana N.R. Birchenough, Malcolm R. Clark, Igor Cristino Silva Cruz, Marina Cunha, Alan Deidun, Judith Gobin, Maruf Hossain, Ana C.M. de Jesus, Carmen Mifsud, Khac Bat Nguyen, Chul Park (chapter lead member), Rachel Przeslawski, Jake Rice, Lennert Schepers, Paul Snelgrove, Natalia Strelkova and Leen Vandepitte.

Keynote points

- As of 2019, 153,434 marine benthic invertebrate species had been described globally.
- Since 2012, researchers have described 10,777 new marine benthic invertebrate species; at the same time, biodiversity is changing globally at rates unprecedented in human history, creating the potential for species extinction before they have been described.
- The deep sea covers 43 per cent of the Earth's surface, with an estimated 95 per cent of marine invertebrate species still undescribed.
- Major pressures on marine invertebrates include temperature increase, ocean acidification, physical impacts on the seabed, the extraction of living and non-living resources, coastal use, invasive species and pollution.
- Large areas of the globe, including areas beyond national jurisdiction, still lack effective and adequate long-term ecosystem monitoring and protection for marine invertebrates.
- Despite new research regarding many important ecosystem processes, functions, goods and services, huge knowledge gaps remain in understanding the impact of reductions in benthic invertebrate biodiversity on human well-being and ecosystem dynamics.

1. Introduction

The present subchapter focuses on benthic shrimps, worms, gastropods, bivalves and other invertebrates living on or in the sea floor that are important food sources for fishes, marine mammals, seabirds and humans, as well as invertebrate species that are targeted by some commercial fisheries. Those taxa form the basis for some of the most productive ecosystems on the planet (e.g., estuaries and coral reefs), rivalling tropical forests (Valiela, 1995) and creating habitats covering more of the Earth's surface than all other habitats combined (Snelgrove and others, 1997). Changes in ocean use, the harvesting of organisms, climate change, pollution and invasive species contribute to global alterations in nature at rates unprecedented in human history. Historically, coastal biota have experienced greater pressures and impacts than the deep sea, but the depletion of coastal marine resources and new technologies create both the capacity and incentive to fish, mine and drill in some of the deepest parts of the ocean (McCauley and others, 2015). Alterations of biodiversity often erode economies, livelihoods, food security, health and quality of life worldwide (Intergovernmental Science-Policy Platform on Biodiversity and Ecosystem Services (IPBES), 2019).

2. Summary of the situation recorded in the first *World Ocean Assessment*

In the first *World Ocean Assessment* (United Nations, 2017b), major drivers and patterns of marine invertebrate biodiversity were identified, from regional to global scales. Complex interactions among drivers, as well as their individual and collective impacts on marine biodiversity at multiple scales of biological organization and observation, limit current capacity to predict regional diversity with confidence. Coastal and oceanic patterns differ globally, and coastal benthic species richness generally peaks near the equator and declines polewards, in contrast to mid-latitude peaks in oceanic species. However, strong longitudinal gradients complicate coastal patterns, with localized hotspots of biodiversity across many taxa in areas such as the tropical Indo-Pacific and the Caribbean.

Areas of low oxygen, bottom instability, variation in ocean chemistry, habitat variables and maritime activities complicate the prediction of marine invertebrate diversity patterns in space and time. The multiple drivers of change, often acting in tandem, make it extremely difficult to disentangle natural changes from human-induced pressures. Biodiversity hotspots often attract and support human extractive activities, directly linking ocean biodiversity and ecosystem services. Moreover, those hotspots also often support important ecosystem functions, such as nutrient recycling, food web support and habitat creation that, in turn, contribute to ecosystem services of direct benefit to humans.

3. Description of environmental changes (2010–2020)

3.1. Marine invertebrate biodiversity

Records in the World Register of Marine Species (WoRMS) (Vandepitte and others, 2018; WoRMS Editorial Board, 2019), indicate that 10,777 new valid marine benthic invertebrate species were described between 2012 and 2019, bringing the total number of such species described globally to 153,434. The taxon Mollusca contain the highest numbers of described marine benthic invertebrate (31 per cent), followed by Arthropoda (24 per cent).

The Ocean Biodiversity Information System (OBIS) contains distribution information for 124,372 marine species, representing 56.4 million distribution records. Among those, WoRMS currently identifies 80,132 species as marine benthic invertebrates, representing 8.1 million distribution records.

According to the data available in OBIS and WoRMS in 2019 (see figure I), the well-sampled North Atlantic Ocean contains the highest numbers of recorded marine benthic invertebrate species (24,214 species), followed by the comparatively undersampled South Pacific Ocean (23,245 species), including the Coral Sea (18,224 species), which will certainly yield many more yet undiscovered species.

A study based on bathymetric zones (see figure I) reveals that the Coral Sea contains the highest number of species recorded at depths shallower than 200 m (11,353 species), followed by the Indian Ocean (9,971), the North Atlantic Ocean (9,915) and the South Pacific Ocean (7,498). In some instances (e.g., Bering Sea, Arctic Ocean and Norwegian Sea) similar latitudes differ in benthic diversity. Below 1,000 m, the better-sampled (relative to other basins) North Atlantic Ocean contains the highest number of species (8,027).[1]

[1] Distribution information is not available for all species described in the World Register of Marine Species (WoRMS). The Ocean Biodiversity Information System (OBIS) constantly receives input from many data providers and shows the exact ocean locations where marine species have been recorded. Because WoRMS documentation of benthic traits is ongoing, some 11,000 of the invertebrate species in OBIS still lack functional group designations, and the overview therefore omits those marine benthic invertebrate species.

Total numbers of recorded marine invertebrate benthic species represented as three depth categories (< 200 m, 200–1,000 m and > 1,000 m)

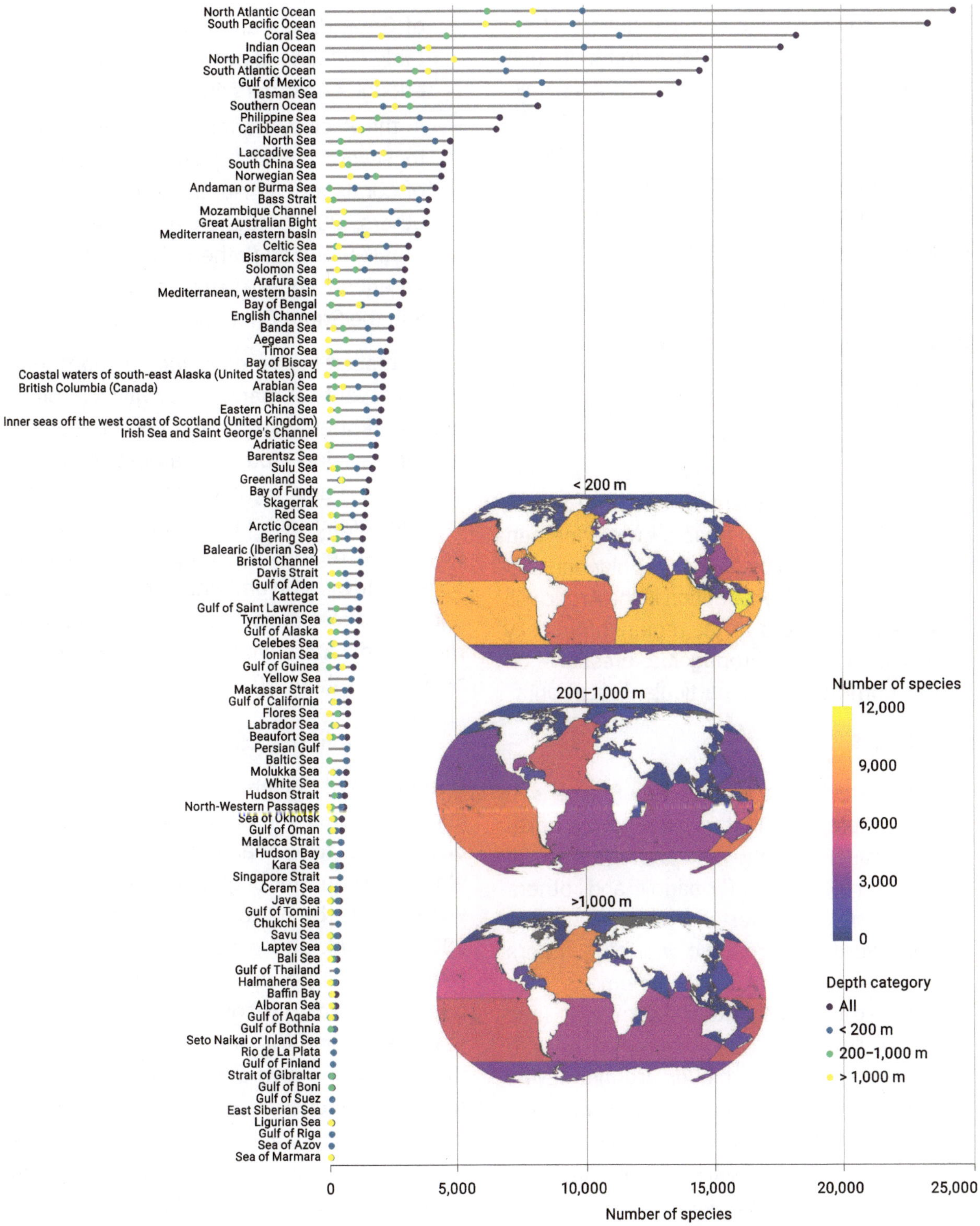

Source: OBIS (2019) for species occurrences; WoRMS for species group information; EMODnet (2016), GEBCO (2015) and Provoost and Bosch (2018) for bathymetry data; and adapted from Marineregions.org (Claus and others, 2014; Flanders Marine Institute (2018)) for sea areas.

3.2. Assessment and state of marine invertebrate biodiversity

Globally, multiple pressures and drivers affect marine benthic invertebrates simultaneously (see table below). While those impacts have been the subject of many studies around the globe, the present section and the table below highlight only some recent targeted or valuable time series studies that illustrate increased understanding since the first Assessment.

3.2.1. Climate warming

Strong evidence indicates unabated warming of the global ocean since 1970, which has taken up more than 90 per cent of the excess heat in the climate system. Since 1993, the rate of ocean warming has probably more than doubled (Intergovernmental Panel on Climate Change (IPCC), 2019). Impacts on marine benthos are particularly profound for polar and subpolar regions. Sea ice reduction in the Arctic will increase ship access to the region, potentially increasing local anthropogenic pressure on benthic communities, in particular in harbours.

Recent findings

- In the Arctic,[2] the Barents Sea (Jørgensen and others, 2019), other seas to the north of Eurasia and the Far Eastern seas in the North Pacific (Lobanov and others, 2014), marine invertebrates are shifting northwards as a result of warming waters (see table). Invertebrate biomass has declined in areas of the Alaska seas (see table) (Grebmeier and others, 2015) with consequences for higher trophic levels (Grebmeier, 2012); native elders link this change to decreased sea ice coverage, the movement of sand bars and alterations in ocean currents (Metcalf and Behe, in Jørgensen and others, 2017).
- In the North Atlantic, climate warming has enabled the arrival of warm-water species in inshore areas of the United Kingdom of Great Britain and Northern Ireland (see table) influenced by the Gulf Stream (Birchenough and others, 2015).
- In the Pacific, marine heatwaves have led to severe bleaching and mass mortality of corals around Australia (Le Nohaïc and others, 2017; Hughes and others, 2018; Stuart-Smith and others, 2018), the Central American coast (Cruz and others, 2018) and the South China Sea (see table).

Some researchers predict increasing frequency and severity of marine heatwaves (Frölicher and Laufkötter, 2018) in the coming decades, even if emission-reduction targets established under the Paris Agreement[3] are met. This warming could eliminate key biogenic habitats in coastal regions of temperate and Arctic seas worldwide (Krumhansl and others, 2016) and affect reef ecosystems located in poorly monitored waters with unknown damage (Genevier and others, 2019).

3.2.2. Bottom trawl fisheries

Bottom trawl fisheries are the most widespread source of anthropogenic physical disturbance to global seabed habitats, and almost one quarter of global seafood landings were caught by bottom trawls from 2011 to 2013 (Hiddink and others, 2017). Trawl gear removes 6–41 per cent of faunal biomass per pass and median recovery times are 1.9–6.4 years (excluding the deep sea), depending on the fishery and environmental context (ibid.). Trawling impact studies demonstrates that decreases in the relative abundance of long-lived fauna (> 10 years) in trawled areas are greater than those of fauna with shorter life spans (1–3 years) (Hiddink and others, 2019).

---

2 See www.arcticbiodiversity.is/index.php/findings/benthos.
3 See FCCC/CP/2015/10/Add.1, decision 1/CP. 21, annex.

Selected national case studies and related natural and anthropogenic drivers and pressures

	Arctic Ocean			North Atlantic Ocean							South Atlantic Ocean	Indian Ocean		North Pacific Ocean			South Pacific Ocean		Indian Ocean-South Pacific Ocean boundary
	Norway and Russian Federation, Barents Sea	Russian Federation, Arctic seas	United States, Arctic	Canada, north-east	Greenland, west and south-east	United Kingdom, North Sea	Portugal, south-west	Greece, bays and gulfs	Malta, coast	Trinidad and Tobago	Brazil, coast and bays	Bangladesh, coast	Australia, west	Viet Nam, coast	South China Sea	Russian Federation, eastern seas	Australia, north-east	New Zealand, east	Australia, south
Climate warming	x	x	x	x	x	x									x	x			x
Temperature events (e.g., El Niño)													x	x			x		
Sedimentation								x		x	x	x		x					
Storms and wave action										x		x	x				x		
Bottom trawl fisheries	x			x	x	x	x	x				x		x	x	x		x	x
Overharvesting of invertebrates											x								
Spreading of new species	x	x	x	x			x		x		x								
Outbreaks of species										x							x		x
Pollution								x			x	x		x	x				
Eutrophication (from agriculture, aquaculture and sewage)								x			x				x				
Oil and gas exploitation and extraction				x		x				x	x				x				x
Offshore wind farms						x													
Large ship-breaking activities												x							
Anchoring								x	x		x								
Coastal infrastructure development								x	x		x				x				
Tourism								x	x		x	x							

Recent findings

- Bottom trawling alters native benthic communities, with impacts characterized as "some modifications" in the North Sea. Studies conducted elsewhere in the North Atlantic and beyond report similar changes in benthic communities resulting from aggregate dredging (Cooper and others, 2017) and experimental trawling (Kenchington and others, 2006), the imposition of "one of the largest footprints per unit of biomass landed" in south-west Portugal (Ramalho and others, 2018) and negative impact on macro-epibenthic composition in southern Greenland (Yesson and others, 2016).
- On bathyal seamounts in the South Pacific, east of New Zealand, the recovery of coral communities after the use of heavy ground gear will likely take many decades (Clark and others, 2019).
- In the North Pacific, negative impacts of bottom trawling on macro-epibenthic composition were reported in the East China Sea (Wang and others, 2018).
- Discarded or lost fishing gear has significant impacts on cold-water coral assemblages (Deidun and others, 2015) at depths of hundreds of metres.
- Invertebrate fishery catches (see also chap. 15) have rapidly expanded globally to more than 10 million tons annually and contribute significantly to global seafood provision, export, trade and local livelihoods. On average, 90 per cent of invertebrate catch can be achieved at a 25 per cent depletion rate, requiring less fishing effort, thereby raising profits, while strongly reducing impacts on other trophic groups (Eddy and others, 2017).
- The harvesting of scallops (*Chlamys islandica*) in the Arctic (Barents Sea) (Nosova and others, 2018) and of sea cucumbers, scallops and crabs in the eastern seas of the Russian Federation (Lysenko and others, 2015) is altering biogenic habitats.

3.2.3. Invasive species

Invasive species (see also chap. 22 and the International Association for Open Knowledge on Invasive Alien Species)[4] occasionally become a dominant pressure on native benthos.

Recent findings

- According to studies on the expanding range of the commercial, predatory snow crab (*Chionoecetes opilio*) in the Arctic, *C. opilio* removes nearly 30,000 tons of macrobenthos in the eastern Barents Sea annually (see table) (Zakharov and others, 2018).
- In the North Atlantic, the invasive green crab (*Carcinus maenas*) has had an impact on seagrasses and sea floor invertebrates in some Canadian coastal areas[5] (see table) (Garbary and others, 2014, Matheson and others, 2016). Extensively invasive *Sargassum* algae (see also chaps. 6E and 6G) now cover beaches and inshore coastal habitats of Trinidad and Tobago and other Caribbean islands (Gobin, 2016). Extensive *Sargassum* beds can alter the abundance of many native marine invertebrates and may provide a suitable habitat for species not previously represented in the local benthic community.
- In the Mediterranean, more than 500 non-indigenous marine invertebrate species have been recorded (Tsiamis and others, 2019), many of which have become established, at least locally, at many sites.
- Outbreaks of the sea urchin *Centrostephanus rodgersii* are degrading kelp forests off the coast of Tasmania, Australia (Ling and Keane, 2018).
- In the South Atlantic, invasive species frequently dominate some Brazilian coastal reefs (Creed and others, 2016, Mantelatto and others, 2018) (see table).

---

4 www.invasivesnet.org/news.

5 Available at: www.dfo-mpo.gc.ca/species-especes/ais-eae/about-sur/index-eng.html.

3.2.4. Consequences of pollution on seabed communities

The consequences of pollution on seabed communities were well documented in the first *World Ocean Assessment* and by IPBES (IPBES, 2019). To assess the environmental state and the resilience of benthic invertebrates, their behaviour, dynamics and multiple interactions with the environment need to be studied (Neves and others, 2013, Pessoa and others, 2019).

Recent findings

- Agricultural run-off and the disposal of municipal waste into the ocean add nutrients that produce algal blooms, which eventually sink to the bottom, creating hypoxic conditions and low pH that typically reduce benthic species diversity. Since the first Assessment, additional algal blooms have been reported by researchers in the Indian Ocean, along the coast of Bangladesh (Kibria and others, 2016; Mallick and others, 2016; Molla and others, 2015), and in the South Atlantic, along the coast of Brazil (Cruz and others, 2018) (see table).
- In the North Atlantic, outflow (sedimentation) from the Orinoco River (Trinidad and Tobago) (see table) increases potential contamination and mortality of benthic invertebrate communities (Gobin, 2016), while a metalliferous discharge caused a multi-year decline in the ecological status of benthic communities along the coast of Greece (Simboura and others, 2014) (see table).

3.2.5. Storms and wave action

Cyclones and tsunamis are among the most critical variables in shaping the biological richness and structure of marine benthic communities and significantly challenging their resilience and stability (Betti and others, 2020). Hurricane frequency and intensity have increased in recent decades along the tropical Atlantic, in close association with climate change-related influences (see references in Hernández-Delgado and others, 2020).

3.2.6. Mining of deep-sea minerals

The mining of deep-sea minerals (see also chap. 18) is a potential new industry that can help to support an expanding "green" economy based on new battery technology for electric vehicles, wind turbines and improved telecommunications and computing technology (Hein and others, 2013). Although no deep-sea mining is currently conducted in the high seas, the International Seabed Authority administers 30 exploration licences (covering an area of 1.5 million km^2) in the Pacific Ocean and the Indian Ocean and along the Mid-Atlantic Ridge. In mining operations, the direct physical removal of sea floor fauna and secondary effects from sediment plumes or the release of ecotoxins will potentially affect benthic environments and will require careful evaluation (Miller and others, 2018). Lack of knowledge of deep-sea biodiversity is a major constraint to ensuring environmental sustainability (Glover and others, 2018).

3.2.7. Human recreational activities, coastal infrastructure development and ship anchoring and bunkering

Human recreational activities, coastal infrastructure development and ship anchoring and bunkering continue to have an impact on vulnerable habitats and associated invertebrate assemblages, as discussed in the first Assessment, with additional records from near Malta (see table) in the Mediterranean (García-March and others, 2007; Mifsud and others, 2006). In addition, ship-breaking activities on the coast of Bangladesh (see table), in the Bay of Bengal, have reduced benthic species diversity (Hossain, 2010).

3.2.8. Crime

The criminal exploitation of marine species occurs globally, as illustrated by the smuggling of abalones out of South Africa by crime groups. A request for assistance from law enforcement agencies in receiving countries may provide a solution (Warchol and Harrington, 2016).

3.2.9. Consequences of changes in marine invertebrate biodiversity on human communities, economies and well-being

Biodiversity changes have both direct and indirect impacts on human well-being (IPBES, 2019). Unfortunately, there is a lack of large-scale and long-term monitoring of large marine areas, even though some Arctic and North Atlantic nations have established long-term monitoring of invertebrate fisheries and by-catch from trawls within existing scientific national fish-assessment surveys (Jørgensen and others, 2017).

Limited publications document specifically how marine benthic invertebrates contribute to human well-being (e.g., Officer and others, 1982; Snelgrove and others, 1997). However, the first and the present Assessments document the importance of benthic invertebrates to marine food webs and the many habitat-forming or habitat-engineering benthic species. Some key issues are summarized below.

- Under a business-as-usual emissions scenario, the United Nations Educational, Scientific and Cultural Organization predicts that the Great Barrier Reef of Australia, along with other World Heritage coral reefs, will have ceased to exist as a functioning coral reef ecosystem by 2100 (Heron and others, 2017).
- Corals, oysters and other living reefs (see also chap. 7F) can dissipate up to 97 per cent of the wave energy reaching them, thus protecting structures and human lives (Ferrario and others, 2014). This is potentially an important mitigation factor as sea level rises. Artificial coastal barriers to protect coastal infrastructure and human communities from climate-related sea level rise will cost an estimated hundreds of billions of dollars by the latter decades of the twenty-first century (IPCC, 2019).
- Increased risk to food security linked to decreases in seafood availability varies greatly on the local and cultural scales. However, for many coastal indigenous peoples and local communities, the harvesting of benthic invertebrates, in particular intertidal species, contributes significantly to their culture and to community-scale food security (IPBES, 2018a, b; IPCC, 2019).
- Elevated sea surface temperatures have contributed to species range extensions globally, including into South Pacific Tasmanian waters (Pecl and others, 2014), which will likely affect fisheries and possibly tourism in the region, as well as ecosystem services.
- Climate-induced changes in the distribution of many benthic invertebrates may cause an increase in food resource species, a decrease, including their local extinction, or even new such species becoming available to dependent coastal communities (IPCC, 2019). Several studies report changes in the poleward range of sessile invertebrates at a slower rate than that of fishes, but also consider benthic invertebrates more likely to respond directly to changes in temperature and pH (IPCC, 2019). Invasive species, such as the snow crab, support increased commercial harvesting in the Arctic Barents Sea (Jørgensen and others, 2019), whereas the crab *Portunus segnis*, a Lessepsian migrant spreading in the Mediterranean, feeds on fish, shelled molluscs, crustaceans and organic matter, thus having a significant impact on trophic processes in native ecosystems, in addition to being the host of a variety of parasites (Rabaoui and others, 2015). In the Africa and Asia-Pacific regions, the impacts of invasive benthic invertebrates increase the risk of failing to meet food security needs (IPBES 2018b, c).
- In the Mediterranean, infrastructure development (e.g., habitat modification for vessels), which has a direct impact on protected species (e.g., *Cladocora caespitosa*) and commercially important species,

decreases the value of marine ecosystem services.

Despite some progress, there remains a need for addressing the huge knowledge gap concerning the effects of biodiversity loss on human communities, economies and well-being. Understanding the underlying causes of change requires repeated time series studies.

4. International and governmental responses

Several ongoing initiatives reflect a growing priority being given to protecting marine biodiversity, in areas both within and beyond national jurisdiction. These initiatives include science processes, such as the World Ocean Assessment, and legal processes, such as the intergovernmental conference on an international legally binding instrument under the United Nations Convention on the Law of the Sea on the conservation and sustainable use of marine biological diversity of areas beyond national jurisdiction, as well as initiatives of Intergovernmental Organizations, such as the International Seabed Authority.

General Assembly resolution 61/105 of 8 December 2006, on sustainable fisheries, in which the Assembly called for fisheries using bottom-contacting gear to avoid significant adverse impacts on vulnerable marine ecosystems, has been particularly influential on marine fisheries. The expert guidance from the Food and Agriculture Organization of the United Nations (FAO) (FAO, 2009) supported States and regional fisheries management organizations in identifying vulnerable marine ecosystems and operating fisheries in ways compliant with the resolution.

Actions taken in line with resolution 61/105 enhanced existing efforts of regional fisheries management organizations to manage the impacts of fisheries on biodiversity. Targeted spatial and temporal closures and move-on rules, triggered by indicators of the presence of vulnerable marine ecosystems, are now applied in combination with a variety of target and limit catch levels spatial management approaches and gear and effort regulations. The aim of these efforts is to keep the impacts of fisheries on target species, by-catch species, seabed habitats and ecological communities within safe ecological levels (Garcia and others, 2014). The performance of regional fisheries management organizations in delivering the mandate to protect seabed habitats and species has been variable over time and among organizations (Gianni and others, 2016), but the frameworks are considered sound and progress is being made (Bell and others, 2019).

4.1. Recent governmental actions

- Some Arctic and North Atlantic nations have established time and cost-efficient, long-term monitoring of invertebrate by-catch from trawls within existing scientific national fish or shrimp assessment surveys (Jørgensen and others, 2017).
- In the South Pacific, New Zealand government policies[6] prohibit bottom trawling and dredging in order to conserve the deep-sea environment in seamount closure areas and benthic protection areas, and there is evidence that benthic species of concern have benefited from those prohibitions (Kelly and others, 2000).
- In the Arctic, in 2019, the Government of Norway closed 442,022 km² to bottom trawling in the Barents Sea (Jørgensen and others, 2020).

6 See www.mpi.govt.nz/dmsdocument/7242-compliance-fact-sheet-7-benthic-protection-areas-and-seamount-closures.

- In the North Pacific Ocean and the Bohai Sea, strict ecological restoration and fishery resources conservation were introduced in 2018.[7]
- In the inlet of the Indian Ocean, despite rules and regulations to protect the marine ecosystem from hazards and destructive activities, actual implementation remains minimal.
- In the Mediterranean, the conservation status of sponges has recently been locally assessed in the Aegean ecoregion (Gerovasileiou and others, 2018).
- Competent authorities in the member States of the European Union are implementing the Marine Strategy Framework Directive.[8] In the areas concerned, among other descriptors, the sea floor integrity shall be kept at a level that safeguards the structure and function of the ecosystems and does not adversely affect benthic ecosystems. The second cycle of the implementation plans under the Directive[9] increases the protection from fishery impacts of seabed features important to benthic invertebrates. This includes, among others, the banning of mobile bottom-contacting gears at depths shallower than 50 m, to protect vulnerable habitats, such as seagrass beds.

The Convention on Biological Diversity Aichi Biodiversity Target 11,[10] another major global policy initiative, has direct relevance for benthic invertebrates. This initiative calls for a robust conservation strategy based on an effectively and equitably managed, ecologically representative and well-connected system of protected areas (see also Kenchington and others, 2019) and other effective area-based conservation measures, integrated into wider seascapes (see also chaps. 26 and 27). Target 11 includes identifying and spatially delineating areas of protection, ensuring scales matching the spatial and temporal needs of the biodiversity features.

This approach is intended to achieve positive and sustained long-term outcomes for the conservation of biodiversity, in particular seabed invertebrate diversity and associated ecosystem functions and services and, where applicable, cultural, spiritual, socioeconomic and other locally relevant values.

Benthic invertebrate biodiversity could particularly benefit from those developments, given that, as documented in the present subchapter, seabed habitats experience pressures and impacts from many sectors and their associated activities and are so diverse that the effectiveness of specific types of conservation measures vary greatly with specific environmental conditions, history and mixes of human pressures, including climate change.

In general, increasing marine protected area network coverage should reduce pressures on benthic invertebrates and facilitate the recovery of negatively affected areas. Aichi Biodiversity Target 11 contributes to a growing awareness that conservation strategies need to move beyond protecting individual, isolated marine areas (Secretariat of the Convention on Biological Diversity, 2011). Marine protected area networks are essential biodiversity conservation tools designed to improve marine biodiversity protection by encompassing spatial scales that better reflect the life history distributions of species. Target 11 also promotes conservation beyond boundaries by recognizing the crucial role of governance and economic, social and ecological factors working in concert to influence ecological outcomes (Meehan and others, 2020).

7 See www.mee.gov.cn/xxgk2018/xxgk/xxgk03/201812/t20181211_684232.html.
8 Available at https://eur-lex.europa.eu/legal-content/EN/TXT/?uri=CELEX:32008L0056.
9 See https://mcc.jrc.ec.europa.eu/main/dev.py?N=24&O=202&titre_chap=D6%20Sea-floor%20integrity&titre_page=Implementation#2016331103713.
10 See www.cbd.int/sp/targets/rationale/target-11.

5. Achievement of relevant Sustainable Development Goals[11] and contribution to Aichi Biodiversity Target 11

Current negative trends in biodiversity and ecosystems will undermine progress towards the achievement of Aichi Biodiversity Target 11, which is aimed at the conservation and integration into the wider landscape and seascape of 10 per cent of coastal and marine areas of particular importance for biodiversity and ecosystem services by 2020, through effectively and equitably managed, ecologically representative and well-connected systems of protected areas and other effective area-based conservation measures.

6. Key remaining knowledge gaps and capacity-building gaps

6.1. Knowledge gaps

- Studies on the effect of protected areas remain limited.
- Reviews do not break down impacts (e.g., climate change, resource exploitation and pollution) on marine biodiversity by species group. This limits knowledge of the value and importance of invertebrates for human well-being.
- Baseline biodiversity studies (for ecoregions or for habitats that are hotspots for biodiversity) are lacking for the mesophotic zone, underwater caves and many of the thousands of global seamounts.

6.2. Capacity-building gaps in the field

- The large-scale protection of the seabed, at both the national and international levels, must continue in order to sustain benthic biodiversity and avoid the extirpation of species before they have even been recorded.
- Listing species with restricted geographical ranges, often arising from specialized habitat requirements, represents the most urgent need. Even describing 100 taxonomic units every year over the next decade would add just 1,000 species before, according to some experts, commercial scale deep-sea mining is expected to begin (Glover and others, 2018).
- To increase knowledge on biodiversity and ecosystem understanding, marine national regular assessment cruises should report both targeted and non-targeted scientific catch.
- Integrated ocean management should be prioritized to coordinate conservation and management among all relevant activities.
- Managers should develop and implement common, well-defined measures to identify and respond to declining benthic habitats in national and international waters.
- Studies are needed to determine the effects on ecosystems of reduced or lost benthos, in particular in the context of food web interactions.
- Studies are needed to determine the effect on food supply if harvested benthic communities disappear.
- The cumulative impact of drivers and pressures that can have a combined effect on marine biodiversity needs to be assessed.

11 See General Assembly resolution 70/1.

References

Betti, F., and others (2020). Effects of the 2018 exceptional storm on the *Paramuricea clavata* (Anthozoa, Octocorallia) population of the Portofino Promontory (Mediterranean Sea). *Regional Studies in Marine Science*, vol. 34, 101037.

Birchenough, Silvana N.R., and others (2015). Climate change and marine benthos: a review of existing research and future directions in the North Atlantic. In *Wiley Interdisciplinary Reviews: Climate Change*, eds. Henning Reiss and others, vol. 6, No. 2, pp. 203–223.

Clark, Malcolm R., and others (2019). Little evidence of benthic community resilience to bottom trawling on seamounts after 15 years. *Frontiers in Marine Science*, vol. 6, art. 63.

Claus, Simon, and others (2014). Marine regions: towards a global standard for georeferenced marine names and boundaries. *Marine Geodesy*, vol. 37, No. 2, pp. 99–125.

Cooper, K.M., and J. Barry (2017). A big data approach to macrofaunal baseline assessment, monitoring and sustainable exploitation of the seabed. *Scientific Reports*, vol. 7, art. 12431.

Creed, Joel C., and others (2016). The invasion of the azooxanthellate coral Tubastraea (Scleractinia: Dendrophylliidae) throughout the world: history, pathways and vectors. *Biological Invasions*, vol. 19, No. 1, pp. 283–305.

Cruz, Igor C.S., and others (2018). Marginal coral reefs show high susceptibility to phase shift. *Marine Pollution Bulletin*, vol. 135, pp. 551–561.

Deidun, Alan, and others (2015). First characterisation of a *Leiopathes glaberrima* (Cnidaria: Anthozoa: Antipatharia) forest in Maltese exploited fishing grounds. *Italian Journal of Zoology*, vol. 82, No. 2, pp. 271–280.

Eddy, Tyler D., and others (2017). Ecosystem effects of invertebrate fisheries. *Fish and Fisheries*, vol. 18, No. 1, pp. 40–53.

EMODnet Bathymetry Consortium (2016). *EMODnet Digital Bathymetry (DTM 2016). EMODnet Bathymetry Consortium.* https://sextant.ifremer.fr/record/c7b53704-999d-4721-b1a3-04ec60c87238.

Food and Agriculture Organization of the United Nations (FAO) (2009). International guidelines for the management of deep-sea fisheries in the high seas. Rome. www.fao.org/in-action/vulnerable-marine-ecosystems/background/deep-sea-guidelines/en/

Ferrario, Filippo, and others (2014). The effectiveness of coral reefs for coastal hazard risk reduction and adaptation. *Nature Communications*, vol. 5, art. 3794.

Flanders Marine Institute (2018). IHO Sea Areas, version 3 (accessed on 25 October 2019). https://doi.org/10.14284/323.

Frölicher, T.L., and Laufkötter, C. (2018). Emerging risks from marine heat waves. *Nature Communications*, vol. 9, art. 650.

Garbary, David J., and others (2014). Drastic decline of an extensive eelgrass bed in Nova Scotia due to the activity of the invasive green crab (*Carcinus maenas*). *Marine Biology*, vol. 161, No. 1, pp. 3–15.

García-March, J.R., and others (2007). Preliminary data on the *Pinna nobilis* population in the marine protected area of Rdum Il-Majjiesa to Ras Ir-Raheb (N.W. Malta). Poster presented at the European Symposium on MPAs as a Tool for Fisheries Management and Ecosystem Conservation. Murcia, Spain.

GEBCO (2015). The GEBCO_2014 Grid, version 20150318 (accessed on 25 October 2019). www.gebco.net.

Genevier, L.G., and others, 2019. Marine heatwaves reveal coral reef zones susceptible to bleaching in the Red Sea. *Global Change Biology*, vol. 25, No. 7, pp. 2338–2351.

Gerovasileiou, V., and others (2018). Assessing the regional conservation status of sponges (Porifera): the case of the Aegean ecoregion. *Mediterranean Marine Science*, vol. 19, No. 3, pp. 526-537. https://doi.org/10.12681/mms.14461.

Glover, Adrian G., and others (2018). Point of View: Managing a sustainable deep-sea 'blue economy' requires knowledge of what actually lives there. *ELife*, vol. 7, e41319.

Gobin, J. (2016). Environmental Impacts on Marine Benthic Communities in an Industrialized Caribbean Island–Trinidad and Tobago. *Marine Benthos: Biology, Ecosystem Functions and Environmental Impact*. New York: Nova Science Publishers.

Grebmeier, Jacqueline M. (2012). Shifting patterns of life in the Pacific Arctic and sub-Arctic seas. *Annual Review of Marine Science*, vol. 4, pp. 63–78.

Grebmeier Jacqueline, and others (2015). Ecosystem characteristics and processes facilitating persistent macrobenthic biomass hotspots and associated benthivory in the Pacific Arctic. *Progress in Oceanography*, vol. 136, pp. 92–114.

Hernández-Delgado, E.A., and others (2020). Hurricane Impacts and the Resilience of the Invasive Sea Vine, *Halophila stipulacea*: a Case Study from Puerto Rico. *Estuaries and Coasts*, pp. 1–21.

Heron, Scott Fraser, and others (2017). *Impacts of Climate Change on World Heritage Coral Reefs: A First Global Scientific Assessment*. Paris: UNESCO.

Hiddink, Jan Geert, and others (2017). Global analysis of depletion and recovery of seabed biota after bottom trawling disturbance. *Proceedings of the National Academy of Sciences*, vol. 114, No. 31, pp. 8301–8306.

Hiddink, J.G., and others (2019). Assessing bottom trawling impacts based on the longevity of benthic invertebrates. *Journal of Applied Ecology*, vol. 56, No. 5, pp. 1075-1084.

Hossain, Maruf Md. M. (2010). *Ship Breaking Activities: Threat to Coastal Environment, Biodiversity and Fishermen Community in Chittagong, Bangladesh*. Publication Cell, Young Power in Social Action.

Hughes, T.P., and others (2018). Large-scale bleaching of corals on the Great Barrier Reef. *Ecology*, vol. 99, No. 2, pp. 501–501.

Intergovernmental Science-Policy Platform on Biodiversity and Ecosystem Services (IPBES) (2018a). *Summary for Policymakers of the Regional Assessment Report on Biodiversity and Ecosystem Services for Africa of the Intergovernmental Science-Policy Platform on Biodiversity and Ecosystem Services*. eds. E . Archer and others. Bonn, Germany: IPBES secretariat.

________ (2018b). *Summary for Policymakers of the Regional Assessment Report on Biodiversity and Ecosystem Services for Asia and the Pacific of the Intergovernmental Science-Policy Platform on Biodiversity and Ecosystem Services*. eds. M. Karki and others. Bonn, Germany: IPBES secretariat.

________ (2018c). *Summary for Policymakers of the Regional Assessment Report on Biodiversity and Ecosystem Services for the Americas of the Intergovernmental Science-Policy Platform on Biodiversity and Ecosystem Services*. eds. J. Rice and others. Bonn, Germany: IPBES secretariat.

________ (2019). *Summary for policymakers of the global assessment report on biodiversity and ecosystem services of the Intergovernmental Science-Policy Platform on Biodiversity and Ecosystem Services*. eds. Sandra Díaz and others. Paris: IPBES secretariat.

Intergovernmental Panel on Climate Change (IPCC) (2019). Summary for Policymakers. In *IPCC Special Report on the Ocean and Cryosphere in a Changing Climate*. https://report.ipcc.ch/srocc/pdf/SROCC_SPM_Approved.pdf.

Jørgensen, Lis L., and others (2017). Benthos. In *State of the Arctic Marine Biodiversity Report*, pp. 85–107. Conservation of Arctic Flora and Fauna (CAFF).

Jørgensen, Lis L., and others (2019). Impact of multiple stressors on sea bed fauna in a warming Arctic. *Marine Ecology Progress Series*, vol. 608, pp. 1–12.

Jørgensen, Lis L., and others (2020). Responding to global warming: new fisheries management measures in the Arctic. *Progress in Oceanography*, vol. 188, art. 102423.

Kelly, S., and others (2000). Spiny lobster, *Jasus edwardsii*, recovery in New Zealand marine reserves. *Biological conservation*, vol. 92, No. 3, pp. 359-369.

Kenchington, Ellen, and others (2006). Effects of experimental otter trawling on benthic assemblages on Western Bank, northwest Atlantic Ocean. *Journal of Sea Research* vol. 56, pp. 249-270.

Kenchington, Ellen, and others (2019). Connectivity modelling of areas closed to protect vulnerable marine ecosystems in the northwest Atlantic. *Deep Sea Research Part I: Oceanographic Research Papers*, vol. 143, pp. 85–103.

Kibria, Golam, and others (2016). Trace/heavy metal pollution monitoring in estuary and coastal area of Bay of Bengal, Bangladesh and implicated impacts. *Marine Pollution Bulletin*, vol. 105, No. 1, pp. 393–402.

Krumhansl, Kira A., and others (2016). Global patterns of kelp forest change over the past half-century. *Proceedings of the National Academy of Sciences*, vol. 113, No. 48, pp. 13785–13790.

Le Nohaïc, Morane, and others (2017). Marine heatwave causes unprecedented regional mass bleaching of thermally resistant corals in northwestern Australia. *Scientific Reports*, vol. 7, art. 14999.

Ling, Scott D., and John P. Keane (2018). Resurvey of the Longspined Sea Urchin (*Centrostephanus rodgersii*) and associated barren reef in Tasmania. Hobart, Australia: University of Tasmania.

Lobanov, V. B., and others (2014). Chapter 5. Impact of climate change on marine natural systems, 5.6: Far-Eastern seas of Russia. In *Second Roshydromet Assessment Report on Climate Change and its Consequences in the Russian Federation*. Moscow: ROSHYDROMET. pp. 684–743.

Lysenko, V.N., and others (2015). The abundance and distribution of the Japanese sea cucumber, *Apostichopus japonicus* (Selenka, 1867) (Echinodermata: Stichopodidae), in nearshore waters of the southern part of the Far Eastern State Marine Reserve. *Russian Journal of Marine Biology*, vol. 41, No. 2, pp. 140–144.

Mallick, Debbrota, and others (2016). Seasonal variability in water chemistry and sediment characteristics of intertidal zone at Karnafully estuary, Bangladesh. *Pollution*, vol. 2, No. 4, pp. 411–423.

Mantelatto, Marcelo Checoli, and others (2018). Invasion of aquarium origin soft corals on a tropical rocky reef in the southwest Atlantic, Brazil. *Marine Pollution Bulletin*, vol. 130, pp. 84–94.

Matheson, K., and others (2016). Linking eelgrass decline and impacts on associated fish communities to European green crab (Linnaeus 1758) invasion. *Marine Ecology Progress Series*, vol. 538, pp. 31–45.

McCauley, Douglas J., and others (2015). Marine defaunation: Animal loss in the global ocean. *Science*, vol. 347, No. 6219, 1255641.

Meehan, Mairi C., and others (2020). How far have we come? A review of MPA network performance indicators in reaching qualitative elements of Aichi Target 11. Conservation Letters, e12746.

Mifsud, C., and others (2006). The distribution and state of health of *Posidonia oceanica* (L.) Delile meadows along the Maltese territorial waters. *Biologia Marina Mediterranea*, vol. 13, No. 4, pp. 255–261.

Miller, Kathryn A., and others (2018). An overview of seabed mining including the current state of development, environmental impacts, and knowledge gaps. *Frontiers in Marine Science*, vol. 4, art. 418.

Molla, H.R., and others (2015). Spatio-temporal variations of microbenthic annelid community of the Karnafuli River Estuary, Chittagong, Bangladesh. *International Journal of Marine Science*, vol. 5, No. 26, pp. 1–11.

Neves, R.A.F., and others (2013). Factors influencing spatial patterns of molluscs in a eutrophic tropical bay. *Marine Biological Association of the United Kingdom. Journal of the Marine Biological Association of the United Kingdom*, vol. 93, No. 3, pp. 577–589.

Nosova, Tatyana, and others (2018). Structure and long-term dynamics of zoobenthos communities in the areas of scallop *Chlamys islandica* beds at Kola Peninsula. *Izvestiya TINRO*, vol. 194, pp. 27–41. https://doi.org/10.26428/1606-9919-2018-194-27-41.

OBIS (2019). Ocean Biogeographic Information System. 2019. www.obis.org.

Pecl, Gretta, and others (2014). Redmap: ecological monitoring and community engagement through citizen science. *Tasmanian Naturalist*, vol. 136, pp. 158–164.

Pessoa, L.A., and others (2019). Intra-annual variation in rainfall and its influence of the adult's *Cyprideis* spp. (Ostracoda, Crustacea) on a eutrophic estuary (Guanabara Bay, Rio de Janeiro, Brazil). *Brazilian Journal of Biology*, (AHEAD).

Provoost, Pieter, and Samuel Bosch (2018). obistools: Tools for data enhancement and quality control. Ocean Biogeographic Information System. Intergovernmental Oceanographic Commission of UNESCO. https://cran.r-project.org/package=obistools.

Rabaoui, Lotfi, and others (2015). Occurrence of the lessepsian species *Portunus segnis* (Crustacea: Decapoda) in the Gulf of Gabes (Tunisia): first record and new information on its biology and ecology. *Cahiers de Biologie Marine*, vol. 56, No. 2, pp. 169–175.

Ramalho, Sofia P., and others (2018). Bottom-trawling fisheries influence on standing stocks, composition, diversity and trophic redundancy of macrofaunal assemblages from the West Iberian Margin. *Deep Sea Research Part I: Oceanographic Research Papers*, vol. 138, pp. 131–145.

Secretariat of the Convention on Biological Diversity (2011). Strategic plan for biodiversity 2011–2020: Provisional technical rationale, possible indicators and suggested milestones for the Aichi Biodiversity Targets. Japan: Nagoya.

Simboura, N., and others (2014). Benthic community indicators over a long period of monitoring (2000–2012) of the Saronikos Gulf, Greece, Eastern Mediterranean. *Environmental Monitoring and Assessment*, vol. 186, No. 6, pp. 3809–3821.

Snelgrove, P.V.R., and others (1997) The importance of marine sediment biodiversity in ecosystem processes, *Ambio*, vol. 26, pp. 578–583.

Officer, C.B., and others (1982). Benthic filter feeding: a natural eutrophication control. *Marine Ecology Progress Series*, vol. 9, pp. 203–210.

Stuart-Smith, Rick D., and others (2018). Ecosystem restructuring along the Great Barrier Reef following mass coral bleaching. *Nature*, vol. 560, pp. 92–96.

Tsiamis, Konstantinos, and others (2019). Non-indigenous species refined national baseline inventories: A synthesis in the context of the European Union's Marine Strategy Framework Directive. *Marine Pollution Bulletin*, vol. 145, pp. 429–435.

United Nations (2017). *The First Global Integrated Marine Assessment: World Ocean Assessment I*. Cambridge: Cambridge University Press.

Valiela, Ivan (1995). *Marine Ecological Processes*. New York, Springer-Verlag, second edition.

Vandepitte, Leen, and others (2018). A decade of the World Register of Marine Species–General insights and experiences from the Data Management Team: Where are we, what have we learned and how can we continue? *PloS One*, vol. 13, No. 4, e0194599.

Wang, H.J., and others (2018). The characteristics and changes of the species and quantity of macrobenthos in Yueqing Bay. *Marine Sciences*, vol. 6, pp. 78–87 (in Chinese with English abstract).

Warchol, Greg, and Michael Harrington (2016). Exploring the dynamics of South Africa's illegal abalone trade via routine activities theory. *Trends in Organized Crime*, vol. 19, No. 1, pp. 21–41.

WoRMS Editorial Board (2019). WoRMS – World Register of Marine Species. www.marinespecies.org. https://doi.org/10.14284/170.

Yesson, Chris, and others (2016). The impact of trawling on the epibenthic megafauna of the west Greenland shelf. *ICES Journal of Marine Science*, vol. 74, No. 3, pp. 866–876.

Zakharov Denis V., and others (2018). Diet of the snow crab in the Barents Sea and macrozoobenthic communities in the area of its distribution. *Trudy VNIRO*. vol. 172, pp. 70–90 (in Russian).

Zalota, Anna K., and others (2018). Development of snow crab *Chionoecetes opilio* (Crustacea: Decapoda: Oregonidae) invasion in the Kara Sea. *Polar Biology*, vol. 41, No. 10, pp. 1983–1994.

Addendum by the Group of Experts of the Regular Process for Global Reporting and Assessment of the State of the Marine Environment, including Socioeconomic Aspects

Status of pelagic invertebrates: cephalopods

Of the 750 species considered by the International Union for Conservation of Nature (IUCN), only one species is classified as Critically Endangered, two as Endangered and another two as Vulnerable, all of which are deep-sea umbrella octopuses (IUCN, 2020).

However, more than 419 species are considered Data Deficient, and they include many deep-sea dwellers (IUCN, 2020). Ten nautilus species were included in appendix II to the Convention on International Trade in Endangered Species of Wild Fauna and Flora in 2017 to regulate international trade therein.

Although information on many deep-sea dwellers is still scarce, recent advances in deep-sea research has increased understanding of the ecology and biology of deep-sea cephalopods. In the central Pacific Ocean, a rare observation of the mating and reproductive behaviours of the deep-sea squids *Chiroteuthis* spp. has been recorded (Vecchione, 2019). A specimen of giant squid, the largest species (up to 13 m) and one of the most enigmatic, was filmed in the Gulf of Mexico in 2019, which was only the second time ever that the species was recorded since it had been first observed in 2012. Analysis of the mitochondrial DNA of 43 specimens from the North Pacific Ocean, the Atlantic Ocean and Oceania supports the hypothesis that giant squids belong to a single species (*Architeuthis dux*) (Winkelmann and others, 2013). Ontogenetic changes in the feeding strategy of the vampire squid (*Vampyroteuthis infernalis*) have been established using stable isotope analyses (Golikov and others, 2019).

Recent work has identified a common multi-decadal increasing trend in the catch rates of dozens of cephalopods species with different biological and ecological strategies (demersal, benthopelagic and pelagic) in diverse oceanic regions (Doubleday and others, 2016). This proliferation has been attributed to their high adaptability and resilience to environmental fluctuations thanks to their rapid growth and flexible development. As an example, shoaling of the oxygen minimum zone in the California Current System has been thought to optimize feeding conditions for the Humboldt squid (*Dosidicus gigas*). This has allowed the species to thrive and expand its distribution northwards up to the Gulf of Alaska (Stewart and others, 2014). In the North Sea, a warming trend from the mid-1980s to the mid-2010s is thought to have been responsible for an increase in overall abundance of several squid species and in an northward expansion of their distribution (van der Kooij and others, 2016). Future warming of the Arctic Ocean may facilitate the trans-Arctic expansion of the European cuttlefish (*Sepia officinalis*) into North Canadian waters by 2300 (Xavier and others, 2016). In Australian waters, warming waters associated with a poleward extension of the Eastern Australian Current are facilitating the expansion of the distribution of the gloomy octopus (*Octopus tetricus*) (Ramos and others, 2018).

References

Doubleday, Zoë A., and others (2016). Global proliferation of cephalopods. *Current Biology*, vol. 26, No. 10, pp. R406–R407.

Golikov, Alexey V., and others (2019). The first global deep-sea stable isotope assessment reveals the unique trophic ecology of vampire squid *Vampyroteuthis infernalis* (Cephalopoda). *Scientific Reports*, vol. 9, No. 1, art. 19099. https://doi.org/10.1038/s41598-019-55719-1.

International Union for Conservation of Nature (IUCN) (2020). *The IUCN Red List of Threatened Species*. www.iucnredlist.org.

Ramos, Jorge E., and others (2018). Population genetic signatures of a climate change driven marine range extension. *Scientific Reports*, vol. 8,art. 9558. https://doi.org/10.1038/s41598-018-27351-y.

Stewart, Julia S., and others (2014). Combined climate- and prey-mediated range expansion of Humboldt squid (*Dosidicus gigas*), a large marine predator in the California current system. *Global Change Biology*, vol. 20, No. 6, pp. 1832–1843. https://doi.org/10.1111/gcb.12502.

Van der Kooij, Jeroen, and others (2016). Climate change and squid range expansion in the North Sea. *Journal of Biogeography*, vol. 43, No. 11, pp. 2285–2298. https://doi.org/10.1111/jbi.12847.

Vecchione, Michael (2019). ROV observations on reproduction by deep-sea cephalopods in the central Pacific Ocean. *Frontiers in Marine Science*, vol. 6, art. 403. https://doi.org/10.3389/fmars.2019.00403.

Winkelmann, Inger, and others (2013). Mitochondrial genome diversity and population structure of the giant squid *Architeuthis*: genetics sheds new light on one of the most enigmatic marine species. *Proceedings of the Royal Society B: Biological Sciences*, vol. 280, No. 1759, 20130273. https://doi.org/10.1098/rspb.2013.0273.

Xavier, José C., and others (2016). Climate change and polar range expansions: could cuttlefish cross the arctic? *Marine Biology*, vol. 163, No. 4, art. 78. https://doi.org/10.1007/s00227-016-2850-x.

Chapter 6C
Fishes

Contributors: Thomas J. Webb (convener), Maria José Juan-Jordá, Hiroyuki Motomura, Francisco Navarrete-Mier, Henn Ojaveer (subchapter lead member), Hazel Oxenford, Chul Park (chapter lead member), Clive Roberts, Mudjekeewis D. Santos, Tracey Sutton and Michael Thorndyke.

Keynote points

- The mobilization of existing data and the development of tools and open, global repositories provide a global picture of the diversity of marine fishes, with 17,762 known species, including 238 species described since the first World Ocean Assessment (United Nations, 2017e).
- While knowledge of the biodiversity of marine fishes exceeds that of many other marine taxa, further improvements will be necessary not only in taxonomic and biosystematic infrastructure but also in the exploration and characterization of the oceans to achieve a complete inventory.
- More than half of the known marine fish species have had their conservation status assessed by the International Union for Conservation of Nature (IUCN), and approximately a third of those assessments having been conducted since the first Assessment.
- Of the fish species with conservation assessments, around 6 per cent of bony fishes, nearly 50 per cent of elasmobranchs, 10 per cent of chimaeras and both species of coelacanths are threatened or near threatened with extinction.
- Capacity for documenting and understanding marine fish diversity continues to grow, but significant gaps remain for certain ecosystem groups (e.g., mesopelagic fishes) and in predicting responses to multiple simultaneous external stressors.

1. Introduction

The present subchapter covers marine fish taxonomy, distribution, habitat and conservation status, emphasizing how the overall state of knowledge has changed since the first *World Ocean Assessment*. The consequences of changing fish diversity for humanity are briefly considered, and perspectives for specific regions are provided. The subchapter concludes with an outlook for fish biodiversity, including continuing gaps in knowledge and capacity. All 17,762 taxonomically valid species within the World Register of Marine Species (WoRMS) superclass Pisces (WoRMS, 2019) are considered, including bony fishes (class Actinopterygii, 16,503 species), sharks and rays (class Elasmobranchii, 1,202 species), chimaeras (class Holocephali, 55 species) and coelacanths (class Coelacanthi, 2 species).

The global biomass of marine fishes is approximately four times the total biomass of all birds and mammals (Bar-On and others, 2018), and fishes constitute an important part of marine biodiversity. Approximately 70 per cent of the marine fish biomass is comprised of mesopelagic fishes, although with wide estimate ranges, found in depths of 200–1,000 m (Irigoien and others, 2014, Hidalgo and Browman, 2019). Fishes occur throughout the world's oceans and in a wide range of depths. For example, the fish seen alive at the greatest depth is the snailfish *Pseudoliparis swirei*, formally described in 2017 and found in depths greater than 8,000 m in the Mariana Trench in the Pacific Ocean (Linley and others, 2016, Gerringer and others, 2017). Fishes play a key role in marine food webs as both predators and prey, often moving through food webs over the course of their lifetimes, such as from planktonic larvae into predatory adults. Fish biodiversity varies between habitats. Habitat affiliations in the Fishbase biodiversity information system for 17,246 species (97 per cent of all known species) show that most bony fishes are demersal or reef-associated, while most species of sharks and rays, chimaeras and coelacanths are demersal or bathydemersal (see table 1).

Fish biodiversity is changing, and fishes are sensitive to environmental changes caused by multiple external pressures (Comte and Olden, 2017) and to exploitation by fisheries (see chap. 15), which has important implications for human well-being (Food and Agriculture Organization of the United Nations (FAO), 2018). The first *Assessment* included chapters on the conservation challenges faced by the 1,088 species of sharks and other elasmobranchs (United Nations, 2017c) and the 25 species of tunas and billfishes (United Nations, 2017d). In addition, overall synthesis chapters revealed that fishes were among the best-known marine groups (United Nations, 2017a, b), with clear latitudinal and depth gradients in diversity. The mechanisms driving fish diversity are complex and include ecosystem stability and age, niche partitioning and predator-mediated dampening of dominance (Rabosky and others, 2018).

Overexploitation and habitat loss and degradation were recently identified as major threats to marine fish biodiversity, and while the impacts of climate change have become more apparent, pollution was not considered a significant threat (Arthington and others, 2016). Subsequently, evidence has emerged that scientific assessment and effective fisheries management can reverse the effects of overexploitation, leading to increases in abundance on average for well-managed stocks constituting half of the reported global fish catch, although overexploitation remains a significant threat in regions with less-developed fisheries management (Hilborn and others, 2020). The impacts of climate change and thermal stress on marine fishes, in particular coral reef fish communities, have become more severe (Robinson and others, 2019), while novel threats, for instance microplastic pollution, are now also attracting increased research interest, even though considerable uncertainty remains about their population-level effects (Villarrubia-Gómez and others, 2018).

Table 1

Number of valid marine species in each taxonomic class of fishes, according to the WoRMS taxonomy, in each broad habitat category

	Class							
	Actinopterygii		*Elasmobranchii*		*Holocephali*		*Coelacanthi*	
Habitat	Existing	Described since 2015	Existing	Described since 2015	Existing	Described since 2015	Existing	Described since 2015
Bathydemersal	1 785 (11%)	4	314 (26%)	2	38 (69%)	–	–	–
Demersal	5 691 (34%)	11	449 (37%)	5	11 (20%)	3	2 (100%)	–
Benthopelagic	1 422 (9%)	18	131 (11%)	13	4 (7%)	–	–	–
Bathypelagic	1 346 (8%)	3	33 (3%)	1	2 (4%)	–	–	–
Pelagic-neritic	807 (5%)	38	34 (3%)	10	–	–	–	–
Pelagic-oceanic	378 (2%)	1	83 (7%)	11	–	–	–	–
Reef-associated	4 618 (28%)	93	98 (8%)	1	–	–	–	–
Unknown	456 (3%)	22	60 (5%)	2	–	–	–	–
Total	16 503	190	1 202	45	55	3	2	–

Source: WoRMS Editorial Board (2019); Froese and Pauly (2019).

2. Documented change in the state of fish biodiversity

Documenting changes in fish biodiversity requires considering fish taxonomy, including the description of new species; spatial distribution, which can be assessed using occurrence records to reveal contractions or expansions in species ranges; and formal assessments of conservation status, to highlight species of conservation concern. Summarizing findings across higher taxonomic groups, and across groups of species occurring in similar habitat zones, is also necessary. Primary data sources used to quantify all those aspects of change are listed in table 2.

Table 2
Major types of global aggregations of data on marine fishes

Data type	Source	Reference
Taxonomy and systematics	World Register of Marine Species (WoRMS)	WoRMS Editorial Board, 2019;
	California Academy of Sciences	Van der Laan and others, 2019
	Eschmeyer's Catalog of Fishes	Fricke and others, 2020
Global occurrence data	Ocean Biodiversity Information System (OBIS)	OBIS, 2018
Habitat affinities	FishBase	Froese and Pauly, 2019
Conservation status	*IUCN Red List of Threatened Species*	IUCN, 2019

2.1. Taxonomy

Since 2015, 238 new marine fish species have been described and added to WoRMS (see table 1). Almost half (49 per cent) of the newly described bony fishes are reef-associated, whereas most of the newly described elasmobranchs are pelagic (see table 1). This rate of description is around 6 to 7 times lower than the one species per day described between 1999 and 2013 (United Nations, 2017a). This taxonomic effort is supplemented by recent phylogenetic studies of bony fishes (Rabosky and others, 2018) and of sharks, rays and chimaeras (Stein and others, 2018).

2.2. Occurrences

Fishes continue to be well represented in global occurrence databases, providing insight into distributions, biogeography and macroecological analyses. Collectively, the Ocean Biodiversity Information System (OBIS) (OBIS, 2018) includes 20,302,222 occurrence records for 15,101 marine fish species, with fishes making up more than a third of all occurrence records. Occurrence records are now available from OBIS for 85 per cent of bony fishes, 84 per cent of elasmobranchs, 78 per cent of chimaeras and one of the two coelacanth species. A total of 306,913 of those occurrence records have been added since the first Assessment, covering 4,099 (23 per cent) fish species, comprising 3,857 (23 per cent) bony fishes (for a total of 241,385 new occurrence records), 233 (19 per cent) sharks and rays (65,480 new records), eight (15 per cent) chimaeras (46 new records) and one of the two coelacanths (two new records). The first ever occurrences of 76 species (68 bony fishes and eight elasmobranchs) have been recorded in OBIS since 2015 (153 occurrence records in total). These species are primarily demersal (32 species) or reef-associated (13 species). Five of the 238 species added to WoRMS since the first Assessment already have occurrence records in OBIS.

2.3. Conservation status

Marine fishes are among the well-assessed marine taxonomic groups in terms of conservation status (Webb and Mindel, 2015). Fifty-three per cent (9,372 species) of all marine fishes have been assessed by IUCN in the 2019 Red List (IUCN, 2019)[1] and 44 per cent (7,756 species) have been assigned to a category other than Data Deficient. Thirty-two per cent (3,008 species) of all assessments of marine fishes have occurred since the first Assessment, in 2015. So far, as no marine fish species has been reassessed since the first Assessment, the IUCN Red List cannot yet be used to assess changes in the status of individual species. However, the proportion of species in each threat category is shown for each taxonomic class in table 3 and that for each habitat affiliation in table 4. Ecological and trait-based methods for predicting the conservation status of species categorized as Data Deficient suggest that, at least with regard to sharks and rays found in European waters, approximately half to two thirds of species in this category should also be considered as at risk of extinction (Walls and Dulvy, 2019). Recent evidence suggests that 24 per cent of the mean monthly space used by sharks falls under the footprint of pelagic longline fisheries and that pelagic sharks have limited spatial refuge from current levels of fishing effort in marine areas beyond national jurisdictions (Queiroz and others, 2019).

2.4. Advances in knowledge and capacity contributing to the evaluation of changes in state

The evaluation of changes in state since the first Assessment is made possible by new data from ongoing long-term monitoring programmes (e.g., international bottom trawl surveys by the International Council for the Exploration of the Sea (ICES)), contributions of fishery observers to scientific data collection, global compilations of fish stock assessments (e.g., the RAM Legacy Stock Assessment Database) and conservation assessments (e.g., the 2019 IUCN Red List), as well as improvements in technology allowing sampling in novel environments (Linley and others, 2016) and the monitoring of individual movements using satellite tagging (Curtis and others, 2018). Dramatic increases in knowledge of fish diversity have also been made possible by increased deep-water fishing (to 1,200 m) by commercial and research vessels, coupled with an increase in shallow-water sampling, enabling the discovery of many cryptic reef fish species in some regions (Gordon and others, 2010). Data infrastructure (e.g., WoRMS, OBIS and ICES data portal) providing the backbone for assessments has been supplemented by new analytical tools enabling users to interact programmatically with the data sources (Boettiger and others, 2012; Chamberlain, 2018; Chamberlain and Salmon, 2018; Provoost and Bosch, 2019; Millar and others, 2019). Those developments and tools have facilitated the use of marine fish data products as indicators of the status of marine ecosystems (ICES, 2018, 2019).

1 Search focused on species in the classes Actinopterygii, Cephalaspidomorphi, Chondrichthyes, Myxini and Sarcopterygii with global scope, in the marine oceanic, marine deep benthic, marine intertidal, marine coastal and supratidal or marine neritic habitats, to ensure that all species within the World Register of Marine Species superclass Pisces were included (see www.iucnredlist.org/search?permalink=c53b-bf34-fec3-4549-8a83-d7630d2bc6bd).

Table 3

Number of marine fish species in each IUCN category by class

	Class											
	Actinopterygii			*Elasmobranchii*			*Holocephali*			*Coelacanthi*		
IUCN category	Pre-first Assessment	Post-first Assessment	Total	Pre-first Assessment	Post-first Assessment	Total	Pre-first Assessment	Post-first Assessment	Total	Pre-first Assessment	Post-first Assessment	Total
Least Concern	4 642	2 071	6 713 (80.6%)	117	201	318 (31.8%)	9	16	25 (54.3%)	-	-	-
Near Threatened	70	27	97 (1.2%)	85	22	107 (10.7%)	2	-	2 (4.3%)	-	-	-
Vulnerable	171	39	210 (2.5%)	80	27	107 (10.7%)	-	1	1 (2.2%)	1		1 (50%)
Endangered	45	18	63 (0.8%)	29	15	44 (4.4%)	-	-	-	-	-	-
Critically Endangered	25	2	27 (0.3%)	14	24	38 (3.8%)	-	-	-	1	-	1 (50%)
Extinct in the Wild or Extinct	2	-	2 (0.02%)	-	-	-	-	-	-	-	-	-
Data Deficient	746	467	1 213 (14.6%)	310	75	385 (38.5%)	15	3	18 (39.1%)	-	-	-
Total	5 701 (34.5%)	2 624 (15.9%)	8 325	635 (52.5%)	364 (30.1%)	999	26 (47.3%)	20 (36.4%)	46	2 (100%)	0 (0%)	2

Note: "Pre-first Assessment" indicates the number of species most recently assessed before 2015, and "Post-first Assessment" the number of species assessed since 2015. Also shown are the total number of species in each IUCN category for each class and the percentage of all assessed species in each IUCN category for each class. The last row shows the total number of species assessed pre- and post-first Assessment in each class, and the percentage of all species in that class that have been assessed.

Table 4
Number of marine fish species in each IUCN category by habitat affiliation

	Not Evaluated		Data Deficient		Not Threatened		Threatened	
	Number of species	Percentage of all known species	Number of species	Percentage of species assessed by IUCN	Number of species	Percentage of species assessed by IUCN	Number of species	Percentage of species assessed by IUCN
Bathydemersal	1 325	61.9	285	34.9	491	60.1	41	5.0
Demersal	3 060	49.7	617	19.9	2 169	69.9	317	10.2
Benthopelagic	936	60.0	124	19.8	440	70.4	61	9.8
Bathypelagic	594	42.7	140	17.6	452	81.9	4	0.5
Pelagic-neritic	351	41.6	120	24.4	335	68.1	37	7.5
Pelagic-oceanic	187	40.5	41	14.9	202	73.5	32	11.6
Reef-associated	1 561	33.0	262	8.3	2 712	85.5	198	6.2
Unknown	425	82.2	27	29.3	55	59.8	10	10.9

Note: The percentages of species categorized as "Not Evaluated" are the percentages of all known species with a given habitat affiliation that have not been assessed by IUCN. Percentages in the other columns are percentages of species assessed by IUCN in each category. The IUCN categories Least Concern and Lower Risk/Least Concern are combined under "Not Threatened", and the categories Near Threatened, Vulnerable, Endangered, Critically Endangered, Extinct in the Wild and Extinct are combined under "Threatened".

3. Consequences of biodiversity change on human communities, economies and well-being

Changes in fish biodiversity have direct and immediate consequences for human communities, economies and well-being through their impacts on commercial, recreational and subsistence fisheries, as well as on alternative sources of income derived from marine ecosystems, including tourism (FAO, 2018). Fishes are integral to the achievement of Sustainable Development Goal 14 of conserving and sustainably using marine resources,[2] with several indicators relating directly to the role that fishes play in sustainable food provision (see chap. 15). Improved knowledge of the distribution and abundance of marine fishes, in particular, is key to monitoring progress toward target 14.4 (effectively regulate harvesting). Increasing the economic benefits to small island developing States and least developed countries through tourism (target 14.7) will involve understanding the distribution and status of charismatic fish species, such as manta rays (Kessel and others, 2017) or fish assemblages, such as coral reef fishes (Wabnitz and others, 2018).

2 See General Assembly resolution 70/1.

4. Key region-specific changes and consequences

4.1. North Atlantic Ocean

In the North Atlantic and adjacent areas, pressure on fish stocks shows an overall downward trend over the period 2003–2017, with the median fishing mortality stabilized at 1.0. The pressure indicator (F/Fmsy) for the Mediterranean and the Black Sea has remained at 2.2. The number of stocks within safe biological limits has almost doubled, from 15 in 2003 to 29 in 2017, with the largest increase in the Bay of Biscay and the Iberian waters – from 2 to 8 stocks. The overall biomass volume has continued to develop positively, increasing by around 36 per cent. In the Mediterranean and the Black Sea, the 2016 spawning stock biomass showed no significant increase compared with 2003. The North-West Atlantic saw a marked change in fish community structure with the collapse of cod and mackerel stocks as a result of overfishing (Shelton and Sinclair, 2008; Van Beveren and others, 2020).

In the Baltic Sea, gradual long-term trends, rather than abrupt changes in functional diversity and multi-trait community composition, were observed between 1971 and 2013 (Törnroos and others, 2018). There are three sub-assemblages along a strong west-east salinity gradient, with low functional redundancy in the Baltic Proper compared with other sub-areas, suggesting an ecosystem more susceptible to external pressures (Frelat and others, 2018). In the North Sea, taxonomic and trait-based indicators provide new evidence of fish assemblage structure and highlight the multifaceted effects of drivers responsible for those changes. Specifically, the central North Sea displayed a decrease in community size structure linked to changes in fishing, and the Norwegian trench region displayed an increase in community size structure primarily linked to climate change, while no change was observed along the eastern Scottish coast where the community size structure was most strongly associated with net primary production (Marshall and others, 2016). In the Mediterranean, the dynamics of small and medium pelagic fish populations exhibit synchrony with climate variability: while the North Atlantic Oscillation is affecting their dynamics in the western and central Mediterranean, anchovy and sardine populations follow the signal of the Atlantic Multidecadal Oscillation in the eastern and central Mediterranean. Thus, there are strong subregional patterns in the temporal dynamics of pelagic fish in the Mediterranean (Tsikliras and others, 2019).

4.2. South Atlantic Ocean

The wider Caribbean region is highly biodiverse and an important region of fish endemism, with approximately 50 per cent of its bony fishes occurring nowhere else (Linardich and others, 2017). Diverse oceanographic and hydrographic features yield an array of subtropical and tropical habitats, including 8 per cent of the world's coral reefs and 6 per cent of seamounts (Oxenford and Monnereau, 2018). Fish biodiversity is negatively affected by overfishing, habitat destruction (in particular of coral reefs) and climate change (Jackson and others, 2014; Oxenford and Monnereau, 2018). Several large-bodied fish species have become commercially extinct or critically endangered (Linardich and others, 2017). The reduction in fish biodiversity is affecting the functioning of Caribbean coral reefs (Lefcheck and others, 2019), with socioeconomic consequences, especially for small island developing States, where up to 22 per cent of the workforce is employed in the fishery sector (Edwards and Yarde, 2019).

A significant emerging phenomenon is the unprecedented bloom across the equatorial Atlantic of pelagic *Sargassum* seaweed, which has been advecting into the Caribbean Sea since 2011 (Wang and others, 2019). This has had a negative impact on critical fish habitats and associated fish biodiversity nearshore (van Tussenbroek and others, 2017; Rodríguez-Martínez and others, 2019), but has had positive effects on some pelagic reef-associated species, the populations of which have increased and are now supporting fisheries (e.g., the yellow jack *Carangoides bartholomaei* and the almaco jack *Seriola rivoliana*) (Ramlogan and others, 2017; Monnereau and Oxenford, 2017). Landings of offshore pelagic species appear to have been disrupted by the presence of *Sargassum*, with some being more readily available, but often as small juveniles (e.g., the dolphinfish *Coryphaena hippurus*), while others (e.g., the flying fish *Hirundichthys affinis*) are more difficult to catch (Oxenford and others, 2019; Caribbean Regional Fisheries Mechanism-Japan International Cooperation Agency, 2019).

4.3. North Pacific Ocean

The North Pacific Ocean, extending from arctic to tropical waters, has the highest fish species diversity in the world, with more than 6,000 species. This rich diversity is derived from and supported by strong water currents flowing northwards and southwards along the north-western continental shelf. Such currents have functioned both to transfer fishes and to isolate fish populations, thereby facilitating speciation (Motomura, 2019). The northern region is a major fishing ground, contributing to about 30 per cent of global catches, mainly targeting the pollack, tunas, sardines and anchovies. The southern region includes the northern part of the Coral Triangle, identified as a marine biodiversity hotspot, and has a higher species richness of shore fishes than any other large marine areas on the globe (Roberts and others, 2002). Most fishes in the southern part are associated with coral reefs and have seen population declines as a result of intense fishing pressure and habitat degradation (Nañola and others, 2011).

4.4. South Pacific Ocean

The South Pacific Ocean includes several highly biodiverse tropical, subtropical and temperate marine ecosystems, modulated directly by the El Niño Southern Oscillation and monsoons. There is high inter-annual variability of primary production that leads to a rich diversity of marine fishes, including reef fishes, pelagic species and highly migratory species (e.g., tuna, sharks and manta rays). Fish biodiversity in this region is affected by fishing (including by-catch) of small pelagics, sharks and tuna, as well as by climate change and pollution, which threaten nursery habitats and drive species from tropical to temperate waters. The destruction of strategic habitats, such as mangroves, can change the distribution and abundance of fish species that use those areas for reproduction and feeding.

Areas of the South-West Pacific that have been explored, including ocean ridges and seamount chains, support a rich marine fish diversity (Clark and Roberts, 2008; Roberts and others, 2015). The fish faunas of the tropical islands of Melanesia and Polynesia in the northern South-West Pacific are predominantly from the Indo-West Pacific in nature, with high diversity but relatively low levels of endemism. In contrast, New Caledonia (France) is a centre of fish endemism, with 107 of 2,341 recorded species endemic to the exclusive economic zone (Fricke and others, 2011; 2015). Off the coast of New Zealand, the number of known marine fish species has grown from around 1,000 species in 1993 to more than 1,294 in 2019 (Roberts and Paulin, 1997; Roberts and others, 2015, 2019), with 22 per cent endemic to the New Zealand region and half of the additional species new to science. Australia is positioned south-west of the aforementioned tropical archipelagos, spans the junctions of two major oceans and is home to some 2,000 known marine fish species.

5. Outlook

Positive outlooks for fish biodiversity come from the evidence that individual fish populations respond positively to effective fisheries management (Hilborn and others, 2020) and that fish diversity and biomass increase within effective marine protected areas (Sala and Giakoumi, 2017). However, the global extinction of the smooth handfish *Sympterichthys unipennis* (Last and others, 2020) is a reminder that fish biodiversity continues to face significant threats as well. Both positive and negative outcomes are known because fishes continue to be among the most systematically studied and monitored components of marine ecosystems, mostly because of their economic value. Nonetheless, considerable fish diversity remains to be discovered: expert estimates indicate that at least another 700 fish species (approximately a 50 per cent increase over the number of currently known species) are yet to be described from the New Zealand exclusive economic zone and extended continental shelf alone (Gordon and others, 2010; Roberts and others, 2019). Further increase in capacity in taxonomy and biosystematics (Taxonomy Decadal Plan Working Group, 2018) and the integration of data from existing biodiversity collections (Nelson and others, 2015) and other sources (Edgar and others, 2016) would pave the way for more comprehensive, synthetic analyses of fish biodiversity over the near to medium term. In addition to improving our understanding of fish biodiversity, improved estimates of fish biomass are needed for some ocean zones, such as the pelagic zone. While it is estimated that mesopelagic fishes dominate global fish biomass, estimates of their biomass span several orders of magnitude and, therefore, the exact contribution that this group makes to global patterns remains poorly understood (Irigoien and others, 2014; Hidalgo and Browman, 2019). In addition, while there are no current estimates of species richness or biomass of bathypelagic fishes, which reside in the world's largest environment (in terms of volume), it is highly likely that those fishes constitute a large portion of global fish biomass (Sutton and others, 2017). Since the first Assessment, the disposal of deep-sea mining water after ore removal has emerged as a significant threat to bathypelagic fishes (Drazen and others, 2019). Key knowledge and capacity gaps in fish biodiversity are summarized in table 5.

Table 5
Key gaps in the understanding of the biodiversity of marine fishes

Knowledge and capacity gaps	Examples of remedial steps taken to address the gaps
Taxonomic and biosystematics infrastructure and capacity	National and international plans to support and develop core taxonomic activities, workforce and infrastructure (e.g., Taxonomy Decadal Plan Working Group, 2018)
Mobilization of existing data into open, global repositories	Historical data rescue, digitization of museum specimens and historical biodiversity literature (e.g., Faulwetter and others, 2016)
Understanding of mesopelagic and deep-sea fish diversity	More and better sampling regimes, employing novel technologies (e.g., Linley and others, 2016; Hidalgo and Browman, 2019)
Response of fishes to multiple simultaneous stressors	Better linking of relevant data across disciplines (e.g., Hodgson and others, 2019)

References

Arthington, Angela H., and others (2016). Fish conservation in freshwater and marine realms: status, threats and management. *Aquatic Conservation: Marine and Freshwater Ecosystems*, vol. 26, No. 5, pp. 838–857.

Bar-On, Yinon M., Rob Phillips, and Ron Milo (2018). The biomass distribution on Earth. *Proceedings of the National Academy of Sciences*, vol. 115, No. 25, pp. 6506–6511. https://doi.org/10.1073/pnas.1711842115.

Boettiger, Carl, and others (2012). rfishbase: exploring, manipulating and visualizing FishBase data from R. *Journal of Fish Biology*, vol. 81, No. 6, pp. 2030–2039.

Caribbean Regional Fisheries Mechanism-Japan International Cooperation Agency (2019). *Fact-Finding Survey Regarding the Influx and Impacts of Sargassum Seaweed in the Caribbean Region*. Belize City: Caribbean Regional Fisheries Mechanism.

Chamberlain, Scott (2018). *worrms: World Register of Marine Species (WoRMS) Client* (version 0.4.0). https://CRAN.R-project.org/package=worrms.

Chamberlain, Scott, and M. Salmon (2018). rredlist: IUCN Red List Client. R Package. 0.6.0. https://CRAN.R-project.org/package=rredlist.

Clark, Malcolm R., and Clive Roberts (2008). *Fish and Invertebrate Biodiversity on the Norfolk Ridge and Lord Howe Rise, Tasman Sea (NORFANZ Voyage, 2003)*. Wellington: Ministry of Fisheries.

Comte, Lise, and Julian D. Olden (2017). Climatic vulnerability of the world's freshwater and marine fishes. *Nature Climate Change*, vol. 7, pp. 718–722.

Curtis, Tobey H., and others (2018). First insights into the movements of young-of-the-year white sharks (*Carcharodon carcharias*) in the western North Atlantic Ocean. *Scientific Reports*, vol. 8, art. 10794.

Drazen, Jeffrey C., and others (2019). Report of the workshop Evaluating the nature of midwater mining plumes and their potential effects on midwater ecosystems. *Research Ideas and Outcomes*, vol. 5, e33527. https://doi.org/10.3897/rio.5.e33527.

Edgar, Graham J., and others (2016). New approaches to marine conservation through the scaling up of ecological data. *Annual Review of Marine Science*, vol. 8, pp. 435–461.

Edwards, Thera, and Thérèse Yarde (2019). *The State of Biodiversity in the Caribbean Community: A Review of Progress Towards the Aichi Biodiversity Targets*. Caribbean Community Secretariat.

Food and Agriculture Organization of the United Nations (FAO) (2018). *The State of World Fisheries and Aquaculture 2018: Meeting the Sustainable Development Goals*. Rome.

Frelat, Romain, and others (2018). A three-dimensional view on biodiversity changes: spatial, temporal, and functional perspectives on fish communities in the Baltic Sea. *ICES Journal of Marine Science*, vol. 75, No. 7, pp. 2463–2475.

Fricke, Ronald, and others (2011). Checklist of the fishes of New Caledonia, and their distribution in the Southwest Pacific Ocean (Pisces). *Stuttgarter Beiträge Zur Naturkunde A, Neue Serie*, vol. 4, pp. 341–463.

Fricke, Ronald, and others (2015). Twenty-one new records of fish species (Teleostei) from the New Caledonian EEZ (south-western Pacific Ocean). *Marine Biodiversity Records*, vol. 8.

Fricke, Ronald, and others (2020). Eschmeyer's Catalog of Fishes: Genera, Species, References. http://researcharchive.calacademy.org/research/ichthyology/catalog/fishcatmain.asp.

Froese, Rainer, and Daniel Pauly (2019). FishBase. Available at www.fishbase.de/search.php.

Gerringer, Mackenzie. E., and others (2017). *Pseudoliparis swirei* sp. nov.: A newly-discovered hadal snailfish (Scorpaeniformes: Liparidae) from the Mariana Trench. *Zootaxa*, vol. 4358, No. 1, pp. 161–177.

Gordon, Dennis P., and others (2010). Marine Biodiversity of *Aotearoa* New Zealand. *PloS One*, vol. 5, No. 8, e 10905.

Hidalgo, Manuel, and Howard I. Browman (2019). Developing the knowledge base needed to sustainably manage mesopelagic resources. *ICES Journal of Marine Science* vol. 76, No. 3.

Hilborn, Ray, and others (2020). Effective fisheries management instrumental in improving fish stock status. *Proceedings of the National Academy of Sciences of the United States of America*, vol. 117, No. 4, pp. 2218–2224.

International Council for the Exploration of the Sea (ICES) (2018). Report of the Working Group on Ecosystem Effects of Fishing Activities (WGECO), 12–19 April 2018, San Pedro del Pinatar, Spain. https://archimer.ifremer.fr/doc/00441/55216.

________ (2019). Working Group on the Ecosystem Effects of Fishing Activities (WGECO). *ICES Scientific Reports*, vol. 1, No. 27. https://doi.org/10.17895/ices.pub.4981.

Irigoien, Xabier, and others (2014). Large mesopelagic fishes biomass and trophic efficiency in the open ocean. *Nature Communications*, vol. 5, art. 3271.

International Union for Conservation of Nature (IUCN) (2019). *The IUCN Red List of Threatened Species*. www.iucnredlist.org/en.

Jackson, J., and others (2014). *Status and Trends of Caribbean Coral reefs: 1970–2012*. Global Coral Reef Monitoring Network. Washington, D.C.: International Union for the Conservation of Nature Global Marine and Polar Program.

Kessel, Steven Thomas, and others (2017). Conservation of reef manta rays (*Manta alfredi*) in a UNESCO World Heritage Site: Large-scale island development or sustainable tourism? *PloS One*, vol. 12, No. 10, e 0185419.

Last, P.R., and others (2020). *Sympterichthys unipennis*. *The IUCN Red List of Threatened Species*. e. T123423283A123424374. https://dx.doi.org/10.2305/IUCN.UK.2020-1.RLTS.T123423283A123424374.en.

Lefcheck, Jonathan S., and others (2019). Tropical fish diversity enhances coral reef functioning across multiple scales. *Science Advances*, vol. 5, No. 3, e aav6420.

Linardich, C., and others (2017). *The Conservation Status of Marine Bony Shorefishes of the Greater Caribbean*. Gland, Switzerland: IUCN.

Linley, Thomas D,. and others (2016). Fishes of the hadal zone including new species, in situ observations and depth records of Liparidae. *Deep Sea Research Part I: Oceanographic Research Papers*, vol. 114, pp. 99–110.

Marshall, Abigail M., and others (2016). Quantifying heterogeneous responses of fish community size structure using novel combined statistical techniques. *Global Change Biology*, vol. 22, No. 5, pp. 1755–1768.

Millar, C., and others (2019). IcesDatras: DATRAS Trawl Survey Database Web Services (version 1.3-0). https://CRAN.R-project.org/package=icesDatras.

Monnereau, I., and H.A. Oxenford (2017). Impacts of climate change on fisheries in the coastal and marine environments of Caribbean Small Island Developing States (SIDS). *Caribbean Marine Climate Change Report Card: Science Review*, vol. 2017, pp. 124–154.

Motomura, H., (2019). Chapter 4. Distribution. In *The Encyclopedia of Ichthyology*, pp. 163–206.

Nañola, Cleto L., and others (2011). Exploitation-related reef fish species richness depletion in the epicenter of marine biodiversity. *Environmental Biology of Fishes*, vol. 90, No. 4, pp. 405–420.

Nelson, Wendy, and others (2015). *National Taxonomic Collections in New Zealand*. Royal Society of New Zealand.

OBIS (2018). *Ocean Biogeographic Information System*. Intergovernmental Oceanographic Commission of UNESCO.

Oxenford, Hazel A., and others (2019). *Report on the Relationships between Sargassum Events, Oceanic Variables and Dolphinfish and Flyingfish Fisheries*. Bridgetown: Centre for Resource Management and Environmental Studies, University of the West Indies, Cave Hill Campus.

Oxenford, Hazel, and Iris Monnereau (2018). Chapter 9: Climate change impacts, vulnerabilities and adaptations: Western Central Atlantic marine fisheries. In *Impacts of Climate Change on Fish and Shellfish in the Coastal and Marine Environments of Caribbean Small Island Developing States (SIDS)*, M. Barange and others, eds. FAO Fisheries Technical Paper, No. 627, pp. 147–68.

Provoost, Pieter, and Samuel Bosch (2019). Robis: R Client to access data from the OBIS API. Ocean Biogeographic Information System. Intergovernmental Oceanographic Commission of UNESCO. https://cran.r-project.org/package=robis.

Queiroz, Nuno, and others (2019). Global spatial risk assessment of sharks under the footprint of fisheries. *Nature*, vol. 572, pp. 461–466.

Rabosky, Daniel L., and others (2018). An inverse latitudinal gradient in speciation rate for marine fishes. *Nature*, vol. 559, pp. 392–395.

Ramlogan, N.R., and others (2017). *Socio-Economic Impacts of Sargassum Influx Events on the Fishery Sector of Barbados*. CERMES Technical Report, No. 81. Bridgetown, Barbados: Centre for Resource Management and Environmental Studies, University of the West Indies, Cave Hill Campus.

Roberts, Callum M., and others (2002). Marine biodiversity hotspots and conservation priorities for tropical reefs. *Science*, vol. 295, No. 5558, pp. 1280–1284.

Roberts, Clive D., and others (2019). *Checklist of the Fishes of New Zealand: Online Version 1.1*. Wellington: Museum of New Zealand Te Papa Tongarewa. https://collections.tepapa.govt.nz/document/10564.

Roberts, Clive D., and Chris D. Paulin (1997). Fish collections and collecting in New Zealand. In *Collection Building in Ichthyology and Herpetology*, Theodore W. Pietsch and William D. Anderson, Jr., eds., ASIH Special Publication 3. pp. 207–229.

Roberts, Clive D., and others (2015). *The Fishes of New Zealand*. Te Papa Press.

Robinson, James P.W., and others (2019). Thermal stress induces persistently altered coral reef fish assemblages. *Global Change Biology*, vol. 25, No. 8, pp. 2739–2750.

Rodríguez-Martínez, R.E., and others (2019). Faunal mortality associated with massive beaching and decomposition of pelagic *Sargassum*. *Marine Pollution Bulletin*, vol. 146, pp. 201–205.

Sala, Enric, and Sylvaine Giakoumi (2017). No-take marine reserves are the most effective protected areas in the ocean. *ICES Journal of Marine Science*, vol. 75. No. 3, pp. 1166–1168. http://doi.org/10.1093/icesjms/fsx059.

Shelton, P.A., and A.F. Sinclair. 2008. It's time to sharpen our definition of sustainable fisheries management. *Canadian Journal of Fisheries and Aquatic Sciences*, vol. 65, pp. 2305–2314.

Stein, R. William, and others (2018). Global priorities for conserving the evolutionary history of sharks, rays and chimaeras. *Nature Ecology & Evolution*, vol. 2, pp. 288–298.

Sutton, Tracey T., and others (2017). A global biogeographic classification of the mesopelagic zone. *Deep Sea Research Part I: Oceanographic Research Papers*, vol. 126, pp. 85–102.

Taxonomy Decadal Plan Working Group (2018). *Discovering Diversity: A Decadal Plan for Taxonomy and Biosystematics in Australia and New Zealand 2018–2028*. Canberra and Wellington: Australian Academy of Science and Royal Society Te Apārangi.

Törnroos, Anna, and others (2019). Four decades of functional community change reveals gradual trends and low interlinkage across trophic groups in a large marine ecosystem. *Global Change Biology*, vol. 25, No. 4, pp. 1235–1246.

Tsikliras, Athanassios C., and others (2019). Synchronization of Mediterranean pelagic fish populations with the North Atlantic climate variability. *Deep Sea Research Part II: Topical Studies in Oceanography*, vol. 159, pp. 143–151.

United Nations (2017a). Chapter 34: Global patterns in marine biodiversity. In *The First Global Integrated Marine Assessment: World Ocean Assessment I*. Cambridge: Cambridge University Press Cambridge.

________ (2017b). Chapter 35: Extent of assessment of marine biological diversity. In *The First Global Integrated Marine Assessment: World Ocean Assessment I*, pp. 525–54. Cambridge: Cambridge University Press.

________ (2017c). Chapter 40: Sharks and other elasmobranchs. In *The First Global Integrated Marine Assessment: World Ocean Assessment I*. Cambridge: Cambridge University Press.

________ (2017d). Chapter 41: Tunas and billfishes. In *The First Global Integrated Marine Assessment: World Ocean Assessment I*. Cambridge: Cambridge University Press.

________ (2017e). *The First Global Integrated Marine Assessment: World Ocean Assessment I*. Cambridge: Cambridge University Press.

Van Beveren, Elisabeth, and others (2020). An example of how catch uncertainty hinders effective stock management and rebuilding. *Fisheries Research*, vol. 224, doi.org/10.1016/j.fishres.2019.105473.

Van der Laan, R., and R. Fricke (2019). Eschmeyer's Catalog of Fishes: Classification, vol. 12. www.calacademy.org/scientists/catalog-of-fishes-classification.

Van Tussenbroek, Brigitta I., and others (2017). Severe impacts of brown tides caused by Sargassum spp. on near-shore Caribbean seagrass communities. *Marine Pollution Bulletin*, vol. 122, Nos. 1 and 2, pp. 272–281.

Villarrubia-Gómez, P., and others (2018). Marine plastic pollution as a planetary boundary threat – The drifting piece in the sustainability puzzle. *Marine Policy*, vol. 96, pp. 213–220.

Wabnitz, Colette C.C., and others (2018). Ecotourism, climate change and reef fish consumption in Palau: Benefits, trade-offs and adaptation strategies. *Marine Policy*, vol. 88, pp. 323–332.

Walls, Rachel H.L., and Nicholas K. Dulvy (2019). Predicting the conservation status of Europe's Data Deficient sharks and rays. *BioRxiv*.

Wang, Mengqiu, and others (2019). The great Atlantic *Sargassum* belt. *Science*, vol. 365, No. 6448, pp. 83–87.

Webb, Thomas J., and Beth L. Mindel (2015). Global patterns of extinction risk in marine and non-marine systems. *Current Biology*, vol. 25, No. 4, pp. 506–511.

World Register of Marine Species (WoRMS) (2019). WoRMS taxon details. Pisces. www.marinespecies.org/aphia.php?p=taxdetails&id=11676.

WoRMS Editorial Board (2019). World Register of Marine Species. www.marinespecies.org.

Chapter 6D
Marine mammals

Contributors: David Lusseau (convener), Luciano Dalla Rosa, Karen Evans (subchapter lead member), André Silva Barreto, Mette Skern-Mauritzen, Chul Park (chapter lead member) and Marta Soeffker.

Keynote points

- Marine mammals play a key role in marine ecosystems in terms of biomass, consumption and energy transfer and continue to provide significant economic and cultural contributions to coastal communities.
- The number of species for which a conservation status is available has increased, with eight species no longer in a status of data deficiency as a result of new information. Thirty-six per cent of baleen whale species are increasing in abundance. Overall, the status of coastal dolphins, sirenians and both marine species of otters is deteriorating, with the vaquita close to extinction. Many species lack population abundance information.
- Fishery by-catch continues to be a dominant conservation threat for many species. Indirect threats, such as habitat alteration, overfishing of prey, land-based pollution, anthropogenic noise, ship strikes and disturbances, are becoming more prevalent, in particular in coastal zones.
- There is apparent increasing consumption of hunted and by-caught small marine mammals in some coastal developing nations.

1. Introduction

There are 132 extant species of marine mammals (cetaceans, pinnipeds, sirenians, otters and the polar bear) with varied habits, ranging from cosmopolitan to comprising multiple discrete local populations (e.g., some dolphin species) or to being endemic to specific ecoregions (e.g., freshwater dolphins). The first *World Ocean Assessment* (United Nations, 2017) recognized direct takes (including commercial and subsistence harvest), fishery interactions, (including entanglement and by-catch) and habitat alteration (including disturbance, coastal and riverine developments and climate change) as key pressures linked to trends in the abundance of marine mammals.

In the present chapter, changes in the global status of marine mammals since the first Assessment are reported on the basis of the International Union for Conservation of Nature (IUCN) Red List assessments carried out by the IUCN Species Survival Commission specialist groups for marine mammals (IUCN, 2019). These assessments are complemented, where needed, using primary literature. Also reported are changes in conservation threats (IUCN, 2019) facing species calculated over two decades, 1999–2008 and 2009–2018.

Overall, fewer marine mammal species have a data deficient status thanks to the increase in information on populations (figure I). Since the first Assessment, the status of eight marine mammal species has improved, while that of four species has declined (figure I). These trends give a picture of cautious optimism, showing that, on a global scale, the individual management measures put in place to reduce well-known conservation threats, together with increased efforts to gather data and information on marine mammal species, are showing signs of effectiveness.

Since the first Assessment, advances in the understanding of the role of marine mammals in the state and productivity of marine systems have been made (Roman and others, 2014), including in nutrient cycling and carbon storage (Doughty and others, 2016), trophic cascades (Estes and others, 2016; Burkholder and others, 2013; Kiszka and others, 2015) and ecosystem engineering. Decreases in sea otter populations have had profound impacts on coastal ecosystems in the eastern Pacific (Estes and others, 1998; Estes and others, 2016). It is likely that the continued recovery of baleen whales, following their overexploitation in the nineteenth and twentieth centuries, will influence marine food webs in multiple ways, through consumption, but also through the vertical (through the water column) and horizontal (between foraging and calving grounds) transfer of nutrients (Roman and others, 2014). As with all predators in marine systems, marine mammal populations are affected by variability in the timing and location of productivity in ocean basins. Some species are likely to be more resilient than others to changes to marine productivity dynamics caused by climate change and overexploitation, owing to more flexible behaviours (Sydeman and others, 2015; Moore and Reeves, 2018).

Intentional takes for subsistence or for commercial harvest and by-catch and entanglement in fishing gear continue to be identified as the main conservation threats for all groups of marine mammals under assessments conducted by IUCN (figure II; IUCN, 2019). The diversification of human activities in the oceans, including for energy production and mining as part of the expanding blue economy in many marine regions (Eikeset and others, 2018), poses new conservation challenges for marine mammals. Climate change and associated changes to marine ecosystem dynamics, anthropogenic noise, ship strikes, habitat modification and behavioural disruptions are now emerging as influencing a broader range of species (figure II; IUCN, 2019). Crucially, individual threats can interact and lead to cumulative effects, compounding their impacts on species (National Academies of Sciences, Engineering and Medicine, 2017; see also chap. 25).

Figure I.A
Changes in the conservation status of marine mammals over three assessment periods (before 1999, 1999–2008 and 2009–2018), on the basis of the IUCN Red List assessments

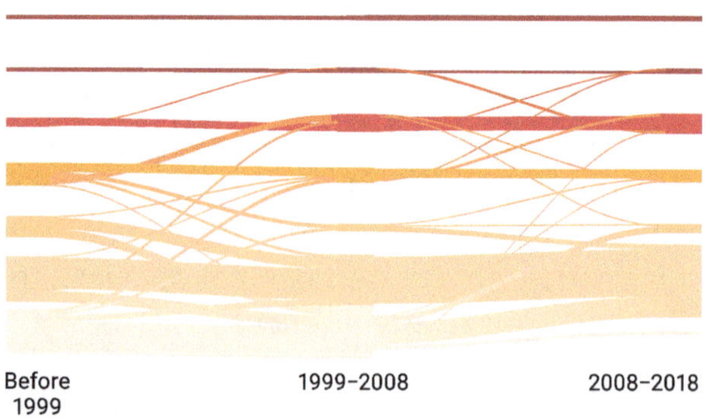

Figure I.B
Composition of the current conservation status of marine mammal species, by group

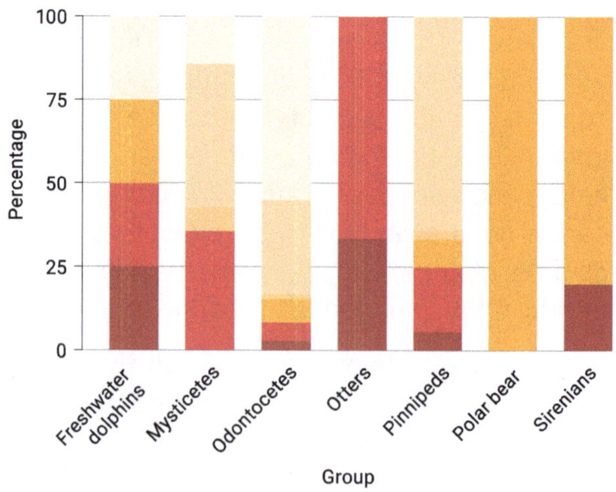

Data Deficient | Least Concern | Near Threatened | Vulnerable | Endangered | Critically Endangered | Extinct

Figure II.A
Currently identified main conservation threats for all marine mammals categorized by the timing of their impact on those species

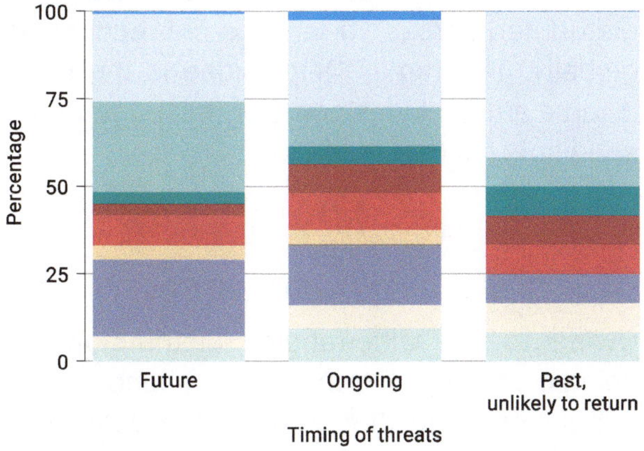

Figure II.B
Conservation threats for specific groups

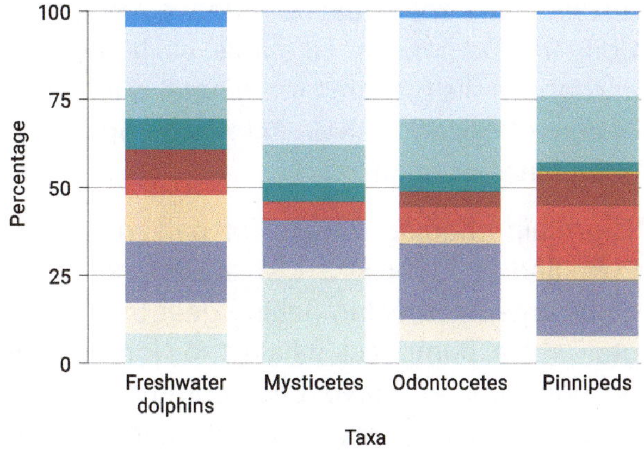

Agriculture and aquaculture
Biological resource use
Climate change and severe weather
Energy production and mining
Geological events
Human intrusion and disturbance
Invasive and other problematic species, genes and diseases
Natural system modifications
Other options
Pollution
Residential and commercial development
Transportation and service corridors

Note: Threat categories are those defined in the IUCN Threats Classification Scheme, where anthropogenic noise is categorized as pollution, and fishing and harvesting aquatic resources as biological resource use (IUCN, 2019).

2. Cetaceans

2.1. Baleen whales (Mysticetes)

2.1.1. Diversity

Fourteen extant species of baleen whales are currently recognized. They are distributed among four families (Balaenidae, Balaenopteridae, Neobalaenidae and Eschrichtiidae).

2.1.2. Abundance and main threats

Of the Balaenidae species, the bowhead whale (*Balaena mysticetus*) and the southern right whale (*Eubalaena australis*) are assessed as of Least Concern, reflecting increasing population trends. However, not all geographically defined populations of the latter are increasing (George and others, 2018). The North Atlantic right whale (*E. glacialis*) has most recently been assessed as Endangered. Although the population of this species was estimated to have increased from 1990 to 2010, it is now estimated to have declined by 16 per cent in the following years (Pettis and others, 2018). No range-wide population size or trend is available for the North Pacific right whale (*E. japonica*) (Endangered).

Of the Balaenopteridae species, new information on the Antarctic minke whale (*Balaenoptera bonaerensis*) and the Bryde's whale (*B. edeni*) has resulted in a change in status from Data Deficient to Near Threatened and of Least Concern, respectively (figure III.A). Globally, increasing population trends have been estimated for the blue whale (*B. musculus*, Endangered), the sei whale (*B. borealis*, Endangered) and the humpback whale (*Megaptera novaeangliae*, of Least Concern), with populations recovering from industrial whaling exploitation (IUCN, 2019). A population increase has also resulted in the status of the fin whale (*B. physalus*) being downlisted from Endangered to Vulnerable. The gray whale (*Eschrichtius robustus*) is listed as of Least Concern and considered stable, while the pygmy right whale (*Caperea marginata*) is listed as of Least Concern with an unknown population abundance or trend.

The main ongoing threats for baleen whales identified in IUCN Red List assessments include entanglement in fishing gear (fin whale, gray whale, humpback whale and North Atlantic right whale), harvesting (common minke whale (*B. acutorostrata*), Antarctic minke whale and sei whale) and ship strike (blue whale, fin whale, gray whale, humpback whale, North Atlantic right whale, North Pacific right whale and southern right whale) (IUCN, 2019). Climate change effects on biological productivity and, consequently, prey availability (Cabrera and others, 2018) are a concern. However, observations for some species are not consistent with projections. For example, the bowhead whale, endemic to the Arctic, is increasing despite current rapid ice loss (Moore and Reeves, 2018) and associated projected declines in prey. Importantly, environmental changes can interact with other anthropogenic threats to cause unforeseen synergistic impacts (Moore and others, 2019; Seyboth and others, 2016). For example, climate-driven shifts in habitat use by the North Atlantic right whale into unprotected shipping and commercial fishing areas have resulted in increases in mortality associated with entanglement and ship strike (Corkeron and others, 2018; Meyer-Gutbrod and Greene, 2018). The direct take of baleen whales through commercial and subsistence harvest are generally within sustainable limits.

2.2. Toothed whales, dolphins and porpoises (Odontocetes)

2.2.1. Diversity

Seventy-five species of toothed cetaceans distributed across 10 families are recognized globally. The family Delphinidae is the most diverse and includes some of the most threatened species (figure III.C).[1]

2.2.2. Abundance and main threats

Pelagic toothed whales, dolphins and porpoises

Owing to the wide-ranging distributions of pelagic cetaceans, trends in the abundance of populations and threats are often difficult to assess. As a result, most oceanic species continue to be listed as of Least Concern, except for the sperm whale (*Physeter macrocephalus*, Vulnerable) and the false killer whale, (*Pseudorca crassidens*, moved from Data Deficient to Near Threatened). A recent global population size or trend estimate for the sperm whale does not exist. Beaked whales consist of 22 pelagic deep-diving species that are still poorly known, with a proposed new species (*Berardius minimus*) currently under consideration (Yamada and others, 2019). The status of all species within this group remains Data Deficient, except for the southern bottlenose whale (*Hyperoodon planifrons*) and the Cuvier's beaked whale (*Ziphius cavirostris*), which are both listed as of Least Concern. The status of the killer whale (*Orcinus orca*), a cosmopolitan species, is Data Deficient globally, but the small coastal southern resident population in the eastern North Pacific is listed as Endangered in the United States of America and Canada as a result of threats associated with prey availability, vessel and acoustic disturbance and the accumulation of contaminants (Southern Resident Orca Taskforce, 2019).

1 It is noted that, with the cessation of whaling operations in the Southern Ocean, this threat has likely been reduced.

Figure III
Changes in conservation status over three assessment periods (before 1999, 1999–2008 and 2009–2018), on the basis of the IUCN Red List assessments

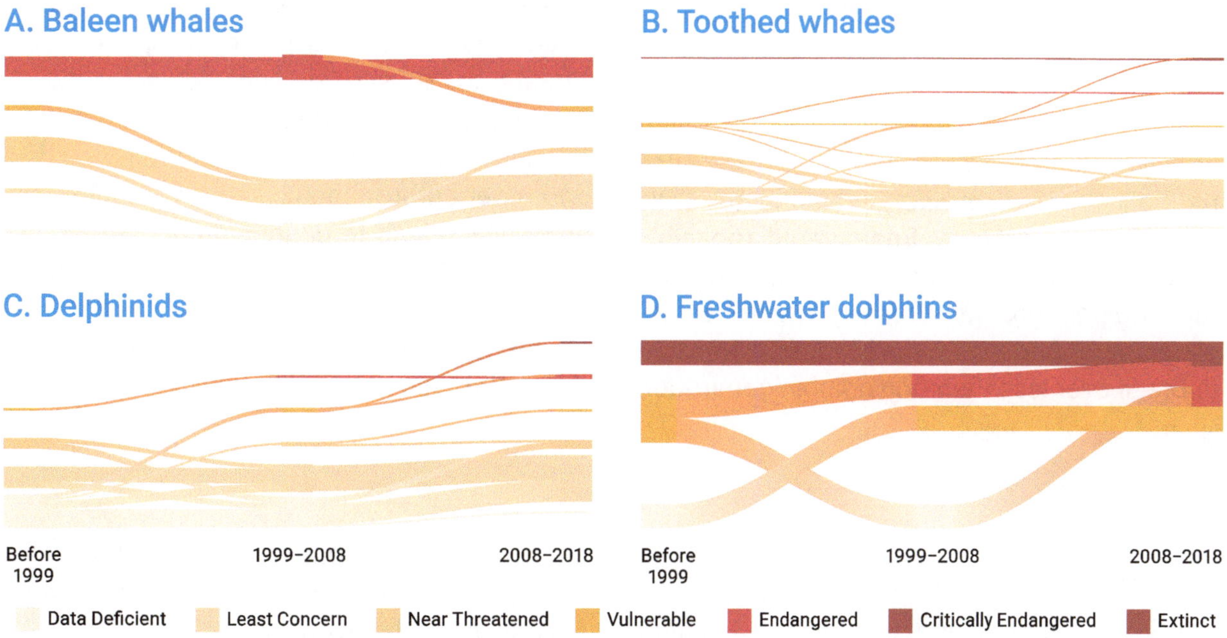

Entanglement in fishing gear is identified as a threat for several oceanic species (IUCN, 2019). Other interactions with fisheries, such as catch depredation and bait stealing by false killer whales, killer whales and sperm whales can result in deterrent actions, such as shooting and subsequent mortality (Tixier and others, 2019; Werner and others, 2015; Hamer and others, 2012). Anthropogenic noise generated, in particular, by mid-frequency active sonar is of concern for deep-diving species, such as beaked whales, *Kogia* spp. and the sperm whale (Pirotta and others, 2018; Harris and others, 2018). Decreasing sea ice and warmer waters have increased interactions between ice-dwelling species such as the narwhal (*Monodon monoceros*) and more boreal mammal species, such as the killer whale, and reduced accessibility to foraging habitat (Breed and others, 2017).

Coastal and estuarine dolphins and porpoises

This group is represented mostly by inshore and regionally restricted species or populations, including endemic species (Möller, 2012), and is consequently the most susceptible to interactions with humans. This susceptibility is reflected in the fact that 10 of the 35 species show decreasing trends, with two Critically Endangered, four Endangered, and four Vulnerable (figure III.B). Regionally restricted populations can result in locally varying states. For example, while the bottlenose dolphin (*Tursiops truncatus*) is assessed as of Least Concern globally, the regional population of Fiordland, New Zealand, is currently assessed as Critically Endangered, the Mediterranean population as Vulnerable and the Black Sea population as Endangered. Populations of coastal and estuarine dolphins are predominantly threatened by intentional and non-intentional capture from fisheries. Despite management plans, failure to reduce captures to sustainable levels have led to severe declines in abundance, in particular for the vaquita (*Phocoena sinus*) (Jaramillo-Legorreta and others, 2019) and the Maui dolphin (*Cephalorhynchus hectori maui*) (Baker and others, 2016). In particular, the vaquita is at a high risk of extinction over the next 10 years (Rojas-Bracho and others, 2019). Other threats to coastal dolphins and porpoises include climate change and associated changes to marine ecosystem dynamics, pollution, ship strike, novel diseases and disturbances caused by industrial and recreational human activities.

Freshwater dolphins

Freshwater dolphins (figure III.D) includes the baiji (*Lipotes vexillifer*), which is currently classified as Critically Endangered (possibly extinct) (IUCN, 2019), as well as the Amazon River dolphin (*Inia geoffrensis*), which has declined by 70 per cent in the Mamirauá Reserve in Brazil over less than one dolphin generation (da Silva and others, 2018), the Ganges dolphin (*Platanista gangetica*) and the Indus dolphin (*P. g. minor*), all of which are currently classified as Endangered. Species-level abundance estimates are lacking for those species. Key threats to all species include water development projects, which fragment habitats, pollution from run-off, by-catch, direct takes and other anthropogenic habitat modification, all of which lead to continuing population declines (IUCN, 2019). Most freshwater dolphin species belong to single species families and their loss, therefore, represents the loss of entire evolutionary lineages.

3. Pinnipeds

3.1. Diversity

Thirty-three extant and two recently extinct species are recognized from three families of pinnipeds (Otariidae, Phocidae and Odobenidae). Most pinnipeds have a limited range, with seven species limited to cold temperate and Arctic waters in the northern hemisphere, and four species limited to Antarctic waters in the southern hemisphere. A further four species are restricted to the Caspian Sea, Lake Baikal, the Hawaiian Islands and the Mediterranean, respectively.

3.2. Abundance and main threats

3.2.1. Phocidae

Global trends are available for eight species of phocids: four are increasing in abundance, including the Mediterranean monk seal (*Monachus monachus*, whose status has been downlisted from Critically Endangered to Endangered); one, the Hawaiian monk seal (*Neomonachus schauinslandi*, Endangered), is decreasing; and three are stable (figure IV.A). The most abundant species in the Antarctic is the crabeater seal (*Lobodon carcinophaga*), with an estimated four million adults, and the most abundant in the Arctic is the harp seal (*Pagophilus groenlandicus*), with an estimated 4.5 million adults (IUCN, 2019).

Threats to phocids are consistent across species and include habitat loss and alteration (loss of pupping and resting areas), interactions with fisheries (intentional killing, entanglement and competition) and disease transfer, potentially from pets and feral terrestrial mammals (figure IV.B; IUCN, 2019). The recent change in trend in the population of the Mediterranean monk seal may be the result of successful local adaptations of the species, due in part to avoidance of human interactions (e.g., utilization of refuges, changes in environmental stewardship and decreasing interactions (Notarbartolo di Sciara and Kotomatas, 2016)).

3.2.2. Otariidae

Currently threatened otariids (e.g., the New Zealand sea lion (*Phocarctos hookeri*), the Australian sea lion (*Neophoca cinereal*) and the Galapagos sea lion (*Zalophus wollebaeki*)) are continuing to decrease in abundance, while those that are of Least Concern are increasing (e.g., the New Zealand fur seal, (*Arctocephalus forsteri*) and the California sea lion (*Zalophus californianus*)). The Steller sea lion (*Eumetopias jubatus*) is an exception, with the species downlisted from Endangered to Near Threatened in 2012 (IUCN, 2019). This improvement has been largely driven by a doubling of the Loughlin's Steller sea lion subspecies (*Eumetopias jubatus monteriensis*) since the 1980s after protection from hunting. The Western Steller sea lion (*E. j. jubatus*), although increasing in parts of its range, continues to decline in the Aleutian Islands. Otariid species currently identified as threatened tend to have more limited ranges and are therefore sensitive to rapid changes in marine productivity caused by climate change (Atkinson and others, 2008; McClatchie and others, 2016) (figure IV.B). Other threats include interactions with fisheries (by-catch and competition for prey) (Chilvers, 2012; Hamer and others, 2013). Although by-catch management has reduced fisheries-related mortality for some species, other factors may be interacting with this conservation threat, leading to reduced survival of particular life stages and cohorts and resulting in a lack of recovery (Hamilton and Baker, 2019).

3.2.3. Odobenidae

The single species in the family Odobenidae, the walrus (*Odobenus rosmarus*), is currently assessed as Vulnerable. There are an estimated 225,000 individuals, although trends are unknown (IUCN, 2019). Although they were unsustainably harvested in the past, current management approaches are yielding sustainable catches. Climate change and associated habitat modification are expected to affect sustainable harvest levels for the species (MacCracken, 2012), with consequences for human food security. The development of human industrial activities in the Arctic linked to sea ice loss further compounds conservation threats for this species (Moore and Reeves, 2018).

Figure IV.A
Changes in the conservation status of pinnipeds over three assessment periods (before 1999, 1999–2008 and 2009–2018), on the basis of the IUCN Red List assessments

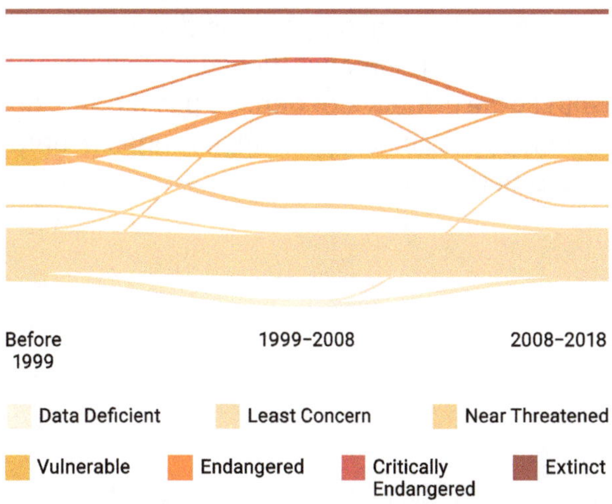

Note: The threat categories are those defined in the IUCN Threats Classification Scheme, where anthropogenic noise is categorized as pollution, and fishing and harvesting aquatic resources as biological resource use (IUCN, 2019).

Figure IV.B
Composition of ongoing and future threats to the three pinniped families

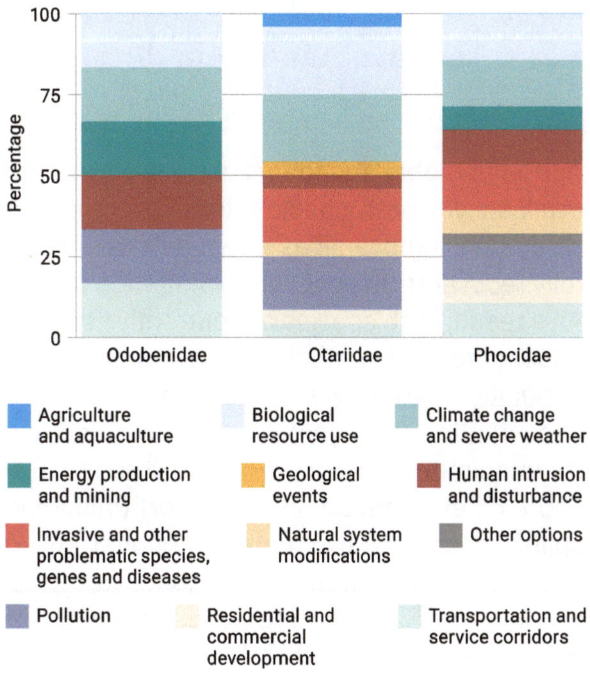

4. Sirenians

4.1. Diversity

There are four extant species in the order Sirenia (the African manatee (*Trichechus senegalensis*), the American manatee (*Trichechus manatus*), the dugong (*Dugong dugon*) and the Amazonian manatee (*Trichechus inunguis*)), with some evidence of genetic partitioning between populations across their ranges (Hunter and others, 2010).

4.2. Abundance and main threats

There continues to be a lack of species abundance estimates for the four species. Indirect evidence indicates decreasing population trends and all species are classified as Vulnerable (IUCN, 2019). Overall, main drivers associated with decreasing populations have been habitat loss, direct and incidental catch and boat collisions (IUCN, 2019). In the north-east of Brazil, high neonate and calf mortality has been connected with decreasing calving habitats associated with the development of shrimp farms and the silting of estuaries (Balensiefer and others, 2017).

5. Otters and polar bear

5.1. Diversity

The family Mustelidae includes two extant marine otter species: the marine otter (*Lontra felina*) and the sea otter (*Enhydra lutris*). The family Ursidae includes one extant marine species: the polar bear (*Ursus maritimus*).

5.2. Abundance and main threats

5.2.1. Marine and sea otters

Although global abundance estimates are not available for both species, they are regarded as decreasing overall as a result of failures to fully recover from past overexploitation for fur. As a result, both species are currently listed as Endangered. Nevertheless, several remnant populations are now increasing as a result of conservation management programmes. New threats limiting recovery include disease, offshore oil exploitation and transport (including spills), poaching, by-catch, intentional killing and disturbance from recreational activities (Duplaix and Savage, 2018). Variability in abundance has been linked to El Niño events and associated effects on Pacific coastal ecosystems. Although projected changes to the El Niño Southern Oscillation are unclear, any changes in occurrence and intensity may have an impact on the marine otter (Vianna and others, 2010).

5.2.2. Polar bear

The polar bear remains listed as Vulnerable, with trends in abundance unknown. A global estimate of 16,000 to 31,000 individuals was recently produced (Hamilton and Derocher, 2019). The most serious threat to this species is the climate change-driven loss of Arctic ice habitat for key demographic functions (Regehr and others, 2016). Pressures from new pathogens, as a result of the reduction of sea ice, and intensifying industrial and recreational activities, as accessibility to the region increases, are having a growing effect on populations (Hamilton and Derocher, 2019).

6. Consequences of changes on human communities, economies and well-being

6.1. Consumption and competition

The recovery of several marine mammal populations is generating the potential for conflicts in some regions and opportunities in others. Marine mammals can learn to associate fishing activities with food availability, leading to the development of behaviours to depredate catches from fishing vessels (Tixier and others, 2019) and the creation of conflict with aquaculture operations (Guerra, 2019).

After an increase in minke whale catches and a resumption of commercial fin whaling prior to the first Assessment, North Atlantic commercial catches of minke whales have decreased and stabilized,[2] and the commercial catch of fin whales was suspended in 2019 and 2020 (small numbers have been taken since the first Assessment as part of regulated subsistence catches). Over the same period, catches of pinnipeds and other cetaceans in the northern hemisphere have remained relatively stable overall (North Atlantic Marine Mammal Commission (NAMMCO), 2019; International Whaling Commission (IWC), 2019). Catches of baleen whales in the western North Pacific have remained broadly stable since the first Assessment (IWC, 2019, catches taken under special permits) and catches in Antarctic waters were suspended in 2019 (IWC, 2019). Regulated subsistence hunting of marine mammals remains stable (NAMMCO, 2019; IWC, 2019). Two intergovernmental organizations continue to provide a forum for discussions on and the assessment and management of catches of marine mammals: IWC, established in 1946, and NAMMCO, established in 1992.

By-caught marine mammals can complement fishery catches for human consumption. This practice can be further complemented by hunting or the use of stranded animals in some countries (Robards and Reeves, 2011). Marine mammals used in such a way has been termed "aquatic wild meat"[3] or "marine bushmeat", the latter as an analogy with terrestrial bushmeat used to support food security in deprived regions (Cosentino and Fisher, 2016; Clapham and Van Waerebeek, 2007). The catch and consumption of coastal species in lower latitudes are likely to have increased (Robards and Reeves, 2011), in particular in South-East Asia and West Africa (Porter and Lai, 2017; Liu and others, 2019; Mintzer and others, 2018; Van Waerebeek and others, 2017), where the sustainability of such practices is often unknown. As habitat change associated with climate change redistributes species and has a potential impact on population abundances (Moore and Reeves, 2018), communities relying on the harvesting of marine mammals for food are also likely to be affected, resulting in future food security challenges (Brinkman and others, 2016).

Marine mammals remain culturally significant, with objects created from body parts and as part of the imagery of coastal traditions and cultures. This cultural heritage is key to community cohesion and identity and includes unique elements, such as cooperative fishing between people and dolphins in Brazil (Daura-Jorge and others, 2012).

---

2 See https://nammco.no.

3 See Convention on the Conservation of Migratory Species of Wild Animals resolution 12.15, on aquatic wild meat.

6.2. Non-lethal activities

Marine mammals continue to be a key feature of marine tourism, which has increased[4] and diversified (Hoyt, 2018). There is anecdotal evidence of the expansion of tourism focused on marine mammals in novel locations and of increased sighting rates in mature touristic activity locations, as a result of distributional shifts associated with population recovery and climate change (e.g., Accardo and others, 2018; Halliday and others, 2018). Tourism is now listed as a conservation threat for 11 cetacean and 13 pinniped species (figure II; IUCN, 2019). Tourism activities offer the opportunity of income generation for coastal communities as long as appropriate management schemes are developed to ensure that marine mammal populations are not overexploited (Christiansen and Lusseau, 2015; Pirotta and Lusseau, 2015), investment is responsible and profits remain in the community (Higham and others, 2016). In an effort to support managers, regulators and operators, IWC and the Convention on the Conservation of Migratory Species of Wild Animals have produced a handbook on whale watching that provides guidelines for the management of whale watching activities. It is reviewed every year, and its contents are regularly updated. Quantification of the socioecological contribution of marine mammal-related tourism to global coastal communities remains unknown.

7. Outlook

Marine mammal conservation successes include an end to overharvesting of large whales and unsustainable by-catch in large-scale pelagic drift net fishing (Reeves and others, 2013). Continuing improvements of stocks can be expected as long as management efforts are maintained (Bejder and others, 2016) and not impaired by climate change (Tulloch and others, 2019).

The number of species listed as Endangered or Critically Endangered (22 species) clearly demonstrates urgent management and conservation challenges. Almost all critically endangered species and subpopulations of species, including the vaquita, the baiji, the Maui dolphin and the Atlantic humpback dolphin (*Sousa teuszii*), have very limited distributions. Despite detailed knowledge of the risks to them, decades of management interventions have not improved their population status (figure III.C). Unless management measures for these species can reduce current threats, current assessments estimate that, within 10 years, those Critically Endangered species or subpopulations will have further deteriorated and may become extinct (Comité Internacional para la Recuperación de la Vaquita, 2019; IUCN 2019).

Global initiatives are required to develop comprehensive management plans for wide-ranging species. Since the first Assessment, the Commission for the Conservation of Antarctic Marine Living Resources (CCAMLR) has established a marine protected area in the Ross Sea (CCAMLR, 2016). The multiple objectives of this marine protected area include protection of core foraging areas of the Weddell seal (*Leptonychotes weddellii*) and the Type C killer whale. Fixed spatial management, such as marine protected areas, has been identified as effective for species conservation (Gormley and others, 2012). However, the current shifts in threats that has been observed (figure II.A), alongside rapid changes in marine ecosystems driven by climate change, make such areas a less flexible tool, especially for ensuring

[4] See Convention on Migratory Species resolution 11.29, on sustainable boat-based marine wildlife watching.

the conservation and sustainable use of highly mobile species (Pinn, 2018; see also chap. 27).

Since the first Assessment, it has become apparent that the cumulative effects of multiple sectors (see also chap. 25) are increasingly influencing the conservation status trajectories of marine mammals (National Academies of Sciences, Engineering and Medicine, 2017). In the coming decades, climate change will influence marine mammals in multiple ways (figure II), including habitat and food web modifications. In addition, increased exposure to human activities and associated stressors will contribute to cumulative effects, potentially curtailing recent recoveries (Tulloch and others, 2019). Trophic amplification may strengthen climate change impacts up the food web, with proportionally stronger impacts on the higher trophic levels occupied by marine mammals (Lotze and others, 2019).

New technological and analytical developments have helped to design frameworks for quantifying the consequences of multiple stressors on populations using observational data (see chap. 25). It is therefore easier to estimate the conservation impact of non-lethal and indirect stressors, such as anthropogenic noise, tourism and offshore renewable energy systems. Ecosystem approaches to risk assessments (Holsman and others, 2017) are increasingly being used as critical elements of integrated ecosystem assessments (see also chap. 27). Those approaches also place risks within the wider socioecological context of the communities that use marine mammals.

8. Key remaining knowledge gaps

The world is changing rapidly, challenging our ability to forecast marine mammal status and exploitation patterns on the basis of retrospective analyses. These rapid changes require new mechanistic approaches to forecast how species and populations will respond to climate change and the sustainability of current and future direct and indirect human impacts. Specifically, there is a need for: (a) developing approaches to assess and forecast how marine mammals respond and adapt to climate change and associated changing marine ecosystems; (b) greater understanding of the cumulative effects of multiple anthropogenic pressures on marine mammals, including ongoing and new exploitations; (c) developing processes to identify and implement management actions that successfully secure critically endangered marine mammal species from extinction; and (d) greater understanding of Data Deficient populations that allows for quantification of their abundances, thereby facilitating their classification and movement out of the Data Deficient category. In addition, further work is needed to better understand the role of marine mammals in ocean processes, including in the spatial transfer of nutrients and carbon.[5]

9. Key remaining capacity-building gaps

A broad understanding of the human dimensions that lead to successful and unsuccessful management interventions and the capacity and resources to engage in them is currently lacking. Where marine mammals are abundant, there is often a gap in institutional capacity for maximizing opportunities for access to those highly valued resources and their

5 See www.imf.org/external/pubs/ft/fandd/2019/12/natures-solution-to-climate-change-chami.htm.

sustainable use. Associated tools are lacking for assessing trade-offs between sectors utilizing marine mammals through incidental by-catch or non-lethal cumulative effects and traditional industries exploiting those species (e.g., fisheries and tourism). Incentivization techniques are being developed for other natural resources, such as forests, to diversify how they are sustainably used, as well as better connect local "nature-wealthy" communities with potential remote markets (Dao, 2018). This could potentially be applied to marine mammals. There is a need for assessing how those approaches could be used to diversify current marine mammal exploitation and offer opportunities to develop derivatives.

References

Accardo, Corey, and others (2018). Sightings of a bowhead whale (*Balaena mysticetus*) in the Gulf of Maine and its interactions with other baleen whales. *Journal of Cetacean Research and Management*, vol. 19, pp. 23–30.

Atkinson, S., and others (2008). Anthropogenic causes of the western Steller sea lion *Eumetopias jubatus* population decline and their threat to recovery. *Mammal Review*, vol. 38, No. 1, pp. 1–18.

Baker, C.S., and others (2016). *Estimating the Abundance and Effective Population Size of Māui Dolphins Using Microsatellite Genotypes in 2015–16, with Retrospective Matching to 2001–16*. Auckland: Department of Conservation.

Balensiefer, Deisi Cristiane, and others (2017). Three decades of Antillean Manatee (*Trichechus manatus manatus*) stranding along the Brazilian coast. *Tropical Conservation Science*, vol. 10. https://doi.org/10.1177/1940082917728375.

Bejder, Michelle, and others (2016). Embracing conservation success of recovering humpback whale populations: evaluating the case for downlisting their conservation status in Australia. *Marine Policy*, vol. 66, pp. 137–141.

Breed, Greg A., and others (2017). Sustained disruption of narwhal habitat use and behavior in the presence of Arctic killer whales. *Proceedings of the National Academy of Sciences*, vol. 114, No. 10, pp. 2628–2633.

Brinkman, Todd J., and others (2016). Arctic communities perceive climate impacts on access as a critical challenge to availability of subsistence resources. *Climatic Change*, vol. 139, Nos. 3 and 4, pp. 413–427.

Burkholder, Derek A., and others (2013). Patterns of top-down control in a seagrass ecosystem: could a roving apex predator induce a behaviour-mediated trophic cascade? *Journal of Animal Ecology*, vol. 82, No. 6, pp. 1192–1202. https://doi.org/10.1111/1365-2656.12097.

Cabrera, Andrea A., and others (2018). Strong and lasting impacts of past global warming on baleen whale and prey abundance. *BioRxiv*, 497388.

CCAMLR (2016). Conservation Measure 91-05 (2016): Ross Sea Region Marine Protected Area.

Chilvers, Barbara (2012). Population viability analysis of New Zealand sea lions, Auckland Islands, New Zealand's sub-Antarctics: assessing relative impacts and uncertainty. *Polar Biology*, vol. 35, No. 10, pp. 1607–1615. https://doi.org/10.1007/s00300-011-1143-6.

Christiansen, Fredrik, and David Lusseau (2015). Linking behavior to vital rates to measure the effects of non-lethal disturbance on wildlife. *Conservation Letters*, vol. 8, No. 6, pp. 424–431.

Clapham, Phil, and Koen Van Waerebeek (2007). Bushmeat and bycatch: the sum of the parts. *Molecular Ecology*, vol. 16, No. 13, pp. 2607–2609.

Comité Internacional para la Recuperación de la Vaquita (2019). Report of the Eleventh Meeting of the Comité Internacional para la Recuperación de la Vaquita (CIRVA). La Jolla, California: Southwest Fisheries Science Center.

Corkeron, Peter, and others (2018). The recovery of North Atlantic right whales, *Eubalaena glacialis*, has been constrained by human-caused mortality. *Royal Society Open Science*, vol. 5, No. 11.

Cosentino, A. Mel, and Sue Fisher (2016). The utilization of aquatic bushmeat from small cetaceans and manatees in South America and West Africa. *Frontiers in Marine Science*, vol. 3, art. 163. https://doi.org/10.3389/fmars.2016.00163.

Dao, David (2018). Decentralized sustainability: beyond the tragedy of the commons with smart contracts + AI. Medium, 21 June. https://medium.com/@daviddao/decentralized-sustainability-9a53223d3001.

Daura-Jorge, F.G., and others (2012). The structure of a bottlenose dolphin society is coupled to a unique foraging cooperation with artisanal fishermen. *Biology Letters*, vol. 8, No. 5, pp. 702–705. https://doi.org/10.1098/rsbl.2012.0174.

Doughty, Christopher E., and others (2016). Global nutrient transport in a world of giants. *Proceedings of the National Academy of Sciences*, vol. 113, No. 4, pp. 868–873.

Duplaix, Nicole, and Melissa Savage (2018). *The Global Otter Conservation Strategy*. IUCN/SSC Otter Specialist Group, Salem, Oregon, United States of America.

Eikeset, Anne Maria, and others (2018). What is blue growth? The semantics of "Sustainable Development" of marine environments. *Marine Policy*, vol. 87, pp. 177–179. https://doi.org/10.1016/j.marpol.2017.10.019.

Estes, James A., and others (1998). Killer whale predation on sea otters linking oceanic and nearshore ecosystems. *Science*, vol. 282, No. 5388, pp. 473–476.

Estes, James A., and others (2016). Megafaunal impacts on structure and function of ocean ecosystems. *Annual Review of Environment and Resources*, vol. 41, pp. 83–116.

da Silva, Vera M., and others (2018). Both cetaceans in the Brazilian Amazon show sustained, profound population declines over two decades. *PLOS ONE*, vol. 13, No. 5, pp. 1–12. https://doi.org/10.1371/journal.pone.0191304.

George, J. Craig, and others (2018). Bowhead Whale: *Balaena mysticetus*. In *Encyclopedia of Marine Mammals*, Elsevier, pp. 133–135.

Gormley, Andrew M., and others (2012). First evidence that marine protected areas can work for marine mammals. *Journal of Applied Ecology*, vol. 49, No. 2, pp. 474–480.

Guerra, Ana Sofía (2019). Wolves of the Sea: managing human-wildlife conflict in an increasingly tense ocean. *Marine Policy*, vol. 99, pp. 369–373.

Halliday, William D., and others (2018). Tourist vessel traffic in important whale areas in the western Canadian Arctic: risks and possible management solutions. *Marine Policy*, vol. 97, pp. 72–81. https://doi.org/10.1016/j.marpol.2018.08.035.

Hamer, D.J., and others (2013). The endangered Australian sea lion extensively overlaps with and regularly becomes by-catch in demersal shark gill-nets in South Australian shelf waters. *Biological Conservation*, vol. 157, pp. 386–400. https://doi.org/10.1016/j.biocon.2012.07.010.

Hamer, Derek J., and others (2012). Odontocete bycatch and depredation in longline fisheries: A review of available literature and of potential solutions. *Marine Mammal Science*, vol. 28, No. 4, pp. E345–374. https://doi.org/10.1111/j.1748-7692.2011.00544.x.

Hamilton, S.G., and A. E. Derocher (2019). Assessment of global polar bear abundance and vulnerability. *Animal Conservation*, vol. 22, No. 1, pp. 83–95.

Hamilton, Sheryl, and G. Barry Baker (2019). Population growth of an endangered pinniped—the New Zealand sea lion (*Phocarctos hookeri*)—is limited more by high pup mortality than fisheries bycatch. *ICES Journal of Marine Science*, vol. 76, No. 6, pp. 1794–1806. https://doi.org/10.1093/icesjms/fsz039.

Harris, Catriona M., and others (2018). Marine mammals and sonar: dose-response studies, the risk-disturbance hypothesis and the role of exposure context. *Journal of Applied Ecology*, vol. 55, No. 1, pp. 396–404.

Higham, James E.S., and others (2016). Managing whale-watching as a non-lethal consumptive activity. *Journal of Sustainable Tourism*, vol. 24, No.,1, pp. 73–90.

Holsman, Kirstin, and others (2017). An ecosystem-based approach to marine risk assessment. *Ecosystem Health and Sustainability*, vol. 3, No. 1, e01256.

Hoyt, Erich (2018). Tourism. In *Encyclopedia of Marine Mammals*, 3rd ed., Bernd Würsig, J.G.M. Thewissen, and Kit M. Kovacs, eds. Academic Press, pp. 1010–1114. https://doi.org/10.1016/B978-0-12-804327-1.00262-4.

Hunter, M.E., and others (2010). Low genetic variation and evidence of limited dispersal in the regionally important Belize manatee. *Animal Conservation*, vol. 13, No. 6, pp. 592–602. https://doi.org/10.1111/j.1469-1795.2010.00383.x.

International Union for Conservation of Nature (IUCN) (2019). *The IUCN Red List of Threatened Species*. www.iucnredlist.org/en.

International Whaling Commission (2019). Total Catches. Available at https://iwc.int/total-catches.

Jaramillo-Legorreta, Armando M., and others (2019). Decline towards extinction of Mexico's vaquita porpoise (*Phocoena sinus*). *Royal Society Open Science*, vol. 6, No. 7. https://doi.org/10.1098/rsos.190598.

Kiszka Jeremy J., and others (2015). Behavioural drivers of the ecological roles and importance of marine mammals. *Marine Ecology Progress Series*, vol. 523, pp. 267–81.

Liu, Mingming, and others (2019). Fishers' experiences and perceptions of marine mammals in the South China Sea: insights for improving community-based conservation. *Aquatic Conservation: Marine and Freshwater Ecosystems*, vol. 29, No. 5, pp. 809–819.

Lotze, Heike K., and others (2019). Global ensemble projections reveal trophic amplification of ocean biomass declines with climate change. *Proceedings of the National Academy of Sciences*, vol. 116, No. 26, pp. 12907–12912.

MacCracken, James G. (2012). Pacific Walrus and climate change: observations and predictions. *Ecology and Evolution*, vol. 2, No. 8, pp. 2072–2090.

McClatchie, Sam, and others (2016). Food limitation of sea lion pups and the decline of forage off central and southern California. *Royal Society Open Science*, vol. 3, No. 3. https://doi.org/10.1098/rsos.150628.

Meyer-Gutbrod, Erin L., and Charles H. Greene (2018). Uncertain recovery of the North Atlantic right whale in a changing ocean. *Global Change Biology*, vol. 24, No. 1, pp. 455–464.

Mintzer, Vanessa Jordan, and others (2018). The use of aquatic mammals for bait in global fisheries. *Frontiers in Marine Science*, vol. 5, art. 191.

Möller, Luciana M. (2012). Sociogenetic structure, kin associations and bonding in delphinids. *Molecular Ecology*, vol. 21, No. 3, pp. 745–764.

Moore, Sue E., and others (2019). Baleen whale ecology in arctic and subarctic seas in an era of rapid habitat alteration. *Progress in Oceanography*, vol. 176.

Moore, Sue E., and Randall R. Reeves (2018). Tracking arctic marine mammal resilience in an era of rapid ecosystem alteration. *PLoS Biology*, vol. 16, No. 10, e2006708.

National Academies of Sciences, Engineering and Medicine, and Medicine (2017). *Approaches to Understanding the Cumulative Effects of Stressors on Marine Mammals*. Washington, D.C.: National Academies Press. https://doi.org/10.17226/23479.

North Atlantic Marine Mammal Commission (NAMMCO) (2019). Marine Mammals. Seals and Walruses (Pinnipeds). https://nammco.no/marinemammals.

Notarbartolo di Sciara, Giuseppe, and S. Kotomatas (2016). Chapter Twelve: are Mediterranean monk seals, *Monachus monachus*, being left to save themselves from extinction? *Advances in Marine Biology*, vol. 75, pp. 359–386. https://doi.org/10.1016/bs.amb.2016.08.004.

Pettis, H., and others (2018). North Atlantic right whale consortium 2018 annual report card. Report to the North Atlantic Right Whale Consortium.

Pinn, Eunice H. (2018). Protected areas: the false hope for cetacean conservation? In *Oceanography and Marine Biology: An Annual Review*, vol. 56. S. J. Hawkins and others, eds. Boca Raton, Florida, United States: CRC Press, pp. 72–104.

[Pirotta, Enrico, and others (2018). Understanding the population consequences of disturbance. *Ecology and Evolution*, vol. 8, No. 19, pp. 9934–9946.

Pirotta, Enrico, and David Lusseau (2015). Managing the wildlife tourism commons. *Ecological Applications*, vol. 25, No. 3, pp. 729–741.

Porter, Lindsay, and Hong Yu Lai (2017). Marine mammals in Asian societies; trends in consumption, bait, and traditional use. *Frontiers in Marine Science*, vol. 4, art. 47.

Reeves Randall R., and others (2013). Marine mammal bycatch in gillnet and other entangling net fisheries, 1990 to 2011. *Endangered Species Research*, vol. 20, No. 1, pp. 71–97.

Regehr, Eric V., and others (2016). Conservation status of polar bears (*Ursus maritimus*) in relation to projected sea-ice declines. *Biology Letters*, vol. 12, No. 12, pp. 20160556.

Robards, Martin D., and Randall R. Reeves (2011). The global extent and character of marine mammal consumption by humans: 1970–2009. *Biological Conservation*, vol. 144, No. 12, pp. 2770–2786.

Rojas-Bracho L., and others (2019). A field effort to capture critically endangered vaquitas *Phocoena sinus* for protection from entanglement in illegal gillnets. *Endangered Species Research*, vol. 38, pp. 11–27.

Roman, Joe, and others (2014). Whales as marine ecosystem engineers. *Frontiers in Ecology and the Environment*, vol. 12, No. 7, pp. 377–385.

Seyboth, Elisa, and others (2016). Southern Right Whale (*Eubalaena australis*) Reproductive Success is Influenced by Krill (*Euphausia superba*) Density and Climate. *Scientific Reports*, vol. 6,art. 28205. https://doi.org/10.1038/srep28205.

Southern Resident Orca Taskforce (2019). *Final Report and Recommendations*.

Sydeman, William J., and others (2015). Climate change and marine vertebrates. *Science*, vol. 350, No. 6262, pp. 772–777.

Tixier, Paul, and others (2019). Commercial fishing patterns influence odontocete whale-longline interactions in the Southern Ocean. *Scientific Reports*, vol. 9, art. 1904.

Tulloch, Vivitskaia J.D., and others (2019). Future recovery of baleen whales is imperiled by climate change. *Global Change Biology*, vol. 25, No. 4, pp. 1263–1281.

United Nations (2017). *The First Global Integrated Marine Assessment: World Ocean Assessment I*. Cambridge: Cambridge University Press.

Van Waerebeek, Koen, and others (2017). New records of Atlantic humpback dolphin in Guinea, Nigeria, Cameroon and Togo underscore fisheries pressure and generalized marine bushmeat demand. *Revue d'Ecologie (Terre et Vie)*, vol. 72, No. 2, pp. 1576–1586.

Vianna, Juliana A., and others (2010). Phylogeography of the marine otter (*Lontra felina*): historical and contemporary factors determining its distribution. *Journal of Heredity*, vol. 101, No. 6, pp. 676–689. https://doi.org/10.1093/jhered/esq088.

Werner, Timothy B., and others (2015). Mitigating bycatch and depredation of marine mammals in longline fisheries. *ICES Journal of Marine Science*, vol. 72, No. 5, pp. 1576–1586.

Yamada, Tadasu K., and others (2019). Description of a new species of beaked whale (*Berardius*) found in the North Pacific. *Scientific Reports*, vol. 9, art. 12723.

Chapter 6E
Marine reptiles

Contributors: Qamar Schuyler (convener), Karen Evans (subchapter lead member), Maximillian Hirschfeld, Carmen Mifsud, Gabriel Hoinsoude Segniagbeto, André Silva Barreto and Vinay Udyawer.

Keynote points

- Changes in the conservation status of marine turtles since the first *World Ocean Assessment* (United Nations, 2017a) are highly variable, with some populations experiencing positive growth rates, while others have experienced catastrophic declines.
- The conservation status of most sea snake and marine iguana populations has not changed since the first Assessment, although huge data gaps remain.
- The main threats to marine reptiles remain similar to those recorded in the first Assessment. By-catch is the most significant threat, although targeted harvesting, marine pollution, habitat loss, coastal development, disease and climate change are also key threatening processes.

1. Introduction

Chapter 39 of the first Assessment outlined the conservation status of marine reptiles, the major threats to those taxa and the most pressing conservation needs as of 2012 (United Nations, 2017b).

The present chapter provides an updated assessment at a global scale and identifies regional trends in the conservation status of marine turtles and sea snakes, with a focus on changes that have occurred since the first Assessment. It also relates to other chapters in the present Assessment, in particular, chapters 4, 7 and 15.

1.1. Assessment frameworks

The primary assessment frameworks used in the first Assessment to assess the status of marine reptiles were the International Union for Conservation of Nature (IUCN) Red List assessments and the IUCN Marine Turtle Specialist Group conservation priorities portfolio (Wallace and others, 2010). The present chapter takes a similar approach and, where updates on conservation status were not available, incorporates updated information from regional reports of the Marine Turtle Specialist Group and peer-reviewed publications.

2. Conservation status of marine reptiles

2.1. Marine turtles

Since the first Assessment, the status of two global populations and four subpopulations of marine turtles has been updated (table 1). The global status of the loggerhead turtle (*Caretta caretta*), based on data from 90 per cent of the total global nesting population (comprising 6 of 10 recognized subpopulations), has improved, moving from Endangered to Vulnerable (table 1). However, the status of each subpopulation varies considerably, ranging from of Least Concern (North-West and South-West Atlantic Ocean, Mediterranean and North Pacific Ocean) to Near Threatened (South-West and South-East Indian Ocean), Endangered (North-East Atlantic Ocean) and Critically Endangered (North-West and North-East Indian Ocean and South Pacific Ocean) (Casale and Tucker, 2017).

The Kemp's ridley turtle (*Lepidochelys kempii*) has been uplisted from Endangered to Critically Endangered, following a reduction in total population of more than 80 per cent from historical levels. This constitutes a reversal of recoveries observed in the 1990s and 2000s, the causes of which are unknown, but may be related to fishery by-catch and the Deepwater Horizon oil spill (Wibbels and Bevan, 2019). Similarly, the North-West Atlantic subpopulation of the leatherback turtle (*Dermochelys coriacea*) has been uplisted from of Least Concern to Endangered (Northwest Atlantic Leatherback Working Group, 2019). This change is primarily the result of updated regional nesting trend analyses.

Subpopulations of the green turtle (*Chelonia mydas*) in the North Indian Ocean (Vulnerable) and South Atlantic Ocean (Least Concern) have been assessed for the first time (Mancini and others, 2019; Broderick and Patricio, 2019), while the Hawaiian subpopulation has been reassessed and remains of Least Concern (Chaloupka and Pilcher, 2019). Although the green turtle has not been reassessed globally by IUCN since the first Assessment, a global assessment under the Endangered Species Act of the United States of America ranked most subpopulations with a very low likelihood of quasi-extinction risk within the next 100 years (Seminoff and others, 2015). Of the subpopulations assessed, that of the Mediterranean was ranked as having the highest risk of extinction.

2.2. Sea snakes

Since the first Assessment, the status of 26 of the 71 currently recognized species of sea snakes has been updated, including that of three of four newly described species. Two species (*Aipysurus apraefrontalis* and *A. foliosquama*), previously assessed as Critically Endangered, were reclassified as Data Deficient owing to changes in their known range (D'anastasi and others, 2016, Udyawer and others, 2020).

Declines in the abundance of the dusky sea snake (*Aipysurus fuscus*) at Ashmore Reef have reduced the known range to three reef systems in the Timor Sea. Currently classified as Endangered, high rates of hybridization with the more common species *A. laevis* across the reduced range of the former species has raised concerns about high levels of introgression of this species (Sanders and others, 2014).

More generally, increased documentation on sea snakes has provided updated records of species assemblages and distribution globally (Rasmussen and others, 2014; Rezaie-Atagholipour and others, 2016; Sarker and others, 2017; Buzás and others, 2018; Ganesh and others, 2019). Expanded genetic evaluation of species across their global range has resulted in the restructuring of the Hydrophiinae phylogeny (Sanders and others, 2013), the reclassification of cryptic species (Sanders and others, 2013; Ukuwela and others, 2013; Ukuwela and others, 2014; Lukoschek, 2018) and the description of four novel species since the first Assessment (Ukuwela and others, 2012; Sanders and others, 2012; Nankivell and others, 2020).

2.3. Marine iguana

The marine iguana (*Amblyrhynchus cristatus*) was reassessed under the IUCN Red List assessment in 2020 and its status remains Vulnerable (MacLeod and others, 2020). A recent taxonomic review of the species based on morphological and genetic information resulted in a reclassification of two subspecies into one subspecies and the addition of five new subspecies, resulting in a total of 11 subspecies (Miralles and others, 2017).

Table 1
Species of marine turtles and sea snakes that have had a change in their IUCN Red List status since the first Assessment

Taxa	Common name	Change in IUCN Red List status
Marine turtles	Loggerhead turtle	Downlisted in 2015 from Endangered to Vulnerable (the subpopulations range from Critically Endangered to of Least Concern)
	Green turtle	The Hawaiian subpopulation was assessed as remaining of Least Concern in 2019. The North Indian and South Atlantic subpopulations were listed as Vulnerable and of Least Concern, respectively, in 2019. The rest of the global population is Endangered (no change, but see Seminoff and others, 2015)
	Kemp's ridley turtle	Uplisted from Endangered to Critically Endangered in 2019
	Leatherback turtle	The North-West Atlantic subpopulation was uplisted from of Least Concern to Endangered in 2019. The global population has not been assessed since 2013 and remains categorized as Vulnerable, although all other subpopulations are either Critically Endangered or Data Deficient (no change)
Sea snakes	Short-nosed sea snake	Listing revised from Critically Endangered to Data Deficient in 2018
	Leaf-scaled sea snake	Listing revised from Critically Endangered to Data Deficient in 2018
	Mosaic sea snake	Assessed as Data Deficient in 2018
	Shark Bay sea snake	Assessed as Data Deficient in 2018
	Rough-scaled sea snake	Assessed as Data Deficient in 2018

Note: The other sea turtle species have not been reassessed since the first Assessment.
Twenty-six Australian sea snake species were reassessed under the IUCN Red List assessment in 2018. Apart from the updates to the five species listed in the table above, listings for the remaining 21 Australian species were unchanged. All other species found outside Australia (45) have not been reassessed since the first Assessment.

3. Regional trends

A variety of sources provide information on local and regional population trends for turtles and sea snakes. Because these sources vary widely in how they report population trends, a brief summary is provided in table 2, with specific references for details on assessment and reporting methods used. Where there are population trend data for a regional management unit as a whole, citations are included for reports on smaller regions within the unit. In some instances, smaller nesting beaches or areas may differ in trend direction from the unit; in such instances, they are reported separately. As table 2 includes only data on trends since the first Assessment, it relies on sources published from 2015 to January 2020. Additional data unknown to the compilers or published after January 2020 may alter the trends reported below.

Table 2
Regional trends in abundance and distribution for sea turtles and sea snakes

Region	Turtles	Sea snakes
North Atlantic Ocean, Caribbean Sea and Mediterranean	**Increasing trends (nesting)** CC: North-West Atlantic Ocean RMU (Ceriani and Meylan, 2017; Mazaris and others, 2017; Nalovic and others, 2018) and Mediterranean RMU (Casale, 2015a; Mazaris and others, 2017; Casale and others, 2018) CM: North-West Atlantic Ocean RMU (Mazaris and others, 2017; Nalovic and others, 2018; Valdivia and others, 2019; National Marine Fisheries Service, 2019), South Atlantic Ocean distinct population segment (Valdivia and others, 2019) and Mediterranean (Casale and others, 2018) EI: West Atlantic Ocean RMU (Mazaris and others, 2017; Nalovic and others, 2018; Valdivia and others, 2019) **Stable trends (nesting)** CC: peninsular Florida (United States) recovery unit (Valdivia and others, 2019) LK: following an exponential recovery after the first Assessment, trends have flattened considerably (Wibbels and Bevan, 2019) **Decreasing trends (nesting)** DC: North-West Atlantic Ocean RMU (Northwest Atlantic Leatherback Working Group, 2019) EI: Mexico (Valdivia and others, 2019)	
South Atlantic Ocean	**Increasing trends (nesting)** CC: South-West Atlantic Ocean RMU (Casale and Marcovaldi, 2015) CM: South Atlantic Ocean RMU (Mazaris and others, 2017; Broderick and Patricio, 2019) DC: Brazil, although variable (Colman and others, 2019) LO: West Atlantic Ocean RMU (Mazaris and others, 2017) **Stable trends (nesting)** LO: French Guiana (France) (Nalovic and others, 2018) **Decreasing trends (nesting)** LO: East Atlantic Ocean RMU (Mazaris and others, 2017)	Potential expansion of distribution with changing climatic conditions (Lillywhite and others, 2017)
Indian Ocean, Arabian Sea and Persian Gulf	**Increasing trends (nesting)** CC: South-West Indian Ocean RMU (Mazaris and others, 2017) CM: South-West Indian Ocean RMU (Mazaris and others, 2017) LO: North-East Indian Ocean RMU (Mazaris and others, 2017) **Stable trends (nesting)** CM: Egypt and Kuwait (Phillott and Rees, 2018) DC: India (Phillott and Rees, 2018) EI: Kuwait and Qatar (Phillott and Rees, 2018) LO: India, two major and one minor nesting sites, either stable or increasing (Phillott and Rees, 2018) **Decreasing trends (nesting)** CC: North-West Indian Ocean RMU (Casale, 2015b) CM: North Indian Ocean RMU (Mancini and others, 2019)	

Region	Turtles	Sea snakes
North Pacific Ocean	**Increasing trends (nesting)** CC: North Pacific Ocean RMU (Casale and Matsuzawa, 2015) CM: North central Pacific Ocean RMU (Mazaris and others, 2017; Chaloupka and Pilcher, 2019) and Northern Mariana Islands (Summers and others, 2018) **Decreasing trends (nesting)** CM: North-West Pacific Ocean RMU (Mazaris and others, 2017) DC: West Pacific Ocean RMU (Tiwari and others, 2013; Mazaris and others, 2017) and East Pacific Ocean RMU (Wallace and others, 2013, Mazaris and others, 2017) **No trend (individuals)** CM: Guam, ocean count of individuals (Valdivia and others, 2019)	Expanding distribution in the north of their geographical range from new data records (Park and others, 2017) Declining trends in unregulated fishery catches in the Gulf of Thailand (Van Cao and others, 2014)
South Pacific Ocean	**Increasing trends (nesting)** CC: Australia (Limpus and others, 2013) **Stable trends (nesting)** ND: northern Australia (Groom and others, 2017) **Decreasing trends** CM: Raine Island (Australia), genetic analysis may indicate a dramatic reduction in hatching success (Jensen and others, 2016) EI: Australia, nesting (Bell and others, 2020) ND: South-West Pacific Ocean RMU (Mazaris and others, 2017)No definable trend (nesting) ND: eastern Australia (Limpus and others, 2017)	Expansion of range upon new data (D'anastasi and others, 2016; Udyawer and others, 2020)

Abbreviations: CC, *Caretta caretta* (loggerhead turtle); CM, *Chelonia mydas* (green turtle); DC, *Dermochelys coriacea* (leatherback turtle); EI, *Eretmochelys imbricata* (hawksbill turtle); LK, *Lepidochelys kempii* (Kemp's ridley turtle); LO, *Lepidochelys olivacea* (Olive ridley turtle); ND, *Natator depressus* (flatback turtle); RMU, regional management unit.

4. Threats

Although many marine reptiles are protected by law and conservation efforts have been undertaken in many regions, globally, threats to marine reptiles remain much the same as those identified in the first Assessment. Mortality from by-catch in fisheries (both regulated and illegal, unreported and unregulated) remains a significant threat to marine turtles and sea snakes (Lewison and others, 2014; Rees and others, 2016; Riskas and others, 2018). Other key threatening factors that affect marine reptiles include unregulated harvesting, marine pollution, habitat loss, coastal development, diseases and climate change. While understanding of the impacts of climate change and marine pollution on marine reptiles has increased since the first Assessment, impacts at the population level are still broadly unknown.

4.1. Sea turtles

While by-catch and the retention of animals are likely to be the most significant threats to turtle populations worldwide, research since the first Assessment has brought about an emerging understanding of the threats posed by climate change and pollution.

The most significant climate change impacts hypothesized for sea turtles include the feminization of the population and an increased embryonic mortality caused by higher nest temperatures (Fuentes and Cinner, 2010). While baseline sex ratios at a number of nesting beaches indicate that the vast majority produce predominantly female hatchlings (e.g., Laloë and others, 2016; Jensen and others, 2018), some models suggest that feminization could actually lead to increased reproductive success in the short term, since males can breed more frequently than females (Hays and others, 2014). Although embryos may be more resilient to high temperatures than previously thought (Howard and others, 2014), rising temperatures do ultimately lead to hatchling mortality (Laloë and others, 2017). There are suggestions that climate change could affect population trends at regional scales, with increased reproductive success in temperate areas perhaps balancing decreased hatchling production in tropical areas (Montero and others, 2018). However, climate change impacts, from sea level rise (and associated habitat loss) to increased incidence of cyclones leading to nest inundation and coastal erosion, are also of concern for turtle populations (Fuentes and Cinner, 2010).

In 2010, the Deepwater Horizon oil spill resulted in hundreds of thousands of green, loggerhead and Kemp's ridley turtles being exposed (at varying levels) to the oil (Deepwater Horizon Natural Resource Damage Assessment Trustees, 2016; Putman and others, 2015). While the long-term impacts of that event on population levels are yet to be quantified, the numbers of loggerhead and Kemp's ridley turtle nests appear to have decreased as a result of both direct mortality (to adult turtles) and deterrence from nesting beaches resulting from beach clean-up activities (Gallaway and others, 2016; Lauritsen and others, 2017). According to global risk assessments of interactions between marine turtles and marine debris, it is estimated that more than 50 per cent of marine turtles are likely to have ingested debris (Schuyler and others, 2016), with hawksbill turtles observed to ingest up to 8.8 g of plastic per kilogram of body weight (Lynch, 2018). Marine debris, like other pollutants, has demonstrably detrimental effects on individuals, but the ecosystem and population-level impacts of pollutants on sea turtle populations still require further study (Nelms and others, 2016; Wilcox and others, 2018).

A new potential threat emerging in the Caribbean is the coastal accumulation of an unprecedented proliferation of *Sargassum*. While open-ocean *Sargassum* mats provide important critical nursery grounds for sea turtles, recent studies suggest that *Sargassum* accumulation on shore may inhibit nesting and impede hatchling dispersal, while decaying mounds may alter oxygen levels and thermal conditions (Maurer and others, 2015). Since this mass seaweed stranding is a new phenomenon, in particular in the eastern Caribbean, its direct impacts on sea turtle nesting remain largely unknown. Nesting beach habitat degradation resulting from coastal development, as identified in the first Assessment, continues to decrease the quantity and quality of nesting areas available to females (Broderick and Patricio, 2019; Casale and Tucker, 2017)

4.2. Sea snakes

In the Gulf of Thailand, high levels of by-catch of sea snakes have been documented in the local squid fishery. Sea snakes are an important commercial by-catch product for Vietnamese fishers who operate in the Gulf of Thailand, the harvest of which is currently unregulated and largely undocumented (Van Cao and others, 2014). Baseline surveys have detected a decline in sea snake harvests between 2008 and 2012 (Van Cao and others, 2014). The development of by-catch reduction devices, for example within the Australian trawl fishery industry, may assist in mitigating high incidental by-catch of sea snakes in tropical fisheries that adopt by-catch reduction measures, but may have limited utility in fisheries that rely on commercial by-catch as a source of income (Lobo and others, 2010).

High concentrations of trace heavy metals recorded in sea snakes in close proximity to mineral extraction operations across their range have highlighted marine pollution as an emerging threat to sea snake populations (Rezaie-Atagholipour and others, 2012; Sereshk and Bakhtiari, 2015; Gillett and others, 2017; Goiran and others, 2017).

4.3. Marine iguana

In the first Assessment, extreme climatic El Niño events, tourism and introduced species were identified as key threats to the marine iguana, together with pollution (Wikelski and others, 2002). However, no information on the direct impact of any of those stressors on population numbers has been published since then. Population size estimates based on molecular approaches suggest that populations of recently proposed subspecies are generally small and have a reduced evolutionary potential, making them vulnerable to threats (Frankham and others, 2014; MacLeod and Steinfartz, 2016).

Further research on threats to the marine iguana has shown that tourism activities cause physiological stress and suppress the immune system (French and others, 2017). Despite the increasing demand on resources to match the growth of the resident human population, tourism and the economy (Benitez-Capistros and others, 2014; Walsh and Mena, 2016; Pizzitutti and others, 2017), which may pose potential threats to marine iguanas, no studies have been undertaken since the first Assessment to assess the impact of pollution through oil spills, agricultural pesticides and plastics on marine iguana populations.

Although programmes to control and eradicate introduced species have been ongoing since the 1980s (Barnett and Rudd, 1983; Carrión, 2016), their effectiveness with regard to marine iguana populations has not been evaluated.

Overall, the improved management and control of immigration, tourism and import of goods since the first Assessment have the potential to reduce the cumulative pressures on marine iguana populations of climate change, pollution, tourism and introduced predators, but they require continuous attention to reduce ongoing population declines (Dirección del Parque Nacional Galápagos, 2014; Asamblea Nacional de la República del Ecuador, 2015; MacLeod and others, 2020).

5. Economic and social consequences of the changes to marine reptile populations

Little has been published on the economic and social consequences of changing marine reptile populations, and there is limited information on the economic and social role of sea snakes, in particular, in many regions of their global range. Balancing economic growth through tourism and the protection of marine reptile populations, especially in the case of the marine iguana, remains a major challenge.

Increases in several populations of the green turtle have resulted in growing interest in exploring whether legal harvesting could be conducted, or expanded, in a sustainable manner, in particular for groups that have cultural or subsistence reasons to harvest (Chaloupka and Balazs, 2007; Rees and others, 2016).

The reliance of fishers from developing countries on income derived from sea snakes caught as by-catch (e.g., Van Cao and others, 2014) is poorly understood. However, high by-catch rates of sea snakes may be a source of income in increasingly unprofitable coastal fisheries in South and South-East Asia (Lobo and others, 2010).

The increased exposure of and research into sea snakes since the first Assessment have led to growing interest from the public in many locations, resulting in the establishment of long-term sea snake data collection programmes based on citizen science (e.g., Goiran and Shine, 2019). The increased reporting by the public of stranded sea snakes has enabled the collection of data on sea snake health that can provide insight into causes of stranding events and changes in distributions (Udyawer and others, 2018).

6. Key knowledge and capacity-building gaps

6.1. Marine turtles

As highlighted in the first Assessment, variability in the demographics of various subpopulations of marine turtles and the threatening processes having an impact on populations highlight the need for ongoing assessments of species and regional subpopulations. A recent review concluded that key knowledge gaps for informing the management of marine turtle populations remain (Rees and others, 2016). Broadly speaking, knowledge is lacking with regard to reproductive biology, including nest selection, hatchling fitness and production; foraging habitats, including connectivity among habitats; demographics; disease pathogenesis; and population-level risks associated with such threats as pollution, by-catch, climate change and the potential unintended consequences of associated mitigation measures.

6.2. Sea snakes

There is a lack of fundamental information on and long-term monitoring of sea snakes across most of their global range. A recent survey of experts identified key knowledge gaps for establishing baselines and progressing the management of sea snake populations (Udyawer and others, 2018). Broadly speaking, knowledge is lacking with regard to geographical distributions, including the movements, dispersal and connectivity of populations; the identification of key habitats, in particular in coastal regions; and quantifying resilience to environmental disturbances (e.g., marine heatwaves and coral bleaching) and the responses to such threats as by-catch and climate change (Fry and others, 2001; Gillett and others, 2014; Heatwole and others, 2016).

It is also currently unclear how emerging threats such as pollutants might be influencing population health (Rezaie-Atagholipour and others, 2012; Sereshk and Bakhtiari, 2015; Goiran and others, 2017).

Increasing reports of deceased turtleheaded sea snakes (*Emydocephalus annulatus*) within protected lagoons in New Caledonia, France, with no obvious cause of death, have highlighted the need for understanding the prevalence of and susceptibility to disease and potential interaction with climate change (Udyawer and others, 2018).

Given the variety of actors likely to have contact with sea snakes (e.g., various industries and recreational beach and ocean users) and potential hazards (e.g., envenomation), potential opportunities for public education and monitoring exist. The collection of data by the public, either opportunistically or as part of citizen science programmes, range from reporting stranded marine snakes (Gillett and others, 2017; Gillett, 2017) to more involved, structured and repeated surveys (Goiran and Shine, 2019).

6.3. Marine iguana

A lack of recent data on the abundance of marine iguana subspecies limits any evaluation of population trends in relation to threats and management actions. Recent advances in population genetics and taxonomy of the marine iguana may guide future conservation research and management aims. Furthermore, building local capacity and allocating resources for comprehensive long-term monitoring could facilitate the assessment of population trends and vulnerability of the marine iguana in the future.

References

Asamblea Nacional de la República del Ecuador (2015). LOREG: Ley Orgánica del Régimen Especial de Galápagos. www.turismo.gob.ec/wp-content/uploads/2016/04/LOREG-11-06-2015.pdf.

Barnett, Bruce D., and Robert L. Rudd (1983). Feral dogs of the Galapagos Islands: impact and control. *International Journal for the Study of Animal Problems*, vol. 4, No. 1.

Bell, I.P., and others (2020). Twenty-eight years of decline: nesting population demographics and trajectory of the north-east Queensland endangered hawksbill turtle (*Eretmochelys imbricata*). *Biological Conservation*, vol. 241, 108376. https://doi.org/10.1016/j.biocon.2019.108376.

Benitez-Capistros, Francisco, and others (2014). Environmental impacts on the Galapagos Islands: identification of interactions, perceptions and steps ahead. *Ecological Indicators*, vol. 38, pp. 113–23. https://doi.org/10.1016/j.ecolind.2013.10.019.

Broderick, A, and Ana Patricio (2019). Green Turtle: *Chelonia mydas (South Atlantic subpopulation)*. In *The IUCN Red List of Threatened Species 2019*. e.T142121866A142086337. https://doi.org/10.2305/IUCN.UK.2019-2.RLTS.T142121866A142086337.en.

Buzás, Balázs, and others (2018). The sea snakes (Elapidae: Hydrophiinae) of Fujairah. *Tribulus*, vol. 26.

Carrión, Víctor (2016). Control y erradicación de animales introducidos: el peligro de las especies invasoras – Parte I: Animales. 4 October 2016. www.carlospi.com/galapagospark/parque_nacional_especies_invasoras_animales.html.

Casale, P. (2015a). *Caretta caretta* (Mediterranean subpopulation). In *The IUCN Red List of Threatened Species 2015*. e.T83644804A83646294.

________ (2015b). *Caretta caretta* (North West Indian Ocean subpopulation). In *The IUCN Red List of Threatened Species 2015*. e.T84127873A84127992.

Casale, P., and M.A. Marcovaldi (2015). *Caretta caretta* (South West Atlantic subpopulation). In *The IUCN Red List of Threatened Species 2015*. e.T84191235A84191397.

Casale, P., and Y. Matsuzawa (2015). *Caretta caretta* (North Pacific subpopulation). In *The IUCN Red List of Threatened Species 2015*. e.T83652278A83652322.

Casale, P., and A.D. Tucker (2017). *Caretta caretta*. In *The IUCN Red List of Threatened Species 2017*. e. 3897A119333622.

Casale, P., and others (2018). Mediterranean sea turtles: current knowledge and priorities for conservation and research. *Endangered Species Research*, vol. 36, pp. 229–267.

Ceriani, S.A., and A. Meylan (2017). *Caretta caretta* (North West Atlantic subpopulation) (amended version of 2015 assessment). In *The IUCN Red List of Threatened Species 2017*. e.T84131194A119339029.

Chaloupka, M., and George Balazs (2007). Using Bayesian state-space modelling to assess the recovery and harvest potential of the Hawaiian green sea turtle stock. *Ecological Modelling*, vol. 205, Nos. 1 and 2, pp. 93–109.

Chaloupka, M., and N.J. Pilcher (2019). *Chelonia mydas (Hawaiian subpopulation)*. In *The IUCN Red List of Threatened Species 2019*. e.T16285718A142098300.

Colman, Liliana P., and others (2019). Thirty years of leatherback turtle *Dermochelys coriacea* nesting in Espírito Santo, Brazil, 1988–2017: reproductive biology and conservation. *Endangered Species Research*, vol. 39, pp. 147–158.

D'anastasi, B.R., and others (2016). New range and habitat records for threatened Australian sea snakes raise challenges for conservation. *Biological Conservation*, vol. 194, pp. 66–70.

Deepwater Horizon Natural Resource Damage Assessment Trustees (2016). *Deepwater Horizon oil spill: Final Programmatic Damage Assessment and Restoration Plan and Final Programmatic Environmental Impact Statement.* Retrieved from www.gulfspillrestoration.noaa.gov/restoration-planning/gulf-plan.

Dirección del Parque Nacional Galápagos (2014). *Plan de Manejo de las Áreas Protegidas de Galápagos para el Buen Vivir.*

Frankham, Richard, and others (2014). Genetics in conservation management: revised recommendations for the 50/500 rules, Red List criteria and population viability analyses. *Biological Conservation*, vol. 170, pp. 56–63.

French, Susannah S., and others (2017). Too much of a good thing? Human disturbance linked to ecotourism has a "dose-dependent" impact on innate immunity and oxidative stress in marine iguanas, *Amblyrhynchus cristatus*. *Biological Conservation*, vol. 210, pp. 37–47.

Fry, G.C., and others (2001). The reproductive biology and diet of sea snake bycatch of prawn trawling in northern Australia: characteristics important for assessing the impacts on populations. *Pacific Conservation Biology*, vol. 7, No. 1, pp. 55–73.

Fuentes, M.M.P.B., and J.E. Cinner (2010). Using expert opinion to prioritize impacts of climate change on sea turtles' nesting grounds. *Journal of Environmental Management*, vol. 91, No. 12, pp. 2511–2518.

Gallaway, Benny J., and others (2016). Evaluation of the status of the Kemp's ridley sea turtle after the 2010 Deepwater Horizon oil spill. *Gulf of Mexico Science*, vol. 33, No. 2, pp. 192–205.

Ganesh, S.R., and others (2019). Marine snakes of Indian coasts: historical resume, systematic checklist, toxinology, status, and identification key. *Journal of Threatened Taxa*, vol. 11, No. 1, pp. 13132–13150.

Gillett, Amber K. (2017). An investigation into the stranding of Australian sea snakes.

Gillett, Amber K., and others (2014). An antemortem guide for the assessment of stranded Australian sea snakes (Hydrophiinae). *Journal of Zoo and Wildlife Medicine*, vol. 45, No. 4, pp. 755–765.

Gillett, Amber K., and others (2017). Postmortem examination of Australian sea snakes (Hydrophiinae): Anatomy and common pathologic conditions. *Journal of Veterinary Diagnostic Investigation*, vol. 29, No. 5, pp. 593–611.

Goiran, C., and others (2017). Industrial melanism in the seasnake *Emydocephalus annulatus*. *Current Biology*, vol. 27, No. 16, pp. 2510–2513.

Goiran, C., and Richard Shine (2019). Grandmothers and deadly snakes: an unusual project in "citizen science". *Ecosphere*, vol. 10, No. 10. e02877.

Groom, Rachel A., and others (2017). Estimating long-term trends in abundance and survival for nesting flatback turtles in Kakadu National Park, Australia. *Endangered Species Research*, vol. 32, pp. 203–211.

Hays, Graeme C., and others (2014). Different male vs. female breeding periodicity helps mitigate offspring sex ratio skews in sea turtles. *Frontiers in Marine Science*, vol. 1, art. 43.

Heatwole, Harold, and others (2016). Physiological, ecological, and behavioural correlates of the size of the geographic ranges of sea kraits (Laticauda; Elapidae, Serpentes): A critique. *Journal of Sea Research*, vol. 115, pp. 18–25.

Howard, Robert, and others (2014). Thermal tolerances of sea turtle embryos: current understanding and future directions. *Endangered Species Research*, vol. 26, No. 1, pp. 75–86.

Jensen, Michael P., and others (2016). Spatial and temporal genetic variation among size classes of green turtles (*Chelonia mydas*) provides information on oceanic dispersal and population dynamics. *Marine Ecology Progress Series*, vol. 543, pp. 241–256.

Jensen, Michael P., and others (2018). Environmental warming and feminization of one of the largest sea turtle populations in the world. *Current Biology*, vol. 28, No. 1, pp. 154–159.

Laloë, Jacques-Olivier, and others (2016). Sand temperatures for nesting sea turtles in the Caribbean: Implications for hatchling sex ratios in the face of climate change. *Journal of Experimental Marine Biology and Ecology*, vol. 474, pp. 92–99.

Laloë, Jacques-Olivier, and others (2017). Climate change and temperature-linked hatchling mortality at a globally important sea turtle nesting site. *Global Change Biology*, vol. 23, No. 11, pp. 4922–4931.

Lauritsen, Ann Marie, and others (2017). Impact of the Deepwater Horizon oil spill on loggerhead turtle *Caretta caretta* nest densities in northwest Florida. *Endangered Species Research*, vol. 33, pp. 83–93.

Lewison, Rebecca L., and others (2014). Global patterns of marine mammal, seabird, and sea turtle bycatch reveal taxa-specific and cumulative megafauna hotspots. *Proceedings of the National Academy of Sciences*, vol. 111, No. 14, pp. 5271–5276.

Lillywhite, Harvey B., and others (2017). Why are there no sea snakes in the Atlantic? *BioScience*, vol. 68, No. 1, pp. 15–24.

Limpus, C.J., and others (2013). Monitoring of Coastal Sea Turtles: Gap Analysis 1. Loggerhead turtles, Caretta caretta, in the Port Curtis and Port Alma Region, Australia. Report produced for the Ecosystem Research and Monitoring Program Advisory Panel as part of Gladstone Ports Corporation's Ecosystem Research and Monitoring Program.

Limpus, C.J., and others (2017). Estimation of population size and comparison of the benefits of mid-season census and whole of breeding season census of flatback turtle reproduction in eastern Australia. Report produced for the Ecosystem Research and Monitoring Program Advisory Panel as part of Gladstone Ports Corporation's Ecosystem Research and Monitoring Program.

Lobo, Aaron Savio, and others (2010). Commercializing bycatch can push a fishery beyond economic extinction. *Conservation Letters*, vol. 3, No. 4, pp. 277–285.

Lukoschek, Vimoksalehi (2018). Congruent phylogeographic patterns in a young radiation of live-bearing marine snakes: Pleistocene vicariance and the conservation implications of cryptic genetic diversity. *Diversity and Distributions*, vol. 24, No. 3, pp. 325–340.

Lynch, Jennifer M. (2018). Quantities of marine debris ingested by sea turtles: global meta-analysis highlights need for standardized data reporting methods and reveals relative risk. *Environmental Science & Technology*, vol. 52, No. 21, pp. 12026–12038.

MacLeod, Amy, and others (2020). *Amblyrhynchus cristatus. The IUCN Red List of Threatened Species* 2020: e.T1086A499235.

MacLeod, Amy, and Sebastian Steinfartz (2016). The conservation status of the Galápagos marine iguanas, *Amblyrhynchus cristatus*: a molecular perspective. *Amphibia-Reptilia*, vol. 37, No. 1, pp. 91–109.

Mancini, A., and others (2019). *Chelonia mydas (North Indian Ocean subpopulation)*. In *The IUCN Red List of Threatened Species 2019*. e.T142121108A142122995.

Maurer, Andrew S., and others (2015). Sargassum accumulation may spell trouble for nesting sea turtles. *Frontiers in Ecology and the Environment*, vol. 13, No. 7, pp. 394–395.

Mazaris, Antonios D., and others (2017). Global sea turtle conservation successes. *Science Advances*, vol. 3, No. 9, e 1600730.

Miralles, Aurélien, and others (2017). Shedding light on the Imps of Darkness: an integrative taxonomic revision of the Galápagos marine iguanas (genus *Amblyrhynchus*). *Zoological Journal of the Linnean Society*, vol. 181, No. 3, pp. 678–710.

Montero, Natalie, and others (2018). Influences of the local climate on loggerhead hatchling production in North Florida: implications from climate change. *Frontiers in Marine Science*, vol. 5, art. 262.

Nalovic, Michel, and others (2018). *Sea Turtles in the North-West Atlantic & Caribbean Region: MTSG Annual Regional Report 2018*. Draft report of the IUCN-SSC Marine Turtle Specialist Group.

Nankivell, J.H., and others (2020). A new species of turtle-headed sea Snake (*Emydocephalus*: Elapidae) endemic to Western Australia. *Zootaxa*. https://doi.org/10.11646/zootaxa.4758.1.6.

Nelms, Sarah E., and others (2016). Seismic surveys and marine turtles: an underestimated global threat? *Biological Conservation*, vol. 193, pp. 49–65.

National Marine Fisheries Service (2019). *Recovering Threatened and Endangered Species, FY 2017-2018. Report to Congress.*

Northwest Atlantic Leatherback Working Group (2019). *Dermochelys coriacea* (Northwest Atlantic Ocean subpopulation). *The IUCN Red List of Threatened Species 2019*. e.T46967827A83327767.

Park, Jaejin, and others (2017). Northward dispersal of sea kraits (*Laticauda semifasciata*) beyond their typical range. *PloS One*, vol. 12, No. 6, e 0179871.

Phillott, A.D., and A. Rees, eds. (2018). Sea Turtles in the Middle East and South Asia Region: MTSG Annual Regional Report 2018. Draft report of the IUCN-SSC Marine Turtle Specialist Group.

Pizzitutti, Francesco, and others (2017). Scenario planning for tourism management: a participatory and system dynamics model applied to the Galapagos Islands of Ecuador. *Journal of Sustainable Tourism*, vol. 25, No. 8, pp. 1117–1137.

Putman, Nathan F., and others (2015). Deepwater Horizon oil spill impacts on sea turtles could span the Atlantic. *Biology Letters*, vol. 11, No. 12, 20150596.

Rasmussen, Arne Redsted, and others (2014). Sea snakes in Australian waters (Serpentes: subfamilies Hydrophiinae and Laticaudinae) – a review with an updated identification key. *Zootaxa*, vol. 3869, No. 4, pp. 351–371.

Rees, A.F., and others (2016). Are we working towards global research priorities for management and conservation of sea turtles? *Endangered Species Research*, vol. 31, pp. 337–382.

Rezaie-Atagholipour, Mohsen, and others (2012). Metal concentrations in selected tissues and main prey species of the annulated sea snake (*Hydrophis cyanocinctus*) in the Hara Protected Area, northeastern coast of the Persian Gulf, Iran. *Marine Pollution Bulletin*, vol. 64, No. 2, pp. 416–421.

Rezaie-Atagholipour, Mohsen, and others (2016). Sea snakes (Elapidae, Hydrophiinae) in their westernmost extent: an updated and illustrated checklist and key to the species in the Persian Gulf and Gulf of Oman. *ZooKeys*, No. 622, pp. 129–164.

Riskas, Kimberly A., and others (2018). Evaluating the threat of IUU fishing to sea turtles in the Indian Ocean and Southeast Asia using expert elicitation. *Biological Conservation*, vol. 217, pp. 232–239.

Sanders, Kate L., and others (2012). *Aipysurus mosaicus*, a new species of egg-eating sea snake (Elapidae: Hydrophiinae), with a redescription of *Aipysurus eydouxii* (Gray, 1849). *Zootaxa*, No. 3431, pp. 1–18.

Sanders, Kate L., and others (2013). Multilocus phylogeny and recent rapid radiation of the viviparous sea snakes (Elapidae: Hydrophiinae). *Molecular Phylogenetics and Evolution*, vol. 66, No. 3, pp. 575–591.

Sanders, Kate L., and others (2014). High rates of hybridisation reveal fragile reproductive barriers between endangered Australian sea snakes. *Biological Conservation*, vol. 171, pp. 200–208.

Sarker, Mohammad Abdur Razzaque, and others (2017). Sea snakes of Bangladesh: a preliminary survey of Cox's Bazar District with notes on diet, reproduction, and conservation status. *Herpetological Conservation and Biology*, vol. 12, No. 2, pp. 384–393.

Schuyler, Qamar A., and others (2016). Risk analysis reveals global hotspots for marine debris ingestion by sea turtles. *Global Change Biology*, vol. 22, No. 2, pp. 567–576.

Seminoff, Jeffrey Aleksandr, and others (2015). Status review of the green turtle (*Chelonia mydas*) under the Endangered Species Act. United States.

Sereshk, Zahra Heydari, and Alireza Riyahi Bakhtiari (2015). Concentrations of trace elements in the kidney, liver, muscle, and skin of short sea snake (*Lapemis curtus*) from the Strait of Hormuz Persian Gulf. *Environmental Science and Pollution Research*, vol. 22, No. 20, pp. 15781–15787.

Summers, Tammy M., and others (2018). Endangered Green Turtles (*Chelonia mydas*) of the Northern Mariana Islands: Nesting Ecology, Poaching, and Climate Concerns. *Frontiers in Marine Science*, vol. 4, art. 428.

Tiwari, M., and others (2013). *Dermochelys coriacea* (West Pacific Ocean subpopulation). In *The IUCN Red List of Threatened Species 2013*. e.T46967817A46967821.

Udyawer, Vinay, and others (2018). Future directions in the research and management of marine snakes. *Frontiers in Marine Science*, vol. 5, art. 399.

Udyawer, Vinay, and others (2020). Prioritising search effort to locate previously unknown populations of endangered marine reptiles. *Global Ecology and Conservation*, vol. 22, e01013.

Ukuwela, Kanishka D.B., and others. (2012). *Hydrophis donaldi* (Elapidae, Hydrophiinae), a highly distinctive new species of sea snake from northern Australia. *Zootaxa*, vol. 3201, No. 1, pp. 45–57.

Ukuwela, Kanishka D.B., and others (2013). Molecular evidence that the deadliest sea snake *Enhydrina schistosa* (Elapidae: Hydrophiinae) consists of two convergent species. *Molecular Phylogenetics and Evolution*, vol. 66, No. 1, pp. 262–269.

Ukuwela, Kanishka D. B., and others (2014). Multilocus phylogeography of the sea snake *Hydrophis curtus* reveals historical vicariance and cryptic lineage diversity. *Zoologica Scripta*, vol. 43, No. 5, pp. 472–484.

United Nations (2017a). *The First Global Integrated Marine Assessment: World Ocean Assessment I*. Cambridge: Cambridge University Press.

________ (2017b). Chapter 39: Marine reptiles. In *The First Global Integrated Marine Assessment: World Ocean Assessment I*. Cambridge: Cambridge University Press.

Valdivia, Abel, and others (2019). Marine mammals and sea turtles listed under the US Endangered Species Act are recovering. *PloS One*, vol. 14, issue 1, e0210164

Van Cao, Nguyen, and others (2014). Sea snake harvest in the Gulf of Thailand. *Conservation Biology*, vol. 28, No. 6, pp. 1677–1687.

Wallace, Bryan P., and others (2010). Regional management units for marine turtles: a novel framework for prioritizing conservation and research across multiple scales. *PloS One*, vol. 5, issue 12, e15465.

Wallace, B., and others (2013). *Dermochelys coriacea* East Pacific Ocean subpopulation. In *The IUCN Red List of Threatened Species*. e.T46967807A46967809.

Walsh, Stephen J., and Carlos F. Mena (2016). Interactions of social, terrestrial, and marine sub-systems in the Galapagos Islands, Ecuador. *Proceedings of the National Academy of Sciences*, vol. 113, No. 51, pp. 14536–14543.

Wibbels, T., and E. Bevan (2019). *Lepidochelys kempii*. In *The IUCN Red List of Threatened Species 2019*. e.T11533A142050590.

Wikelski, Martin and others (2002). Galapagos islands: marine iguanas die from trace oil pollution. *Nature*, vol. 417, pp. 607–608.

Wilcox, Chris and others (2018). A quantitative analysis linking sea turtle mortality and plastic debris ingestion. *Scientific Reports*, vol. 8, art. 12536.

Chapter 6F
Seabirds

Contributors: Martin Cryer (convener), Igor Debski, Maria Dias, Karen Evans (subchapter lead member), Carolina Hazin, Chul Park (chapter lead member), Cleo Small and Graeme Taylor.

Keynote points

- Since the first *World Ocean Assessment* (United Nations, 2017), the global conservation status of seabirds has worsened, continuing a long-term trend.
- Thirty-one per cent of species are now threatened with extinction, up from 28 per cent in 2010.
- Pressures related to fishing (by-catch and prey depletion) are now affecting more species, while pollution is affecting fewer species (although marine debris, especially plastics, is an emerging threat with poorly understood consequences).
- Invasive alien species and climate change also remain major causes of seabird decline and affect a number of species similar to that in 2010.
- Current capacity and resources limit the ability to assess consequences at the population level and the implications for ecosystem services of existing and emerging threats.

1. Introduction

Seabirds (defined as bird species of which a large proportion of the population relies on the marine environment for at least part of the year (Croxall and others, 2012)) play an important role in the world's marine ecosystems, being top predators that consume about the same amount of biomass as all fisheries combined (Brooke, 2004). Seabirds occur across all oceans, from coastal areas to the high seas, and many species are highly migratory, connecting different marine systems or ocean basins (Croxall and others, 2005; Shaffer and others, 2006; Egevang and others, 2010; Dias and others, 2011).

A total of 359 species are identified under current taxonomy, representing six orders and 12 families. The worldwide distribution of species (by countries) was summarized by Croxall and others (2012) in terms of species richness, number of endemics and number of threatened species. Seabirds are relatively well studied, compared with most other marine taxa, and several assessments documenting the status and recent trends of specific taxonomic groups have been conducted since the first Assessment (Trathan and others, 2015; Phillips and others, 2016; Rodríguez and others, 2019).

The first Assessment reported that 97 species of seabirds were classified as threatened, to varying degrees (namely, those species that were classified as Critically Endangered, Endangered or Vulnerable in the 2010 International Union for Conservation of Nature (IUCN) *Red List of Threatened Species*), which represented 28 per cent of the 346 species evaluated at the time. The first Assessment also highlighted that pelagic seabird species were particularly threatened and that albatrosses (family Diomedeidae), gadfly petrels (family Procellariidae, genera *Pterodroma* and *Pseudobulweria*) and penguins (family Spheniscidae) were the groups with the highest percentages of species in threatened categories in the IUCN Red List. The first Assessment concluded that the decline in seabird populations was the result of 10 primary pressures. At sea, those included: incidental by-catch (in longline, gill-net and trawl fisheries); pollution (from oil spills and marine debris, including plastics); depletion of prey species by fishing; and offshore energy production and mining. On land, the main threats were invasive alien species; problematic native species (e.g., those that have become superabundant); human disturbance; industrial and residential development; and hunting and trapping. Climate change and severe weather were identified as affecting seabirds both on land and at sea.

2. Description of environmental changes between 2010 and 2020

The number of species in each IUCN Red List category in 2018, by seabird order, is shown in table 1. Since then, a quantitative review of the threats affecting all seabird species globally has been conducted using data collated from more than 900 publications and a standardized assessment approach based on the IUCN Red List Threats Classification Scheme (IUCN, 2019) (Dias and others, 2019).

In their review, Dias and others (2019) adopted an approach similar to that used by Croxall and others (2012), so that their results can be used to evaluate changes in the status of, and threats to, seabirds since the first Assessment. Since the first Assessment, 28 species of seabirds have been uplisted (that is, their conservation status has worsened) and 11 species have been downlisted (that is, their conservation status has improved) (table 2). Particularly noticeable has been the declining status of species within the orders Anseriformes (sea ducks), with 5 of 18 species uplisted, and Procellariiformes (tubenoses), with 11 of 131 species uplisted and 4 downlisted. Procellariiformes (in particular albatrosses and gadfly petrels) and penguins remain the groups with the highest proportions of threatened species (see table 1). The downlisting of species since the first Assessment has resulted from improved knowledge (e.g., new colonies discovered and taxonomy revision) rather than a genuine improvement in their status.

Table 1
Number of seabird species (346), by order, in each IUCN Red List category in 2018 (considered by Croxall and others (2012) and Dias and others (2019))

Seabird order	EX	CR	EN	VU	NT	LC	DD	Total
Procellariiformes (tubenoses)	2	13	20	27	19	47	3	131
Sphenisciformes (penguins)	0	0	5	5	3	5	0	18
Charadriiformes (gulls and auks)	1	1	4	10	11	93	0	120
Anseriformes (sea ducks)	0	0	0	4	2	12	0	18
Suliformes (gannets and boobies)	0	2	5	8	3	26	0	44
Gaviiformes (loons)	0	0	0	0	1	4	0	5
Phaethontiformes (tropicbirds)	0	0	0	0	0	3	0	3
Pelecaniformes (pelicans)	0	0	0	0	1	2	0	3
Podicipediformes (grebes)	0	0	0	1	0	3	0	4
Total	3	16	34	55	40	195	3	346

Abbreviations: EX, Extinct; CR, Critically Endangered; EN, Endangered; VU, Vulnerable; NT, Near Threatened; LC, Least Concern; DD, Data Deficient.

A comparison of the reviews by Dias and others (2019) and Croxall and others (2012) also identified that such factors as invasive alien species, climate change and severe weather continue to affect a number of globally threatened species similar to that in 2010 (figure I). Dias and others (2019) identified fishery by-catch and prey depletion by fishing as having an impact on more species of seabirds in 2018 than in 2010 (50 globally threatened species, an increase of 10 from 2010, and 22 globally threatened species, an increase of 12 from 2010, respectively). These increases are at least partly the result of a better understanding of the impacts of by-catch on seabirds, in particular in gill-net fisheries (Crawford and others, 2015; see also below) and the effects of competition for prey species between fisheries and seabirds (Crawford and others, 2015; Grémillet and others, 2018; Trathan and others, 2015). Declines in prey species can also be caused by factors other than fishing, including climate change (Mitchell and others, 2020). The major threats affecting threatened species are the same as those affecting all seabird species (Dias and others, 2019).

Climate change has been reported to have already caused declines in almost 100 seabird species (Dias and others, 2019). For example, changes in sea surface temperature during late winter were associated with declines in the population growth rate of the black-browed albatross (*Thalassarche melanophris*), primarily through effects on prey availability and subsequent juvenile survival (Jenouvrier and others, 2018). Similarly, Carroll and others (2015) found that the breeding success at 11 colonies of the black-legged kittiwake (*Rissa tridactyla*) in Ireland and the United Kingdom of Great Britain and Northern Ireland was higher when stratification was weaker before breeding and when sea surface temperatures were lower during the breeding season.

Figure I
Number of globally threatened seabird species assessed as being affected by marine, terrestrial and overarching threatening processes in 2010 and 2018

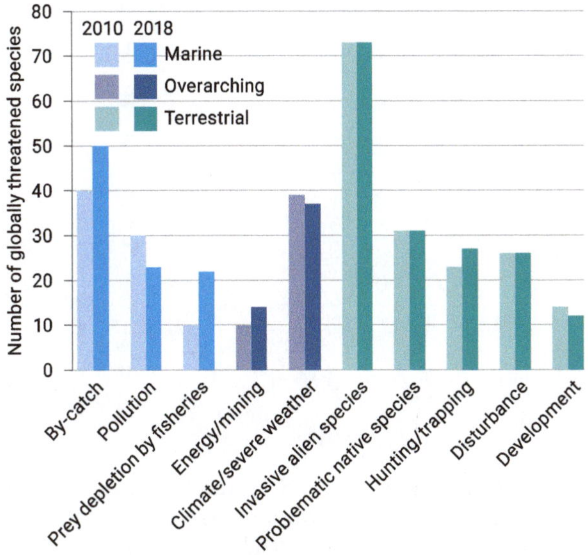

Source: Croxall and others (2012), for 2010 data; Dias and others (2019), for 2018 data.
Note: Only processes considered in both assessments are shown.

Invasive alien species continue to pose a threat to 73 species, the same number as in 2010, with rats and cats being a particular threat to small petrels, such as gadfly petrels and storm petrels (Rodríguez and others, 2019). In their global review, Jones and others (2016) identified 122 species (202 populations) of seabirds that had benefited (in terms of an increase in population size or colonization of a location) from the eradication of invasive mammals from islands. Other programmes to restore or improve habitat for seabirds have also been found to be beneficial, including planting and reforestation, weed control, enhancing or providing nesting opportunities and erosion control (Beck and others, 2015; Bried and Neves, 2015; Buxton and others, 2016).

Table 2
Summary of changes in the IUCN Red List status of seabird species between 2010 and 2018[a]

Seabird order	Uplisted	Unchanged	Downlisted	Data deficient	Total
Procellariiformes	11	112	4	4	131
Sphenisciformes	1	15	2	0	18
Charadriiformes	8	108	4	0	120
Anseriformes	5	13	0	0	18
Suliformes	2	41	1	0	44
Gaviiformes	0	5	0	0	5
Phaethontiformes	0	3	0	0	3
Pelecaniformes	0	3	0	0	3
Podicipediformes	1	3	0	0	4
Total	28	303	11	4	346

Source: Croxall and others (2012) for 2010; Dias and others (2019) for 2018.
Note: "Uplisting" means a worse conservation status in 2018 than in 2010, while "downlisting" means a better conservation status. Species are considered data deficient if they were so categorized in the IUCN Red List in either 2010 or 2018 because no meaningful assessment of change in status could be made.
[a] The comparison is limited to the 346 species of seabirds considered by both authors.

Fishery by-catch remains the biggest at-sea threat to Procellariiformes seabirds, affecting mostly albatrosses, large petrels and shearwaters (Phillips and others, 2016; Rodríguez 2019). Target 14.2 of the Sustainable Development Goals[1] provides that marine and coastal ecosystems should be sustainably managed and protected by 2020 to avoid significant adverse impacts; and impacts on biodiversity will need to be taken into account in order to make progress towards Goal 12. Efforts to reduce the by-catch of seabirds in fisheries have been increasing, including through the adoption or updating of national plans of action by some nations using longline, trawl or gill-net fishing methods, in which by-catch is most commonly an issue. Mandatory mitigation measures have also been introduced within some areas of national jurisdiction and some parts of the high seas, including, for example, line weighting, night setting, bird-scaring lines and area closures (Brothers and others, 1999; Abraham and others, 2017). There is evidence that the number of threatened seabird species affected by the depletion of their prey by fisheries has more than doubled in the past decade (Croxall and others, 2012; Dias and others, 2019), although this increase is at least in part due to increased understanding in that area.

In contrast, the most recent assessment identified that threats associated with marine pollution have decreased, with pollution now affecting 23 globally threatened species (7 fewer than in 2010). This decrease is primarily driven by overall reductions in pollution associated with oil spills in recent decades (Roser, 2013). Pollution in the form of marine plastics has been documented as affecting seabird species on a wide scale (e.g., Wilcox and others, 2015). Despite Sustainable Development Goal target 14.1, which provides for the prevention and significant reduction of marine

[1] See General Assembly resolution 70/1.

pollution of all kinds by 2025, plastic in the marine environment is expected to continue to affect many seabird species in the coming decades (Kühn and others, 2015; Ryan and others, 2009; Wilcox and others, 2015). Although this form of pollution has not yet been identified as a direct cause of many declines at the population level (but see Auman and others (1997) and Lavers and others (2014)), small, highly pelagic species such as storm petrels, prions and auklets (Roman and others, 2019; Wilcox and others, 2015) are most likely to be at risk. Light pollution, both at the colonies (Rodríguez and others, 2017; Rodríguez and others, 2019) and from oil platforms, vessels and other artificial structures at sea, poses a threat to small petrels (Montevecchi, 2006; Rodríguez and others, 2019), although with poorly understood impacts on populations. This threat was not considered in the first Assessment.

Other emerging threats identified by Dias and others (2019) include energy production, in particular offshore wind farms, deep-sea mining (Green and others, 2016) and light pollution, including from such marine infrastructure as platforms and vessels (Rodríguez and others, 2017; Rodríguez and others, 2019). Understanding the consequences of those threats at the population level remains limited, but juvenile seabirds and birds near colonies appear to be particularly susceptible to light pollution (Rodríguez and others, 2015). Such adverse effects as collision and mortality have been described for at least 21 species of Procellariiformes, including as a result of birds being attracted to the artificial light of fishing and offshore oil and gas facilities (Montevecchi, 2006). Development in those areas is relevant to the achievement of Sustainable Development Goal target 15.9, which provides that ecosystem and biodiversity values should be integrated into national and local planning and development processes by 2020 (noting also the relevance of Goals 7, 9 and 11 in this area).

Although the population sizes of some penguin species are increasing, climate change has been identified as a major threat to many species in the group, with declines primarily associated with changes in habitat conditions, increased frequency of floods and storms and extreme temperatures (Trathan and others, 2015; Dias and others, 2019). Sustainable Development Goal target 13.2 provides that climate change measures should be integrated into national policies, strategies and planning, even though it does not include a target date. By-catch, competition with fisheries, pollution, invasive alien species and disturbance at the colonies are also now known to be important stressors for penguin species (Trathan and others, 2015; Crawford and others, 2017; Dias and others, 2019).

3. Consequences of changes in seabird populations on human communities, economies and well-being

Changes in seabird populations, in particular substantial declines, have an impact on biodiversity and the associated functioning of marine systems and the ecosystem services that they provide (Wenny and others, 2011; Burdon and others, 2017; Tavares and others, 2019). For example, seabirds feeding at sea and nesting ashore can contribute a substantial proportion of nutrients entering the latter system, enhancing the productivity of local fauna and flora and adjacent coastal systems (Graham and others, 2018). Changes in this nutrient transfer would profoundly affect those systems. The potential consequences of changes in seabird populations for ecosystem services, although poorly understood, are likely to be varied and complex. Thus, the consequential impacts on ecosystem services of changes in seabird populations are directly relevant to many of the Sustainable Development Goals, including Goals 7 (affordable and clean energy), 9 (industry, innovation and infrastructure), 11 (sustainable cities and communities), 12 (responsible consumption and production), 13 (climate action), 14 (life below water) and 15 (life on land).

4. Outlook

A continuing long-term decline in the status of seabird populations, especially for pelagic species over the last decade (figure II), and the persistence of major threats do not provide a positive outlook for seabirds in the near future.

Current efforts to mitigate the impacts of fishery by-catch and invasive species, in particular in island habitats, will likely continue, as their importance for the conservation of biodiversity and seabirds is increasingly recognized and prioritized (Buxton and others, 2016; Jones and others, 2016). However, if fishing pressure on forage fish intensifies, competition between fisheries and seabirds may increase with associated potential detrimental impacts on some seabird populations, even though the empirical evidence for a consistent effect is not strong (Hilborn and others, 2017). The impacts of potentially increasing competition may be exacerbated by any decreases in prey abundance related to changes in oceanographic conditions driven by climate change (Grémillet and Boulinier, 2009). In this context, the transition of fisheries to lower trophic levels, especially mesopelagic species (St. John and others, 2016), may be particularly problematic, because mesopelagic fishes are an important part of the diet of many pelagic seabirds (Watanuki and Thiebot, 2018).

Climate change is projected to have profound implications for many seabird populations through potential redistribution of prey and changes to the composition of marine communities. Direct climate change effects are likely to: increase heat stress at breeding and roosting colonies; increase disturbance to breeding and roosting colonies through increasing storm frequency and intensity, in particular in low-lying colony regions; and increase inundation of low-lying nesting and feeding areas through sea level rise (Grémillet and Boulinier 2009). Groups such as penguins are particularly vulnerable to the negative consequences of climate change (Dias and others, 2019), especially those species dependent on pack ice or particular habitat conditions that are likely to decline under climate change (Ainley and others, 2010). Although many more species are projected to be affected negatively by climate change (BirdLife International and National Audubon Society, 2015), some species are also projected to be positively affected through increases in range or population size. For example, the populations of the king penguin (*Aptenodytes patagonicus*) and the black-browed albatross nesting on Heard Island (Australia) are expected to increase (Chambers and others, 2011) as a result of an increase in the extent of breeding grounds, as glaciers retreat.

As the use of renewable energy (wind, water and waves) technology continues to expand globally, the risk of seabirds interacting with related structures is likely to increase. These interactions are likely to occur predominantly with coastal species, such as divers, scoters, terns and shags (Garthe and Hüppop, 2004) and may be more problematic for highly mobile species, such as shearwaters (Busch and Garthe, 2018). The potential for increasing impacts could be reduced by locating facilities in areas less favoured by seabirds, identified through the use of observation and tracking data on habitat use (e.g., Busch and others, 2013; Winship and others, 2018).

Although the number of major oil spills has decreased over the past decades, other aspects of pollution, including marine debris, are likely to increase, in particular in view of predicted large increases in the amount of plastic waste in the marine environment (Jambeck and others, 2015). Impacts associated with light pollution are also expected to grow, largely associated with continued growth in maritime traffic, which is expanding at a rate of about 4 per cent per year (United Nations Conference on Trade and Development, 2018). Endangered species and species with small population sizes are more likely to be affected (Rodríguez and others, 2019). Growing maritime traffic may also increase the risk of introducing alien predators or pathogens to additional locations, potentially affecting seabird populations (Renner and others, 2018).

New tracking technology for seabirds (see, e.g., Sansom and others, 2018; Zhang and others, 2019) and sophisticated mapping and analytical software will make it possible to estimate the exposure of seabird populations to individual threats with increasing precision. This should enable the identification of the species, life stages, threats, places and times where mitigation will bring the most benefit for seabird populations, as well as the pinpointing of critical knowledge gaps. Quantitative spatial overlap approaches have been used mostly to assess impacts and risks posed by fishing (see, e.g., Tuck and others, 2011; Abraham and others, 2017; Clay and others, 2019), but it could be used in future assessments of all threats with a spatial component (e.g., as reported by Currey and others, 2012; Redfern and others, 2013).

Figure II
Red List Index of species survival for seabirds for the period 1988–2017

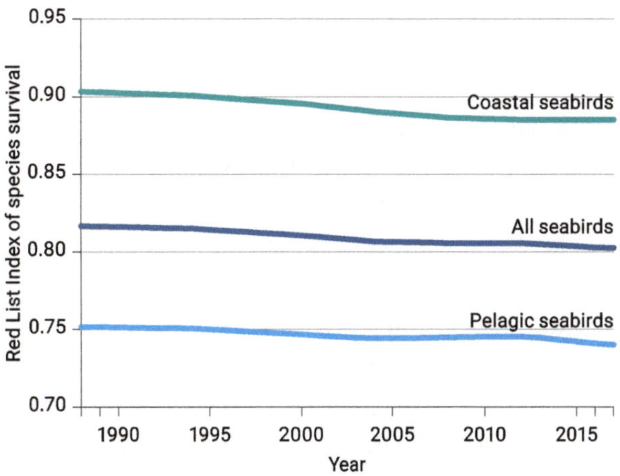

Note: The Red List Index uses information from the IUCN Red List to measure the projected overall survival rates of species within groups (Butchart and others, 2007). It is based on changes in the proportion of species in each threat classification category resulting from genuine improvement or deterioration in the status of each species. The revised Index is scaled so that, for any given group, a value of 1 indicates that all species are classified as of Least Concern, while a value of 0 indicates that all species have gone extinct.

5. Key remaining knowledge gaps

Despite the fact that seabirds are relatively well-studied, several knowledge gaps remain in the demography, status, at-sea distribution and population trends of small species, such as storm petrels, gadfly petrels, prions and auklets. In addition, the at-sea distributions of most seabird species at young life-history stages are poorly understood compared with those of adults. Perhaps the biggest gaps, however, concern the likely population-level consequences, and resultant changes to ecosystem services (and progress towards the Sustainable Development Goal targets), of the impacts of such emerging threats as marine debris (especially plastics), coastal and offshore wind and tidal energy facilities, deep-sea mining and light pollution.

6. Key remaining capacity-building gaps

The key remaining gaps in capacity relate to the knowledge gaps identified above. Current capacity and resources limit our ability to monitor population trends, understand the at-sea distribution of birds at different life stages and estimate the demographics and productivity of all but the most intensively studied species. Those gaps greatly limit our ability to assess population-level consequences and implications for ecosystem services of existing and emerging threats.

References

Abraham, Edward R., and others (2017). Assessment of the risk of southern hemisphere surface longline fisheries to ACAP Species. In ACAP – Eighth Meeting of the Seabird Bycatch Working Group. SBWG8-Doc-07. Wellington.

Ainley, David, and others (2010). Antarctic penguin response to habitat change as Earth's troposphere reaches 2° C above preindustrial levels. *Ecological Monographs*, vol. 80, No. 1, pp. 49–66.

Auman, Heidi J., and others (1997). Plastic ingestion by Laysan Albatross chicks on Sand Island, Midway Atoll, in 1994 and 1995. *Albatross Biology and Conservation*, vol. 239244.

Beck, Jessie, and others (2015). Año Nuevo State Park Seabird Conservation and Habitat Restoration: Report 2015. Oikonos–Ecosystem Knowledge (2015).

BirdLife International, and National Audubon Society (2015). *The Messengers: What Birds Tell Us about Threats from Climate Change and Solutions for Nature and People*. Cambridge: BirdLife International.

Bried, Joël, and Verónica C. Neves (2015). Habitat restoration on Praia islet, Azores archipelago, proved successful for seabirds, but new threats have emerged. *Airo*, vol. 23, pp. 25–35.

Brooke, de L.M. (2004). The food consumption of the world's seabirds. *Proceedings of the Royal Society of London. Series B: Biological Sciences*, vol. 271, suppl. No. 4, pp. S246–S248.

Brothers, Nigel P., and others (1999). *The Incidental Catch of Seabirds by Longline Fisheries: Worldwide Review and Technical Guidelines for Mitigation*. FAO Fisheries Circular, No. 937, pp. 1–100.

Burdon, Daryl, and others (2017). The matrix revisited: a bird's-eye view of marine ecosystem service provision. *Marine Policy*, vol. 77, pp. 78–89.

Busch, Malte, and Stefan Garthe (2018). Looking at the bigger picture: The importance of considering annual cycles in impact assessments illustrated in a migratory seabird species. *ICES Journal of Marine Science*, vol. 75, No. 2, pp. 690–700.

Busch, Malte, and others (2013). Consequences of a cumulative perspective on marine environmental impacts: offshore wind farming and seabirds at North Sea scale in context of the EU Marine Strategy Framework Directive. *Ocean & Coastal Management*, vol. 71, pp. 213–224.

Butchart, Stuart H.M., and others (2007). Improvements to the Red List Index. *PloS One*, vol. 2, No. 1, e 140.

Buxton, Rachel T., and others (2016). Deciding when to lend a helping hand: a decision-making framework for seabird island restoration. *Biodiversity and Conservation*, vol. 25, pp. 467–484.

Carroll, Matthew, and others (2015). Effects of sea temperature and stratification changes on seabird breeding success. *Climate Research*, vol. 66, pp. 75–89.

Chambers, Lynda E., and others (2011). Observed and predicted effects of climate on Australian seabirds. *Emu-Austral Ornithology*, vol. 111, pp. 235–251.

Clay, Thomas A., and others (2019). A comprehensive large-scale assessment of fisheries by-catch risk to threatened seabird populations. *Journal of Applied Ecology*, vol. 56, pp. 1882–1893.

Crawford, Robert J.M., and others (2015). A changing distribution of seabirds in South Africa: the possible impact of climate and its consequences. *Frontiers in Ecology and Evolution*, vol. 3, art. 10.

Crawford, Rory, and others (2017). Tangled and drowned: a global review of penguin by-catch in fisheries. *Endangered Species Research*, vol. 34, pp. 373–396.

Croxall, John P., and others (2005). Global circumnavigations: tracking year-round ranges of nonbreeding albatrosses. *Science*, vol. 307, No. 5707, pp. 249–250.

Croxall, John P., and others (2012). Seabird conservation status, threats and priority actions: a global assessment. *Bird Conservation International*, vol. 22, No. 1, pp. 1–34.

Currey, Rohan J.C., and others (2012). *A Risk Assessment of Threats to Maui's Dolphins*. New Zealand Ministry for Primary Industries and Department of Conservation.

Dias, Maria P., and others (2011). Breaking the routine: individual Cory's shearwaters shift winter destinations between hemispheres and across ocean basins. *Proceedings of the Royal Society B: Biological Sciences*, vol. 278, No. 1713, pp. 1786–1793.

Dias, Maria P., and others (2019). Threats to seabirds: a global assessment. *Biological Conservation*, vol. 237, pp. 525–537.

Egevang, Carsten, and others (2010). Tracking of Arctic terns *Sterna paradisaea* reveals longest animal migration. *Proceedings of the National Academy of Sciences*, vol. 107, No. 5, pp. 2078–2081.

Garthe, Stefan, and Ommo Hüppop (2004). Scaling possible adverse effects of marine wind farms on seabirds: developing and applying a vulnerability index. *Journal of Applied Ecology*, vol. 41, No. 4, pp. 724–734.

Graham, Nicholas A.J., and others (2018). Seabirds enhance coral reef productivity and functioning in the absence of invasive rats. *Nature*, vol. 559, pp. 250–253.

Green, Rhys E., and others (2016). Lack of sound science in assessing wind farm impacts on seabirds. *Journal of Applied Ecology*, vol. 53, No. 6, pp. 1635–1641.

Grémillet, David, and others (2018). Persisting worldwide seabird-fishery competition despite seabird community decline. *Current Biology*, vol. 28, No. 24, pp. 4009–4013.

Grémillet, David, and Thierry Boulinier (2009). Spatial ecology and conservation of seabirds facing global climate change: a review. *Marine Ecology Progress Series*, vol. 391, pp. 121–137.

Hilborn, Ray, and others (2017). When does fishing forage species affect their predators? *Fisheries Research*, vol. 191, pp. 211–221.

International Union for Conservation of Nature and Natural Resources (IUCN) (2019). Threats Classification Scheme (version 2019-3). *The IUCN Red List of Threatened Species*. www.iucnredlist.org/en.

Jambeck, Jenna R., and others (2015). Plastic waste inputs from land into the ocean. *Science*, vol. 347, No. 6223, pp. 768–771.

Jenouvrier, Stéphanie, and others (2018). Climate change and functional traits affect population dynamics of a long-lived seabird. *Journal of Animal Ecology*, vol. 87, pp. 906–920.

Jones, Holly P., and others (2016). Invasive mammal eradication on islands results in substantial conservation gains. *Proceedings of the National Academy of Sciences*, vol. 113, pp. 4033–4038.

Kühn, Susanne, and others (2015). Deleterious effects of litter on marine life. In *Marine Anthropogenic Litter*. Cham, Switzerland: Springer, pp. 75–116.

Lavers, Jennifer L., and others (2014). Plastic ingestion by Flesh-footed Shearwaters (*Puffinus carneipes*): implications for fledgling body condition and the accumulation of plastic-derived chemicals. *Environmental Pollution*, vol. 187, pp. 124–129.

Mitchell, Ian, and others (2020). Impacts of climate change on seabirds, relevant to the coastal and marine environment around the United Kingdom. *MCCIP Science Review 2020*, pp. 382–399. https://doi.org/10.14465/2020.arc17.sbi.

Montevecchi, William A. (2006). Influences of Artificial Light on Marine Birds. In *Ecological Consequences of Artificial Night Lighting*, pp. 94–113.

Phillips, Richard A., and others (2016). The conservation status and priorities for albatrosses and large petrels. *Biological Conservation*, vol. 201, pp. 169–183.

Redfern, J.V., and others (2013). Assessing the risk of ships striking large whales in marine spatial planning. *Conservation Biology*, vol. 27, No. 2, pp. 292–302.

Renner, Martin, and others (2018). The risk of rodent introductions from shipwrecks to seabirds on Aleutian and Bering Sea islands. *Biological Invasions*, vol. 20, No. 9, pp. 2679–2690.

Rodríguez, Airam, and others (2015). GPS tracking for mapping seabird mortality induced by light pollution. *Scientific Reports*, vol. 5, art. 10670. https://doi.org/10.1038/srep10670.

Rodríguez, Airam, and others (2017). Seabird mortality induced by land-based artificial lights. *Conservation Biology*, vol. 31, No. 5, pp. 986–1001.

Rodríguez, Airam, and others (2019). Future directions in conservation research on petrels and shearwaters. *Frontiers in Marine Science*, vol. 6, art. 94.

Roman, Lauren, and others (2019). Ecological drivers of marine debris ingestion in Procellariiform Seabirds. *Scientific Reports*, vol. 9, art. 916.

Roser, Max (2013). Our World in Data. Oil Spills. Available at https://ourworldindata.org/oil-spills.

Ryan, Peter G., and others (2009). Monitoring the abundance of plastic debris in the marine environment. *Philosophical Transactions of the Royal Society B: Biological Sciences*, vol. 364, No. 1526, pp. 1999–2012.

Sansom, Alex, and others (2018). Comparing marine distribution maps for seabirds during the breeding season derived from different survey and analysis methods. *PloS one*, vol. 13, No. 8, e0201797.

Shaffer, Scott A., and others (2006). Migratory shearwaters integrate oceanic resources across the Pacific Ocean in an endless summer. *Proceedings of the National Academy of Sciences*, vol. 103, No. 34, pp. 12799–12802.

St. John, Michael A., and others (2016). A dark hole in our understanding of marine ecosystems and their services: perspectives from the Mesopelagic community. *Frontiers in Marine Science*, vol. 3, art. 31.

Tavares, Davi Castro, and others (2019). Traits shared by marine megafauna and their relationships with ecosystem functions and services. *Frontiers in Marine Science*, vol. 6, art. 262.

Trathan, Phil N., and others (2015). Pollution, habitat loss, fishing, and climate change as critical threats to penguins. *Conservation Biology*, vol. 29, No. 1, pp. 31–41.

Tuck, Geoffrey N., and others (2011). An assessment of seabird: fishery interactions in the Atlantic Ocean. *ICES Journal of Marine Science*, vol. 68, No. 8, pp. 1628–1637.

United Nations (2017). *The First Global Integrated Marine Assessment: World Ocean Assessment I*. Cambridge: Cambridge University Press.

United Nations Conference on Trade and Development (2018). *Review of Maritime Transport*. United Nations publication.

Watanuki, Yutaka, and Jean-Baptiste Thiebot (2018). Factors affecting the importance of myctophids in the diet of the world's seabirds. *Marine Biology*, vol. 165, No. 4, art. 79.

Wenny, Daniel G., and others (2011). The Need to Quantify Ecosystem Services Provided by Birds. *The Auk*, vol. 128, No. 1, pp. 1–14.

Wilcox, Chris, and others (2015). Threat of plastic pollution to seabirds is global, pervasive, and increasing. *Proceedings of the National Academy of Sciences*, vol. 112, No. 38, pp. 11899–11904.

Winship, Arliss J., and others (2018). *Modeling At-Sea Density of Marine Birds to Support Atlantic Marine Renewable Energy Planning: Final Report*. OCS Study BOEM 2018-010. Sterling, Virginia: United States Department of the Interior, Bureau of Ocean Energy Management, Office of Renewable Energy Programs.

Zhang, Jingjing, and others (2019). GPS telemetry for small seabirds: using hidden Markov models to infer foraging behaviour of common diving petrels (*Pelecanoides urinatrix urinatrix*). *Emu — Austral Ornithology*, vol. 119, No. 2, pp. 126–137.

Chapter 6G
Marine plants and macroalgae

Contributors: Hilconida Calumpong, Hugh Kirkman and Nair Sumie Yokoya (joint conveners), Jason M. Hall-Spencer, Nahid Abdel Rahim Osman, Chul Park (chapter lead member), Franciane Pellizzari and Elizabeth Sinclair.

Keynote points

- Some 90 per cent of mangrove, seagrass and marsh plant species have been assessed as being at risk of extinction; 19 per cent of mangroves, 21 per cent of seagrass species and one marsh plant species are on the International Union for Conservation of Nature (IUCN) *Red List of Threatened Species*.
- Among macroalgae, 1 species of red seaweed from Australia (*Vanvoorstia bennettiana*) is listed as Extinct, 10 species (six red algae and four brown algae) are listed as Critically Endangered, 1 species of brown alga is listed as Endangered, and 4 species (three red algae and one brown alga) are listed as Vulnerable. The number of macroalgal species assessed and reported in the IUCN Red List is less than 1 per cent of the total number of species listed in the Ocean Biodiversity Information System (OBIS). All 15 threatened species are endemic to the Galapagos Islands, and 47 species have been assessed as at a risk of extinction in the Mediterranean. This highlights the knowledge gap with regard to macroalgae.
- In terms of macroalgal endemism, Antarctica ranks highest, with 27 per cent endemics, followed by South America (22 per cent) and the Red Sea Large Marine Ecosystem (9 per cent).
- New techniques such as genomics have been developed for species identification and for elucidating phylogenetic relationships. As a result, the number of species is expected to rise, especially for macroalgae; however, owing to uneven human and infrastructure capacities among regions, some regions will be less studied than others.

1. Introduction

The present chapter considers the taxonomy, conservation status and population trends of marine plants and focuses specifically on mangroves, salt marsh plants, seagrasses and macroalgae (or seaweeds), which comprise red, green and brown macroalgae. Although mangroves, salt marshes and seagrass meadows were considered individually in chapters 48, 49 and 47, respectively, of the first *World Ocean Assessment* (United Nations, 2017), and are covered in chapters 7H, 7I and 7G, respectively, of the present Assessment, they were considered at the ecosystem scale, not at the species level. Macroalgae were also covered in the first Assessment, but as a source of food (in chap. 14) and as ecosystems (in chap. 47, on kelp forests, and chap. 50, on the Sargasso Sea).

2. Mangroves

Mangroves consist of shrubs and trees that grow in the coastal belt in tropical and subtropical areas worldwide. They have developed characteristic features that allow them to survive in brackish and shallow marine habitats, such as: (a) short lateral root extensions, called pneumatophores, that grow upwards from muddy and anoxic (oxygen-free) substrates and enable the absorption of oxygen from the air; (b) a branched stem system, known as aerial roots or stilt roots, and buttress roots for better anchorage in soft substrates and to withstand strong winds or waves; (c) succulent leaves with internal water storage tissues;

(d) salt-secreting or exclusion mechanisms, such as salt glands in the leaves of some species, for osmotic balance; and (e) viviparous reproduction, that is, the seed germinates into a seedling, called a propagule, while still attached to the parent plant, producing an elongated crude root eventually dropping directly into the substrate, thus, "planting" itself. Undisturbed forests are marked by distinct zonation, with each zone dominated by characteristic species.

2.1. Taxonomic treatment

The World Mangroves Database (Dahdouh-Guebas, 2020) lists 65 "valid" or correct names of mangrove taxa in 14 families, inclusive of five hybrids and exclusive of three species of the fern mangrove *Acrostichum* and two species of the leguminous mangrove *Cynometra*. No new species of mangroves has been described over the past decade, although new hybrids have been identified using molecular methods, bringing the current number to eight (Ragavan and others, 2017; Ono and others, 2016). Relative to other plant groups such as ferns or grasses, the number of species may not be considered high, but mangrove taxa are present in a broad cross section of 16 flowering plant families, except for three species that belong to the fern family. Of the 16 families, only two (Pellicieraceae and Rhizophoraceae) contain exclusively marine species

2.2. Current state and trends

It was reported in chapter 48 of the first Assessment that, depending on the criteria used in defining strict or "true" mangrove species, the number of species ran between 70 and 73, inclusive of hybrids. It was also emphasized that the groupings of species changed as a result of taxonomic studies. For example, the genus *Sonneratia*, which used to belong to the family Sonneratiaceae, is now classified in the family Lythraceae (Little and others, 2004), while Sonneratiaceae has been relegated to the subfamily level.

IUCN assessed 64 species for risk of extinction between 1998 and 2018. As at 19 November 2019, 3 had been categorized as Critically Endangered (*Sonneratia griffithii* (Duke and others, 2010a), *Bruguiera hainesii*[1] (Duke and others, 2010b) and *Sonneratia hainanensis* (World Conservation Monitoring Centre, 1998)); 3 as Endangered, all belonging to the family Malvaceae (*Camptostemon philippinensis* (Duke and others, 2010c), *Heritiera fomes* (Kathiresan and others, 2010) and *H. globosa* (Sukardjo, 2010)); 5 as Near Threatened (*Aegialitis rotundifolia* (Ellison and others, 2010a), *Aegiceras floridum* (Ellison and others, 2010b), *Ceriops decandra* (Duke and others, 2010d), *Sonneratia ovata* (Salmo and others, 2010) and *Rhizophora samoensis* (Ellison and Duke, 2010)), 5 as Vulnerable (*Avicennia lanata* (Chua, 1998), *A. integra* (Duke, 2010a), *A. rumphiana* (Duke and others, 2010e), *A. bicolor* (Duke, 2010b) and *Pelliciera rhizophorae* (Ellison and others, 2010c)); 47 as of Least Concern; and only 1 (*Excoecaria indica* (Ellison and others, 2010d)) as Data Deficient.

The distribution of those species across marine regions as defined by IUCN is shown in figure I. No mangroves are found in the North Atlantic Ocean. All globally Critically Endangered species are found in the eastern Indian Ocean and the north-eastern and eastern central Pacific Ocean, while all the globally Endangered species are found in the eastern Indian Ocean and the western central Pacific Ocean. Those that are in the Near Threatened category are found in the eastern Indian Ocean and the western central Pacific Ocean. Major threats are residential and commercial development, aquaculture and agriculture, biological

[1] In 2016, using molecular markers, Ono and others identified this species as a hybrid of *Bruguiera cylindrica* and *B. gymnorhiza*.

resource use, such as logging for construction materials and fuel, climate change causing habitat shifting and alteration, pollution, sand mining and replacement by invasive species.

At the regional level, however, some species of Least Concern may be threatened by various factors. *Avicennia marina* in the Red Sea Large Marine Ecosystem (Sherman and Hempel, 2008), for example, is particularly threatened, as it is highly grazed and harvested for forage, being palatable to such livestock as camels, goats and cattle (Nawata, 2013) owing to its lower content of soluble tannins. Goods and services and changes in mangrove ecosystems were treated in detail in chapter 48 of the first Assessment.

The global distribution, diversity and abundance of mangroves are affected by climate change, such as alterations in temperatures and rainfall regimes (Donato and others, 2011; Ward and others, 2016; Friess and Webb, 2013). Warmer winters and sea level rise are expected to allow for expansion polewards at the expense of salt marshes. However, dispersal and habitat availability constraints may hinder expansion near certain range limits. Along arid and semi-arid coasts, decreases or increases in rainfall are expected to lead to mangrove contraction or expansion, respectively.

Figure I
Distribution of mangrove species by IUCN categories and marine regions

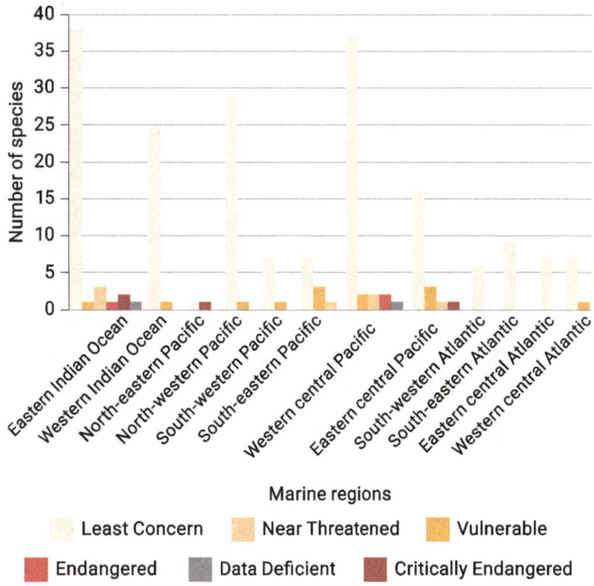

3. Salt marsh plants

Salt marsh plants constitute the major vegetation in the intertidal zones and inland saline areas of the temperate zones. They constitute the lush and highly productive "intertidal grasslands" or salt marshes, with distinct zonation in response to steep physical gradients, such as salt concentrations exceeding 500 millimolars (Yuan and others, 2019). They have a high ability to filter nutrients, which increases water quality in nearby coastal systems affected by urban, aquaculture and agricultural run-off, and to store carbon. They protect coastal communities from storm and erosion and enhance human well-being by acting as critical nursery habitats for fish and other marine organisms harvested for food. The essential ecosystem goods and services provided by salt marshes benefit millions of people and are valued at $10,000 per hectare per year (Barbier and others, 2011; Hopkinson and others, 2012; Möller and others, 2014).

Unlike mangroves, which are dominated by trees, salt marsh plants are usually grasses or shrubs. Just like mangroves, they have developed and adapted to high saline environments, flooding and desiccation, and anoxic conditions.

3.1. Taxonomic treatment

Globally, plant species richness in marshes is surprisingly high, with more than 500 marsh plant species known. However, the majority of those are found in freshwater lakes and rivers, extending into brackish aquatic environments and occupying an estimated 45,000 km^2 worldwide (Greenberg and others, 2006). The salt-tolerant species inhabiting salt marshes belong to three major families: Chenopodiaceae, with two species of chenopods (*Salicornia veneta* and *S. rubra*); Poaceae, with three species of *Spartina* grasses (*S. alterniflora, S. gracilis and S. maritima*) and many species of *Phragmites*; and Juncaceae (many species). However, only two species are reported to inhabit exclusively the coastal zone: the chenopod *Salicornia veneta* and the grass *Spartina alterniflora*. The others inhabit freshwater lakes, rivers and swamplands and extend into "artificial" brackish and marine habitats, such as aquaculture ponds and canals. *Salicornia* and *Spartina* are found throughout temperate North America, with *Salicornia* extending to Mexico. *Salicornia* is also found in some parts of Europe and northern parts of Asia.

3.2. Current state and trends

At the global level, up to half of coastal wetlands have been lost due to agriculture, aquaculture and other anthropogenic changes in land use (Pendleton and others, 2012). Climate change, declining water quality and changes in sediment delivery rates associated with human activity continue to affect the world's remaining wetlands, such as salt marshes (Kirwan and Megonigal, 2013).

Salicornia veneta is the only threatened species listed in the IUCN Red List (Foggi and others, 2011) and is categorized as Vulnerable. It is found along the coast of Italy in the Adriatic Sea, and occupies an area of less than 500 km^2. Although common in its distribution area, its population is reported to be decreasing as a result of coastal development, settlements and tourism. It is protected under national legislation, and one site is a protected area. *Spartina alterniflora* is categorized as of Least Concern (Maiz-Tome, 2016).

4. Seagrasses

Seagrasses are marine flowering plants that inhabit tidal and subtidal marine environments. They require high levels of light and are normally abundant in shallow waters, where they are productive components of nearshore environments, providing food and shelter for many economically important species (Heck and Orth, 1980).

Seagrasses are among some of the oldest plants on Earth, with fossil deposits believed to date back to the Pliocene (Tuya and others, 2017). They have developed adaptations to survive in their particular niche (Papenbrock, 2012), such as mostly thin, flattened, elongated or strap-like leaves that allow for flexibility in waters with strong waves and currents and the diffusion of gases (since they do not have stomata); an extensive system of roots and rhizomes that enable them to anchor in muddy and sandy substrates (e.g., the temperate species in the genus *Phyllospadix* have hooks that allow them to attach to rocks); an adaptability to survive in high and often varying salinities; pollen in gelatinous tubes or floating packets for submarine or pollination occurring on the water surface; and, in some species, viviparous or cryptic viviparous reproduction that enables them to compete with other species (Green and Short, 2003).

The distribution of seagrasses depends in part on the dispersal of fruits, seeds, seedlings and vegetative propagules by ocean currents. Using a combination of population genetic assignment procedures and dispersal predictions from a hydrodynamic model, it was predicted that 60 per cent of *Posidonia australis* fruits dispersed within a 20 km range (Sinclair and others, 2018). That study provided insight into the role of physical transport for long-distance dispersal of fruits and the consequences for the spatial genetic structuring of seagrass meadows.

4.1. Taxonomic treatment

Seagrasses are flowering plants belonging to the class Liliopsida. As of 2011, 72 species in 6 families and 15 genera had been recognized (Short and others, 2011). They are distributed throughout the world, except in Antarctica. So far, at least two species have been described using molecular characters: the new species *Thalassodendron johnsonii* (Duarte and others, 2012), and one species (*Halophila major*) separated from the *Halophila ovalis* complex (Nguyen and others, 2014). Subpopulations have also been identified using barcoding techniques (Nguyen and others, 2015).

4.2. Current state and trends

It was reported in chapter 47 of the first Assessment that, as of 2011, 31 per cent (22 of 72 species) of the world's total number of species had declining populations, 5 per cent had increasing population trends and 22 per cent had an unknown status. The same report stated that seagrasses had been disappearing at a rate of 110 km^2 a year since 1980 and that 29 per cent of the known areal extent had disappeared since seagrass areas had been initially recorded in 1879, with rates of decline accelerating from a median of 0.9 per cent a year before 1940 to 7 per cent a year since 1990.

On a global scale, no further species extinction risk assessments have been conducted since 2011. Of the 72 species, three, all in the family Zosteraceae, remain in the IUCN Endangered category, all with decreasing populations (Short and Waycott, 2010a, b, c), namely, *Phyllospadix japonicus, Zostera chilensis* and *Z. geojeensis*. *P. japonicus* and *Z. geojeensis* are both found in the Pacific North-West, while *Z. chilensis* occurs in the Pacific South-East (figure II).

Figure II
Distribution of seagrass species by IUCN categories and marine regions

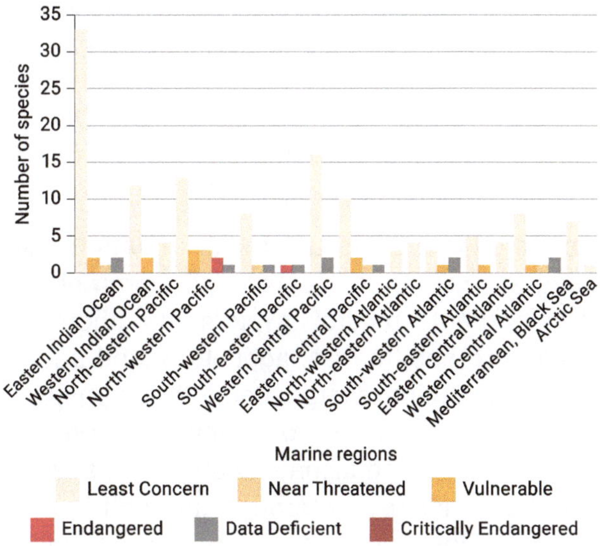

Five species are listed as Near Threatened, all with decreasing populations (Short and Waycott, 2010d, e, f). *Posidonia australis* (Short and others, 2010a) occurs in both the eastern Indian Ocean and the south-western Pacific Ocean, while *Zostera asiatica* (Short and Waycott, 2010d) occurs in the north-western and eastern central Pacific Ocean. All others occur only in one marine region each: *Halophila engelmanni* (Short and others, 2010b) in the western central Atlantic Ocean and *H. nipponica* (Short and others, 2010c) and *Zostera caulescens* (Short and Waycott, 2010e) in the north-western Pacific Ocean (figure II).

Seven species are listed as Vulnerable, with decreasing populations. These belong to three families: Posidoniaceae (*Posidonia sinuosa* (Short and others, 2010d)), Hydrocharitaceae (*Halophila baillonii* (Short and others, 2010e), *H. beccarii* (Short and others, 2010f) and *H. hawaiiana* (Short and others, 2010g)) and Zosteraceae (*Phyllospadix iwatensis, Zostera caespitosa* (Short and Waycott, 2010f, g) and *Z. capensis* (Short and others, 2010h)). Except for *H. beccarii* and *Z. capensis*, which are found in two marine regions, the Indian Ocean and the Pacific

Ocean for the former and the Indian Ocean and the Atlantic Ocean for the latter, all are restricted in their distribution: *Posidonia sinuosa* in the eastern Indian Ocean, *Zostera caespitosa* in the north-eastern Pacific Ocean, *Halophila hawaiiana* in the central Pacific Ocean, *Phyllospadix iwatensis* in the north-eastern Pacific Ocean and *Halophila baillonii* in the south-western and western central Atlantic Ocean.

As stated in the Red List, major threats to the seagrasses are residential and commercial development, natural system modification leading to habitat loss, agriculture and aquaculture, pollution, energy production, transportation and service corridors, invasive species and diseases, and climate change and severe weather leading to alteration and shifting of habitats. *Posidonia oceanica*, which is endemic to the Mediterranean, is an example of a species that has been negatively affected by those threats. The poleward expansion of tropical species has been observed in the South Atlantic, with *Halophila decipiens* (Gorman and others, 2016) and *Halodule wrightii* (Ferreira and others, 2015).

Nine species listed in the IUCN Red list as Data Deficient (Short and Waycott, 2010n, o, p, q, r, s, t, u, v) have remained unstudied to this day: two species in the family Zannichelliaceae (*Lepilaena australis* and *L. marina* (Short, 2010a, b), the former found in the eastern Indian Ocean and the south-western and western central Pacific Ocean and the latter in the eastern Indian Ocean only; four in the family Cymodoceaceae, with *Halodule beaudettei* (Short and others, 2010i) and *H. emarginata* (Short and others, 2010j) occurring in the eastern Indian Ocean, *H. bermudensis* (Coates and others, 2010) in the south-western and western central Atlantic Ocean and the eastern Indian Ocean, and *H. ciliata* (Short, 2010c) in the eastern central Pacific Ocean; two in the family Hydrocharitaceae (*Halophila euphlebia* (Short and others, 2010k) and *H. sulawesii* (Short, 2010d)) found in the north-western and western central Pacific Ocean, respectively; and one in the family Ruppiaceae (*Ruppia filifolia*) occurring in the south-eastern Pacific Ocean and the south-western Atlantic Ocean (figure II).

At the regional level, some species may be threatened by various factors. For example, *Enhalus acoroides* has a very limited distribution in the Red Sea Large Marine Ecosystem (El Shaffai, 2016) and may be vulnerable to grazing, especially since the area is home to a significant dugong population (Shawky, 2019; Nasr and others, 2019). Dugongs graze selectively, changing not only the community and population structure but also the species composition of seagrass beds. In the Atlantic coast of Canada, the European green crab (*Carcinus maenas*), which is invasive and non-indigenous to the region, is having a negative impact on eelgrass (Matheson and others, 2016).

General decreases or increases in seagrass beds in some parts of their distribution have been reported, such as decreases in *Zostera marina* in Nova Scotia and in the Gulf of Saint Lawrence and increases in Newfoundland (Canada) (Bernier and others, 2018). The non-indigenous *Halophila stipulacea* is reportedly expanding westwards from the eastern Mediterranean (Sghaier and others, 2011), while *H. decipiens* was newly recorded in a locality of the Aegean Sea (Gerakaris and others, 2020).

5. Macroalgae

The term "macroalgae" or "seaweeds" refers to non-flowering plant-like organisms that grow anchored in nearshore areas, except for some species of *Sargassum* that grow floating, mostly in the Sargasso Sea (see chap. 7Q). They have developed many adaptations that have enabled them to colonize various habitats, from polar to equatorial regions, and from shallow to very deep areas up to the limit of the euphotic zone. These adaptations include various pigments to trap light, diverse life-history patterns and morphologies to increase survival, and production of anti-herbivory compounds to evade grazing. These adaptations are used to characterize and identify groups and species.

Seaweeds form the most extensive and productive vegetated coastal habitats, as kelps and other algal beds, in global coastal environments such as rock shores and biogenic reefs, estimated to cover about 3.4 million km^2 and support a global net primary production of about 173 teragrams of carbon per year (Krause-Jensen and Duarte, 2016), and are harvested and farmed for food and other uses (see chap. 17 for more details). They are often used as indicators of water quality and reef health. For example, species of the green seaweed *Ulva* are used as indicators of heavy metal pollution and eutrophication (Alp and others, 2012).

A group of red algae that contain calcium carbonate in their cell walls (called coralline algae, from their resemblance to hard corals) can cover more sublittoral rocky substrata than any other group of macroorganisms in the photic zone, from intertidal habitats to 270 m deep, and are the deepest recorded macroalgae. Most of those heavily calcified red algae encrust rock or other substrata, but some species grow unattached to form important complex habitats that build up over thousands of years and are known as "maerl beds" or "rhodolith beds" (Riosmena-Rodríguez, 2017). These free-living coralline algae cover extensive areas of coastal seabed and are common in fossil marine carbonate deposits. The largest continuous latitudinal distribution of rhodolith beds occurs off Brazil, contributing to the formation of mesophotic reefs on extensive areas of the continental shelf and seamounts tops and around oceanic islands and atolls (Amado-Filho and others, 2017). Free-living coralline algal thalli grow slowly (a few millimetres per year) and can be long-lived (> 100 years). They provide a three-dimensional calcareous habitat attracting recruits to the benthos and providing refuge for the juveniles of commercially important shellfish.

5.1. Taxonomic treatment

Macroalgae are currently classified as protists (kingdom Protista). However, recent phylogenetic studies using plastids suggested that red and green algae share a common ancestor of the kingdom Plantae, while brown algae share a common ancestor of the kingdom Chromista (Delwiche, 2007). They are a diverse group, consisting of three major divisions based on their dominant pigmentation: red algae (Rhodophyta); brown algae, which used to be classified in the division Phaeophyta but were recently placed in their own class in the division Ochrophyta; and green algae (Chlorophyta). They contain chlorophyll and conduct photosynthesis. Many of them are "plant-like" in appearance but have simple bodies called thalli and lack the water-conducting system observed in terrestrial plants. Unlike seagrasses, they do not bear flowers.

As of 2012, Guiry (2012) had listed 12,471 species of algae belonging to the three divisions, of which 6,131 were red, 1,792 were brown and 4,548 were green, but estimated that there were about 27,000 species yet undescribed, including macroalgae and microalgae living in non-marine habitats. The Ocean Biodiversity Information System (OBIS, 2020; see also the table below), which records only marine species, lists 3,065 species of red algae (Rhodophyta); 879 species of brown algae (Phaeophyceae) and 844 species of green algae (Chlorophyta). For lower taxa (i.e., subspecies or taxa of unknown rank), the numbers are higher: 3,406, 1,070 and 1,164, respectively (see table below). The number and records of green algal species are lower than those of red and brown algae because the majority of green algae are found in freshwater environments.

Records of red (Rhodophyta), brown (Phaeophyceae) and green algae (Chlorophyta) in the Ocean Biodiversity Information System

Data	Rhodophyta	Phaeophyceae	Chlorophyta
Occurrence records	614 096	568 806	392 594
Species level records	449 392	477 331	209 396
Species	3 065	879	844
Lower taxa	3 406	1 070	1 164
Datasets	266	234	371
Time range	1865–2019	1869–2019	1778–2019

Source: OBIS, 2020.

5.2. Current state and trends

5.2.1. Red algae (Rhodophyta)

Red algae comprise a higher number of species than brown and green algae (see table above). They are distributed mainly from tropical to temperate marine waters (figure III.A), with very few species in freshwater ecosystems. They are found in areas with sea surface temperatures ranging from 5 to 30°C, a salinity of 5–35 practical salinity units (PSU) and mostly at depths of 0–20 m (figure III.B), although some rhodoliths have been recorded at much greater depths.

5.2.2. Brown algae (Phaeophyceae)

Brown algae, which comprise the least species (see table above), are purely marine and have a wide distribution, mainly in the cold and temperate waters of the Pacific Ocean, the Atlantic Ocean and the Indian Ocean and in the Southern Ocean (figure IV.A). They typically grow in areas with sea surface temperatures ranging from 5 to 30°C, although they can tolerate low temperatures (from -5 to 5°C) and a salinity of 5–35 PSU, and are mostly found at depths of 0–20 m (figure IV.B).

5.2.3. Green algae (Chlorophyta)

The number of green algae is between those of red and brown algae (see table above). They are distributed widely, but mainly in the northern hemisphere (figure V.A). They grow in diverse environments, from land to ocean and in seas with sea surface temperatures ranging from 5 to 30°C, although they can tolerate low temperatures (from -5 to 5°C), and a salinity of 0-35 PSU, and are mostly found at depths of 0–20 m (figure V.B).

Figure III.A
Global distribution of red algae (Rhodophyta)

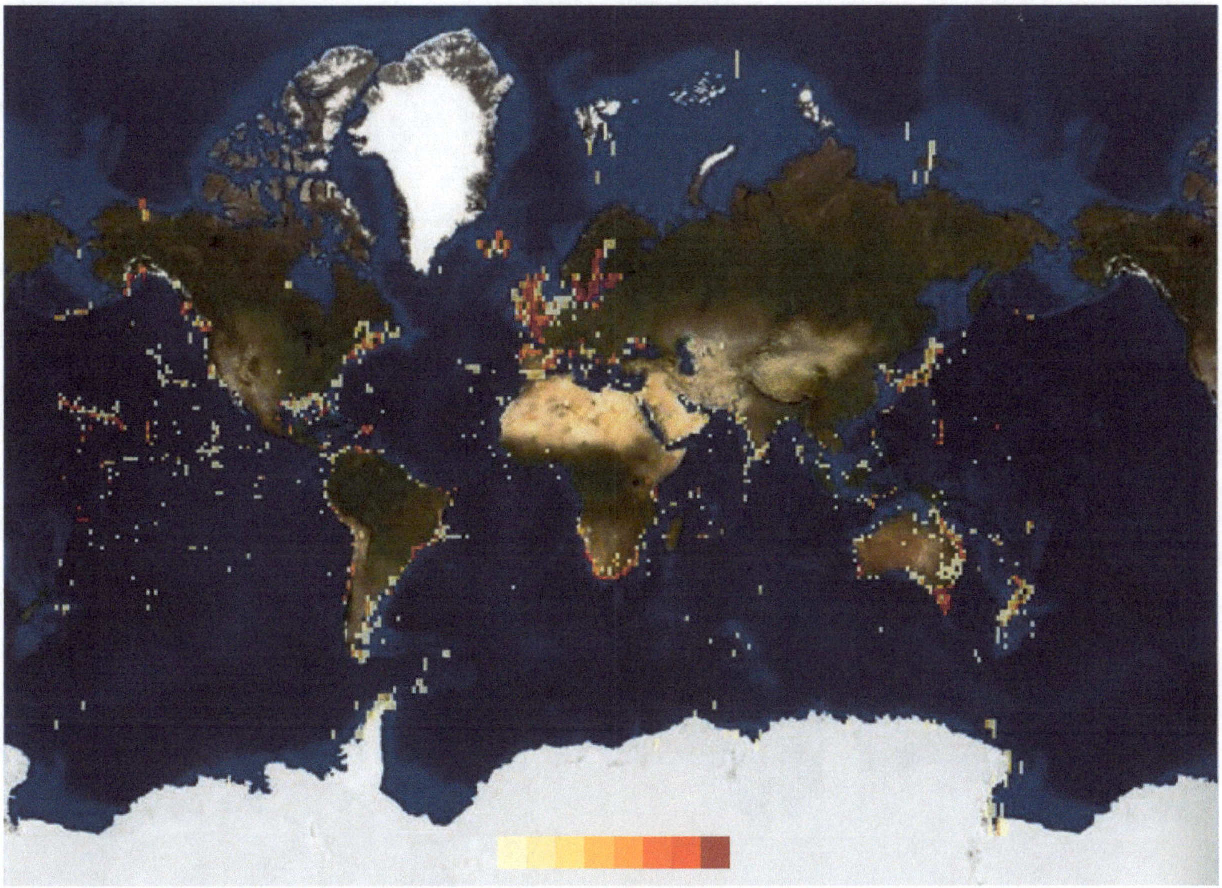

Figure III.B
Environmental conditions (temperature, sea surface salinity and depths) of red algae (Rhodophyta) occurrence around the world

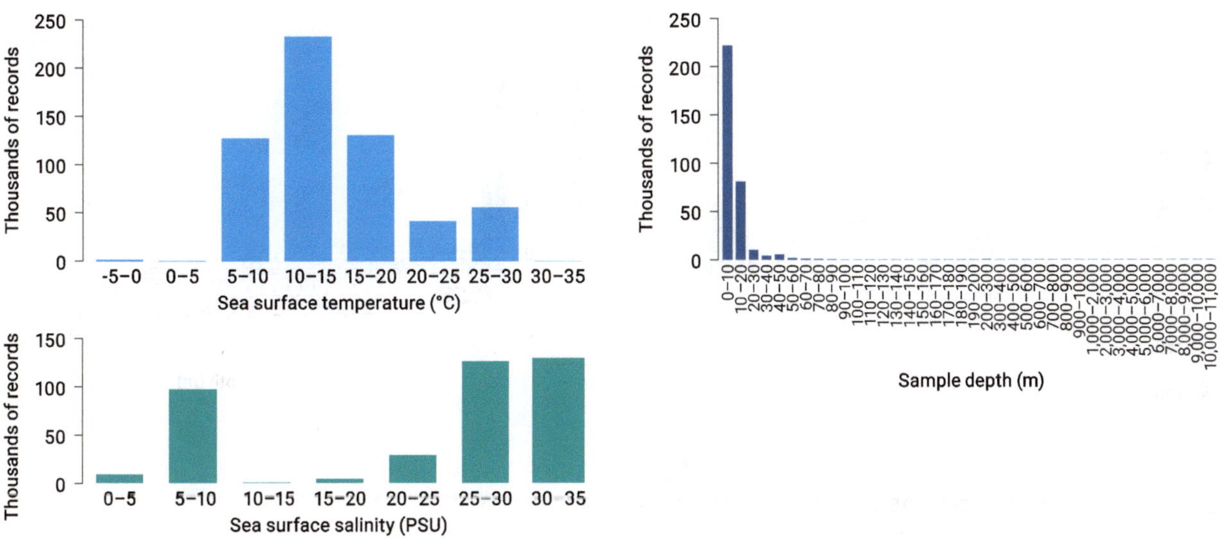

Source: OBIS (2020), https://mapper.obis.org/?taxonid=852.

Figure IV.A
Global distribution of brown algae (Phaeophyceae)

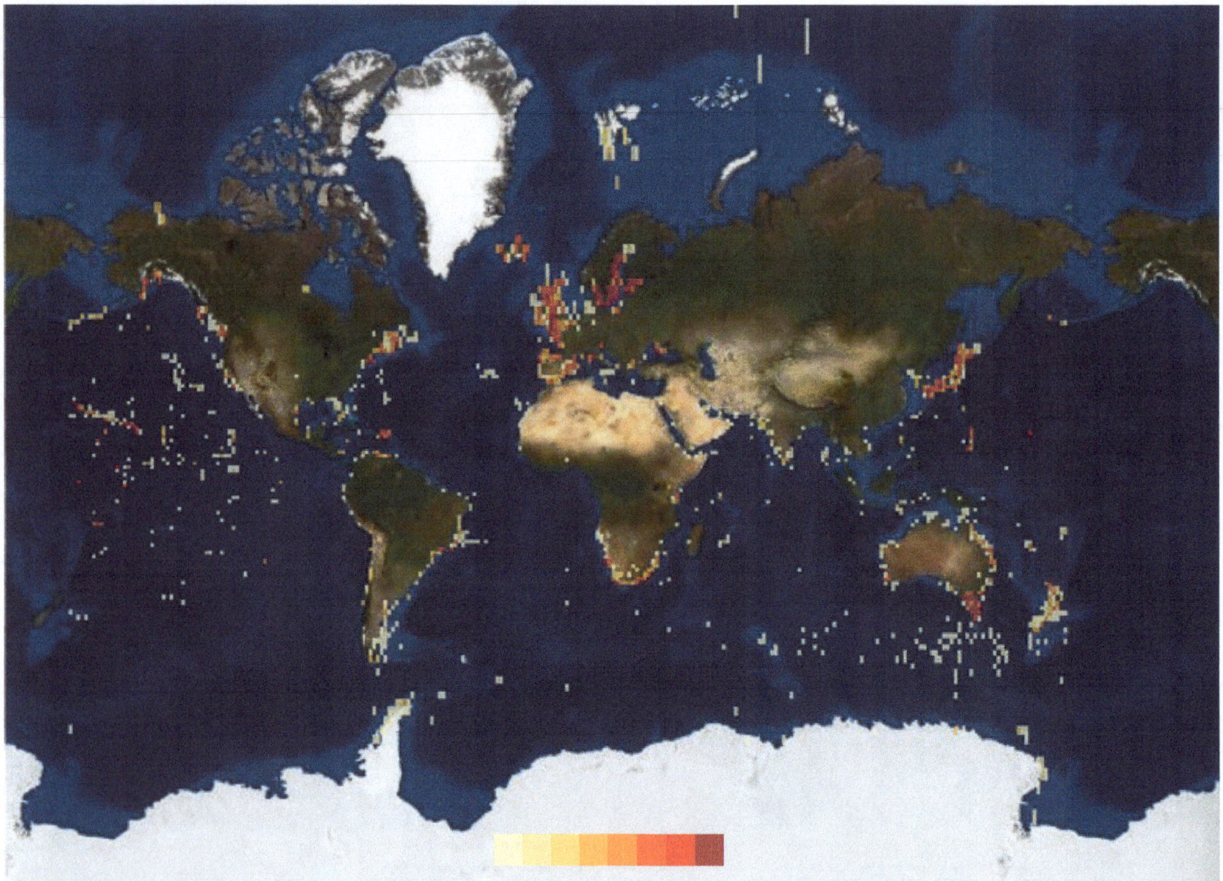

Figure IV.B
Environmental conditions (temperature, sea surface salinity and depths) of brown algae (Phaeophyceae) occurrence around the world

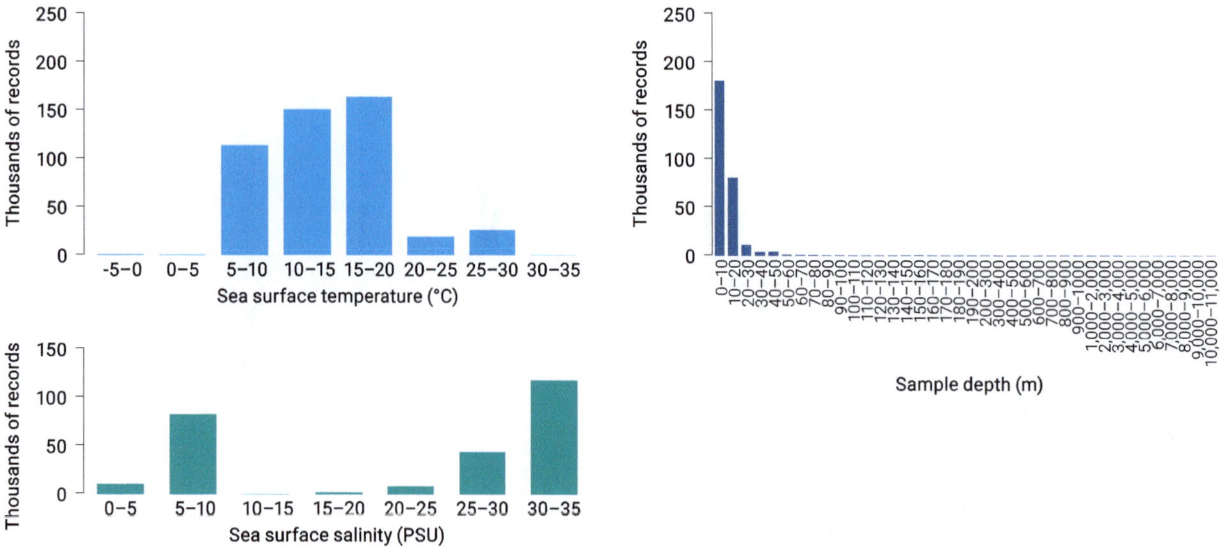

Source: OBIS (2020), https://mapper.obis.org/?taxonid=830.

Figure V.A
Global distribution of green algae (Chlorophyta)

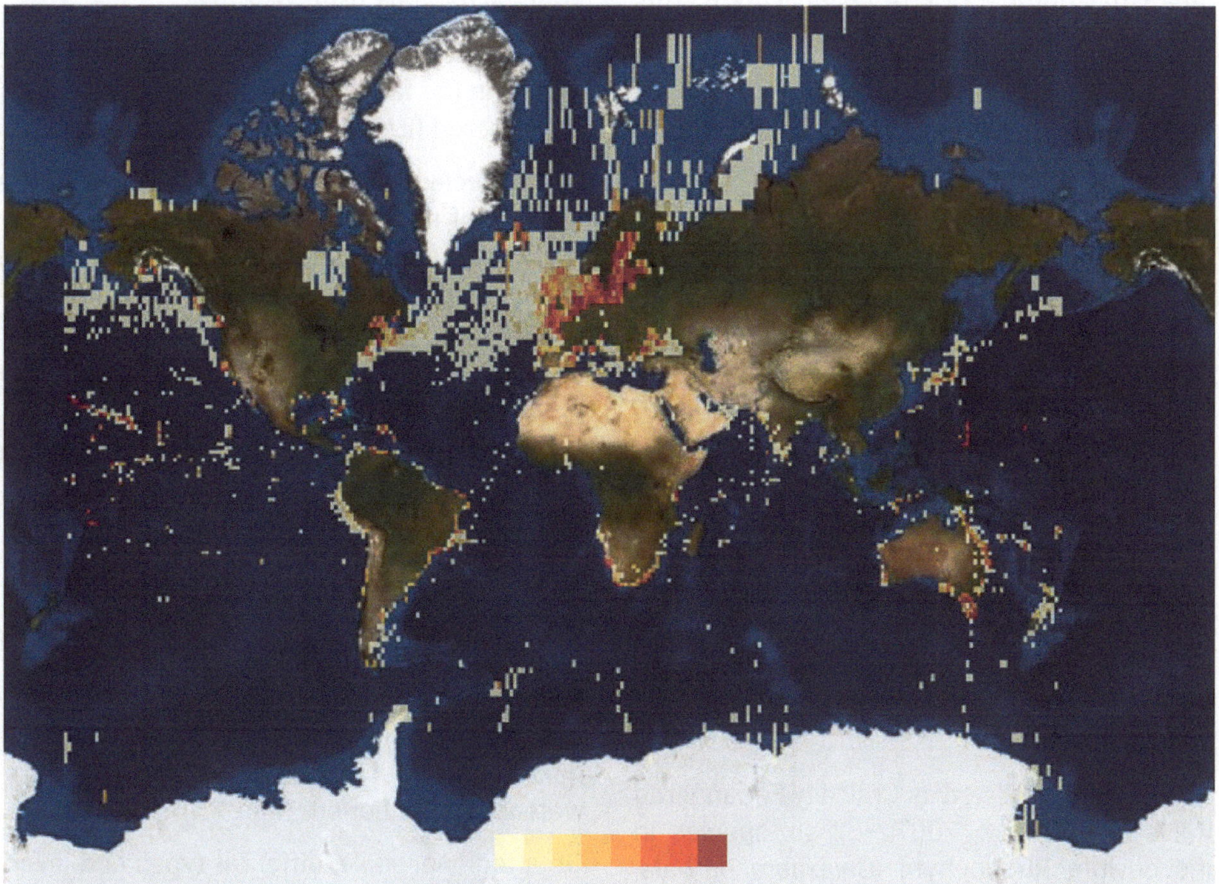

Figure V.B
Environmental conditions (temperature, sea surface salinity and depths) of green algae (Chlorophyta) occurrence around the world

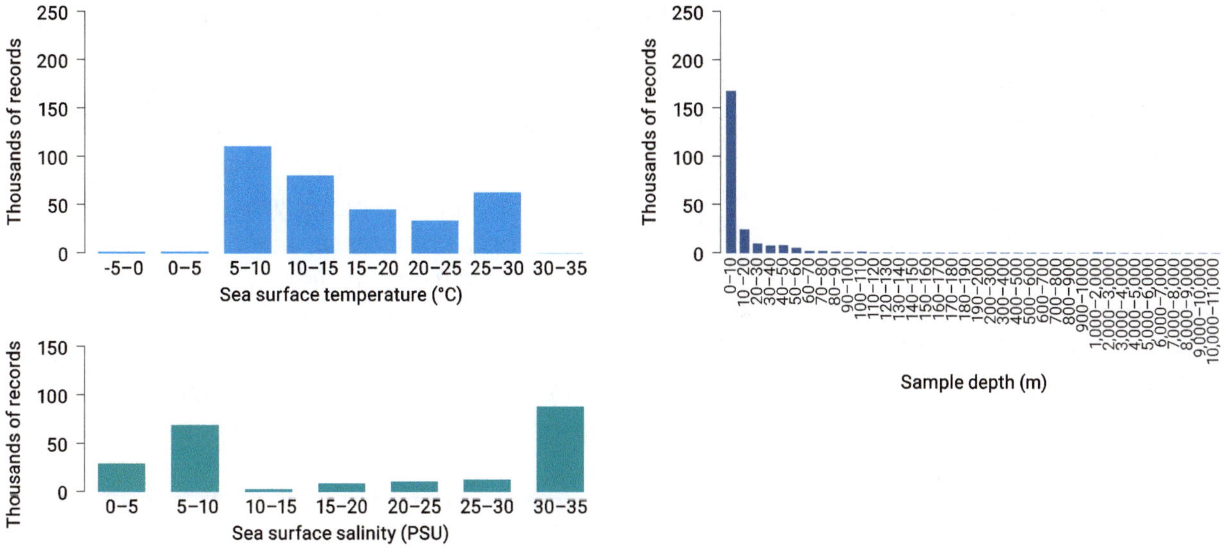

Source: OBIS (2020), https://mapper.obis.org/?taxonid=801.

An extinction risk assessment of algae was conducted in 2007 (Guiry and Guiry, 2020) and on a small number of species only (125). This does not even cover 1 per cent of the total species inventoried to date.

One red algal species is listed as Extinct and 15 red and brown algae are in the threatened list, all occurring in the south-eastern Pacific Ocean and all endemic to the Galapagos Islands (IUCN, 2019).

Vanvoorstia bennettiana,[2] in the family Delesseriaceae (Rhodophyta), is listed as Extinct (Millar, 2003). It was first discovered by William Harvey in 1855 growing in Port Jackson, Sydney Harbour, Australia. It was re-collected in 1886 at about eight kilometres east of the type locality (Millar, 2001). Despite intense searching, no specimens have been seen or collected in more than a century, and it is thought that habitat loss through human activities caused the extinction of that species (Guiry and Guiry, 2020).

Ten species are listed as Critically Endangered (Miller and others, 2007a–o): six species in the phylum Rhodophyta (*Galaxaura barbata, Gracilaria skottsbergii, Laurencia oppositocladia, Myriogramme kylinii, Phycodrina elegans* and *Schizymenia ecuadoreana*); and four species in the class Phaeophyceae (*Bifurcaria galapagensis, Desmarestia tropica, Dictyota galapagensis* and *Spatoglossum schmittii*). The Critically Endangered species of red algae belong to the families Galaxauraceae, Gracilariaceae, Delesseriaceae and Schizymeniaceae, while those of brown algae belong to the families Sargassaceae, Desmarestiaceae and Dictyotaceae. The period since 1970 has been marked by significant changes in the populations of macroalgae in the Galapagos during the El Niño events of 1982–1983 and 1997–1998, affecting *Bifurcaria galapagensis*, an endemic brown macroalga of the Galapagos, which inhabits the shallow intertidal and subtidal habitats (Garske 2002).

One species of brown algae, *Sargassum setifolium*, in the family Sargassaceae, is listed as Endangered (IUCN, 2019).

Four species are listed as Vulnerable: three species of red algae (*Austrofolium equatorianum, Acrosorium papenfussii* and *Pseudolaingia hancockii*); and one species of brown alga (*Eisenia galapagensis*) (IUCN, 2019).

Four species, all of red algae, are listed as of Least Concern and 54 species as Data Deficient (IUCN, 2019). None of the green algae assessed are listed as threatened, and only *Rhizoclonium robustum* is listed as Data Deficient.

The major threats mentioned in the IUCN reports are "climate change and severe weather", followed by "invasive and other problematic species". "Residential and commercial development", "transportation and service corridors", "biological resource use" and "pollution" were least mentioned.

On a regional scale, different types and levels of seaweed biodiversity assessments have been conducted.

In the Mediterranean, an extinction risk assessment was conducted under the Mediterranean Action Plan of the Convention for the Protection of the Marine Environment and the Coastal Region of the Mediterranean[3] (wherein 47 species were listed as threatened). Among the threatened species, typical examples are the "habitat-forming" *Cystoseira* species, except for *C. compressa*, which are declining and have even become locally extinct (Mancuso and others, 2018; Thibaut and others, 2015). However, Verlaque and others (2019) recommend that this list be reviewed on case-by-case basis, as it includes species that are far

2 Images of and taxonomical information on *Vanvoorstia bennettiana* are available at www.algaebase.org/search/species/detail/?species_id=23738.

3 United Nations, *Treaty Series*, vol. 1102, No. 16908.

from threatened and even considered invasive (e.g., *Caulerpa prolifera*), and a reassessment is suggested. The seaweeds of the Mediterranean, especially the perennial slow growers, are mostly threatened by commercial and industrial development (United Nations Environment Programme, 2015; Mansour and others, 2007; Husain and Khalil, 2013), coastal discharges (Mohorjy and Khan, 2006; Peña-García and others, 2014; Fabbrizzi and others, 2020), climate change (Piñeiro-Corbeira and others, 2018), and the introduction of exotic and invasive species through the Suez Canal (Galil and others, 2019). Israel and others (2020) reported that 16 per cent of the marine flora of Israel are regarded as invasive or exotic.

Miloslavich and others (2011) analysed the marine biodiversity of South America and found that species richness was higher in the tropical East Pacific than the tropical West Atlantic, and that the Humboldt Current System was richer than the Patagonian Shelf. Endemism analyses showed that 22 per cent of South American species are endemic and that 75 per cent of species are reported within only one of the South American subregions. In the South Atlantic, local stressors and coastal urbanization are causing substantial loss of seaweed biodiversity. Seaweed richness is 26 per cent less in urban areas than in areas with higher vegetation cover (Scherner and others, 2013). Among global stressors, heatwaves deserve major attention, as it is an important threat to temperature-sensitive species, such as the ecologically and biotechnologically important red alga *Laurencia catarinensis*. This species lost around 50 per cent of its total biomass during a heatwave that lasted from 8 October to 13 November 2014, when temperatures reached 2.66 degrees above the threshold calculated for those calendar days (Gouvêa and others, 2017).

In the Red Sea, macroalgal endemism level is about 9 per cent (Persga, 2003) and this is likely to increase with future research, as the Red Sea macroalgae are currently one of the least studied, despite the long history of scientific exploration dating back to the seventeenth century (Sheppard and others, 1992). Previous records (Walker, 1987) show that the Red Sea contained about 485 macroalgal species that were circumtropical and subtropical in distribution, occurring over extensive parts of the Indo-Pacific, Mediterranean and Caribbean regions. The composition, distribution and diversity of the Red Sea macroalgae seem to follow the natural latitudinal gradient in salinity, temperature and nutrient richness (Kürten and others, 2014), where diversity was greater in the northern and southern parts than in the central part (Walker, 1987; Sheppard, 1992).

Antarctic macroalgae are characterized by low species richness when compared with other regions of the world, with a high endemism originally estimated at approximately 33 per cent (Wiencke and Clayton, 2002; Wiencke and others, 2014), but since then reduced to 27 per cent (Oliveira and others, 2020). The highest endemism is found in brown algae (35.3 per cent), followed by red algae (29.4 per cent) and green algae (12.5 per cent) (Oliveira and others, 2020).

An inverse relationship between species diversity and latitude is observed in Antarctic macroalgae (Wiencke and Clayton, 2002). A total of 104 taxa were identified in the South Shetland Islands (Pellizzari and others, 2017), which is higher than in Adelaide Island, with 41 taxa (Cormaci and others, 1992), and Terra Nova Bay (Ross Sea, above 70° south), with 17 taxa (Mystikou and others, 2014).

The impact of oceanic warming on the distribution of Antarctic seaweeds was evaluated by Müller and others (2009), who concluded that the temperature increase may not directly affect the latitudinal distribution of some Antarctic seaweeds. However, Pellizzari and others (2020) suggested that macroalgal diversity in Antarctica, mainly in the surroundings of the Antarctic Peninsula, needed to be monitored, since the area is susceptible to species introduction and meteorological and oceanographic changes (Hughes and Ashton, 2017).

6. Consequences of changes on human communities, economies and well-being

The loss of species comprising the major coastal and ocean ecosystems, such as mangroves, salt marsh plants, seagrasses and seaweeds or those harvested for use and other applications, could have major health and economic repercussions on society.

A direct impact of the loss of species on economies and well-being is demonstrated by the loss of the kelp species, which are brown algae forming huge forests in the temperate oceans. These species are harvested for food and industrial, cosmetic, medical and other applications. Kelps are deeply affected by rising ocean temperatures, as they require cold waters to reproduce and grow. The effect is most apparent at the northernmost and southernmost limits of their distribution (Reed and others, 2016). As ocean temperatures increase, the distribution of some kelp populations have moved south in the southern latitudes, and north in the northern latitudes, and the associated grazer populations, such as urchins, have also shifted geographically in recent decades (Wahl and others, 2015). Wernberg and others (2019) and others have shown trajectories of change in kelp abundance based on long-term biomass records (figure VI).

Figure VI
Trajectory of change in records of kelp abundance

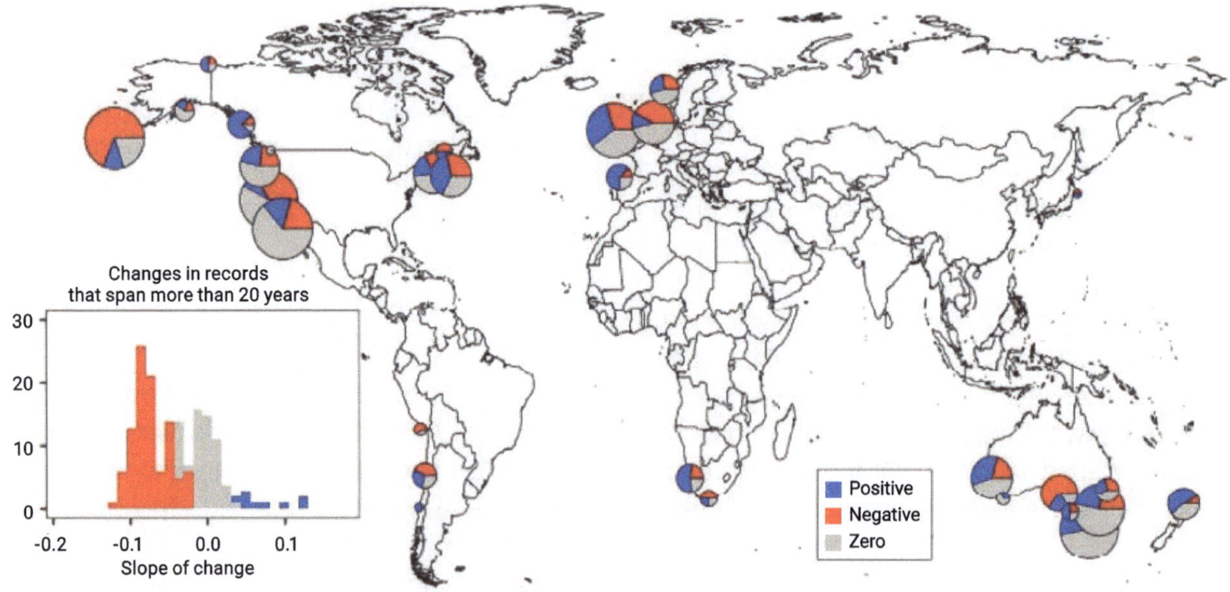

Source: Wernberg and others, 2019, reprinted with permission.

The loss to fisheries as a result of climate change impacts on reef-forming coralline algae and maerl-bed forming rhodoliths is also significant. Some publications (Barberá and others, 2003; Riosmena-Rodríguez and others, 2010) on the conservation status of maerl and rhodolith habitats in the Atlantic, Mediterranean and Gulf of California waters show that the health of those habitats is decreasing in many parts of the world. Activities such as dredging (e.g., for soil conditioner or shipping channels), destructive fishing (e.g., with dredges or trawls) and fish farming can reduce the complexity and biodiversity of those habitats, as can the spread of invasive species, such as the gastropod *Crepidula fornicata* (Peña and others, 2014). In addition to those direct impacts, maerl beds have been reported to face pressures from ocean warming and acidification, since slow growing coralline algae are highly vulnerable to anthropogenic CO_2 emissions (Martin and Hall-Spencer, 2017; Cornwall and others, 2019). European maerl beds are protected because they provide a wide range of ecological niches for associated flora and fauna.

Pioneering species such as *Halophila ovalis, Halodule uninervis* and *Cymodocea rotundata* are now being used as indicators of seagrass meadow resiliency in vulnerability assessments. Several seagrass species are used as bioindicators for heavy metal pollution, for example, *Halophila ovalis* and *H. minor* (Ahmad and others, 2015), while *Thalassia hemprichii, Enhalus acoroides* and *Cymodocea rotundata* are potential bioindicators for cadmium content in sediments and zinc content in seawater (Li and Xiaoping, 2012).

7. Key remaining knowledge and capacity-building gaps

Although new techniques such as genomics have been developed for species identification and for elucidating phylogenetic relationships, human and infrastructure capacities are still lacking in many regions. Few people study to become systematists, and even fewer to become phycologists (algal taxonomists). Taxonomic and systematic studies are important tools to monitor marine biodiversity, which is the foundation for development, in particular in small island States and archipelagic countries. As such, they respond to Sustainable Development Goal 14, specifically target 14.a.[4] With the development of new techniques for identifying species, the number of species is expected to rise, especially for macroalgae. However, some regions will still be less studied than others, depending on the capacities available. In addition, the vulnerability of a majority of plant species, including macroalgae, to changing climate and ocean conditions has not been assessed.[5]

8. Outlook

Climate change is now recognized as a major pressure on populations. It could be an opportunity for some species to expand their distribution, as in the case of some mangrove or marsh plant species, or cause others to become more restricted or even go extinct, as in the case of some kelp species. For example, Pergent and others (2014) projected that, in the Mediterranean, the endemic seagrass *Posidonia oceanica,* which has low tolerance to changes in salinity and temperature, would probably decline, mainly in the Levantine Sea, where sea surface

4 See General Assembly resolution 70/1.
5 See www.fisheries.noaa.gov/national/climate/climate-vulnerability-assessments.

temperature and salinity are expected to increase. *Zostera marina,* which grows in colder temperatures, could first be more confined and isolated in the northernmost parts of the Mediterranean, and then become extinct. Those two species could also be outcompeted by species such as *Cymodocea nodosa* and *Halophila stipulacea*, which grow well in warmer climates, which could lead to a decrease in the structural complexity of habitats.

According to some projections of seaweed species loss by 2100 made with habitat modelling based on the Representative Concentration Pathways (RCP) of greenhouse gas emissions, the current distribution of 15 prominent species of kelp and canopy-forming temperate species of seaweeds in Australia is expected to decrease by an average of 62 per cent (with decreases in range comprised between 27 and 100 per cent) under the most conservative emission scenario of RCP2.6, while eight kelp species in the North Atlantic are expected to lose 50 per cent of their range. On the other hand, it is projected that some species will expand their limits of distribution, such as three of the eight species in the North Atlantic into the Arctic, replace another species or form new forests. The change in populations of herbivores as a result of climate change is also expected to have an impact on macroalgal populations (see Wernberg and others, 2019).

References

Ahmad F., and others (2015). Tropical seagrass as a bioindicator for metal accumulation. *Sains Malaysiana*. vol. 44, No. 2, pp. 203–210. 10.17576/JSm-2015-4402-06.

Alp Mehmet Tahir, and others . Determination of heavy metal levels in sediment and macroalgae (*Ulva* sp. and *Enteromorpha* sp.) on the Mersin Coast (2012). *Ekoloji*, vol. 21, No. 82, pp. 47–55.

Amado-Filho, Gilberto M., and others (2017). South Atlantic rhodolith beds: latitudinal distribution, species composition, structure and ecosystem functions, threats and conservation status. In *Rhodolith/Maërl Beds: A Global Perspective*, Rafael Riosmena-Rodríguez, Wendy Nelson, and Julio Aguirre, eds. Cham, Switzerland: Springer International Publishing, pp. 299–317. https://doi.org/10.1007/978-3-319-29315-8_12.

Barberá, C., and others (2003). Conservation and management of northeast Atlantic and Mediterranean maerl beds. *Aquatic Conservation: Marine and Freshwater Ecosystems*, vol. 13, No. S1, pp. S65–S76. https://doi.org/10.1002/aqc.569.

Barbier, Edward B., and others (2011). The value of estuarine and coastal ecosystem services. *Ecological Monographs*, vol. 81, No. 2, pp. 169–93. https://doi.org/10.1890/10-1510.1.

Bernier, R.Y., and others, eds. (2018). *State of the Atlantic Ocean Synthesis Report*. Canadian Technical Report. of Fisheries and Aquatic Sciences, No. 3167. Ottawa: Department of Fisheries and Oceans Canada.

Chua, L.S.L. (1998). *Avicennia lanata. The IUCN Red List of Threatened Species*. https://dx.doi.org/10.2305/IUCN.UK.1998.RLTS.T31819A9662485.en. Coates, K., and others, 2010. *Halodule bermudensis. The IUCN Red List of Threatened Species 2010*. e.T173374A7002336. https://dx.doi.org/10.2305/IUCN.UK.2010-3.RLTS.T173374A7002336.en.

Cormaci, M., and others (1992). Observations taxonomiques et biogéographiques sur quelques espèces du genre *Cystoseira* C. Agardh. *Bulletin de l'Institut océanographique (Monaco)*, pp. 21–35.

Cornwall, Christopher E., and others (2019). Impacts of ocean warming on coralline algal calcification: meta-analysis, knowledge gaps, and key recommendations for future research. *Frontiers in Marine Science*, vol. 6, art. 186. https://doi.org/10.3389/fmars.2019.00186.

Dahdouh-Guebas, F., ed. (2020). World Mangroves database and Herbarium. Mangroves: the forgotten habitat in the middle of everywhere. Available at www.vliz.be/vmdcdata/mangroves.

Delwiche, Charles F. (2007). Algae in the warp and weave of life: bound by plastids. In *Unravelling the Algae. The Past, Present, and Future of Algal Systematics*. Juliet Brodie and Jane Lewis, eds. The Systematics Association Special Volume Series, No. 75. Boca Raton, Florida: CRC Press, pp. 7–20.

Donato, Daniel C., and others (2011). Mangroves among the most carbon-rich forests in the tropics. *Nature Geoscience*, vol. 4, No. 5, pp. 293–297. https://doi.org/10.1038/ngeo1123.

Duarte, M.C., and others (2012). Systematics and ecology of a new species of seagrass (*Thalassodendron*, Cymodoceaceae) from Southeast African Coasts. *Novon: A Journal for Botanical Nomenclature*, vol. 22, No. 1, pp. 16–24.

Duke, N. (2010a). *Avicennia integra. The IUCN Red List of Threatened Species*, e.T178844A7624677. http://dx.doi.org/10.2305/IUCN.UK.2010-2.RLTS.T178844A7624677.en.

________(2010b). *Avicennia bicolor. The IUCN Red List of Threatened Species*, e.T178847A7625682. http://dx.doi.org/10.2305/IUCN.UK.2010-2.RLTS.T178847A7625682.en.

Duke, N., and others (2010a). *Sonneratia griffithii. The IUCN Red List of Threatened Species*, e.T178799A7609832. http://dx.doi.org/10.2305/IUCN.UK.2010-2.RLTS.T178799A7609832.en.

Duke, N., and others (2010b). *Bruguiera hainesii. The IUCN Red List of Threatened Species*, e.T178834A7621565. http://dx.doi.org/10.2305/IUCN.UK.2010-2.RLTS.T178834A7621565.en.

Duke, N., and others (2010c). *Camptostemon philippinense. The IUCN Red List of Threatened Species*, e.T178808A7612909. http://dx.doi.org/10.2305/IUCN.UK.2010-2.RLTS.T178808A7612909.en.

Duke, N., and others (2010d). *Ceriops decandra. The IUCN Red List of Threatened Species*, e.T178853A7627935. http://dx.doi.org/10.2305/IUCN.UK.2010-2.RLTS.T178853A7627935.en.

Ellison, J., and others (2010a). *Aegialitis rotundifolia. The IUCN Red List of Threatened Species 2010*, e.T178839A7623021. http://dx.doi.org/10.2305/IUCN.UK.2010-2.RLTS.T178839A7623021.en.

Ellison, J., and others (2010b). *Aegiceras floridum. The IUCN Red List of Threatened Species 2010*, e.T178856A7628795. http://dx.doi.org/10.2305/IUCN.UK.2010-2.RLTS.T178856A7628795.en.

Ellison, J., and others (2010c). *Pelliciera rhizophorae. The IUCN Red List of Threatened Species 2010*, e.T178833A7621318. http://dx.doi.org/10.2305/IUCN.UK.2010-2.RLTS.T178833A7621318.en.

Ellison, J., and others (2010d). *Excoecaria indica. The IUCN Red List of Threatened Species 2010*, e.T178836A7622053. http://dx.doi.org/10.2305/IUCN.UK.2010-2.RLTS.T178836A7622053.en.

Ellison, J., and J. Duke (2010). *Rhizophora samoensis. The IUCN Red List of Threatened Species*, e.T178831A7620672. http://dx.doi.org/10.2305/IUCN.UK.2010-2.RLTS.T178831A7620672.en.

El Shaffai, A. (2016). *Field Guide to Seagrasses of the Red Sea*. 2nd ed. Anthony Rouphael and Ameer Abdulla, eds. Gland, Switzerland: IUCN.

Fabbrizzi, E., and others (2020). Modeling macroalgal forest distribution at Mediterranean scale: present status, drivers of changes and insights for conservation and management. *Frontiers in Marine Science*, vol. 7, art. 20. https://doi.org/10.3389/fmars.2020.00020.

Ferreira, Chirle, and others (2015). Anatomical and ultrastructural adaptations of seagrass leaves: an evaluation of the southern Atlantic groups. *Protoplasma*, vol. 252, No. 1, pp. 3–20. https://doi.org/10.1007/s00709-014-0661-9.

Foggi, B., and others (2011). *Salicornia veneta. The IUCN Red List of Threatened Species*, e.T164320A5824288. http://dx.doi.org/10.2305/IUCN.UK.2011-1.RLTS.T164320A5824288.en.

Galil, B.S., and others (2019). Invasive biota in the deep-sea Mediterranean: an emerging issue in marine conservation and management. *Biological Invasions*, vol. 21, pp. 281–88.

Garske L.E. (2002). Macroalgas marinas. In *Reserva Marina de Galápagos: Línea Base de la Biodiversidad*. Eva Danulat and Graham J. Edgar, eds. Santa Cruz, Galápagos, Ecuador: Fundación Charles Darwin/Servicio Parque Nacional Galápagos, Pp. 419–439.

Gerakaris, V., and others (2020). First record of the tropical seagrass species *Halophila decipiens* Ostenfeld in the Mediterranean Sea. *Aquatic Botany*, vol. 160, 103151. https://doi.org/10.1016/j.aquabot.2019.103151.

Gouvêa, L.P., and others (2017). Interactive effects of marine and eutrophication on the ecophysiology of a widespread and ecologically important macroalga. *Limnology and Oceanography*, vol. 62, No. 5, pp. 2056–2075. https://doi.org/10.1002/lno.10551.

Gorman, Daniel, and others (2016). Population expansion of a tropical seagrass (*Halophila decipiens*) in the southwest Atlantic (Brazil). *Aquatic Botany*, vol. 132, pp. 30–36. https://doi.org/10.1016/j.aquabot.2016.04.002.

Green, E.P., and F.T. Short (2003). *World Atlas of Seagrasses*. Berkeley, California: University of California Press, p. 324.

Greenberg, Russell, and others (2006). Tidal marshes: a global perspective on the evolution and conservation of their terrestrial vertebrates. *BioScience*, vol. 56, No. 8, pp. 675–85. https://doi.org/10.1641/0006-3568(2006)56[675:TMAGPO]2.0.CO;2.

Guiry, M. D. (2012). How many species of algae are there? *Journal of Phycology*, vol. 48, No. 5, pp. 1057–1063. https://doi.org/10.1111/j.1529-8817.2012.01222.x.

Guiry, M.D., and Guiry, G.M. (2020). AlgaeBase. World-wide electronic publication. Galway: Ireland: National University of Ireland. www.algaebase.org.

Heck, Kenneth L., and Robert J. Orth (1980). Seagrass habitats: the roles of habitat complexity, competition and predation in structuring associated fish and motile macroinvertebrate assemblages. In *Estuarine Perspectives*. V. S. Kennedy, ed. New York: Academic Press, pp. 449–464.

Hopkinson, Charles S., and others (2012). Carbon sequestration in wetland dominated coastal systems — a global sink of rapidly diminishing magnitude. *Current Opinion in Environmental Sustainability*, vol. 4, No. 2, pp. 186–194. https://doi.org/10.1016/j.cosust.2012.03.005.

Hughes, Kevin A., and Gail V. Ashton (2017). Breaking the ice: the introduction of biofouling organisms to Antarctica on vessel hulls. *Aquatic Conservation: Marine and Freshwater Ecosystems*, vol. 27, No. 1, pp. 158–164.

Husain, Tahir, and Ahmed Abdulwahab Khalil (2013). Environment and sustainable development in the Kingdom of Saudi Arabia: current status and future strategy. *Journal of Sustainable Development*, vol. 6, No. 12, pp. 14–30.

International Union for Conservation of Nature (IUCN) (2019). *The IUCN Red List of Threatened Species*. www.iucnredlist.org.

Israel, Alvaro, and others (2020). The seaweed resources of Israel in the Eastern Mediterranean Sea. *Botanica Marina*, vol. 63, No. 1, pp. 85–95. https://doi.org/10.1515/bot-2019-0048.

Kathiresan, K., and others (2010). *Heritiera fomes. The IUCN Red List of Threatened Species*, e. T178815A7615342. http://dx.doi.org/10.2305/IUCN.UK.2010-2.RLTS.T178815A7615342.en.

Kirwan, Matthew L., and J. Patrick Megonigal (2013). Tidal wetland stability in the face of human impacts and sea-level rise. *Nature*, vol. 504, pp. 53–60. https://doi.org/10.1038/nature12856.

Krause-Jensen, Dorte, and Carlos M. Duarte (2016). Substantial role of macroalgae in marine carbon sequestration. *Nature Geoscience*, vol. 9, No. 10, pp. 737–42. https://doi.org/10.1038/ngeo2790.

Kürten, Benjamin, and others (2014). Ecohydrographic constraints on biodiversity and distribution of phytoplankton and zooplankton in coral reefs of the Red Sea, Saudi Arabia. *Marine Ecology*, vol. 36, No. 4, pp. 1195–1214. https://doi.org/10.1111/maec.12224.

Li, Lei, and Xiaoping Huang (2012). Three tropical seagrasses as potential bio-indicators to trace metals in Xincun Bay, Hainan Island, South China. *Chinese Journal of Oceanology and Limnology*, vol. 30, No. 2, pp. 212–224. https://doi.org/10.1007/s00343-012-1092-0.

Little, Stefan A., and others (2004). Duabanga-like leaves from the Middle Eocene Princeton chert and comparative leaf histology of Lythraceae sensu lato. *American Journal of Botany*, vol. 91, No. 7, pp. 1126–1139.

Maiz-Tome, L., ed. (2016). *Spartina alterniflora. The IUCN Red List of Threatened Species*, e.T13491788A13491792. https://dx.doi.org/10.2305/IUCN.UK.2016-1.RLTS.T13491788A13491792.en.

Mancuso, F.P., and others (2018). Status of vulnerable *Cystoseira* populations along the Italian infralittoral fringe, and relationships with environmental and anthropogenic variables. *Marine Pollution Bulletin*, vol. 129, No. 2, pp. 762–771. https://doi.org/10.1016/j.marpolbul.2017.10.068.

Mansour, Abbas M., and others (2007). Sedimentological and environmental impacts of development projects along the coast of Hurghada, Red Sea, Egypt. *Egyptian Journal of Aquatic Research*, vol. 33, No. 1, pp. 59–84.

Martin, Sophie, and Jason M. Hall-Spencer (2017). Effects of Ocean Warming and Acidification on Rhodolith/Maërl Beds. In *Rhodolith/Maërl Beds: A Global Perspective,* Rafael Riosmena-Rodríguez, Wendy Nelson, and Julio Aguirre, eds. Cham, Switzerland: Springer International Publishing, pp. 55–85. https://doi.org/10.1007/978-3-319-29315-8_3.

Matheson, K., and others (2016). Linking eelgrass decline and impacts on associated fish communities to European green crab *Carcinus maenas* invasion. *Marine Ecology Progress Series*, vol. 548, pp. 31–45.

Millar, A.J.K. (2003). The world's first recorded extinction of a seaweed. In *Proceedings of the XVIIth International Seaweed Symposium*. Anthony Chapman and others. New York: Oxford University Press, pp. 313–318.

Miller, K.A., and others (2007a). *Acrosorium papenfussii. The IUCN Red List of Threatened Species 2007*, e.T63609A12696272. https://dx.doi.org/10.2305/IUCN.UK.2007.RLTS.T63609A12696272.en.

Miller, K.A., and others (2007b). *Austrofolium equatorianum. The IUCN Red List of Threatened Species 2007*, e.T63610A12696491. https://dx.doi.org/10.2305/IUCN.UK.2007.RLTS.T63610A12696491.en.

Miller, K.A., and others (2007c). *Bifurcaria galapagensis. The IUCN Red List of Threatened Species 2007*, e.T63593A12686056. https://dx.doi.org/10.2305/IUCN.UK.2007.RLTS.T63593A12686056.en.

Miller, K.A., and others (2007d). *Desmarestia tropica. The IUCN Red List of Threatened Species 2007*, e.T63585A12684515. https://dx.doi.org/10.2305/IUCN.UK.2007.RLTS.T63585A12684515.en.

Miller, K.A., and others (2007e). *Dictyota galapagensis. The IUCN Red List of Threatened Species 2007*, e.T63587A12684867. https://dx.doi.org/10.2305/IUCN.UK.2007.RLTS.T63587A12684867.en.

Miller, K.A., and others (2007f). *Eisenia galapagensis. The IUCN Red List of Threatened Species 2007*, e.T63598A12686906. https://dx.doi.org/10.2305/IUCN.UK.2007.RLTS.T63598A12686906.en.

Miller, K.A., and others (2007g). *Galaxaura barbata. The IUCN Red List of Threatened Species 2007*, e.T63651A12703033. https://dx.doi.org/10.2305/IUCN.UK.2007.RLTS.T63651A12703033.en.

Miller, K.A., and others (2007h). *Gracilaria skottsbergii. The IUCN Red List of Threatened Species 2007*, e.T63646A12702413. https://dx.doi.org/10.2305/IUCN.UK.2007.RLTS.T63646A12702413.en.

Miller, K.A., and others (2007i). *Laurencia oppositocladia. The IUCN Red List of Threatened Species 2007*, e.T63622A12699120. https://dx.doi.org/10.2305/IUCN.UK.2007.RLTS.T63622A12699120.en.

Miller, K.A., and others (2007j). *Myriogramme kylinii. The IUCN Red List of Threatened Species 2007*, e.T63612A12696918. https://dx.doi.org/10.2305/IUCN.UK.2007.RLTS.T63612A12696918.en.

Miller, K.A., and others (2007k). *Phycodrina elegans. The IUCN Red List of Threatened Species 2007*, e.T63614A12697346. https://dx.doi.org/10.2305/IUCN.UK.2007.RLTS.T63614A12697346.en.

Miller, K.A., and others (2007l). *Pseudolaingia hancockii. The IUCN Red List of Threatened Species 2007*, e.T63615A12697574. https://dx.doi.org/10.2305/IUCN.UK.2007.RLTS.T63615A12697574.en.

Miller, K.A., and others (2007m). *Sargassum setifolium. The IUCN Red List of Threatened Species 2007*, e.T63596A12686555. https://dx.doi.org/10.2305/IUCN.UK.2007.RLTS.T63596A12686555.en.

Miller, K.A., and others (2007n). *Schizymenia ecuadoreana. The IUCN Red List of Threatened Species 2007*, e.T63653A12703293. https://dx.doi.org/10.2305/IUCN.UK.2007.RLTS.T63653A12703293.en.

Miller, K.A., and others (2007o). *Spatoglossum schmittii. The IUCN Red List of Threatened Species 2007*, e.T63591A12685707. https://dx.doi.org/10.2305/IUCN.UK.2007.RLTS.T63591A12685707.en.

Miloslavich, Patricia, and others (2011). Marine biodiversity in the Atlantic and Pacific coasts of South America: knowledge and gaps. *PLOS ONE*, vol. 6, No. 1, pp. 1–43. https://doi.org/10.1371/journal.pone.0014631.

Möller, Iris, and others (2014). Wave attenuation over coastal salt marshes under storm surge conditions. *Nature Geoscience*, vol. 7, No. 10, pp. 727–731. https://doi.org/10.1038/ngeo2251.

Mohorjy, Abdullah M., and Ahmed M. Khan (2006). Preliminary assessment of water quality along the Red Sea coast near Jeddah, Saudi Arabia. *Water International*, vol. 31, No. 1, pp. 109–115. https://doi.org/10.1080/02508060608691920.

Müller, Ruth, and others (2009). Impact of oceanic warming on the distribution of seaweeds in polar and cold-temperate waters. *Botanica Marina*, vol. 52, No. 6, pp. 617–638.

Mystikou, Alexandra, and others (2014). Seaweed biodiversity in the south-western Antarctic Peninsula: surveying macroalgal community composition in the Adelaide Island/Marguerite Bay region over a 35-year time span. *Polar Biology*, vol. 37, No. 11, pp. 1607–1619.

Nasr, Dirar, and others. (2019). Status of Red Sea dugongs. In *Oceanographic and Biological Aspects of the Red Sea*. Najeeb M. A. Rasul and Ian C. F. Stewart, eds. Cham, Switzerland: Springer International Publishing, pp. 327–354. https://doi.org/10.1007/978-3-319-99417-8_18.

Nawata, H. (2013). Relationship between humans and camels in arid tropical mangrove ecosystems on the Red Sea coast. *Global Environmental Research*, vol. 17, pp. 233–246.

Nguyen, V.X., and others (2014). Genetic species identification and population structure of *Halophila* (Hydrocharitaceae) from the Western Pacific to the Eastern Indian Ocean. *BMC Evolutionary Biology*, vol. 14, No. 1, pp. 92. https://doi.org/10.1186/1471-2148-14-92.

Nguyen, V.X., and others (2015). New insights into DNA barcoding of seagrasses. *Systematics and Biodiversity*, vol. 13, No. 5, pp. 496–508.

Ocean Biodiversity Information System (OBIS) (2020). Intergovernmental Oceanographic Commission of UNESCO. Available at https://obis.org (accessed on 10 April 2020).

Oliveira, M.C., and others (2020). Diversity of Antarctic seaweeds. In *Antarctic Seaweeds: Diversity, Adaptation and Ecosystem Services*. Iván Gómez and Pirjo Huovinen, eds. Springer, pp. 23–42. https://doi.org/10.1007/978-3-030-39448-6_2.

Ono, Junya, and others (2016). *Bruguiera hainesii*, a critically endangered mangrove species, is a hybrid between *B. cylindrica* and *B. gymnorhiza* (Rhizophoraceae). *Conservation Genetics*, vol. 17, No. 5, pp. 1137–1144. https://doi.org/10.1007/s10592-016-0849-y.

Papenbrock, Jutta (2012). Highlights in seagrasses' phylogeny, physiology, and metabolism: what makes them special? *ISRN Botany*, vol. 2012, art. 103892. https://doi.org/10.5402/2012/103892.

Peña, V., and others (2014). The diversity of seaweeds on maerl in the NE Atlantic. *Marine Biodiversity*, vol. 44, No. 4, pp. 533–551. https://doi.org/10.1007/s12526-014-0214-7.

Peña-García, David, and others (2014). Input and dispersion of nutrients from the Jeddah Metropolitan Area, Red Sea. *Marine Pollution Bulletin*, vol. 80, Nos. 1 and 2, pp. 41–51.

Pellizzari, F., and others (2017). Diversity and spatial distribution of seaweeds in the South Shetland Islands, Antarctica: an updated database for environmental monitoring under climate change scenarios. *Polar Biology*, vol. 40, No. 8, pp. 1671–1685.

Pellizzari F., and others (2020). Biogeography of Antarctic seaweeds facing climate changes. In *Antarctic Seaweeds: Diversity, Adaptation and Ecosystem Services*. Iván Gómez and Pirjo Huovinen, eds. Springer, pp. 83–102. https://doi.org/10.1007/978-3-030-39448-6_5.

Pendleton, Linwood, and others (2012). Estimating global "blue carbon" emissions from conversion and degradation of vegetated coastal ecosystems. *PLOS ONE*, vol. 7, No. 9, e43542. https://doi.org/10.1371/journal.pone.0043542.

Pergent, G., and others (2014). Climate change and Mediterranean seagrass meadows: a synopsis for environmental managers, *Mediterranean Marine Science*, vol. 15, No. 2. http://dx.doi.org/10.12681/mms.621.

Piñeiro-Corbeira, Cristina, and others (2018). Seaweed assemblages under a climate change scenario: functional responses to temperature of eight intertidal seaweeds match recent abundance shifts. *Scientific Reports*, vol. 8,art. 12978. https://doi.org/10.1038/s41598-018-31357-x.

Ragavan, P., and others (2017). Natural hybridization in mangroves – an overview. *Botanical Journal of the Linnean Society*, vol. 185, No. 2, pp. 208–224. https://doi.org/10.1093/botlinnean/box053.

Reed, Daniel, and others (2016). Extreme warming challenges sentinel status of kelp forests as indicators of climate change. *Nature Communications*, vol. 7, art. 13757. https://doi.org/10.1038/ncomms13757.

Riosmena-Rodríguez, Rafael (2017). Natural history of rhodolith/maërl beds: their role in near-shore biodiversity and management. In *Rhodolith/Maërl Beds: A Global Perspective*, Rafael Riosmena-Rodríguez, Wendy Nelson, and Julio Aguirre, eds. Cham, Switzerland: Springer International Publishing, pp. 3–26.

Riosmena-Rodríguez, Rafael, and others, "Reefs that rock and roll: biology and conservation of rhodolith beds in the Gulf of California", in *The Gulf of California: biodiversity and conservation*, R. C. Bursca, ed. (Tucson, University of Arizona and Arizona-Sonora Desert Museum Press, 2010).

Salmo III, S.G., and others (2010). *Sonneratia ovata. The IUCN Red List of Threatened Species*, e.T178814A7615033. http://dx.doi.org/10.2305/IUCN.UK.2010-2.RLTS.T178814A7615033.en.

Scherner, F., and others (2013). Coastal urbanization leads to remarkable seaweed species loss and community shifts along the SW Atlantic. *Marine Pollution Bulletin*, vol. 76, Nos .1 and 2, pp. 106–115.

Sghaier, Y.R., and others (2011). Occurrence of the seagrass *Halophila stipulacea* (Hydrocharitaceae) in the southern Mediterranean Sea. *Botanica Marina*, vol. 54, No. 6, pp. 575–582. https://doi.org/10.1515/BOT.2011.061.

Shawky, A.M. (2019). Evidence of the occurrence of a large dugong in the Red Sea, Egypt. *Egyptian Journal of Aquatic Research*, vol. 45, No. 3, pp. 247–250.

Sheppard, Charles, and others (1992). *Marine Ecology of the Arabian Region: Pattern and Processes in Extreme Tropical Environments*. London: Academic Press. Sherman, Kenneth, and Gotthilf Hempel, eds. (2008). The UNEP large marine ecosystem report: a perspective on changing conditions in LMEs of the world's regional seas. UNEP Regional Seas Report and Studies, No. 182. Nairobi, United Nations Environment Programme.

Sinclair, Elizabeth A., and others (2018). Seeds in motion: genetic assignment and hydrodynamic models demonstrate concordant patterns of seagrass dispersal. *Molecular Ecology*, vol. 27, No. 24, pp. 5019–5034.

Short, F.T. (2010a). *Lepilaena australis. The IUCN Red List of Threatened Species 2010*. e.T173353A6997857. https://dx.doi.org/10.2305/IUCN.UK.2010-3.RLTS.T173353A6997857.en.

________(2010b). *Lepilaena marina. The IUCN Red List of Threatened Species 2010*. e.T173359A6998923. https://dx.doi.org/10.2305/IUCN.UK.2010-3.RLTS.T173359A6998923.en.

________(2010c). *Halodule ciliata. The IUCN Red List of Threatened Species 2010*. e.T173334A6993582. https://dx.doi.org/10.2305/IUCN.UK.2010-3.RLTS.T173334A6993582.en.

________(2010d). *Halophila sulawesii. The IUCN Red List of Threatened Species 2010*. e.T173326A6991316. https://dx.doi.org/10.2305/IUCN.UK.2010-3.RLTS.T173326A6991316.en.

Short, F.T., and M. Waycott (2010a). *Phyllospadix japonicus. The IUCN Red List of Threatened Species*, e.T173341A6994909. https://dx.doi.org/10.2305/IUCN.UK.2010-3.RLTS.T173341A6994909.en.

________(2010b). *Zostera chilensis. The IUCN Red List of Threatened Species*, e.T173322A6990689. https://dx.doi.org/10.2305/IUCN.UK.2010-3.RLTS.T173322A6990689.en.

________(2010c). *Zostera geojeensis. The IUCN Red List of Threatened Species 2010*. e.T173345A6995781. https://dx.doi.org/10.2305/IUCN.UK.2010-3.RLTS.T173345A6995781.en.

________(2010d). *Zostera asiatica. The IUCN Red List of Threatened Species 2010*. e.T173339A6994461. https://dx.doi.org/10.2305/IUCN.UK.2010-3.RLTS.T173339A6994461.en.

Short, F.T., and M. Waycott (2010e). *Zostera caulescens. The IUCN Red List of Threatened Species 2010*. e.T173335A6993689. https://dx.doi.org/10.2305/IUCN.UK.2010-3.RLTS.T173335A6993689.en.

________(2010f). *Phyllospadix iwatensis. The IUCN Red List of Threatened Species 2010*. e.T173344A6995596. https://dx.doi.org/10.2305/IUCN.UK.2010-3.RLTS.T173344A6995596.en.

________(2010g). *Zostera caespitosa. The IUCN Red List of Threatened Species 2010*. e.T173357A6998463. https://dx.doi.org/10.2305/IUCN.UK.2010-3.RLTS.T173357A6998463.en.

Short, F.T., and others (2010a). *Posidonia australis. The IUCN Red List of Threatened Species 2010*. e.T173333A6993340. https://dx.doi.org/10.2305/IUCN.UK.2010-3.RLTS.T173333A6993340.en.

Short, F.T., and others (2010b). *Halophila engelmanni. The IUCN Red List of Threatened Species 2010*. e.T173337A6994043. https://dx.doi.org/10.2305/IUCN.UK.2010-3.RLTS.T173337A6994043.en.

Short, F.T., and others (2010c). *Halophila nipponica. The IUCN Red List of Threatened Species 2010*. e.T173381A7004341. https://dx.doi.org/10.2305/IUCN.UK.2010-3.RLTS.T173381A7004341.en.

Short, F.T., and others (2010d). *Posidonia sinuosa. The IUCN Red List of Threatened Species 2010*. e.T173349A6996688. https://dx.doi.org/10.2305/IUCN.UK.2010-3.RLTS.T173349A6996688.en.

Short, F.T., and others (2010e). *Halophila baillonii. The IUCN Red List of Threatened Species 2010*. e.T173382A7004500. https://dx.doi.org/10.2305/IUCN.UK.2010-3.RLTS.T173382A7004500.en.

Short, F.T., and others (2010f). *Halophila beccarii. The IUCN Red List of Threatened Species 2010*. e.T173342A6995080. https://dx.doi.org/10.2305/IUCN.UK.2010-3.RLTS.T173342A6995080.en.

Short, F.T., and others (2010g). *Halophila hawaiiana. The IUCN Red List of Threatened Species 2010*. e.T173338A6994270. https://dx.doi.org/10.2305/IUCN.UK.2010-3.RLTS.T173338A6994270.en.

Short, F.T., and others (2010h). *Zostera capensis. The IUCN Red List of Threatened Species 2010*. e.T173370A7001305. https://dx.doi.org/10.2305/IUCN.UK.2010-3.RLTS.T173370A7001305.en.

Short, F.T., and others (2010i). *Halodule beaudettei. The IUCN Red List of Threatened Species 2010*. e.T173329A6992218. https://dx.doi.org/10.2305/IUCN.UK.2010-3.RLTS.T173329A6992218.en.

Short, F.T., and others (2010j). *Halodule emarginata. The IUCN Red List of Threatened Species 2010*. e.T173347A6996342. https://dx.doi.org/10.2305/IUCN.UK.2010-3.RLTS.T173347A6996342.en.

Short, F.T., and others (2010k). *Halophila euphlebia. The IUCN Red List of Threatened Species 2010*. e.T173325A6991162. https://dx.doi.org/10.2305/IUCN.UK.2010-3.RLTS.T173325A6991162.en.

Short, F.T., and others (2010l). *Ruppia filifolia. The IUCN Red List of Threatened Species 2010*. e.T173362A6999534. https://dx.doi.org/10.2305/IUCN.UK.2010-3.RLTS.T173362A6999534.en.

Short, F.T., and others (2011). Extinction risk assessment of the world's seagrass species. *Biological Conservation*, vol. 144, No. 7, pp. 1961–1971.

Sukardjo, S. (2010). *Heritiera globosa. The IUCN Red List of Threatened Species*, e.T178807A7612712. http://dx.doi.org/10.2305/IUCN.UK.2010-2.RLTS.T178807A7612712.en.

Thibaut T., and others (2015). Decline and local extinction of Fucales in the French Riviera: the harbinger of future extinctions? *Mediterranean Marine Science*, vol. 16, No. 1, pp. 206–224.

Tuya, Fernando, and others (2017). Seagrass paleo-biogeography: fossil records reveal the presence of *Halodule* cf. in the Canary Islands (eastern Atlantic). *Aquatic Botany*, vol. 143, pp. 1–7. https://doi.org/10.1016/j.aquabot.2017.08.002.

United Nations (2017). *The First Global Integrated Marine Assessment: World Ocean Assessment I*. Cambridge: Cambridge University Press.

United Nations Environment Programme (2015). Regional Coordination Mechanism (RCM): issues brief for the Arab Sustainable Development Report. Marine Resources in the Arab Region.

Verlaque, Marc, and others (2019). Mediterranean seaweeds listed as threatened under the Barcelona Convention: a critical analysis. In *Scientific Reports of Port-Cros National Park*, , vol. 33, pp. 179–214.

Wahl, Martin, and others (2015). The responses of brown macroalgae to environmental change from local to global scales: direct versus ecologically mediated effects. *Perspectives in Phycology*, vol. 2, No. 1, pp. 11–29.

Walker, Diana I. (1987). Chapter 8: benthic algae. In *Red Sea*, Alasdair J. Edwards and Stephen M. Head, eds. Key Environment Series. Amsterdam: Pergamon. pp. 152–168. https://doi.org/10.1016/B978-0-08-028873-4.50013-X.

Ward, R.D., and others (2016). Impacts of climate change on mangrove ecosystems: a region by region overview. *Ecosystem Health and Sustainability*, vol. 2, No. 4. e01211. https://doi.org/10.1002/ehs2.1211.

Wernberg, T., and others (2019). Chapter 3: status and trends for the world's kelp forests. In *World Seas: An Environmental Evaluation*, 2nd ed., pp. 57–78. London: Academic Press. Wiencke, Christian, and others (2014). Macroalgae. In *Biogeographic Atlas of the Southern Ocean*. Claude de Broyer and others, eds. Cambridge, United Kingdom: Scientific Committee on Antarctic Research, pp. 66–73.

Wiencke C., and Clayton M.N. (2002). Antarctic Seaweeds. In *Synopses of the Antarctic Benthos*, vol. 9. Johann-Wolfgang Wägele, ed. Rugell, Liechtenstein.

World Conservation Monitoring Centre, World Conservation Monitoring Centre (1998). *Sonneratia hainanensis. The IUCN Red List of Threatened Species*, e.T32472A9709212. https://dx.doi.org/10.2305/IUCN.UK.1998.RLTS.T32472A9709212.en.

Yuan F., and others (2019) Reproductive physiology of halophytes: current standing. *Frontiers in Plant Science*, vol. 9, art. 1954. https://doi.org/10.3389/fpls.2018.01954.

Chapter 7
Trends in the state of biodiversity in marine habitats

Chapter lead member: Hilconida Calumpong.

Introduction

The present chapter is composed of 17 subchapters that detail the state of coastal and marine habitats, from the coast to the deepest abyssal plains. The change in state since the first *World Ocean Assessment* is provided on mangrove forests, salt marshes, estuaries and deltas, seagrass meadows, cold water corals, tropical and subtropical coral reefs, the Sargasso Sea, high-latitude ice, hydrothermal vents and cold seeps, and submarine habitats, such as seamounts, submarine canyons and trenches.

The subchapter on submarine canyons has been expanded to include continental slopes, the subchapter on seamounts includes pinnacles and the subchapter on trenches includes ridges and plateaux. Kelp forests, which were included with seagrasses in the first Assessment, are now integrated into the subchapter on marine plants and macroalgae. New assessments are provided on sandy and muddy substrates, intertidal zones, atolls and island lagoons, abyssal plains and the open ocean.

Where a baseline of the state of the habitat was available in the first Assessment, it was used as a basis for looking at change over the last decade. Key threats to habitats have been identified and their influence on changes observed has been discussed. When available, specific regional changes have been highlighted, and an outlook for habitats over the near to medium term has been provided.

Chapter 7A
Intertidal zone

Contributors: Julia Sigwart (convener), Hilconida Calumpong (chapter lead member), Ronaldo Adriano Christofoletti, Karen Evans (subchapter lead member), Judith Gobin and Patricia Miloslavich.

Keynote points

- The intertidal zone encompasses many different habitats on the world's coasts.
- A large proportion of humans live in proximity to the intertidal zone.
- Human activities affect the intertidal zone directly, through coastal modification, and indirectly, through climate change.
- Despite our close relationship with intertidal habitats, key knowledge gaps remain and taxonomic infrastructure is needed in developing countries to resolve baseline data.

1. Introduction

The intertidal zone stands at the vanguard of human influence on the oceans – where the seas meet the land. The intertidal region worldwide encompasses the diverse habitats that occur at the shore, and those environments are unified by special properties in that they are not continuously covered by water, but are regularly exposed by the waning tide. The interface between terrestrial and marine factors creates a spectrum of increasing saltwater influence, with species and habitats occupying different points along a gradient. That is clearly seen in the stacked bands or zones on rocky intertidal shores (figure I) or in the succession from dunes to salt marsh to tidal flats (figure II). The intertidal zone further encompasses sandy beaches, mangroves, coral rubble and shallow reefs (figures III and IV), and intertidal areas provide the primary habitats for macrofauna considered to have special significance, such as marine reptiles (see chap. 6D). Species that inhabit intertidal zones are characterized by special adaptations that enable them to tolerate the periodic transitions between air and water. The intertidal zone is the most accessible part of the oceans and, therefore, is of particular importance for subsistence and small-scale fisheries and harvesting. Because of that accessibility, the intertidal zone is the marine realm most closely connected to human activities and interactions.

Figure I
Exposed intertidal area on a rocky outcrop showing horizontal bands formed by mussels (black band, closest to sand), barnacles and lichen

Photo: J. Sigwart.
Note: Ucluelet, British Columbia, Canada.

Figure II
Exposed intertidal area of mudflats fringed by rural development and rock armour

Photo: J. Sigwart.
Note: Newtownards, Northern Ireland, United Kingdom.

Figure III
Vegetative intertidal in the form of mangroves, adjacent to rocky substrata

Photo: J. Sigwart.
Note: Phuket, Thailand.

Figure IV
Intertidal live coral, shown below the water surface, in an area of shallow coral reef and coral rubble infrequently exposed on extreme natural tides

Photo: J. Sigwart.
Note: Port Dickson, Malaysia.

In the period since the first *World Ocean Assessment* (United Nations, 2017), the most important changes to intertidal habitats have been those resulting from climate change and human alteration to coastlines. In the context of the second *World Ocean Assessment*, the "intertidal" or "coastal" zone is not a single habitat, but includes aspects of many other habitats covered in chapters 6 and 7 (see table below), including sandy beaches, rock, high-latitude ice, mangroves, coral rubble and shallow reefs. Similarly, it is important to clarify that coastal habitats and communities are primarily benthic, but the benthos is a broader category of sea bottom, from the intertidal to the deep sea.

The action of tides differs considerably across the regions of the world, and those dynamics influence the flora and fauna, as well as human activities in the oceans. In many enclosed water bodies, such as the Mediterranean, the flux of the tides is almost negligible. Areas of very high tidal flux are target sites for tidal energy extraction, such as in the Strangford Lough in Northern Ireland, United Kingdom, and Sihwa Lake in the Republic of Korea (Leary and Esteban, 2009). Areas with lower tidal currents, including protected estuaries, are often sites of port development linked to the world's major cities and global shipping centres. The introduction of artificial substrates and built marine structures is increasing with the development of coastal regions worldwide. Artificial substrates occur on all coasts and broadly incorporate reclaimed land and built islands, as well as maritime infrastructure and human-created habitats, such as artificial reefs.

Source table for information on habitats with an intertidal aspect

Habitat type	Intertidal aspects	Major changes and threats	Source
Sand and mud substrates (soft bottom)	Sand beaches and mudflats	Sand extraction for building artificial islands; erosion and redistribution of sediment from increased wave action	Chapter 7B
Rocky substrates and reefs	Rocky shores	Increased thermal range and wave impacts from storms, and invasive species, all decrease local biodiversity	Chapter 7B
Atoll and island lagoons	Shallow fringing reefs, coral rubble	Rising sea levels, ocean warming, increased wave heights and coastal erosion	Chapter 7C
Tropical and subtropical coral reefs	Coral rubble and intertidal hard and soft corals	Coral bleaching and physiological responses to ocean warming; coastal erosion and nutrient run-off	Chapter 7D
Estuaries and deltas	Tidal estuaries	Rising sea levels, terrestrial pollutants and run-off	Chapter 7F
Kelp forests and algal beds	Intertidal algae	Increased thermal range and wave impacts from storms, and invasive species, all decrease local biodiversity	Chapter 6G
Seagrass meadows	Intertidal seagrass	Physical disturbance from anchoring or development; warming sea temperatures	Chapter 7G
Mangroves	Coastal mangroves	Logging and clearance	Chapter 7H
Salt marshes	Tidal marshes	Rising sea levels, terrestrial pollutants and run-off	Chapter 7I
High-latitude ice	Polar coastal environments	Climate-related loss of winter ice cover increases disturbance from temperature fluctuations and physical impact of broken sea ice and iceberg scouring; loss of ice also opens pathways for species to invade new areas	Chapter 7K
Artificial substrates and built environment		Invasive species, pollutants	Chapter 7A

2. Description of the environmental changes between 2010 and 2020

Coastal and near-coastal marine environments are the marine habitats most affected by climate change (Hoegh-Guldberg and Bruno, 2010). Many intertidal species are harvested or cultivated worldwide, depending on access to coastal habitats and the condition of those habitats in the context of species distributions, physical disturbance, development, traffic and pollution. The impacts of degradation of intertidal environments are worst for island and coastal countries, where the intertidal zone represents a larger proportion of their territorial area, but all nations are affected, directly or indirectly (Curran and others, 2002). Vegetated intertidal areas, including salt marshes and mangroves, have been removed or severely degraded by coastal development, with over 50 per cent of wetlands and mangroves removed during the last century (Burke and others, 2000). Coastal environments are also affected by pollution run-off from terrestrial sources. The combined impacts can alter fresh water as well as marine resources. Human activities continue to modify the physical shape of coasts both directly and indirectly through construction that alters or creates shorelines and consequent modifications to hydrodynamics and sediment transportation, which all change habitat conditions.

The overall effect of changes to coastal environments is to reduce available intertidal habitats and the quality of the remaining habitat. Intertidal organisms and ecosystems are typically at the edge of tolerance for one stressor and may have unexpected reactions to additional stresses from environmental change, meaning that local responses are often unpredictable (Hewitt and others, 2016). That limits the distribution and sustainability of fished species. The impacts of climate change include temperature change, but also sea level rise and changes in wave heights and increasing storm events. Sea level rise reduces the availability of intertidal habitats as the natural environment is confronted by mitigation efforts, such as the construction of sea walls and coastal defences. That results in a type of "coastal squeeze" whereby, when sea level rises, marine influences move inland to space already occupied by human activities (Pontee, 2013).

Human alteration of coastlines also includes urbanization and development, the construction of urban and maritime infrastructure and the development of recreational activities. Physical infrastructure includes bridges, roads, sea walls, dams and flood gates, as well as energy infrastructure, such as wind and tidal energy converters. Such structures create substrata, a potential reef-like hard bottom that can be occupied by rocky intertidal species. However, while that may create a local-scale increase in species diversity, the overall effect is habitat loss. In recent years, the scale of human alteration to the coasts has accelerated dramatically, with major projects to build artificial islands and peninsular structures to increase coastal housing. Those projects affect intertidal communities in as yet unknown ways, as the physical material, rock and sand, imported from outside the local region will bring additional alien biological material. Construction of new islands smothers the habitat that previously occupied the same space and changes local hydrodynamic conditions and sedimentation that will smother further adjacent habitats. New structures are also inhabited by high-density human populations that introduce additional environmental impacts.

3. Economic and social consequences

The ocean and, in particular, coastal areas have an impact on all Sustainable Development Goals.[1] Intertidal habitats provide the most common examples of marine ecosystem goods and services, and coastal habitats have value for biodiversity as well as being of service to humanity. There is also a strong gender element to the exploitation of marine resources, with some fisherwomen's organizations being developed in Europe, starting in the 1990s (Frangoudes and others, 2014).

Species appropriate to a local context may not be the most efficient example of a particular service but, more importantly, they contribute to regional biodiversity. For example, mussels and oysters provide water filtration and food, but there are over 300 species in those taxonomic families of bivalves (World Register of Marine Species (WoRMS) Editorial Board, 2017), many of which fill discrete niches or ecosystem functions. Many of those species are widely cultivated and consumed. It is not sustainable in the long term to select a single species for large-scale aquaculture on all global coasts. Additional species in their local native habitats support biodiversity and the diversification of human resources.

The current consequences of human alterations to coastlines have both positive and negative impacts on intertidal biodiversity. The relative impacts of the various anthropogenic pressures mentioned in section 2 differ between developed and developing economies. Rock armour, or riprap, are hard structures built to control the erosion of coastlines. The materials include structures designed to support habitat space for intertidal organisms that can provide important local-scale mitigation to habitat loss and improve delivery of additional benefits to humans (Chapman and Underwood, 2011). Structures engineered to increase habitat space are sometimes called "living sea walls" and can reduce some impacts of coastal hardening. Artificial substrates also apparently favour non-native and invasive marine species that outcompete native fauna in rocky substrata (Tyrrell and Byers, 2007). Such eco-engineering approaches can limit, if not mitigate, habitat loss through coastal sprawl. Another form of coastal modification is land construction ("reclaimed" land), which, although beneficial to humans in the short term, reduces the ability of natural systems to deliver other benefits, including natural wave and storm defences. Coastal communities are at risk from changes to physical safety and access to food, with implications for profoundly important issues related to the Sustainable Development Goals, such as poverty, education and the availability of food.

4. Key region-specific changes and consequences

Specific habitat types have greater prevalence in some regions, depending on local coastal morphology. For example, rocky intertidal habitats have very high biodiversity in temperate latitudes in the North Atlantic and the North Pacific, while the coast of Brazil in the South Atlantic is considered a hotspot for macroalgae (Miloslavich and others, 2016). Mangroves and corals (extending to subtidal depths) are characteristic habitats in tropical and subtropical coastlines where the threat from sea level rise is the greatest.

1 See General Assembly resolution 70/1.

We know more about intertidal than subtidal habitats almost everywhere in the world, except where coasts are inhospitable or pose significant dangers (e.g., areas dominated by saltwater crocodiles) and in high latitudes where human population density is sparse to non-existent. The Antarctic and Arctic regions contain areas that remain completely unsampled for coastal flora and fauna. Tropical regions, in particular in South-East Asia, contain disproportionate numbers of new species that have not been described, but are increasingly being recognized, in particular through molecular genetic analysis. Species under pressure in areas that have been less studied are at higher risk from potential extinction, as appropriate conservation measures cannot be assessed.

Artificial habitats also vary in their distribution according to local conditions. Built islands are features that are found mainly in the shallow, sheltered seas of the Arabian Gulf and South-East Asia. Rock armour is found worldwide, but in terms of its contribution to habitat space has been most studied in Australia, North America and Europe. Energy converters, such as offshore wind turbines are found especially in Europe and increasingly in North America (see chap. 21). There is a growing demand for coastal development, both for housing and urban development, but also for coastal resources such as aquaculture and energy converters, with increasing detrimental effects on vegetated habitats. In regions with intensive coastal urbanization, such as Australia, the Middle East, Asia, Europe and the United States, more than half of the available shoreline in some regions has been modified through engineering and built coastal structures (Dafforn and others, 2015). Climate change is increasing coastal erosion, which prompts construction of additional hard engineering defences, such as sea walls, and accelerates coastal modification (Asif and Muneer, 2007).

5. Outlook

The outlook for the knowledge base on intertidal habitats is good in many respects, and marine research has a natural emphasis on intertidal and coastal regions because of their accessibility in most regions and their importance to human activities. Intertidal areas have been included in some marine protected areas.

The socioeconomic consequences of continued change in intertidal habitats are potentially severe. In countries where extensive tidal flats are well developed and local populations are highly dependent on marine ecosystem services, such as in many Asian coastal areas, a reduction in intertidal space through coastal squeeze will have severe impacts, reducing both the area and the supply of resources. Physical degradation of coastlines through climate change will decimate local economies. Biotic degradation through altered hydrodynamics, invasive species and overexploitation has dramatic and complex impacts. The removal of mangroves and biotic reefs eliminates natural coastal defences that protect human settlements. Invasive species reduce local biodiversity. Overfishing or overdependency on monoculture aquaculture species, mainly cultivated in intertidal regions, lowers nutritional quality and puts human prosperity at risk. Coastal areas house key public infrastructure, such as power, wastewater treatment and transportation facilities (e.g., airports), which are also at risk from sea level rise. Protection of natural local coastal areas and intertidal biodiversity are crucial for human sustainability.

6. Key remaining knowledge gaps

Several topics require critical attention to ensure the sustainability of intertidal habitats. Slow, cumulative systemic shifts are often not recognized until the change becomes catastrophic. Assessments of damaged environments set the targets for conservation, meaning that the environment will never recover to a truly robust and sustainable condition (Plumeridge and Roberts, 2017). Even in European seas, where there is perhaps the longest continuous history of recorded observations, pre-industrial "baseline" data are already influenced by human impacts, and the problem is far worse in understudied systems and in many developing nations.

The physical parameters and coastline alterations associated with anthropogenic impacts and sea level rise require additional studies to enable predictive models for hydrodynamic impacts and small, local-scale models using analogous systems, where the behaviour of one well-studied physical system can be applied to predict impacts in another place. The first *World Ocean Assessment* highlighted a need for more information on the succession of habitat types and species ranges with alterations in coastlines, but that area remains a key knowledge gap. Finally, underpinning all of those issues is an urgent need for more detailed studies of biodiversity in understudied areas, especially in regions where significant knowledge gaps exist and scientific infrastructure is less developed, but where there is great species diversity (Lira-Noriega and Soberón, 2015). Many species, even in the well-studied intertidal zones, remain unnamed and undescribed by science. Without species identification, the biodiversity of the habitats cannot be accurately quantified or monitored.

7. Key remaining capacity-building gaps

"Coastal squeeze" reduces intertidal environments, with sea level rise on one side and human urbanization on the other side. Human development should include future planning to provide coastal and intertidal habitats space for retreat from increasingly frequent storm events and climate disruption to maintain those important protective buffers.

The countries that are the most extremely diverse in terms of important areas of biodiversity and species richness are, for the most part, developing economies (Lira-Noriega and Soberón, 2015). There is an urgent need to support baseline studies and monitoring to develop and sustain the same kind of long-term data sets in developing countries that are available in developed nations, especially in Europe and North America. In developed countries, citizen science approaches can be effective tools to expand monitoring; with further capacity development for taxonomy, that approach could be applied more widely.

There is an urgent need also to build taxonomic infrastructures, which support such emerging technologies as environmental DNA (eDNA), through specimen collections and barcode catalogues, and to develop human capacity through training and technology transfer and access to the latest scientific resources and to data and scientific literature in the country of origin. It is not possible to use emerging technologies, such as eDNA barcoding, where there is no taxonomic infrastructure in place. Barcodes can recognize only what is already in a database. Taxonomic infrastructure must also include specialist skills, literature and the resources to support fundamental science. They are essential to enable robust environmental impact assessment processes, especially in developing countries. Further, those types of fundamental science underpin the ability to further develop capacity to conduct climate vulnerability assessments with respect to key marine species and habitats.

Although the intertidal zone includes the most accessible (and most vulnerable) habitats, a high proportion of undescribed invertebrate and algal species occur in tropical shallow marine ecosystems. The lack of taxonomic ability to identify local species obscures potential loss, encroachment of species with shifting ranges and indicators of disturbance and makes it difficult to correctly recognize invasive species and take appropriate measures to protect local resources (Sigwart, 2018). Scientific infrastructure underpins downstream economic growth and the protection of environmental resources.

References

Asif, M., and T. Muneer (2007). Energy supply, its demand and security issues for developed and emerging economies. *Renewable and Sustainable Energy Reviews*, vol. 11, No. 7, pp. 1388–1413. https://doi.org/10.1016/j.rser.2005.12.004.

Burke, Lauretta, and others (2000). *Pilot Analysis of Global Ecosystems: Coastal Ecosystems*. Washington, D.C.: World Resources Institute. www.wri.org/publication/pilot-analysis-global-ecosystems-coastal-ecosystems.

Chapman, M.G., and A.J. Underwood (2011). Evaluation of ecological engineering of "armoured" shorelines to improve their value as habitat. *Journal of Experimental Marine Biology and Ecology*, vol. 400, Nos. 1–2, pp. 302–313. https://doi.org/10.1016/j.jembe.2011.02.025.

Curran, Sara, and others (2002). Interactions between Coastal and Marine Ecosystems and Human Population Systems: Perspectives on How Consumption Mediates this Interaction. *AMBIO: A Journal of the Human Environment*, vol. 31, No. 4, pp. 264–268. https://doi.org/10.1579/0044-7447-31.4.264.

Dafforn, Katherine A., and others (2015). Marine urbanization: an ecological framework for designing multifunctional artificial structures. *Frontiers in Ecology and the Environment*, vol. 13, No. 2, pp. 82–90. https://doi.org/10.1890/140050.

Frangoudes, Katia, and others (2014). Women's organisations in fisheries and aquaculture in Europe: History and future prospects. *MARE Publication Series*, vol. 9, pp. 215–231. https://doi.org/10.1007/978-94-007-7911-2_12

Hewitt, Judi E., and others (2016). Multiple stressors, nonlinear effects and the implications of climate change impacts on marine coastal ecosystems. *Global Change Biology*, vol. 22, No. 8, pp. 2665–75. https://doi.org/10.1111/gcb.13176.

Hoegh-Guldberg, Ove, and John F. Bruno (2010). The Impact of Climate Change on the World's Marine Ecosystems. *Science*, vol. 328, No. 5985, pp. 1523–1528. https://doi.org/10.1126/science.1189930.

Leary, David, and Miguel Esteban (2009). Renewable energy from the ocean and tides: a viable renewable energy resource in search of a suitable regulatory framework. *Carbon & Climate Law Review*, No. 4, pp. 417–25.

Lira-Noriega, Andrés, and Jorge Soberón (2015). The relationship among biodiversity, governance, wealth, and scientific capacity at a country level: Disaggregation and prioritization. *Ambio*, vol. 44, No. 5, pp. 391–400.

Miloslavich, Patricia, and others (2016). Chapter 3: Benthic Assemblages in South American Intertidal Rocky Shores: Biodiversity, Services, and Threats. In *Marine Benthos: Biology, Ecosystem Functions and Environmental Impact*, ed. Rafael Riosmena-Rodríguez. Nova Science Publisher.

Plumeridge, Annabel A., and Callum M. Roberts (2017). Conservation targets in marine protected area management suffer from shifting baseline syndrome: A case study on the Dogger Bank. *Marine Pollution Bulletin*, vol. 116, Nos. 1–2, pp. 395–404.

Pontee, Nigel (2013). Defining coastal squeeze: A discussion. *Ocean & Coastal Management*, vol. 84, pp. 204–7. https://doi.org/10.1016/j.ocecoaman.2013.07.010.

Sigwart, Julia D. (2018). *What Species Mean: A User's Guide to the Units of Biodiversity.* CRC Press.

Tyrrell, Megan C., and James E. Byers (2007). Do artificial substrates favor nonindigenous fouling species over native species? *Journal of Experimental Marine Biology and Ecology*, vol. 342, No. 1, pp. 54–60.

United Nations (2017). *The First Global Integrated Marine Assessment: World Ocean Assessment I*. Cambridge: Cambridge University Press.

World Register of Marine Species (WoRMS) Editorial Board (2017). *World Register of Marine Species*. VLIZ. www.marinespecies.org.

Chapter 7B
Biogenic reefs and sandy, muddy and rocky shore substrates

Contributors: Ronaldo Christofoletti and Judith Gobin (joint conveners), Frédéric Guichard, Sergiy Medinets, Evangelina Schwindt and Julia Sigwart.

Keynote points

- Biogenic reefs and sandy, muddy and rocky shores support high biodiversity and a wide range of ecosystem services that benefit human populations.
- They are under pressure from multiple stressors as a result of climate change, urbanization and the use of resources; storms, land reclamation, contaminants and pollutants have emerged as the main drivers.
- There is a gap in interdisciplinary research and participative governance to promote resilience and provide for the sustainable development of these habitats.
- Owing to their cultural significance and importance for tourism worldwide, they are in a unique position to serve as flagship habitats to promote the role of the ocean in the implementation of the 2030 Agenda for Sustainable Development, including Sustainable Development Goal 14.[1]

1. Introduction

Coastal environments are home to a variety of valuable natural resources, including sandy, muddy and rocky shores and reefs. All of those habitats present a high biodiversity (see chap. 6 of the present Assessment), and an increasing number of studies examine patterns, processes and impacts associated with them. Biodiversity and the human impact on rocky shores were recently reviewed at regional scales (Hawkins and others, 2019). However, a gap remains in the understanding of rocky and muddy shores at a global scale. For sandy shores, recent trends show that species richness worldwide is related to ecoregions, where temperature and latitude predict an increase in species richness from temperate to tropical shores (Barboza and Defeo, 2015; figure I). Reefs form biogenic habitats that are ubiquitous in coastal systems worldwide but vary in extent and species composition across biogeographic regions (Firth and others, 2016). While in tropical regions they are typically formed by the calcium carbonate secreted by reef-building corals and calcareous red algae in shallow-water settings (Huang and Roy, 2015), in temperate regions they are formed by invertebrates, including oysters, mussels and annelid worms (Barbier and others, 2008; Dubois and others, 2009; Firth and others, 2016; figure II).

The present chapter provides an integrated overview of biogenic reefs and sandy, muddy and rocky shores, in both intertidal and subtidal zones, which are connected by their locations in the marine and terrestrial interface. The habitats cover all coastlines worldwide (Firth and others, 2016; Luijendijk and others, 2018; chap. 7A of the present Assessment) and are linked to different ecosystems, including atoll and island lagoons (chap. 7C), coral reefs (chaps. 7D and 7E), estuaries and deltas (chap. 7F), kelp forests and algal beds (chap. 6G), seagrass meadows (chap. 7G), mangroves (chap. 7H) and salt marshes (chap. 7I). They are influenced by many drivers and oceanographic dynamics that emerge from local to global scales (chaps. 4, 5 and 22). To minimize overlap and highlight their interactions, there will be a focus on biogenic reefs and sandy, muddy and rocky shores. The link with other chapters in which close habitats are detailed (e.g., chaps. 7A and 7G), in particular coral reefs in chapter 7D (the tropical and subtropical zone) and cold water corals in chapter 7E, should be noted.

1 See General Assembly resolution 70/1.

Figure I
Sandy beach species richness in temperate and tropical ecoregions

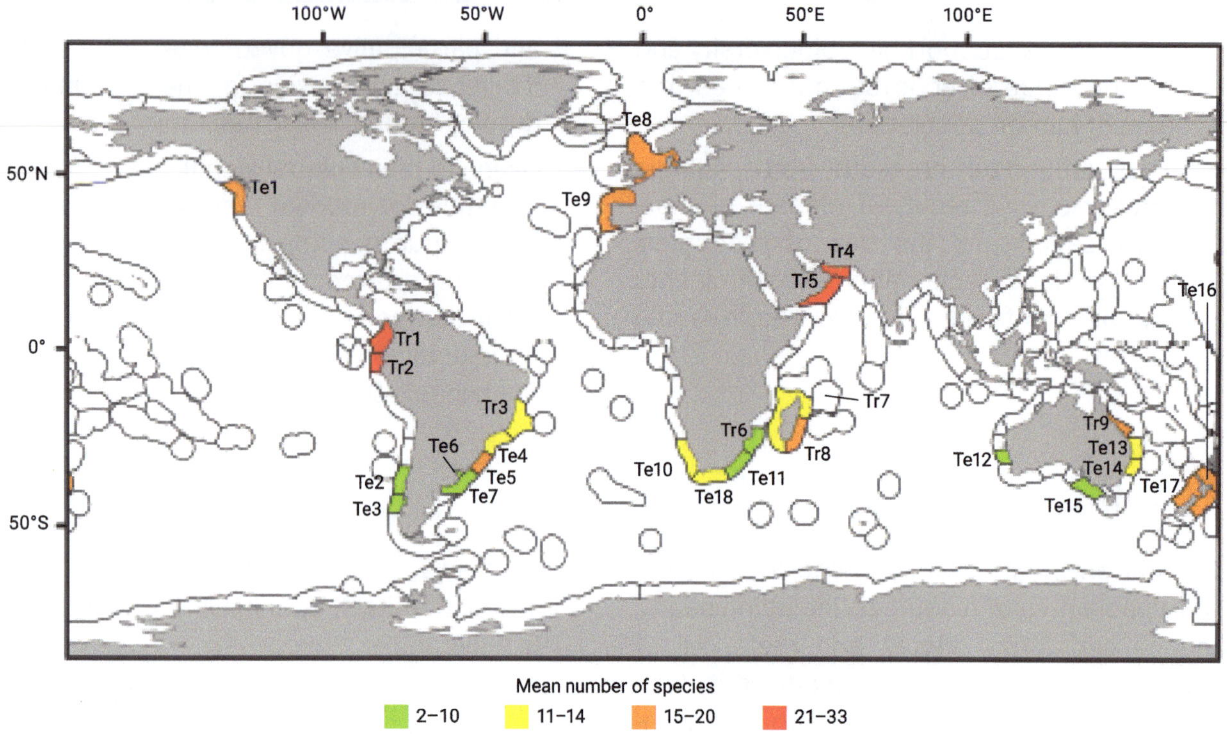

Source: Reprinted from Barboza and Defeo, 2015; Spalding, M.D. and others, "Marine ecoregions of the world: a bioregionalization of coastal and shelf areas", *Bioscience*, vol. 57 (2007), pp. 573–583.
Note: The map containing ecoregions was downloaded from http://maps.tnc.org/gis_data.html. The final map was generated using gvSIG 1.12 (www.gvsig.org).
Abbreviations: Te, temperate; Tr, tropical; Te1, Oregon, Washington, Vancouver coast and Shelf; Te2, Araucanian; Te3, Chiloense; Te4, South-eastern Brazil; Te5, Rio Grande; Te6, Rio de la Plata; Te7, Uruguay-Buenos Aires Shelf; Te8, North Sea; Te9, South European Atlantic Shelf; Te10, Namaqua; Te11, Natal; Te12, Houtman; Te13, Tweed-Moreton; Te14, Manning-Hawkesbury; Te15, Western Bassian; Te16, North-eastern New Zealand; Te17, Central New Zealand; Te18, Agulhas Bank; Tr1, Panama Bight; Tr2, Guayaquil; Tr3, Eastern Brazil; Tr4, Gulf of Oman; Tr5, Western Arabian Sea; Tr6, Delagoa; Tr7, Western and northern Madagascar; Tr8, South-eastern Madagascar; Tr9, Central and southern Great Barrier Reef.

Biogenic reefs and sandy, muddy and rocky shores are characterized by high biodiversity (chap. 6) and the provision of ecosystem services (chaps. 8 and 21), including water filtration and nutrient cycling (chaps. 10 and 11). There is a strong connection between their ecosystem services and urbanization (chap. 8), as about 60 per cent of the world's populations live and derive livelihoods in coastal areas (Nicholls and others, 2007). Such coastal environments, where key activities include boating, fishing, surfing, swimming and bird watching (Everard and others, 2010; Rodríguez-Revelo and others, 2018), are economically relevant for tourism, for recreational, artisanal and commercial fishing and also for aesthetic and recreational value (chaps. 8 and 15). Such environments are also interconnected with many aspects of development, including urbanization, aquaculture and infrastructures (chaps. 8, 14 and 16).

As a consequence of the wide range of ecosystem services provided by such habitats and their association with coastal urbanization and coastal protection, they are vulnerable to impacts from multiple stressors (chap. 25). They continue to be adversely affected by pollutants and contaminants, such as excessive nutrients from fertilizers, toxic chemicals and heavy metals, sewage, waste and plastic (chaps. 10, 11 and 12) and also by mining, oil and gas exploration and exploitation activities (chaps. 11, 18 and 19) and the more recently documented threats of invasive species (chap. 22). At the same time, sedimentation and changes in coastal erosion (chap. 13) are long-term processes that are increasing under the pressure of climate change impacts (chap. 9) that contribute to changes in the formation of coastlines and may also pose a threat to life and property (Rangel-Buitrago and Anfuso, 2009; Le Duff and others, 2017).

The coastal zone is the most urbanized region in the world, hosting 15 of the 20 megacities (with populations of over 10 million people), in which there is a diversity of ecosystem services provided by the habitats and a conflict with increasing urbanization, as stated in the first *World Ocean Assessment* (United Nations, 2017). The present chapter will cover the changes recognized since the first Assessment, including advances in knowledge or policy. It will also highlight the global analysis of sandy shores and the lack of information at the global level on biogenic reefs and muddy and rocky shores needed to support coastal management and marine spatial planning (chaps. 26 and 27). Despite the utility and economic benefits that coasts provide and the advances in global-scale studies on sandy shores, there is no reliable global-scale assessment of historical trends in shoreline changes for rocky and muddy substrates and there are still regions with very little information and data available on their ecosystems.

Figure II
Global distribution of coastal biogenic reefs (coral, mussels, oysters, worms)

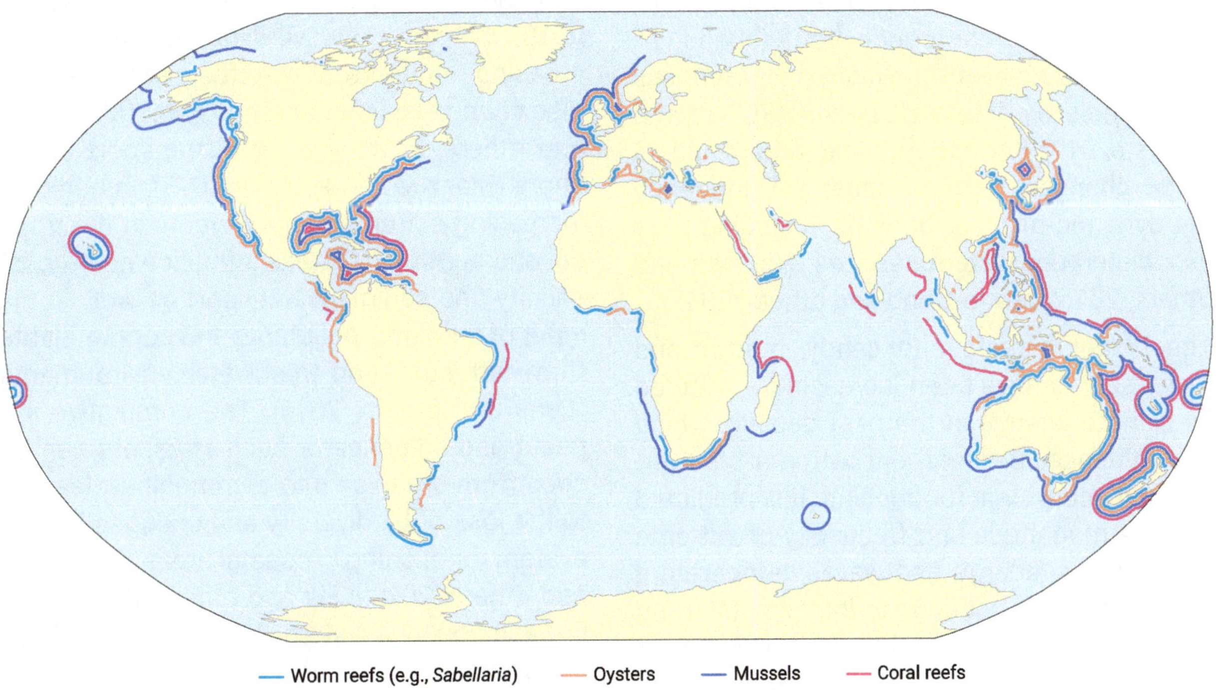

Source: Reprinted from Firth and others, 2016. Data extracted from the Global Biodiversity Information Facility, available at www.gbif.org/, and the United Nations Environment Programme World Conservation Monitoring Centre Ocean Data viewer, available at https://data.unep-wcmc.org.

Note: Maps created by Shaun Lewin, Plymouth University.

2. Documented change in state of biogenic reefs and sandy, muddy and rocky shore substrates

Threats to biogenic reefs and sandy, muddy and rocky shores can be from multiple environmental (some extreme, such as storm surges, hurricanes, earthquakes, tsunamis, heatwaves and floods) and anthropogenic drivers, as stated in the introduction. Such drivers emerge from local to regional and global scales, while the anthropogenic factors dominate change at multiple scales (Mentaschi and others, 2018).

Changes in biogenic reefs and sandy, muddy and rocky shores are all influenced by different components of the seascape, as mangroves, rhodoliths and algae beds, deep zones, coral reefs and seagrass meadows all respond to stressors differently. There is also atmospheric deposition on coastal zones (Medinets and Medinets, 2010, 2012; Medinets, 2014) and discharge of sediments and nutrients on coastal marine systems by estuaries and freshwater (Teixeira and others, 2018; Oelsner and Stets, 2019). Those natural connections (chaps. 7A to 7I) clearly highlight the interconnectedness and complexity of coastal systems (Elliott and others, 2019; Kermagoret and others, 2019) since changes in one habitat will influence the dynamic of other habitats, including their associated ecosystem services (Narayan and others, 2016; Osorio-Cano and others, 2019).

The biogenic reefs and sandy, muddy and rocky shores have been increasingly affected by climate change in the past decade, which has influenced environment patterns, biodiversity and ecosystem functioning. It is predicted that the magnitude and frequency of extreme events (wave energy, heatwaves, temperature and rainfall) will continue to intensify (Herring and others, 2018; Intergovernmental Panel on Climate Change (IPCC), 2018). Changes in the number of days that exceed the temperature thresholds specific to each species, or changes in rain and drought regimes, can lead to sublethal stress, owing to physiological and behavioural changes in organisms, in particular those in intertidal and shallow zones (Pinsky and others, 2019; Rilov and others, 2019). The changes in frequency and intensity of the events can lead to physiological lethal levels, increasing mortality and altering the biodiversity, the range of distribution of organisms and the ecosystem services that such habitats provide (Poloczanska and others, 2013).

In the environmental context, high-latitude, intertidal rocky substrates are affected by ice scour (Scrosati and Ellrich, 2018; chaps. 7A and 7K). Sandy, muddy and rocky shores worldwide are also under increasing wave and extreme rainfall disturbances (Mentaschi and others, 2018), which influence sediment dynamics, erosion, boulder movements and landslides that can change biological communities in sandy and rocky shores in wave-exposed areas (Petrovic and Guichard, 2008; Castelle and others, 2018). Changes in wave dynamics and the increasing frequency of extreme weather events also change sediment composition (Masselink and others, 2016) and larval transport to the shore (Mazzuco and others, 2015). In addition, increasing extreme rainfall events in the tropical and subtropical areas influence near-coast salinity and sediment transport as well as the input of nutrients, pollutants and contaminants from terrestrial and freshwater environments (Lana and others, 2018). The cumulative impacts and influence of such stressors can be seen from organisms to communities, leading to the loss of biodiversity and changes in ecosystem functioning in coastal areas (O'Gorman and others, 2012; Ellis and others, 2017), with impacts on ecosystem services, as well as commercial, recreational and aesthetic values.

In addition to the changes in coastal oceanographic dynamics and the increase in the frequency of extreme events owing to the effects of climate change (chap. 9), other drivers, such as seabed exploration (chap. 18), urbanization (chaps. 8 and 14) and artificial coastal infrastructures (chaps. 7A and 14), are influencing reefs and sandy, muddy and rocky shores owing to contamination (chaps. 10, 11 and 12) and changes in the processes of erosion and sedimentation (chap. 13). Sandy beaches are present in the coastline worldwide, varying from 22 per cent in Europe to 66 per cent in Africa, with their relative occurrence increasing in the subtropics and lower middle latitudes (20°–40°) but decreasing (<20 per cent) in the humid tropics, where mud and mangroves are most abundant as a result of high temperatures and rainfall (figure III; Luijendijk and others, 2018). The erosion of sandy beaches has been increasing over time and with the intensity of greenhouse gas emissions (Vousdoukas and others, 2020). Erosive and accretive tendencies have been interchanging across regions and along nearby coastal segments (Vousdoukas and others, 2020), with more than 50 per cent of the world's sandy shores suffering chronic and severe rates of change over the period 1984–2016, and with 24 per cent of shores eroding at a rate exceeding 0.5 m per year, while 27 per cent are accreting (Luijendijk and others, 2018; figure IV). From a continental perspective, Oceania and Africa present net erosion, while all other continents show net accretion, with the highest accretion rate (1.27 m/year) in Asia (Luijendijk and others, 2018), likely owing to land reclamation and artificial structures (Luijendijk and others, 2018; chap. 14). On a global scale, a relatively high percentage of sandy shorelines recorded in the World Database on Protected Areas are experiencing erosion, bearing in mind that 32 per cent of all marine protected shorelines are sandy, and 37 per cent of those protected sandy shorelines are eroding at a rate of above 0.5 m per year, while 32 per cent are accreting (Luijendijk and others, 2018).

Changes in erosion and sedimentation and the presence of artificial structures can directly influence biodiversity and ecosystem services at different scales. The increase in coastal infrastructure to avoid erosion requires the strengthening of blue engineering approaches for sustainable development (Firth and others 2016; Strain and others, 2018). Although coral concrete has been suggested for the development of marine infrastructure and land reclamation (Wang and others, 2018; Liu and others, 2018), to achieve sustainable development, it is critical to understand the source, the amount of the coral material needed and the impact of its extraction from the environment, as coral reefs have biological, chemical and physical importance in the dynamics of coastal areas and in the climate change scenario (chaps. 7D and 7E).

The impacts of urbanization on coastal zones warrant consideration of several key stressors, including anthropogenic drivers, that exist in the coastal zone. Invasive species (chap. 22) have increased worldwide (Seebens and others, 2017), thus affecting all types of substrates. Biological invasions are expected to increase as a result of maritime transport and also the increase in coastal infrastructure in coastal areas as a new substrate to rocky and reef species (Ivkić and others, 2019; Sardain and others, 2019). Furthermore, recent data sets worldwide show land-based pollution (nutrient pollution, agrochemicals, sewage discharge, chemical contamination by persistent organic pollutants in the form of pharmaceuticals, pesticides and heavy metals), coastal urbanization, land reclamation and oil spills that change habitats, increase contamination and cause sublethal to lethal processes that affect muddy, sandy and rocky shore biodiversity and ecosystem health (Kovalova and others, 2010; Snigirov and others, 2012; Environmental Monitoring of the Black Sea (EMBLAS), 2019; Martinez and others, 2019; Zhai and others, 2020). Many impacts on the shores have their origin offshore, such as oil spills (Escobar, 2019; Soares and others, 2020), or inland, such as mine tailings, with accidental discharge in the coastal zone, through riverine inputs, reaching and affecting the biodiversity and ecosystem services of sandy, muddy and rocky shores on large spatial and temporal scales on the coast (Queiroz and others, 2018) and affecting local and indigenous communities that depend on those ecosystem services for their survival (Dadalto and others, 2019).

Finally, another consequence of coastal urbanization relates to the negative impacts of tourism and human exploitation on biogenic reefs and sandy, muddy and rocky shores at local scales (Mendez and others, 2017). Artificial light at night has been shown to change macroinvertebrate community structure on sandy shores (Garratt and others, 2019) and influence trophic interactions on rocky shores (Underwood and others, 2017; Maggi and Benedetti-Cecchi, 2018). Similarly, shading by artificial infrastructure can influence biodiversity and ecosystem functioning on rocky shores (Pardal-Souza and others, 2017). Trampling has been shown to negatively affect biodiversity on sandy, muddy and rocky shores (Leite and others, 2012; Schlacher and Thompson, 2012; Kim and others, 2018), in addition to other factors, such as littering, noise and extraction (EMBLAS, 2019). Plastic and chemical pollution has become a global threat to the marine environment, especially to sandy shores, where plastic input has increased as a result of transport during oceanographic and meteorological events (Krelling and Turra, 2019) and direct contamination by locals and tourists (EMBLAS, 2019).

Figure III
Global distribution of sandy shorelines

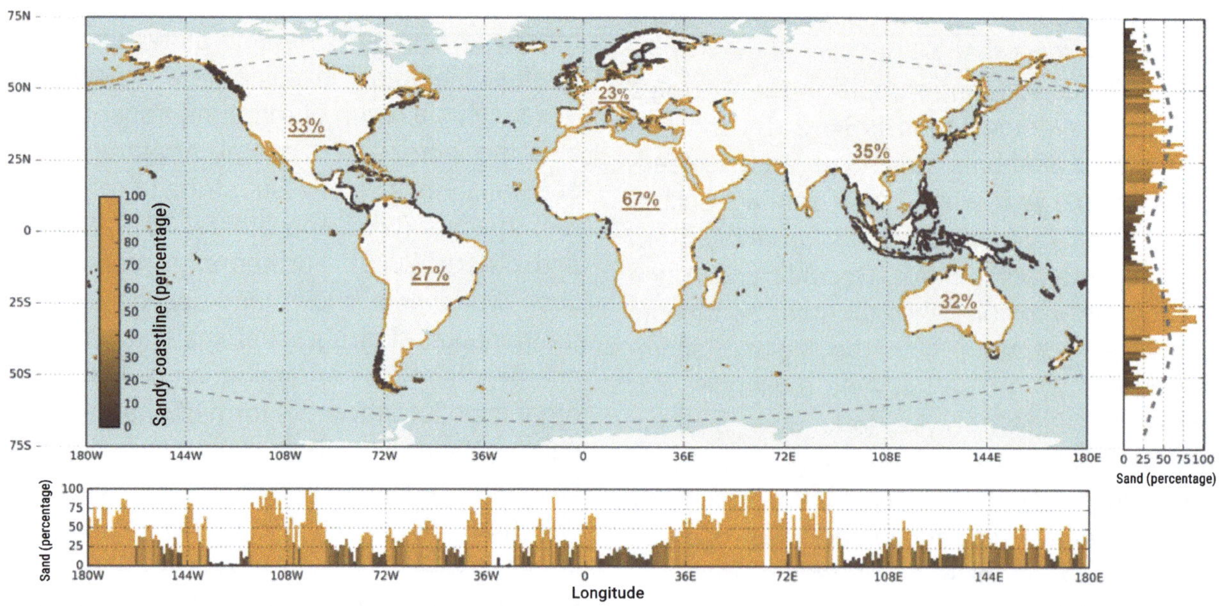

Source: Reprinted from Luijendijk and others, 2018.

Notes: The dots along the world's shoreline represent the local percentage of sandy shorelines (yellow is sand, dark brown is non-sand). The subplot to the right presents the relative occurrence of sandy shorelines per degree latitude, where the dashed line shows the latitudinal distribution of sandy shorelines (see Hayes, M.O., "Relationship between coastal climate and bottom sediment type on the inner continental shelf", *Marine Geology*, vol. 5, No. 2 (1967), pp. 111–132). The lower subplot presents the relative occurrence of sandy shorelines per degree longitude. The curved, dashed grey lines in the main plot represent the boundaries of the ice-free shorelines considered in the analysis. The underlined percentages indicate the percentages of sandy shorelines averaged per continent. The map was created with Python 2.7.12, available at www.python.org, using Cartopy, v.0.15.1. Met Office UK, available at https://pypi.python.org/pypi/Cartopy/0.15.1, and Matplotlib (Hunter, J.D., "Matplotlib: a 2D graphics environment". *Computing in Science & Engineering* vol. 9, No. 3 (2007)).

Figure IV
Global hotspots of beach erosion and accretion

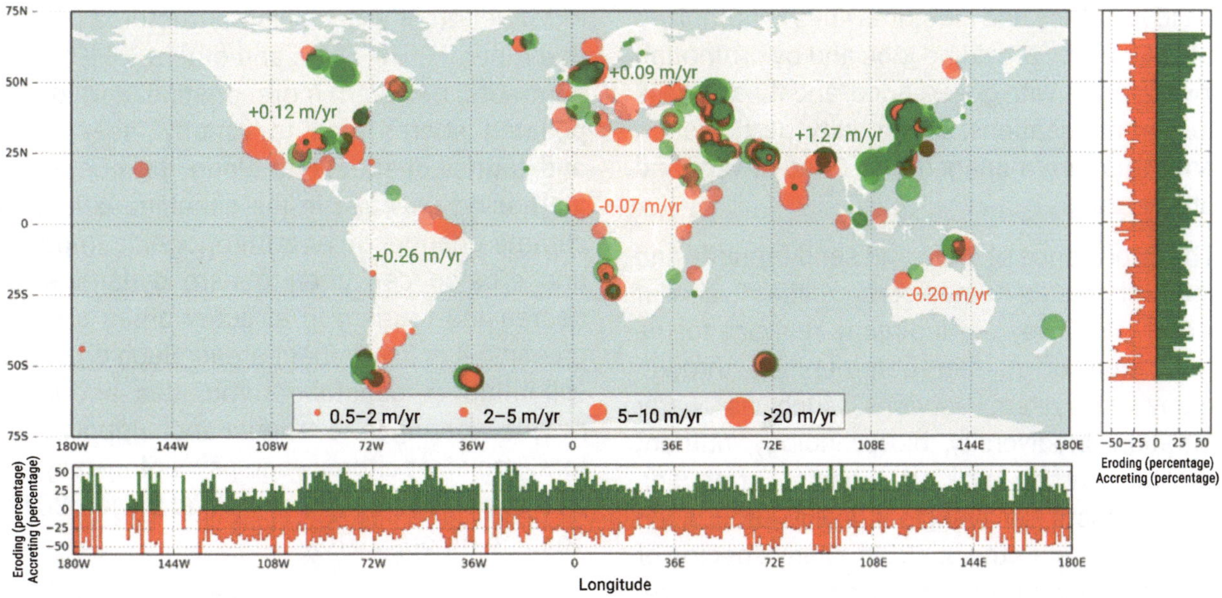

Source: Reprinted from Luijendijk and others, 2018.
Notes: The red (green) circles indicate erosion (accretion) for the four relevant shoreline dynamic classifications (see legend). The bar plots to the right and at the bottom present the relative occurrence of eroding (accreting) sandy shorelines per degree latitude and longitude, respectively. The numbers presented in the main plot represent the average change rate for all sandy shorelines per continent. The map was created with Python 2.7.12, available at www.python.org, using Cartopy, v.0.15.1. Met Office UK, available at https://pypi.python.org/pypi/Cartopy/0.15.1, and Matplotlib (Hunter, J.D., "Matplotlib: a 2D graphics environment", *Computing in Science & Engineering*, vol. 9, No. 3 (2007)).

3. Consequences of the changes on human communities, economies and well-being

Coastal habitats are the first point of contact between human society and the ocean. They provide many direct and indirect services, including space for leisure and sports, environmental chemical-physical processes, biological and fish stocks resources and coastline protection. From intertidal to subtidal areas, reefs and sandy, muddy and rocky shores are explored in many ways by human populations, thus creating a relationship that has been affected by changes over centuries, but at an accelerating pace in recent decades (Biedenweg and others, 2016; Zhai and others, 2020).

Sandy and rocky shores provide space and natural resources for leisure, sports, educational and scientific studies, and traditional, religious and cultural practices for indigenous peoples and traditional communities, and a place to visit for residents from urban areas and tourists (Everard and others, 2010). There are many physical, mental and spiritual health benefits related to time spent in coastal environments and to the health of that environment (Gascon and others, 2017; Marselle and others, 2019). The human-environment link develops psychologically through a sense of place and identity, a connection with nature and belonging, with feelings of pride in one's environment and the revitalizing properties of aesthetic landscapes, while the environment also physically influences people by providing tangible services, such as food (Biedenweg and others, 2016). Finally, muddy, sandy and rocky shores also benefit society as a result of economic aspects related to business and industrial links in urbanized coastlines, by providing jobs and by promoting involvement with governance, and thus access to communications, community participation and trust in management (Biedenweg and others, 2016).

The importance of reefs and sandy, muddy and rocky shores for human communities is similar worldwide. They are indirectly relevant to the well-being of human communities through many ecosystem services, such as water filtration, biodiversity, biotechnology, nutrient cycling, carbon sinking, coastal protection and impacts on pelagic primary production (Hoerterer and others, 2020). Many species of biotechnological interest have been studied recently, thus indicating a potential for scientific and economic development (Park and others, 2019; Girão and others, 2019). Biogenic reefs and sandy, muddy and rocky shores present species of economic interest, mainly molluscs, crustaceans and fishes, that are particularly important for traditional communities as sources of proteins and income, based on traditional fishing (e.g., Gelcich and others, 2019).

The importance of services from sandy, muddy and rocky shores to human well-being and the economy explains the high urbanization and tourism at coastal regions, which have an important economic contribution (Nitivattananon and Srinonil, 2019). However, the higher the population at the coast, the bigger the coastal impacts are. The contrast between more pristine and more polluted areas changes tourism and recreational and artisanal fishing values (Qiang and others, 2019). Low-impact, natural sandy and rocky shores are an important tourist attraction, both for leisure and snorkelling (Drius and others, 2019). Clean and healthy shores attract many tourists, thus triggering the development of a region's tourist sector. Meanwhile, sandy and rocky shores become very vulnerable under the pressure of recreational and touristic activities, as a result of contamination and the alteration of the natural habitats of many organisms owing to the introduction of artificial infrastructures (Strain and others, 2018; Drius and others, 2019). The contrasts between more pristine and more polluted shores lead to variable seascapes and tourist destinations along the coastline. Increasing changes in the coastline owing to climate change, other anthropogenic impacts and erosion can alter tourism dynamics by decreasing tourism in affected areas and increasing it in low-impact areas. Such changes will influence coastal communities, and local people working in the tourist and support sectors, socially, culturally and psychologically, as well as economically (Jarratt and Davies, 2019; You and others, 2018).

Investment in the sustainable development plans of coastal areas brings multiple economic, social and environmental benefits. Urbanization pressure is increasing as people look to improve their well-being and take advantage of the benefits of coastal environments. Rocky shores tend not to be heavily exploited directly but, as one of the most conspicuous and favoured landscapes, local communities grow up around them, with houses often within 100 m of the shore. Muddy shores are challenging but lucrative areas for real estate developers, investors and private builders. For that reason, there is often excessive construction in such areas, and many existing building codes and standards are ignored. The natural shores are then damaged, including as a result of huge landslides that negatively affect the marine ecosystem. Sandy shores are adversely affected by urbanization owing to the removal of coastal vegetation, land cleaning and increased instability during extreme events and erosion, leading to weaker coastline protection (Defeo and others, 2009). The construction of artificial islands has been driven by demand for coastal housing. Yet the development of such sites in many regions has accelerated without any detailed consideration of the ecological impacts, both on the source areas, where sand is extracted on a large scale as construction material, and on the local ecosystems that are disturbed or displaced by construction activities (Rahman, 2017a, 2017b).

Changes in coastlines, sea level rise, extreme events and tourism activities lead to a change in environmental perception by human populations and increase social conflicts in the coastal zone (Robinson and others, 2019; Whitney and Ban, 2019). Wave height is arguably beneficial for wave-power electricity, as it provides increasing power, but ongoing increases in average wave height present a threat to the physical generator infrastructure that is already pushing the limits of engineering endurance (Penalba and others, 2018).

The complexity of ecosystem services from sandy, muddy and rocky shores, the drivers influencing those environments and the conflict between use and conservation, with the benefits for and impacts from human population, highlight the importance of sustainable development. The complex system that integrates the habitats and the potential and challenges for its governance show the importance of marine spatial planning to support and regulate the use of those environments, and of the 2030 Agenda for Sustainable Development and its Sustainable Development Goals, including Goal 14, as well as the inclusion of goals to decrease impacts on sandy, muddy and rocky shores (Kidd and others, 2020; Borja and others, 2020).

4. Key region-specific changes and consequences

There is limited information from the Arctic and Southern regions of the Ocean, but other regions have recorded changes. Along the shores of the North-West Atlantic, North-East Pacific, North Sea and Black Sea, oceanographic drivers are an important issue owing to wave disturbance (Voorhies and others, 2018), ice scour (Scrosati and Ellrich, 2010) and the increasing frequency of extreme weather events (Smale and Wernberg, 2013) that impact intertidal rocky shores, thus changing sediment transportation, which affects muddy and sandy shores (Masselink and others, 2016), and increasing erosion and boulder movements, which can change biological communities in wave-exposed areas (Petrovic and Guichard, 2008; Castelle and others, 2018). Changes in wave dynamics in those regions also influence benthic-pelagic coupling and ecosystem functioning (Griffiths and others, 2017) by influencing larval supply (Mazzuco and others, 2018), organic matter (Massé Jodoin and Guichard, 2019), temperature and hypoxic events (Vaquer-Sunyer and Duarte, 2011).

In the north-west part of the Black Sea, sandy shores have become narrower in recent years in most parts of the coastline (Allenbach and others, 2015), where vegetation has expanded (Allenbach and others, 2015). Meanwhile, the depth of shelf waters near the mouths of the Danube and Dniester rivers has decreased because of riverine sediment input (Anton and others, 2017), which has also led to the formation of shore dunes in some areas. However, in the eastern part of the Black Sea, coastline erosion is associated with sediment starvation in riverine water discharge as a result of dams and engineering works (Kosyan and Velikova, 2016). There has been significant erosion of muddy and rocky shores as a result of landslides caused by both climatic and anthropogenic factors over the past decades (Freiberg and others, 2010, 2011; Goryachkin, 2013; Tătui and others, 2019). Increased erosion rates have been registered in coastlines adjacent to rural areas, where there are no wave breakers, and also around Serpent's Island (Cherkez and others, 2006, 2020; Goryachkin, 2013). The impacts of all those factors, together with socioeconomic drivers related to the over-exploitation of shores for construction, and recreational and touristic activities for higher revenues, have affected the coastline in many ways (Goryachkin, 2013; Stanchev and others, 2013, 2018; Kucuksezgin and others, 2019).

Oceanographic drivers also cause coastal erosion and a reduction in the surface area of sandy shores on the coasts of Argentina and Brazil in the South-West Atlantic, thus also influencing wave energy and larval supply as a result of the increasing frequency of extreme events and cold fronts (Mazzuco and others, 2015, 2018). Changes in erosion and coastline impacts affect the economy of local communities and change the way they perceive natural coastal ecosystems on the Atlantic coast of South America as a whole (Bunicontro and others, 2015). In addition to a constant driver, such as changes in the oceanographic dynamics and their influence on coastal habitats, environmental disasters have been a key issue in the South-West Atlantic (Gil and others, 2019; Marcovecchio and others, 2019). In the past five years, there have been two inland mine tailing disasters as a result of which discharge reached the coastal zone, thus impacting different habitats, including reefs, sandy, muddy and rocky shores and communities in Brazil, and one oil spill that affected more than 3,000 km of coastline (Escobar, 2019; Soares and others, 2020). Such disasters have a high temporal and spatial impact on the environment, ecosystem services and human communities, in particular in view of the cumulative effects of oceanic-climatic drivers that might resuspend the chemicals in the sediments from sandy and muddy shores (Queiroz and others, 2018; Dadalto and others, 2019).

In the Indian Ocean region, the construction of artificial islands has created new local navigational hazards from unsecured installations intended to prevent erosion, as well as the illegal dumping of waste material (Rahman, 2017a), with new structures that have changed the routes to fishing grounds (Rahman, 2017b). The results of the initial detailed environmental impact assessment of a project in Malaysia led to a revision of the planned layout of new islands to prevent destruction by smothering of a diverse seagrass meadow (Williams, 2016; chap. 7G). However, further and long-term environmental impacts require ongoing review.

A global survey shows that the coasts of the western Pacific and the eastern Atlantic are hotspots of concentrations of several pollutants and that they are affected by the warming climate (Lu and others, 2018). Although many of the drivers from other regions also influence the Pacific coast, climate change and oceanic-climatic events emerge as a key issue in the eastern Pacific (Xiu and others, 2018). The eastern Pacific coast is one of the most productive marine ecosystems owing to the presence of upwelling systems, which are considered to be the most important driving factor for changes in sandy, muddy and rocky shores (Randall and others, 2020). In the North Pacific, it is expected that increased upwelling intensity associated with stronger alongshore winds in the coastal region (Xiu and others, 2018) will change the ecosystem functioning on sandy, muddy and rocky shores owing to changes in nutrient input and oceanographic conditions. In the South Pacific, changes in the Humboldt Current System are influencing different countries in different manners (e.g., an increase in winds that are favourable to upwelling off Chile and a decrease off Peru) (Bertrand and others, 2019). The Pacific coast is strongly affected by El Niño and extreme weather events that may become more frequent and influence the coastline and coastal ecosystem services from sandy, muddy and rocky shores (Bertrand and others, 2019). Climate change and associated impacts affect the natural coastal dynamics of sandy and rocky shores and their services, including fishing, aquaculture, erosion and tourism, owing to the increased frequency of extreme events that lie outside the realm of present-day experience (Aguilera and others, 2019).

5. Outlook

When considering a business-as-usual scenario, reefs and sandy, muddy and rocky shores worldwide will be affected, with a serious loss of ecosystem services. It is expected that in the medium term (approximately 20 years) all issues will be aggravated significantly, and substantial parts of natural shores may be lost, with negative socioeconomic and cultural consequences. There is an increase in human population on the coasts and, as a consequence, an increase in the pollutants, waste and other factors influencing the sandy, muddy and rocky shores. The increase in coastal infrastructures and land reclamation will accelerate that process and, to date, little is known of the long-term impacts on coastlines from changes in hydrodynamics, biodiversity and the source of materials to build those infrastructures. On the other hand, if coastal urbanization develops based on blue engineering, it will present an opportunity to increase sustainable initiatives (chap. 7A; Strain and others, 2018) and increase public awareness of the value of coastal ecosystems and socioecological coastal systems based on ocean literacy (Santoro and others, 2017; Fleming and others, 2019).

There is a direct connection between coastal populations and ecosystem services from biogenic reefs and sandy, muddy and rocky shores, where the increase in the population and use of environmental resources might be higher than the shore resilience. At the same time, climate change will increase the frequency and intensity of storms reaching the coast (IPCC, 2018). From an ocean perspective, there are expected changes in oceanic-climatic drivers affecting the coastline by increasing wave energy, erosion, sediment transport and sea level rise, including a reduction in the intertidal range of some shores (Herring and others, 2018). From an inland perspective, increasing rainfall will disturb sediment transport and increase the input of nutrients, contaminants and pollutants from terrestrial and freshwater environments to coastal habitats (Lana and others, 2018). It is expected that 13.6–15.2 per cent (36,097–40,511 km) of the world's sandy beaches could face severe erosion by 2050, and 35.7–49.5 per cent (95,061–131,745 km) by the end of the century. Therefore, a number of countries could face extensive sandy beach erosion issues by the end of the century (Vousdoukasand others, 2020).

The cumulative effects of climate change and other anthropogenic influences will continue to affect biodiversity, ecosystem services and environmental health. Those continuous multiple stressors, according to the One Health concept of the World Health Organization, will influence human well-being and health (Fleming and others, 2019). Increasing coastal populations and infrastructures can also be expected to lead to an increase in cultural conflict between traditional and indigenous communities with the advent of larger cities and industrial activities. In different areas, the increase in the contamination of resources, the loss of biodiversity, changes to the coastline and an increase in conflicts, with the loss of indigenous and local knowledge and traditions, will have a negative economic impact as a result of a decrease in tourism and an increase in investments needed in the health, economic and infrastructure sectors for the citizens of the region. On the other hand, studies in some regions have shown the potential of including traditional ecological knowledge in the governance processes for decreasing conflicts and strengthening a positive sustainable development (Stori and others, 2019; Van Assche and others, 2019). Also, there is potential for the use of integrated catchment management as a key tool in helping to manage coastal marine systems (Henderson and others, 2020).

6. Key remaining knowledge and capacity-building gaps

There have been advances in recent decades in knowledge relating to biogenic reefs and sandy, muddy and rocky shores, which allow us to better understand their importance and critical impacts. New satellite imaging and modelling also provide important data to visualize change and identify areas at high risk, involving multiple scientific areas (Sagar and others, 2017; Mentaschi and others, 2018). However, there remain some gaps in knowledge. Despite recent scientific advances, we have too little information available to anticipate mid- or long-term scenarios with accuracy. Also, regional knowledge and the volume of data available are unbalanced across many regions of the world, such as the South Atlantic, the wider Caribbean and the western Pacific. Most of the data available worldwide are from local and regional analyses and very few global results allow for a critical review of the situation of coastal habitats. In that context, however, there is a clearer global assessment of sandy shores, so plans of action can be established to mitigate impacts (Luijendijk and others, 2018; Vousdoukas and others, 2020). A global analysis of biodiversity and impacts relating to biogenic reefs and muddy and rocky shores remains outstanding. Considering the increasing impacts on such shores and the lack of data sets, it is important to improve scientific protocols, capacity-building and databases for the standardized monitoring of indicators for biodiversity, ecosystem functioning and environmental drivers to be applied with regard to biogenic reefs and sandy, muddy and rocky shores worldwide. At the moment, many scientific data are collected at local levels using different protocols, which precludes any integrative regional or global analysis.

Interdisciplinary science needs to be promoted in order to strengthen the joint development of natural and social sciences in order to generate scientific data on the human dimension of the environment (McKinley and others, 2020), in particular with regard to biogenic reefs and sandy, muddy and rocky shores. Considering the range and interlinkages of disciplines related to biogenic reefs and sandy, muddy and rocky shores, owing to their high biodiversity and ecosystem services, including the human presence on intertidal shores, as well as all the related economic and health services, it is necessary to integrate natural and social science to promote nature-based solutions, blue engineering, ecosystem resilience and human well-being (McKinley and others, 2020; Stepanova and others, 2020). It is necessary to increase the knowledge on multiple stressors in such habitats in order to support a better understanding of the impact of such threats on those habitats, both as individual drivers and in relation to the synergy among multiple stressors. The increasing knowledge on the effect of multiple stressors will support better science-based decision-making.

It is important to build capacity in the development of science through multisectoral cooperation, where scientific questions are looked at not only on the basis of where there are scientific gaps but also where there are social, management and economic gaps (Lubchenco and others, 2019; Urban and others, 2020). Decision makers and policymakers need sound research to solve practical problems when managing resources and biodiversity. Marine spatial planning is a key issue and a great example of how conflicts in the coastal zone and impacts on biogenic reefs and sandy, muddy and rocky shores can be managed on the basis of a multi-stakeholder and interdisciplinary approach for the benefit of sustainable development (Kidd and others, 2020). Also, we need to try to understand and include the human dimension in research on biogenic reefs and sandy, muddy and rocky shores, and increase communication and awareness through ocean literacy (Santoro and others, 2017). The habitats can be used as a flagship to promote the role of science in the implementation of the 2030 Agenda for Sustainable Development, including Goal 14. By increasing scientific knowledge of how to integrate human and natural dimensions into studies on the impacts and conservation of biogenic reefs and sandy, muddy and rocky shores, science that can support the best coastal management practices based on a multi-stakeholder partnership and understanding of the importance of the ocean, the coastal habitats and the multiple stressors they are enduring will be enhanced.

References

Aguilera, Moisés A., and others (2019). Chapter 29 – Chile: environmental status and future perspectives. In *World Seas: An Environmental Evaluation*, ed. Charles Sheppard, pp. 673–702. Elsevier.

Allenbach, Karin, and others (2015). Black Sea beaches vulnerability to sea level rise. *Environmental Science & Policy*, vol. 46, pp. 95–109.

Anton, Catalin, and others (2017). An analysis of the coastal risks in the Romanian nearshore. *Mechanical Testing and Diagnosis*, vol. 7, No. 1, pp. 18–27.

Barbier, Edward B., and others (2008). Coastal ecosystem-based management with nonlinear ecological functions and values. *Science*, vol. 319, No. 5861, pp. 321–323.

Barboza, Francisco Rafael, and Omar Defeo (2015). Global diversity patterns in sandy beach macrofauna: a biogeographic analysis. *Scientific Reports*, vol. 5, No. 1, pp. 1–9.

Bertrand, Arnaud, and others (2019). Climate change impacts, vulnerabilities and adaptations: Southwest Atlantic and Southeast Pacific marine fisheries. *Impacts of Climate Change on Fisheries and Aquaculture*, p. 325.

Biedenweg, Kelly, and others (2016). A holistic framework for identifying human wellbeing indicators for marine policy. *Marine Policy*, vol. 64, pp. 31–37.

Borja, Angel, and others (2020). Moving Toward an Agenda on Ocean Health and Human Health in Europe. *Frontiers in Marine Science*, vol. 7, art. 37.

Bunicontro, M. Paula, and others (2015). The effect of coastal defense structures (mounds) on southeast coast of Buenos Aires province, Argentine. *Ocean & Coastal Management*, vol. 116, pp. 404–413.

Castelle, Bruno, and others (2018). Increased winter-mean wave height, variability, and periodicity in the Northeast Atlantic over 1949–2017. *Geophysical Research Letters*, vol. 45, No. 8, pp. 3586–3596.

Cherkez, E.A., and others (2020). Using of Landsat Space Images to Study the Dynamic of Coastline Changes in the Black Sea North-Western Part in 1983-2013. In XIXth International Conference Geoinformatics: Theoretical and Applied Aspects (11-14 May 2020), EAGE and AUAG, Kyiv, Ukraine.

Cherkez, E.A., and others (2006). Landslide protection of the historical heritage in Odessa (Ukraine). *Landslides*, vol. 3, No. 4, pp. 303–309.

Dadalto, Maria Cristina, and others (2019). Changes perceived by traditional fishing communities after a major dam disaster in Brazil. *International Journal of Environmental Studies*, 1–9.

Defeo, Omar, and others (2009). Threats to sandy beach ecosystems: a review. *Estuarine, Coastal and Shelf Science*, vol. 81, No. 1, pp. 1–12.

Drius, Mita, and others (2019). Tackling challenges for Mediterranean sustainable coastal tourism: An ecosystem service perspective. *Science of the Total Environment*, vol. 652, pp. 1302–1317.

Dubois, Stanislas, and others (2009). Feeding response of the polychaete *Sabellaria alveolata* (*Sabellariidae*) to changes in seston concentration. *Journal of Experimental Marine Biology and Ecology*, vol. 376, No. 2, pp. 94–101.

Elliott, Michael, and others (2019). A synthesis: what is the future for coasts, estuaries, deltas and other transitional habitats in 2050 and beyond? In *Coasts and Estuaries*, pp. 1–28. Elsevier.

Ellis, J.I., and others (2017). Multiple stressor effects on marine infauna: responses of estuarine taxa and functional traits to sedimentation, nutrient and metal loading. *Scientific Reports*, vol. 7, No. 1, art. 12013. https://doi.org/10.1038/s41598-017-12323-5.

Environmental Monitoring of the Black Sea (EMBLAS) (2019). 12-Months National Pilot Monitoring Studies in Georgia, Russian Federation and Ukraine, 2016-2017. In *Final Scientific Report*, J. Slobodnik and others, eds. European Commission and UNDP. http://emblasproject.org/wp-content/uploads/2019/07/EMBLAS-II_NPMS_12_months-2016_2017_FinDraft2.pdf.

Escobar, Herton (2019). Mystery oil spill threatens marine sanctuary in Brazil. *Science*, vol. 366, No. 6466, pp. 672–672. https://doi.org/10.1126/science.366.6466.672.

Everard, Mark, and others Watts (2010). Have we neglected the societal importance of sand dunes? an ecosystem services perspective. *Aquatic Conservation: Marine and Freshwater Ecosystems*, vol. 20, No. 4, pp. 476–487.

Firth, Louise B., and others (2016). Ocean sprawl: challenges and opportunities for biodiversity management in a changing world. In *Oceanography and Marine Biology*, pp. 201–278. CRC Press.

Fleming, Lora E., and others (2019). Fostering human health through ocean sustainability in the 21st century. *People and Nature*, vol. 1, No. 3, pp. 276–83. https://doi.org/10.1002/pan3.10038.

Freiberg, E., and others (2010). Some Peculiarities and Results of Explorations of Deformation Processes of The Rocks of Adzhalykskiy Firth Valley Slopes. In *ISRM International Symposium-6th Asian Rock Mechanics Symposium*. International Society for Rock Mechanics and Rock Engineering.

Freiberg, E., and others (2011). The Impact of Structural-Tectonic and Lithogenous Peculiarities of the Rock Mass on the Formation and Development of Geo-Deformation Processes. In *12th ISRM Congress*. International Society for Rock Mechanics and Rock Engineering.

Garratt, Matthew J., and others (2019). Mapping the consequences of artificial light at night for intertidal ecosystems. *Science of The Total Environment*, vol. 691, pp. 760–768.

Gascon, Mireia, and others (2017). Outdoor blue spaces, human health and well-being: a systematic review of quantitative studies. *International Journal of Hygiene and Environmental Health*, vol. 220, No. 8, pp. 1207–1221.

Gelcich, Stefan, and others (2019). Comanagement of small-scale fisheries and ecosystem services. *Conservation Letters*, vol. 12, No. 2, e12637. https://doi.org/10.1111/conl.12637.

Gil, Mónica Noemí, and others (2019). Southern Argentina: The Patagonian Continental Shelf. In *World Seas: An Environmental Evaluation*, pp. 783–811. Elsevier.

Girão, Mariana, and others (2019). Actinobacteria isolated from *Laminaria ochroleuca*: A source of new bioactive compounds. *Frontiers in Microbiology*, vol. 10, art. 683.

Goryachkin, Yuri N. (2013). Ukraine. In *Coastal Erosion and Protection in Europe*, Enzo Pranzini and Allan Williams, eds., pp. 413–426. London: Routledge.

Griffiths, Jennifer R., and others (2017). The importance of benthic–pelagic coupling for marine ecosystem functioning in a changing world. *Global Change Biology*, vol. 23, No. 6, pp. 2179–96. https://doi.org/10.1111/gcb.13642.

Hawkins, Stephen J., and others (2019). *Interactions in the Marine Benthos*. vol. 87. Cambridge University Press.

Henderson, C.J., and others (2020). Landscape transformation alters functional diversity in coastal seascapes. *Ecography*, vol. 43, pp. 138–148.https://doi.org/10.1111/ecog.04504.

Herring, Stephanie C., and others (2018). Explaining extreme events of 2016 from a climate perspective. *Bulletin of the American Meteorological Society*, vol. 99, No. 1, pp. S1–S157.

Hoerterer, Christina, and others (2020). Stakeholder perspectives on opportunities and challenges in achieving sustainable growth of the blue economy in a changing climate. *Frontiers in Marine Science*.

Huang, D.W., and K. Roy (2015). The future of evolutionary diversity in reef corals. *Philosophical Transactions of the Royal Society B: Biological Sciences*, 370, 20140010.

Intergovernmental Panel on Climate Change (IPCC) (2018). *Special Report on Global Warming of 1.5°C* (SR1.5).

Ivkić, Angelina, and others (2019). The potential of large rafting objects to spread Lessepsian invaders: the case of a detached buoy. *Biological Invasions*, vol. 21, No. 6, pp. 1887–1893.

Jarratt, David, and Nick J. Davies (2019). Planning for climate change impacts: coastal tourism destination resilience policies. *Tourism Planning & Development*, 1–18.

Kermagoret, Charlène, and others (2019). How does eutrophication impact bundles of ecosystem services in multiple coastal habitats using state-and-transition models. *Ocean & Coastal Management*, vol. 174, pp. 144–153.

Kidd, Sue, and others (2020). Marine spatial planning and sustainability: examining the roles of integration-scale, policies, stakeholders and knowledge. *Ocean & Coastal Management*, vol. 191, p. 105182.

Kim, Tae Won, and others (2018). Effect of Mudflat Trampling on Activity of Intertidal Crabs. *Ocean Science Journal*, vol. 53, No. 1, pp. 101–6. https://doi.org/10.1007/s12601-018-0004-4.

Kosyan, R. D., and V. N. Velikova (2016). Coastal zone – Terra (and aqua) incognita – Integrated Coastal Zone Management in the Black Sea. *Estuarine, Coastal and Shelf Science*, vol. 169, pp. A1–16. https://doi.org/10.1016/j.ecss.2015.11.016.

Kovalova, N., and others (2010). Long-term changes of bacterioplankton and chlorophyll a as indicators of changes of north-western part of the Black Sea ecosystem during the last 30 years. *Journal of Environmental Protection and Ecology*, vol. 11, No. 1, pp. 191–198.

Krelling, Allan Paul, and Alexander Turra (2019). Influence of oceanographic and meteorological events on the quantity and quality of marine debris along an estuarine gradient. *Marine Pollution Bulletin*, vol. 139, pp. 282–98. https://doi.org/10.1016/j.marpolbul.2018.12.049.

Kucuksezgin, Filiz, and others (2019). Chapter 12 - The Coasts of Turkey. In *World Seas: An Environmental Evaluation (Second Edition)*, ed. Charles Sheppard, pp. 307–32. Academic Press. https://doi.org/10.1016/B978-0-12-805068-2.00015-2.

Lana, Paulo da Cunha, and others (2018). Benthic estuarine assemblages of the Southeastern Brazil Marine Ecoregion (sbme). In *Brazilian Estuaries: A Benthic Perspective*, Paulo da Cunha Lana and Angelo Fraga Bernardino, eds., pp. 117–75. Cham, Switzerland: Springer International Publishing. https://doi.org/10.1007/978-3-319-77779-5_5.

Le Duff, M., and others (2017). Coastal Erosion Monitoring on Ouvea Island (New Caledonia): Involving the Local Community in Climate Change Adaptation. In *Climate Change Adaptation in Pacific Countries*, pp. 255–268. Springer.

Leite, Lucas G., and others (2012). Abundance of biofilm on intertidal rocky shores: Can trampling by humans be a negative influence? *Marine Environmental Research*, vol. 79, pp. 111–115.

Liu, Jinming, and others (2018). Literature review of coral concrete. *Arabian Journal for Science and Engineering*, vol. 43, No. 4, pp. 1529–1541.

Lu, Yonglong, and others (2018). Major threats of pollution and climate change to global coastal ecosystems and enhanced management for sustainability. *Environmental Pollution*, vol. 239, pp. 670–80. https://doi.org/10.1016/j.envpol.2018.04.016.

Lubchenco, Jane, and others (2019). Connecting science to policymakers, managers, and citizens. *Oceanography*, vol. 32, No. 3, pp. 106–115.

Luijendijk, Arjen, and others (2018). The State of the World's Beaches. *Scientific Reports*, vol. 8, No. 1, art. 6641. https://doi.org/10.1038/s41598-018-24630-6.

Maggi, Elena, and L. Benedetti-Cecchi (2018). Trophic compensation stabilizes marine primary producers exposed to artificial light at night. *Marine Ecology Progress Series*, vol. 606, pp. 1–5.

Marcovecchio, Jorge E., and others (2019). The Northern Argentine Sea. In *World Seas: An Environmental Evaluation*, ed. Charles Sheppard, pp. 759–781. Elsevier.

Marselle, Melissa R., and others (2019). Review of the Mental Health and Well-being Benefits of Biodiversity. In *Biodiversity and Health in the Face of Climate Change*, Melissa R. Marselle and others, eds., pp. 175–211. Cham, Switzerland: Springer International Publishing. https://doi.org/10.1007/978-3-030-02318-8_9.

Martinez, Aline S., and others (2019). Functional responses of filter feeders increase with elevated metal contamination: Are these good or bad signs of environmental health? *Marine Pollution Bulletin*, vol. 149, p. 110571.

Massé Jodoin, Julien, and Frédéric Guichard (2019). Non-resource effects of foundation species on meta-ecosystem stability and function. *Oikos*, vol. 128, No. 11, pp. 1613–1632.

Masselink, Gerd, and others (2016). Extreme wave activity during 2013/2014 winter and morphological impacts along the Atlantic coast of Europe. *Geophysical Research Letters*, vol. 43, No. 5, pp. 2135–43. https://doi.org/10.1002/2015GL067492.

Mazzuco, Ana Carolina de Azevedo, and others (2015). Temporal variation in intertidal community recruitment and its relationships to physical forcings, chlorophyll-a concentration and sea surface temperature. *Marine Biology*, vol. 162, No. 9, pp. 1705–1725.

Mazzuco, Ana Carolina de Azevedo, and others (2018). The influence of atmospheric cold fronts on larval supply and settlement of intertidal invertebrates: Case studies in the Cabo Frio coastal upwelling system (SE Brazil). *Journal of Sea Research*, vol. 137, pp. 47–56.

McKinley, E., and others (2020). Marine social sciences: looking towards a sustainable future. *Environmental Science & Policy.*

Medinets, S., and V. Medinets (2010). Results of investigations of atmospheric pollutants fluxes in Zmeiny Island in Western part of the Black Sea in 2003-2007 years. *Journal of Environmental Protection and Ecology*, vol. 11, No. 3, pp. 1030–1036.

Medinets, Sergiy (2014). The black sea nitrogen budget revision in accordance with recent atmospheric deposition study. *Turkish Journal of Fisheries and Aquatic Sciences*, vol. 14, No. 5, pp. 981–992.

Medinets, Sergiy, and Volodymyr Medinets (2012). Investigations of atmospheric wet and dry nutrient deposition to marine surface in western part of the Black Sea. *Turkish Journal of Fisheries and Aquatic Sciences*, vol. 12, No. 5, pp. 497–505.

Mendez, María M., and others (2017). Effects of recreational activities on Patagonian rocky shores. *Marine Environmental Research*, vol. 130, pp. 213–220.

Mentaschi, Lorenzo, and others (2018). Global long-term observations of coastal erosion and accretion. *Scientific Reports*, vol. 8, No. 1, pp. 1–11.

Narayan, Siddharth, and others (2016). The effectiveness, costs and coastal protection benefits of natural and nature-based defences. *PloS One*, vol. 11, No. 5.

Nicholls, R., and others (2007). Coastal systems and low-lying areas. In *Climate Change 2007: Impacts, Adaptation, and Vulnerability*, Martin Parry and others, eds., pp. 315–357. Cambridge, United Kingdom: Cambridge University Press.

Nitivattananon, Vilas, and Sirinapha Srinonil (2019). Enhancing coastal areas governance for sustainable tourism in the context of urbanization and climate change in eastern Thailand. *Advances in Climate Change Research*, vol. 10, No. 1, pp. 47–58.

O'Gorman, Eoin J., and others (2012). Multiple anthropogenic stressors and the structural properties of food webs. *Ecology*, vol. 93, No. 3, pp. 441–48. https://doi.org/10.1890/11-0982.1.

Oelsner, Gretchen P., and Edward G. Stets (2019). Recent trends in nutrient and sediment loading to coastal areas of the conterminous us: insights and global context. *Science of the Total Environment*, vol. 654, pp. 1225–1240.

Osorio-Cano, Juan D., and others (2019). Ecosystem management tools to study natural habitats as wave damping structures and coastal protection mechanisms. *Ecological Engineering*, vol. 130, pp. 282–295.

Pardal-Souza, André Luiz, and others (2017). Shading impacts by coastal infrastructure on biological communities from subtropical rocky shores. *Journal of Applied Ecology*, vol. 54, No. 3, pp. 826–835.

Park, Hae-Ryung, and others (2019). Transcriptomic response of primary human airway epithelial cells to flavoring chemicals in electronic cigarettes. *Scientific Reports*, vol. 9, No. 1, pp. 1–11.

Penalba, Markel, and others (2018). Wave energy resource variation off the west coast of Ireland and its impact on realistic wave energy converters' power absorption. *Applied Energy*, vol. 224, pp. 205–219.

Petrovic F., and F. Guichard (2008). Scales of *Mytilus* spp. population dynamics: importance of adult displacement and aggregation. *Marine Ecology Progress Series*, vol. 356, pp. 203–14.

Pinsky, Malin L., and others (2019). Greater vulnerability to warming of marine versus terrestrial ectotherms. *Nature*, vol. 569, No. 7754, pp. 108–11. https://doi.org/10.1038/s41586-019-1132-4.

Poloczanska, Elvira S., and others (2013). Global imprint of climate change on marine life. *Nature Climate Change*, vol. 3, No. 10, pp. 919–25. https://doi.org/10.1038/nclimate1958.

Qiang, Mengmeng, and others (2019). Loss of tourism revenue induced by coastal environmental pollution: a length-of-stay perspective. *Journal of Sustainable Tourism*, vol. 28, No. 4, pp. 550–567.

Queiroz, Hermano M., and others (2018). The Samarco mine tailing disaster: a possible time-bomb for heavy metals contamination? *Science of the Total Environment*, vol. 637, pp. 498–506.

Rahman, Serina (2017a). *Johor's Forest City Faces Critical Challenges*. Trends in Southeast Asia 3. ISEAS Yusof Ishak Institute.

_________(2017b). *The Socio-Cultural Impacts of Forest City*. ISEAS Yusof Ishak Institute. http://hdl.handle.net/11540/7217.

Randall, Carly J., and others (2020). Upwelling buffers climate change impacts on coral reefs of the eastern tropical Pacific. *Ecology*, vol. 101, No. 2. e02918. https://doi.org/10.1002/ecy.2918.

Rangel-Buitrago, N., and G. Anfuso (2009). Assessment of coastal vulnerability in La Guajira Peninsula, Colombian Caribbean Sea. *Journal of Coastal Research*, pp. 792–796.

Rilov, Gil and others (2019). Adaptive marine conservation planning in the face of climate change: what can we learn from physiological, ecological and genetic studies? *Global Ecology and Conservation*, vol. 17, e00566. https://doi.org/10.1016/j.gecco.2019.e00566.

Robinson, Danielle, and others (2019). Community perceptions link environmental decline to reduced support for tourism development in small island states: a case study in the Turks and Caicos Islands. *Marine Policy*, vol. 108, art. 103671.

Rodríguez-Revelo, Natalia, and others (2018). Environmental services of beaches and coastal sand dunes as a tool for their conservation. In *Beach Management Tools - Concepts, Methodologies and Case Studies*, Camilo M. Botero, and others, eds., pp. 75–100. Cham, Switzerland: Springer International Publishing. https://doi.org/10.1007/978-3-319-58304-4_5.

Sagar, Stephen, and others (2017). Extracting the intertidal extent and topography of the Australian coastline from a 28 year time series of Landsat observations. *Remote Sensing of Environment*, vol. 195, pp. 153–169.

Santoro, Francesca and others (2017). *Ocean Literacy for All - A Toolkit.*

Sardain, Anthony, and others (2019). Global forecasts of shipping traffic and biological invasions to 2050. *Nature Sustainability*, vol. 2, No. 4, pp. 274–282.

Schlacher, Thomas A., and Luke Thompson (2012). Beach recreation impacts benthic invertebrates on ocean-exposed sandy shores. *Biological Conservation*, vol. 147, No. 1, pp. 123–132.

Scrosati, Ricardo A., and Julius A. Ellrich (2018). Benthic-pelagic coupling and bottom-up forcing in rocky intertidal communities along the Atlantic Canadian coast. *Ecosphere*, vol. 9, No. 5, e02229. https://doi.org/10.1002/ecs2.2229.

Seebens, Hanno, and others (2017). No saturation in the accumulation of alien species worldwide. *Nature Communications*, vol. 8, No. 1, art. 14435. https://doi.org/10.1038/ncomms14435.

Smale, Dan A., and Thomas Wernberg (2013). Extreme climatic event drives range contraction of a habitat-forming species. *Proceedings of the Royal Society B: Biological Sciences*, vol. 280, No.1754. https://doi.org/10.1098/rspb.2012.2829.

Snigirov, Sergey, and others (2012). The fish community in Zmiinyi Island waters: structure and determinants. *Marine Biodiversity*, vol. 42, No. 2, pp. 225–239.

Soares, Marcelo de Oliveira, and others (2020). Oil spill in South Atlantic (Brazil): Environmental and governmental disaster. *Marine Policy*, vol. 115, art. 103879.

Stanchev, Hristo, and others (2018). Analysis of shoreline changes and cliff retreat to support Marine Spatial Planning in Shabla Municipality, Northeast Bulgaria. *Ocean & Coastal Management*, vol. 156, pp. 127–40. https://doi.org/10.1016/j.ocecoaman.2017.06.011.

Stanchev, Hristo, and others (2013). Integrating GIS and high resolution orthophoto images for the development of a geomorphic shoreline classification and risk assessment—a case study of cliff/bluff erosion along the Bulgarian coast. *Journal of Coastal Conservation*, vol. 17, No. 4, pp. 719–28. https://doi.org/10.1007/s11852-013-0271-2.

Stepanova, Olga, and others (2020). Understanding mechanisms of conflict resolution beyond collaboration: an interdisciplinary typology of knowledge types and their integration in practice. *Sustainability Science*, vol. 15, No. 1, pp. 263–279.

Stori, Fernanda Terra, and others (2019). Traditional ecological knowledge supports ecosystem-based management in disturbed coastal marine social-ecological systems. *Frontiers in Marine Science*, vol. 6, art. 571. https://doi.org/10.3389/fmars.2019.00571.

Strain, Elisabeth M.A., and others (2018). Eco-engineering urban infrastructure for marine and coastal biodiversity: Which interventions have the greatest ecological benefit? *Journal of Applied Ecology*, vol. 55, No. 1, pp. 426–41. https://doi.org/10.1111/1365-2664.12961.

Tătui, Florin, and others (2019). The Black Sea coastline erosion: index-based sensitivity assessment and management-related issues. *Ocean & Coastal Management*, vol. 182, art. 104949.

Teixeira, I.G., and others (2018). Response of phytoplankton to enhanced atmospheric and riverine nutrient inputs in a coastal upwelling embayment. *Estuarine, Coastal and Shelf Science*, vol. 210, pp. 132–141.

Underwood, Charlotte N., and others (2017). Artificial light at night alters trophic interactions of intertidal invertebrates. *Journal of Animal Ecology*, vol. 86, No. 4, pp. 781–789.

United Nations (2017). *The First Global Integrated Marine Assessment: World Ocean Assessment I.* Cambridge: Cambridge University Press.

Urban, Edward R., and others (2020). The importance of bottom-up approaches to international cooperation in ocean science. *Oceanography*, vol. 33, No. 1, pp. 11–15.

Van Assche, Kristof, and others (2019). Governance and the coastal condition: towards new modes of observation, adaptation and integration. *Marine Policy*, vol. 112. https://doi.org/10.1016/j.marpol.2019.01.002.

Vaquer-Sunyer, Raquel, and Carlos M. Duarte (2011). Temperature effects on oxygen thresholds for hypoxia in marine benthic organisms. *Global Change Biology*, vol. 17, No. 5, pp. 1788–97. https://doi.org/10.1111/j.1365-2486.2010.02343.x.

Voorhies, Kristen J., and others (2018). Longstanding signals of marine community structuring by winter storm wave-base. *Marine Ecology Progress Series*, vol. 603, pp. 135–146.

Vousdoukas, Michalis I., and others (2020). Sandy coastlines under threat of erosion. *Nature Climate Change*, vol. 10, No. 3, pp. 260–63. https://doi.org/10.1038/s41558-020-0697-0.

Wang, Aiguo, and others (2018). The development of coral concretes and their upgrading technologies: a critical review. *Construction and Building Materials*, vol. 187, pp. 1004–1019.

Whitney, Charlotte K., and Natalie C. Ban (2019). Barriers and opportunities for social-ecological adaptation to climate change in coastal British Columbia. *Ocean & Coastal Management*, vol. 179, art. 104808.

Williams, Joseph Marcel R. (2016). Evaluating the diverse impacts of megaprojects: the case of Forest City in Johor, Malaysia. PhD Thesis, Massachusetts Institute of Technology. https://dspace.mit.edu/handle/1721.1/105036.

Xiu, Peng, and others (2018). Future changes in coastal upwelling ecosystems with global warming: The case of the California Current System. *Scientific Reports*, vol. 8, No. 1, art. 2866.

You, Soojin, and others (2018). Coastal landscape planning for improving the value of ecosystem services in coastal areas: using system dynamics model. *Environmental Pollution*, vol. 242, pp. 2040–2050.

Zhai, Tianlin, and others (2020). Assessing ecological risks caused by human activities in rapid urbanization coastal areas: towards an integrated approach to determining key areas of terrestrial-oceanic ecosystems preservation and restoration. *Science of The Total Environment*, vol. 708, art. 135153.

Chapter 7C
Atoll and island lagoons

Contributors: Colin D. Woodroffe (convener), Karen Evans (subchapter lead member), David Obura, Fernanda de Oliveira Lana and Arthur P. Webb.

Keynote points

- The health of atolls and island lagoons and the sustainability of the communities that depend on them for their livelihoods are challenged by many environmental limitations and stressors, often exacerbated by human actions.
- Islands on atolls and on other coral reefs are low-lying and very vulnerable to the impacts of climate change, in particular sea level rise; individual islands are likely to respond in different ways.
- Climate change threatens coral reef ecosystems, with implications for the habitability of islands. Coral bleaching, the erosion and inundation of islands, carbonate dissolution and the effects of extreme events, such as tropical storms, are of particular significance.
- Developed urban atoll islands are increasingly dependent on engineering solutions, which need to integrate "hard" and "green/blue" options to avoid unintended impacts, whereas less populated rural island communities rely on the health, productivity and function of surrounding marine and coastal ecosystems.

1. Introduction

Low-lying tropical coral reef and atoll islands, with their associated lagoon systems, are geologically young features formed over the past few millenniums. Their formation and maintenance is constrained by sea level, the biological production of calcium carbonate sediments, and oceanic and atmospheric conditions that rework, transport and redeposit those sediments. Islands are scattered, and often very isolated, across reef-forming seas. Common attributes are that they are low-lying, small in extent and exposed to surrounding marine conditions. They offer marginal agronomic potential and limited fresh groundwater resources to the subsistence communities that live on them and that have an inextricable dependence on the surrounding reefs for daily food security. Associated marine and coastal ecosystems, including seagrasses, mangroves and terrestrial vegetation, are also important to the agroecological stability and services on which local communities depend.

Atoll islands and lagoons were not evaluated specifically in the first *World Ocean Assessment* (United Nations, 2017b), although its chapter 7 (United Nations, 2017a) described carbonate production and contribution to coastal sediments, and atolls were mentioned in other chapters. In the present Assessment, chapter 7D, on tropical and subtropical coral reefs, is particularly relevant, and there is complementary material in chapter 7G, on seagrass meadows.

A recent review concluded that 439 features could be classified as atolls worldwide (Goldberg, 2016). There are 268 atolls with associated reef islands, but the present chapter includes consideration of sand cays and shingle motu on other coral reefs, which behave similarly to islands that form on atoll rims. Atolls are most numerous in the Pacific Ocean (84 per cent of atolls occur in the Pacific, including the South China Sea, the Philippines and Indonesian archipelagos); approximately 13 per cent occur in the Indian Ocean and less than 3 per cent occur in the Caribbean (Goldberg, 2016). French Polynesia contains 83 atolls (nearly 20 per cent of the global total). Tabiteuea, a single atoll system, has some 50 named islands along its eastern rim. Rocas Atoll is the only atoll in the South Atlantic and it is a fully protected Brazilian conservation unit

(Pereira and others, 2010; Soares and others, 2011). The Maldives and Lakshadweep archipelagos contain most of the atoll formations in the Indian Ocean, with only isolated atolls elsewhere (e.g., Aldabra and Glorieuses).

Atoll and associated island formation was first explained by Charles Darwin (Darwin, 1874). An emergent volcanic island that subsequently subsides, eventually below sea level, grows a carbonate reef rim constructed by corals and associated organisms, eventually leaving behind just an atoll reef rim and a central lagoon. Reef islands can form where atoll rims are emergent above highest tides and conditions are conducive. They are entirely composed of the calcareous skeletal remains of reef organisms, including coral broken down into boulders, cobbles, shingle and sand, and other organisms, such as foraminifers, molluscs and coralline algae. Island expansion and persistence depends on ongoing reef sediment production to counterbalance continual erosion by waves, currents and wind.

Reef islands appear to have formed on many atolls, as a consequence of a slight fall in sea level from a mid-Holocene highstand, enabling accumulation of reef sediments. Inhabited reef islands tend to be those that are larger, have interior areas where more complex soils and vegetation have developed and have more reliable fresh groundwater resources. Colonization by people may have occurred soon after islands formed (Nunn, 2016; Allen and others, 2016); intensive human use of such complex ecosystems leads to degradation. Atolls with no islands or uninhabitable islands have considerable ecological value and represent the healthiest, least disturbed and most resilient coral reef habitats (Riegl and others, 2012; Donner and Carilli, 2019). A decline in reef health or productivity threatens the persistence of such ecosystems and the communities that depend on them.

2. Documented changes in state of atolls and island lagoons

2.1. Description of environmental changes between 2010 and 2020

Erosion of islands. Erosion of island shorelines has often been attributed to rising sea level, although proving a link between erosion and sea level rise is challenging. Studies demonstrate variability of response around individual islands and across entire atolls (Ford and Kench, 2015), and short-term fluctuations often mask any genuine long-term signals (Mann and others, 2016; Ryan and others, 2016; Nunn and others, 2017, 2019). In Solomon Islands, several islands have disappeared, but displacement of others in response to wave activity indicates that the islands are not simply drowning by sea level rise (Albert and others, 2016, 2017). Variations in shoreline position have been recorded on many islands in the Indian Ocean (Hamylton and East, 2012; Purkis and others, 2016; Testut and others, 2016) and many have grown, and seem likely to continue to grow in the future (Beetham and others, 2017). In the Pacific Ocean, relatively insignificant changes have been observed on Tuamotu atolls where the role of sea level seems to have been relatively minor, compared with that of climate regime, sediment supply and anthropogenic impacts on shoreline stability (Le Cozannet and others, 2013; Duvat and Pillet, 2017). In Tuvalu, reef islands on Funafuti Atoll have increased in area, in part as a consequence of tropical cyclones (Kench and others, 2015; McLean and Kench, 2015), and shoreline change for all 101 islands in Tuvalu have indicated a land area increase in eight of the nine atolls over the past four decades (Kench and others, 2018). However, recent shoreline impacts following Tropical

Cyclone Pam in March 2015 have caused significant shoreline recession in some islands in Tuvalu, highlighting the significant variability in the effects of tropical disturbances on the islands. An analysis of available data, covering 30 Pacific and Indian Ocean atolls, including 709 islands, revealed that no atoll lost overall land area and that 88.6 per cent of islands within them were either stable or had increased in area, while only 11.4 per cent had contracted (Duvat, 2018).

Inundation of reef islands. There has been disproportionately less focus on the inundation of reef islands and the increased recurrence of nuisance flooding (Ford and others, 2018). A detailed digital elevation model of atoll topography developed for Majuro Atoll, using drone-derived photography, enabled comprehensive consideration of potential errors when mapping atoll vulnerability to future flooding (Gesch and others, 2020).

Altered wave climate. Wave conditions may change as sea level rises (Esteban and others, 2018; Costa and others, 2019). Increased overwash is foreshadowed for many reef islands (Storlazzi and others, 2015), and recent analysis has inferred that most atolls may be uninhabitable by the 2050s (Storlazzi and others, 2018). Reef islands are likely to become thinner and longer, with increased run-up and inland flooding (Shope and others, 2016, 2017). Impacts can be affected by dimensions of adjacent reefs and patterns of longshore sediment transport (Quataert and others, 2015; Shope and Storlazzi, 2019), with salinization of groundwater (Oberle and others, 2017).

Reef degradation. It is estimated that coral reefs cover 0.5 per cent of the oceans, equivalent to approximately 1,500,000 km^2 (Leão and others, 2008). It is also estimated that more than 30 per cent of reefs are already severely damaged and that approximately 60 per cent of reef areas will be totally degraded in coming decades, as a result of human actions, especially overfishing, marine pollution and global climate change (Gherardi and Bosence, 2005; Pereira and others, 2010). Warming of tropical surface waters is causing widespread and more frequent coral bleaching globally (Eakin and others, 2019) and is discussed in chapter 7D. The period 2014–2017 was marked by an unprecedented succession of record-breaking hot years, coinciding with the most severe, widespread and longest-lasting global-scale coral bleaching event ever recorded (Eakin and others, 2019). Bleaching of atoll reefs has been recorded across the tropics (Marshall and others, 2017; Head and others, 2019).

Lagoon pollution. Intensive use of lagoon ecosystems leads to water pollution and ecosystem degradation. Studies in Tuvalu have identified domestic wastewater as the primary pollution source and recorded heavy metal contamination of sediments (Fujita and others, 2013, 2014).

Ecological implications. Remote uninhabited or sparsely inhabited atolls can be locations of unique ecological value. Climate change and sea level rise pressures threaten the persistence and unique ecology of such islands as well as potentially threatened and endangered species (Gillespie and others, 2008). For example, in the remote and mostly uninhabited Phoenix Islands, cycles of coral mortality and recovery from increasingly strong heatwaves may eventually fail, in spite of active management (Rotjan and others, 2014). Cyclones may have devastating effects on critical habitats on small islands, posing a critical challenge to vulnerable species over the long term (Huang and others, 2017). Climate threats may also intensify local pressures on islands that are more exposed to human pressures, such as invasive species, which may result in those areas facing reduced resistance under climate impacts (Russell and others, 2017), and the spread of diseases, such as the stony coral tissue loss disease in the Caribbean (Aeby and others, 2019).

2.2. Factors associated with the changes: drivers, pressures, impacts and response

Interactions between islands and the dominant currents and wave regimes affecting them, together with geomorphological characteristics of subsidence and/or uplift, impose overarching controls on island morphology and change. Modelling studies of Rocas Atoll imply that increased wave action owing to refraction following slight sea level rise may explain planimetric and volumetric changes to reef islands (Costa and others, 2017, 2019).

Coral reefs in warm shallow seas accrete vertically and, under some circumstances, accretion rates can exceed current rates of sea level rise (Perry and others, 2015a, 2015b). However, a gradual fall of sea level over the past 2,000 years has caused coral growth to cease on Indo-Pacific reef flats (Harris and others, 2015). Individual islands will be subject to the pattern of relative sea level change at that location, with subtle variations owing to oceanographic and geophysical drivers (Pfeffer and others, 2017). Insights into marine climatic and environmental reconstructions over different timescales, interpreted from massive, long-lived corals that contain retrospective geochemical archives, are becoming possible (Dassié and Linsley, 2015; Evangelista and others, 2018).

Quantification of rates of carbonate production, together with estimates of erosion and sediment removal, provide insights into the budget of reef sediments (Perry and others, 2016, 2017a; Hamylton and others, 2016; Morgan and Kench, 2017). Sediment production contributes to gradual lagoon infilling. For example, the grazing of reefs by parrotfish produces fine sediment (Perry and others, 2015b; Yarlett and others, 2018), which is augmented close to continental shores by terrigenous sediment (Perry and others, 2017b). Sediment budgets have rarely been calculated for reef islands; they are dependent on biogenic production by a range of reef organisms (Morgan and Kench, 2016). Different reef islands can be in one of several stages of development, namely nucleation, growth, stable, decay, relict or endangered (Garcin and others, 2016). Small sand islands composed of freshly deposited coral fragments lack soil and are less able to support human livelihoods than older, more established islands (Connell, 2015).

Reef islands are fragile systems prone to devastation by extreme climatic events, in particular tropical storms. In 2017, Hurricanes Maria and Irma caused major damage and casualties across numerous Caribbean islands and, in 2018, Tropical Cyclone Gita struck the Pacific islands of Eua and Tongatapu, affecting 80 per cent of the population of Tonga and causing destruction of buildings, crops and infrastructure (Magnan and others, 2019). Cyclone Idai in the western Indian Ocean was one of the strongest cyclones ever recorded in the region, resulting in the second highest death toll. Such high-energy events leave lasting morphological impacts for several years on reef islands (Jeanson and others, 2014; Kayanne and others, 2016). The proportion of very intense cyclones has increased since 1975, which is attributed to warming (Holland and Bruyere, 2014) and is expected to continue into the future (Walsh and others, 2016). Add to that the growing populations and increased infrastructure on islands, thus higher exposure, and severe impacts to islands from cyclones are bound to increase. Islands are also vulnerable to unusually high tides and water levels attributed to distant-source wind waves, as occurred in 1987 in Maldives (Wadey and others, 2017) and in several Pacific islands in December 2008 (Hoeke and others, 2013; Smithers and Hoeke, 2014).

Geochemical changes in the ocean, in particular ocean acidification, may lead to the dissolution of lagoonal sediments, the decreased availability of sand to replenish reef islands and reduced potential for reefs to keep up with future sea level rise (Perry and others, 2018). Recent studies have shown that the dissolution of reef sediments is negatively correlated with the aragonite saturation state of seawater and is 10 times more sensitive to ocean acidification than coral calcification (Cyronak and Eyre, 2016; Eyre and others, 2018).

3. Consequences of the changes on human communities, economies and well-being

Communities that live on reef islands contend with many pressures, and the outcomes of those multiple stressors remain highly uncertain. Despite the widespread perception of susceptibility to various impacts of climate change, there is little evidence that can be directly attributed to it. Many problems facing small reef islands result from other pre-existing pressures (Birk, 2014; Duvat and others, 2017), in particular anthropogenic causes that have exacerbated their vulnerability (Connell, 2015; McCubbin and others, 2015).

In its recent *Special Report on the Ocean and Cryosphere in a Changing Climate*, the Intergovernmental Panel on Climate Change considers the consequences of climate change for low-lying islands (Oppenheimer and others, 2019). It distinguishes urban atoll islands from the many outlying smaller islands, which include capital islands (or groups of islands), such as Fongafale (Tuvalu), South Tarawa (Kiribati) and Malé (Maldives). The future of urban atoll islands is important because they concentrate human populations (about 3,200 people per km^2 in South Tarawa; about 65,700 people per km^2 in Malé), economic activities and critical infrastructure (airports, harbourfronts) in low-lying areas that are exposed to marine flooding and coastal erosion. The populous islands are more dependent on imported food than local cultivation (McCubbin and others, 2017). There is also heavy dependence on hard engineering protection. In some cases, relocation of people and critical infrastructure to another island is being considered (Oppenheimer and others, 2019). However, there are many barriers to migration (Birk and Rasmussen, 2014), including a lack of will to move (Jamero and others, 2017, 2019).

A range of hard and soft engineering options can be used to protect vulnerable island shorelines (Wong, 2018), many of which can be considered adaptive responses. Hard defences protecting Malé have proven successful in preventing further damage. However, the impacts of hard engineering coastal defences on natural shoreline and ecosystem processes can be severe and maladaptive, with long-term negative impacts that may overshadow earlier benefits (Donner and Webber, 2014; David and others, 2019). The expense of hard engineering options has led to greater interest in "soft" ecosystem-based resilience measures (Naylor, 2015). With greater experience of coastal erosion and extreme events since 2004 and the Indian Ocean tsunami, the values of natural reef and coastal vegetation ecosystems are becoming clearer and leading to design principles that harness natural as well as artificial structures to reduce coastal vulnerability.

Inhabitants of atolls do not identify climate change as their prime concern. For example, more than 50 per cent of Maldivians questioned perceive future sea level rise to be a serious national challenge, but many other cultural, religious, economic and social factors also play an important role in making decisions about whether or not to migrate (Stojanov and others, 2017). Similarly, most Tuvalu residents are not anticipating outmigration (Mortreux and Barnett, 2009). In Kiribati, a "sinking nation paradigm" has politicized decision-making, with "adaptation" becoming a metaphor for economic development (Mallin, 2018), and social science counter-narratives are challenging notions of islander outmigration (Barnett, 2017; Kelman, 2018; Yamamoto and Esteban, 2017). Coarse-scale assessments downplay variations in community experiences and local knowledge of environmental change (Leon and others, 2015; Owen and others, 2016). There may be reluctance to move from low-lying islands, based on strong cultural traditions and, in some low-lying islands, inhabitants have opted for in situ adaptation strategies, such as raised houses on stilts, in response to flooding rather than migration to the mainland (Jamero and others, 2017). Such options are seen as preferable even if it seems likely that livelihoods in such circumstance might not be sustainable in the long term (Duce and others, 2010; McNamara and others, 2017).

Sea level rise is often inferred as the primary explanation for harmful, unusual or unprecedented environmental changes on small islands, when other factors are actually driving that change. Contemporary environmental changes on most Pacific islands are more likely to be responses to local stresses, including cyclones, sea wall construction, pollution, overfishing, habitat degradation and sand mining. It has been suggested that small islands offer sites where global climate change narratives can be made tangible and visible, and attributed to distant sources (Connell, 2015).

4. Key region-specific changes and consequences

Although most atolls are in the Pacific Ocean, with several archipelagos in the Indian Ocean and very few in the Atlantic Ocean (Intergovernmental Science-Policy Platform on Biodiversity and Ecosystem Services (IPBES), 2018), no gross differences were apparent on the basis of ocean basin in a recent study of atoll island changes in size (Duvat, 2018). The study indicated that Maldives, in the Indian Ocean, appeared more affected by erosion than Pacific islands, with 23.3 per cent decreasing in the former, compared with 7.5 per cent in the latter. Most notably, the study found high variability within and across atolls and archipelagos. Differences have been noted between urban and outlying rural atoll islands, both in demographic trends and in their response to climate change, sea level rise and other threats. Variability between islands suggests that attention should be paid to the specific context of individual archipelagos, atolls and islands in understanding change and its consequences, which appear to mask regional patterns.

5. Outlook

Atolls and island lagoons remain vulnerable to varied environmental hazards, but it is the synergies and interactions between the hazards, the specificity of how they play out in a local geographic and geomorphological setting, and interactions with social and economic factors that may determine the outlook for islands (Duvat and Magnan, 2019b). Because of their small size and vulnerability, climate change may affect islands through increased magnitude of oscillations in major climatic systems, such as the El Niño Southern Oscillation, as seen in stronger and longer thermal stress and coral bleaching events throughout coral reef island systems globally (Eakin and others, 2019; Hughes and others, 2018).

Key local and climate change pressures include:

a) Warming of ocean temperatures that increases coral bleaching;

b) Sea level rise, which threatens to drown islands and perhaps increase erosion, and may increase wave processes across reefs;

c) Ocean acidification, which may lead to weaker calcareous skeletons and appears likely to reduce lagoon and reef island sediments owing to alkalinity changes;

d) Storms and rare wave events, which play an important role in transporting sediment, as any increase in storm frequency or intensity may have consequences for reefs and reef islands;

e) Overfishing and mismanagement of natural resources, in particular those that play a key role in island and habitat structure, such as coral reefs and mangroves;

f) Demographics and human population density, and their influence on pollution and impacts on local island systems, and on the exposure and vulnerability of people and infrastructure to environmental and climate threats.

Duvat and Magnan (2019a) identify five key adaptation pathways for addressing those interacting challenges on atoll islands, which are: to focus on ecosystem resilience; to minimize the risk of maladaptation; to facilitate internal relocation; to ensure appropriate shoreline protection with respect to ground elevation; and to consider and support permanent international migration.

The outlook for islands depends heavily on policy dimensions, both at the national level between individual island States and other countries, and at global levels through the United Nations and other forums. The former is key in identifying island-specific future options, whether they involve investment in adaptive infrastructure (e.g., to build resilience to sea level rise) or relocation.

6. Key remaining knowledge gaps

The response of reefs, lagoon habitats and reef islands to the combination of local and global threats they now face is inadequately understood. There is little information on how reef processes will respond to changes in individual, and compounded, climate drivers as they change. Geographical variability in shoreline erosion and inundation is observed, but the causes of those spatial patterns are poorly understood, which largely precludes any forecasting of how particular locations will behave. Observational analyses of shoreline changes over past decades are being augmented by attempts to model how reef islands may respond. Recent attempts at modelling include large wave tank experiments (Tuck and others, 2018, 2019; Masselink and others, 2019, 2020), as well as hydrodynamic and shoreline response models (Costa and others, 2019; Ortiz and Ashton, 2019; Shope and Storlazzi, 2019). The vulnerability of small islands composed of calcareous bioclastic sand and shingle that occur on atoll rims or in other reef and lagoon environments can be examined in greater detail using more recent sophisticated remote-sensing technologies to monitor reef island shorelines, including high-resolution satellite imagery, airborne light detection and ranging, and drone-acquired imagery (Casella and others, 2016; Lowe and others, 2019; Gesch and others, 2020).

Whereas most atolls are not experiencing net overall erosion, their physical viability is under increased pressure. Atolls and the reef islands around them and lagoons within them are the product of calcifying organisms that contribute to sediment budgets that determine the trajectory of individual islands. There is insufficient knowledge about the productivity of major organisms contributing bioclastic sediments, the breakdown and transport of derived sand and gravel, and the dissolution and removal of that material.

A further aspect that requires more detailed investigation is the fate of the groundwater lens beneath small islands upon which populations depend, as it seems likely that it will contract if shorelines erode or waves overtop islands (Terry and Chui, 2012; Gulley and others, 2016; Bailey and others, 2016; Deng and Bailey, 2017; Ford and others, 2018). Resilience of the fresh groundwater lens to changing natural and demographic factors, in particular during drought conditions, has only recently emerged as an area of active research (Werner and others, 2017; Oberle and others, 2017), and it requires further examination, in particular as it appears dependent on the morphological adjustments of islands, which are incompletely understood.

From a social-ecological perspective, how island communities and nations will adapt to, and influence, the responses of island systems to the aforementioned threats is important (Duvat and Magnan, 2019b). Government, citizens and community institutions, foreign aid and investment partners, and non-governmental actors all play a role in determining how islands will respond to future challenges, and in anticipating and diminishing crises. The Sustainable Development Goals[1] provide a framework, both for national and international policy, but also for the integrated planning and action that will be needed at multiple levels (Obura, 2020).

7. Key remaining capacity-building gaps

There is clearly a lack of adequately trained and resourced personnel within small island communities to monitor changes at the local level, to undertake site-specific research and assessment and to implement adaptation and other programmes, despite the efforts of international agencies to build capacity within small island nations. Skills addressing the multiple threats and emerging issues mentioned above will be needed. The adaptive capacity of small island communities appears limited, and many are overly dependent on international aid for sponsorship of major projects. Global frameworks, such as the Sustainable Development Goals, the SIDS Accelerated Modalities of Action (SAMOA) Pathway[2] and the United Nations Decade of Ocean Science for Sustainable Development (2021–2030),[3] provide multiple opportunities for identifying critical needs and channelling resources for capacity-building to meet those needs.

1 See General Assembly resolution 70/1.
2 General Assembly resolution 69/15, annex.
3 See General Assembly resolution 72/73.

References

Aeby, Greta, and others (2019). Pathogenesis of a tissue loss disease affecting multiple species of corals along the Florida Reef Tract. *Frontiers in Marine Science*, vol. 6, art. 678.

Albert, Simon and others (2016). Interactions between sea level rise and wave exposure on reef island dynamics in the Solomon Islands. *Environmental Research Letters*, vol. 11, No.5, 054011. https://doi.org/10.1088/1748-9326/11/5/054011.

Albert, Simon and others (2017). Winners and losers as mangrove, coral and seagrass ecosystems respond to sea level rise in Solomon Islands. *Environmental Research Letters*, vol. 12, No. 9, 094009. https://doi.org/10.1088/1748-9326/aa7e68.

Allen, Melinda S., and others (2016). Timing, magnitude and effects of late Holocene sea level drawdown on island habitability, Aitutaki, Cook Islands. *Archaeology in Oceania*, vol. 51, No. 2, pp. 108–121.

Bailey, Ryan T., and others (2016). Predicting Future Groundwater Resources of Coral Atoll Islands. *Hydrological Processes*, vol. 30, No. 13, pp. 2092–2105.

Barnett, Jonathon (2017). The dilemmas of normalising losses from climate change: Towards hope for Pacific atoll countries. *Asia Pacific Viewpoint*, vol. 58, No. 1, pp. 3–13.

Beetham, Edward, and others (2017). Future reef growth can mitigate physical impacts of sea level rise on atoll islands. *Earth's Future*, vol. 5, No. 10, pp. 1002–1014.

Birk, Thomas (2014). Assessing vulnerability to climate change and socioeconomic stressors in the Reef Islands group, Solomon Islands. *Geografisk Tidsskrift-Danish Journal of Geography*, vol. 114, No. 1, pp. 59–75.

Birk, Thomas, and Kjeld Rasmussen (2014). Migration from atolls as climate change adaptation: Current practices, barriers and options in Solomon Islands. In *Natural Resources Forum*, vol. 38, pp. 1–13. Wiley Online Library.

Casella, Elisa, and others (2016). Mapping coral reefs using consumer-grade drones and structure from motion photogrammetry techniques. *Coral Reefs*, vol. 36, No. 1, pp. 269–275.

Connell, John (2015). Vulnerable islands: climate change, tectonic change, and changing livelihoods in the Western Pacific. *The Contemporary Pacific*, pp. 1–36.

Costa, Mirella B., and others (2017). Planimetric and volumetric changes of reef islands in response to wave conditions. *Earth Surface Processes and Landforms*, vol. 42, No. 15, pp. 2663–2678.

Costa, Mirella B., and others (2019). Wave refraction and reef island stability under rising sea level. *Global and Planetary Change*, vol. 172, pp. 256–267.

Cyronak, Tyler, and Bradley D. Eyre (2016). The synergistic effects of ocean acidification and organic metabolism on calcium carbonate (CaCO3) dissolution in coral reef sediments. *Marine Chemistry*, vol. 183, pp. 1–12.

Darwin, Charles (1874). *The Structure and Distribution of Coral Reefs* (2nd ed.), London: Smith Elder and Co.

Dassié, Emilie P., and Braddock K. Linsley (2015). Refining the sampling approach for the massive coral *Diploastrea heliopora* for δ18O-based paleoclimate applications. *Palaeogeography, Palaeoclimatology, Palaeoecology*, vol. 440, pp. 274–282.

David, Gabriel, and others (2019). Coastal Infrastructure on Reef Islands–the Port of Fuvahmulah, the Maldives as Example of Maladaptation to Sea level Rise? *Coastal Structures 2019*, pp. 874–885.

Deng, Chenda, and Ryan T. Bailey (2017). Assessing groundwater availability of the Maldives under future climate conditions. *Hydrological Processes*, vol. 31, No. 19, pp. 3334–3349.

Donner, Simon D., and Jessica Carilli (2019). Resilience of Central Pacific reefs subject to frequent heat stress and human disturbance. *Scientific Reports*, vol. 9, No. 1, pp. 1–13.

Donner, Simon D., and Sophie Webber (2014). Obstacles to climate change adaptation decisions: a case study of sea level rise and coastal protection measures in Kiribati. *Sustainability Science*, vol. 9, No. 3, pp. 331–345.

Duce, Stephanie J., and others (2010). *A Synthesis of Climate Change and Coastal Science to Support Adaptation in the Communities of Torres Strait*. Townsville: Reef and Rainforest Research Centre.

Duvat, Virginie K.E. (2018). A global assessment of atoll island planform changes over the past decades. *Wiley Interdisciplinary Reviews: Climate Change*, vol. 10, No. 1, e557.

Duvat, Virginie K.E., and A.K. Magnan (2019a). Contrasting potential for nature-based solutions to enhance coastal protection services in atoll islands. In *Dealing with climate change on small islands: Towards effective and sustainable adaptation?* Klöck, C. and Fink, M., eds., pp. 45–75. Göttingen: Göttingen University Press.

_________ (2019b). Rapid human-driven undermining of atoll island capacity to adjust to ocean climate-related pressures. *Scientific Reports*, vol. 9, art. 15129.

Duvat, Virginie K.E., and Valentin Pillet (2017). Shoreline changes in reef islands of the Central Pacific: Takapoto Atoll, Northern Tuamotu, French Polynesia. *Geomorphology*, vol. 282, pp. 96–118.

Eakin, C. Mark, and others (2019). The 2014–2017 global-scale coral bleaching event: insights and impacts. *Coral Reefs*, vol. 38, No. 4, pp. 539–545.

Esteban, Miguel, and others (2018), Adaptation to sea level rise on low coral islands: lessons from recent events. *Ocean & Coastal Management*, vol. 168, pp. 35–40.

Evangelista, H., and others (2018). Climatic constraints on growth rate and geochemistry (Sr/Ca and U/Ca) of the coral *Siderastrea stellata* in the Southwest Equatorial Atlantic (Rocas Atoll, Brazil). *Geochemistry, Geophysics, Geosystems*, vol. 19, No. 3, pp. 772–786.

Eyre, Bradley D., and others (2018). Coral reefs will transition to net dissolving before end of century. *Science*, vol. 359, No. 6378, pp. 908–911.

Ford, Murray R., and others (2018). Inundation of a low-lying urban atoll island: Majuro, Marshall Islands. *Natural Hazards*, vol. 91, No. 3, pp. 1273–1297.

Ford, Murray R., and Paul S. Kench (2015). Multi-decadal shoreline changes in response to sea level rise in the Marshall Islands. *Anthropocene*, vol. 11, pp. 14–24.

Fujita, Masafumi, and others (2013). Anthropogenic impacts on water quality of the lagoonal coast of Fongafale Islet, Funafuti Atoll, Tuvalu. *Sustainability Science*, vol. 8, No. 3, pp. 381–390.

Fujita, Masafumi, and others (2014). Heavy metal contamination of coastal lagoon sediments: Fongafale Islet, Funafuti Atoll, Tuvalu. *Chemosphere*, vol. 95, pp. 628–634.

Garcin, Manuel, and others (2016). Lagoon islets as indicators of recent environmental changes in the South Pacific–The New Caledonian example. *Continental Shelf Research*, vol. 122, pp. 120–140.

Gesch, Dean, and others (2020). Inundation Exposure Assessment for Majuro Atoll, Republic of the Marshall Islands Using A High-Accuracy Digital Elevation Model. *Remote Sensing*, vol. 12, No. 1, art. 154.

Gherardi, D.F.M., and D.W.J. Bosence (2005). Late Holocene reef growth and relative sea level changes in Atol das Rocas, equatorial South Atlantic. *Coral Reefs*, vol. 24, No. 2, pp. 264–272.

Gillespie, Rosemary G., and others (2008). Biodiversity dynamics in isolated island communities: interaction between natural and human-mediated processes. *Molecular Ecology*, vol. 17, No. 1, pp. 45–57.

Goldberg, Walter M. (2016). Atolls of the world: Revisiting the original checklist. *Atoll Research Bulletin*, vol. 610, pp. 1–47.

Gulley, J.D., and others (2016). Sea level rise and inundation of island interiors: Assessing impacts of lake formation and evaporation on water resources in arid climates. *Geophysical Research Letters*, vol. 43, No. 18, pp. 9712–9719.

Hamylton, Sarah, and Holly East (2012). A Geospatial Appraisal of Ecological and Geomorphic Change on Diego Garcia Atoll, Chagos Islands (British Indian Ocean Territory). *Remote Sensing*, vol. 4, No. 11, pp. 3444–3461.

Hamylton, Sarah M., and others (2016). Linking pattern to process in reef sediment dynamics at Lady Musgrave Island, southern Great Barrier Reef. *Sedimentology*, vol. 63, No. 6, pp. 1634–1650.

Harris, Daniel L., and others (2015). Late Holocene sea level fall and turn-off of reef flat carbonate production: Rethinking bucket fill and coral reef growth models. *Geology*, vol. 43, No. 2, pp. 175–178.

Head, Catherine E.I., and others (2019). Coral bleaching impacts from back-to-back 2015–2016 thermal anomalies in the remote central Indian Ocean. *Coral Reefs*, vol. 38, No. 4, pp. 605–618.

Hoeke, Ron K., and others (2013). Widespread inundation of Pacific islands triggered by distant-source wind-waves. *Global and Planetary Change*, vol. 108, pp. 128–138.

Holland, Greg, and Cindy Bruyere (2014). Recent intense hurricane response to global climate change. *Climate Dynamics*, vol. 42, Nos. 3–4, pp. 617–627.

Huang, Ryan M., and others (2017). Sooty tern (*Onychoprion fuscatus*) survival, oil spills, shrimp fisheries, and hurricanes. *PeerJ*, vol. 5, e3287.

Hughes, Terry P., and others (2018). Spatial and temporal patterns of mass bleaching of corals in the Anthropocene. *Science*, vol. 359, No. 6371, pp. 80–83.

Intergovernmental Science-Policy Platform on Biodiversity and Ecosystem Services (IPBES) (2018). *The IPBES regional assessment report on biodiversity and ecosystem services for Asia and the Pacific*. Karki, M., and others (eds). Secretariat of the Intergovernmental Science-Policy Platform on Biodiversity and Ecosystem Services, Bonn, Germany. 612 pages.

Jamero, Laurice, and others (2017). Small island communities in the Philippines prefer local measures to relocation in response to sea level rise. *Nature Climate Change*, vol. 7, pp. 581–586

Jamero, Laurice, and others (2019). In-situ adaptation against climate change can enable relocation of impoverished small islands. *Marine Policy*, vol. 108, art. 103614,

Jeanson, Matthieu, and others (2014). Morphodynamic characterization of beaches on a Pacific atoll island: Tetiaroa, French Polynesia. *Journal of Coastal Research*, vol. 70, special issue No. 1, pp. 176–181.

Kayanne, Hajime, and others (2016). Eco-geomorphic processes that maintain a small coral reef island: Ballast Island in the Ryukyu Islands, Japan. *Geomorphology*, vol. 271, pp. 84–93.

Kelman, Ilan (2018). Islandness within climate change narratives of small island developing states (SIDS). *Island Studies Journal*, vol. 13, No. 1, pp. 149–166.

Kench, Paul S., and others (2015). Coral islands defy sea level rise over the past century: Records from a central Pacific atoll. *Geology*, vol. 43, No. 6, pp. 515–518.

Kench, Paul S., and others (2018). Patterns of island change and persistence offer alternate adaptation pathways for atoll nations. *Nature Communications*, vol. 9, No. 1, pp. 1–7.

Le Cozannet, Gonéri, and others (2013). Exploring the relation between sea level rise and shoreline erosion using sea level reconstructions: an example in French Polynesia. *Journal of Coastal Research*, vol. 65, special issue No. 2, pp. 2137–2142.

Leão, Z.M.A.N., and others (2008). Coral bleaching in Bahia reefs and its relation with sea surface temperature anomalies. *Biota Neotropica*, vol. 8, No. 3, pp. 1–14.

Leon, Javier X., and others (2015). Supporting local and traditional knowledge with science for adaptation to climate change: lessons learned from participatory three-dimensional modeling in BoeBoe, Solomon Islands. *Coastal Management*, vol. 43, No. 4, pp. 424–438.

Lowe, Meagan K., and others (2019). Assessing Reef-Island Shoreline Change Using UAV-Derived Orthomosaics and Digital Surface Models. *Drones*, vol. 3, No. 2, p. 44.

Magnan, A.K., and others (2019). Cross-Chapter Box 9: Integrative cross-chapter box on low-lying islands and coasts. In *Special Report on the Ocean and Cryosphere in a Changing Climate*, ed. IPCC, pp. 657–74.

Mallin, Marc-Andrej Felix (2018). From sea level rise to seabed grabbing: The political economy of climate change in Kiribati. *Marine Policy*, vol. 97, pp. 244–252.

Mann, Thomas, and others (2016). A geomorphic interpretation of shoreline change rates on reef islands. *Journal of Coastal Research*, vol. 32, No. 3, pp. 500–507.

Marshall, Paul, and others (2017). *Maldives Coral Bleaching Response Plan*. Marine Research Centre.

Masselink, Gerd, and others (2019). Physical and Numerical Modeling of Infragravity Wave Generation and Transformation on Coral Reef Platforms. *Journal of Geophysical Research: Oceans*, vol. 124, No. 3, pp. 1410–1433.

Masselink, Gerd, and others (2020). Coral reef islands can accrete vertically in response to sea level rise. *Science Advances*, vol. 6, No. 24, eaay3656.

McCubbin, Sandra G., and others (2017). Social-ecological change and implications for food security in Funafuti, Tuvalu. *Ecology and Society*, vol. 22, No. 1.

McCubbin, Sandra, and others (2015). Where does climate fit? Vulnerability to climate change in the context of multiple stressors in Funafuti, Tuvalu. *Global Environmental Change*, vol. 30, pp. 43–55.

McLean, Roger, and Paul Kench (2015). Destruction or persistence of coral atoll islands in the face of 20th and 21st century sea level rise? *Wiley Interdisciplinary Reviews: Climate Change*, vol. 6, No. 5, pp. 445–463.

McNamara, Karen E., and others (2017). Identification of limits and barriers to climate change adaptation: case study of two islands in Torres Strait, Australia. *Geographical Research*, vol. 55, No. 4, pp. 438–455.

Morgan, Kyle M., and Paul S. Kench (2016). Reef to island sediment connections on a Maldivian carbonate platform: using benthic ecology and biosedimentary depositional facies to examine island-building potential. *Earth Surface Processes and Landforms*, vol. 41, No. 13, pp. 1815–1825.

________ (2017). New rates of Indian Ocean carbonate production by encrusting coral reef calcifiers: Periodic expansions following disturbance influence reef-building and recovery. *Marine Geology*, vol. 390, pp. 72–79.

Mortreux, Colette, and Jon Barnett (2009). Climate change, migration and adaptation in Funafuti, Tuvalu. *Global Environmental Change*, vol. 19, No. 1, pp. 105–112.

Naylor, Alexander K. (2015). Island morphology, reef resources, and development paths in the Maldives. *Progress in Physical Geography*, vol. 39, No. 6, pp. 728–749.

Nunn, Patrick D. (2016). Sea levels, shorelines and settlements on Pacific reef islands. *Archaeology in Oceania*, vol. 51, No. 2, pp. 91–98.

Nunn, Patrick D., and others (2017). Identifying and assessing evidence for recent shoreline change attributable to uncommonly rapid sea level rise in Pohnpei, Federated States of Micronesia, Northwest Pacific Ocean. *Journal of Coastal Conservation*, vol. 21, No. 6, pp. 719–730.

Nunn, Patrick D., and others (2019). Origin, development and prospects of sand islands off the north coast of Viti Levu Island, Fiji, Southwest Pacific. *Journal of Coastal Conservation*, vol. 23, No. 6, pp. 1005–1018.

Oberle, Ferdinand K.J., and others (2017). Atoll groundwater movement and its response to climatic and sea level fluctuations. *Water*, vol. 9, No. 9, art. 650.

Obura, David O. (2020). The Sustainable Development Goals as an ocean narrative. *Marine Policy Journal*, submitted.

Oppenheimer, Michael, and others (2019). Sea level rise and implications for low lying islands, coasts and communities.

Ortiz, Alejandra C., and Andrew D. Ashton (2019). Exploring carbonate reef flat hydrodynamics and potential formation and growth mechanisms for motu. *Marine Geology*, vol. 412, pp. 173–186.

Owen, S.D., and others (2016). Improving understanding of the spatial dimensions of biophysical change in atoll island countries and implications for island communities: A Marshall Islands' case study. *Applied Geography*, vol. 72, pp. 55–64.

Pereira, N.S., and others (2010). Mapeamento geomorfológico e morfodinâmica do Atol das Rocas, Atlântico Sul. *Revista de Gestão Costeira Integrada-Journal of Integrated Coastal Zone Management*, vol. 10, No. 3, pp. 331–345.

Perry, Chris T., and others (2015a). Linking reef ecology to island building: Parrotfish identified as major producers of island-building sediment in the Maldives. *Geology*, vol. 43, No. 6, pp. 503–506.

Perry, Chris T., and others (2015b). Remote coral reefs can sustain high growth potential and may match future sea level trends. *Scientific Reports*, vol. 5, art. 18289.

Perry, Chris T., and others (2016). Sediment generation by *Halimeda* on atoll interior coral reefs of the southern Maldives: A census-based approach for estimating carbonate production by calcareous green algae. *Sedimentary Geology*, vol. 346, pp. 17–24.

Perry, Chris T., and others (2017a). Terrigenous sediment-dominated reef platform infilling: an unexpected precursor to reef island formation and a test of the reef platform size–island age model in the Pacific. *Coral Reefs*, vol. 36, No. 3, pp. 1013–1021.

Perry, Chris T., and others (2017b). Reef habitat type and spatial extent as interacting controls on platform-scale carbonate budgets. *Frontiers in Marine Science*, vol. 4, art. 185.

Perry, Chris T., and others (2018). Loss of coral reef growth capacity to track future increases in sea level. *Nature*, vol. 558, No. 7710, pp. 396–400.

Pfeffer, Julia, and others (2017). Decoding the origins of vertical land motions observed today at coasts. *Geophysical Journal International*, vol. 210, No. 1, pp. 148–165.

Purkis, Sam J., and others (2016). A half-century of coastline change in Diego Garcia–The largest atoll island in the Chagos. *Geomorphology*, vol. 261, pp. 282–298.

Quataert, Ellen, and others (2015). The influence of coral reefs and climate change on wave-driven flooding of tropical coastlines. *Geophysical Research Letters*, vol. 42, No. 15, pp. 6407–6415.

Riegl, Bernhard M., and others (2012). Human impact on atolls leads to coral loss and community homogenisation: a modeling study. *PloS One*, vol. 7, No. 6.

Rotjan, Randi, and others (2014). Establishment, management, and maintenance of the phoenix islands protected area. *Advances in Marine Biology*, vol. 69, pp. 289–324. Elsevier.

Russell, James C., and others (2017). Invasive alien species on islands: impacts, distribution, interactions and management. *Environmental Conservation*, vol. 44, No. 4, pp. 359–370.

Ryan, Emma J., and others (2016). Multi-scale records of reef development and condition provide context for contemporary changes on inshore reefs. *Global and Planetary Change*, vol. 146, pp. 162–178.

Shope, James Brandon, and Curt Storlazzi (2019). Assessing morphologic controls on atoll island alongshore sediment transport gradients due to future sea level rise. *Frontiers in Marine Science*, vol. 6, art. 245.

Shope, James B., and others (2016). Changes to extreme wave climates of islands within the Western Tropical Pacific throughout the 21st century under RCP 4.5 and RCP 8.5, with implications for island vulnerability and sustainability. *Global and Planetary Change*, vol. 141, pp. 25–38.

Shope, James B., and others (2017). Projected atoll shoreline and run-up changes in response to sea level rise and varying large wave conditions at Wake and Midway Atolls, Northwestern Hawaiian Islands. *Geomorphology*, vol. 295, pp. 537–550.

Smithers, S.G., and R.K. Hoeke (2014). Geomorphological impacts of high-latitude storm waves on low-latitude reef islands—Observations of the December 2008 event on Nukutoa, Takuu, Papua New Guinea. *Geomorphology*, vol. 222, pp. 106–121.

Soares, Marcelo de Oliveira, and others (2011). Aspectos biogeomorfológicos do Atol das Rocas, Atlântico Sul Equatorial. *Brazilian Journal of Geology*, vol. 41, No. 1, pp. 85–94.

Stojanov, Robert, and others (2017). Local perceptions of climate change impacts and migration patterns in Malé, Maldives. *The Geographical Journal*, vol. 183, No. 4, pp. 370–385.

Storlazzi, Curt D., and others (2018). Most atolls will be uninhabitable by the mid-21st century because of sea level rise exacerbating wave-driven flooding. *Science Advances*, vol. 4, No. 4, eaap9741.

Storlazzi, Curt D., and others (2015). Many atolls may be uninhabitable within decades due to climate change. *Scientific Reports*, vol. 5, art. 14546.

Terry, James P., and Ting Fong May Chui (2012). Evaluating the fate of freshwater lenses on atoll islands after eustatic sea level rise and cyclone-driven inundation: A modelling approach. *Global and Planetary Change*, vol. 88, pp. 76–84.

Testut, Laurent, and others (2016). Shoreline changes in a rising sea level context: The example of Grande Glorieuse, Scattered Islands, Western Indian Ocean. *Acta Oecologica*, vol. 72, pp. 110–119.

Tuck, Megan E., and others (2018). Physical modelling of reef platform hydrodynamics. *Journal of Coastal Research*, vol. 85, special issue No. 1, pp. 491–495.

Tuck, Megan E., and others (2019). Physical modelling of the response of reef islands to sea level rise. *Geology*, vol. 47, No. 9, pp. 803–806.

United Nations (2017a). Chapter 7: Calcium carbonate production and contribution to coastal sediments. In *The First Global Integrated Marine Assessment: World Ocean Assessment I*. Cambridge: Cambridge University Press.

________(2017b). *The First Global Integrated Marine Assessment: World Ocean Assessment I*. Cambridge: Cambridge University Press.

Wadey, Matthew, and others (2017). Coastal flooding in the Maldives: an assessment of historic events and their implications. *Natural Hazards*, vol. 89, No.1, pp. 131–159.

Walsh, Kevin J.E., and others (2016). Tropical cyclones and climate change. *Wiley Interdisciplinary Reviews-climate Change*, vol. 7, pp. 65–89.

Werner, Adrian D., and others (2017). Hydrogeology and management of freshwater lenses on atoll islands: Review of current knowledge and research needs. *Journal of Hydrology*, vol. 551, pp. 819–844.

Wong, Poh Poh (2018). Coastal Protection Measures–Case of Small Island Developing States to Address Sea level Rise. *Asian Journal of Environment & Ecology*, vol. 6, pp. 1–14.

Yamamoto, Lilian, and Miguel Esteban (2017). Migration as an adaptation strategy for atoll island states. *International Migration*, vol. 55, No. 2, pp. 144–158.

Yarlett, Robert T., and others (2018). Constraining species-size class variability in rates of parrotfish bioerosion on Maldivian coral reefs: Implications for regional-scale bioerosion estimates. *Marine Ecology Progress Series*, vol. 590, pp. 155–169.

Chapter 7D
Tropical and subtropical coral reefs

Contributors: Ian Butler (convener), Karen Evans (subchapter lead member), Hazel Oxenford and Hiroya Yamano.

Keynote points

- Global declines in coral cover continue, primarily owing to increasing ocean temperatures associated with climate change, as well as extractive activities, pollution and sedimentation, novel coral diseases and the physical destruction of coral reefs.
- The frequency of disturbances caused by heatwaves, storms, flooding and crown-of-thorns starfish outbreaks has increased, resulting in a decrease in recovery time between disturbances.
- The understanding of the value of ecosystem services provided by coral reefs is improving, not only in terms of direct economic benefits (market use value), but also through less tangible use, such as aesthetic value.
- Substantial knowledge gaps remain, in particular with regard to responses of coral reef communities to climate change and how those responses might influence human use of coral reefs.
- Projections of future states suggest continued decreases in coral abundance, reef-associated fishes and the architectural complexity of reef frameworks.

1. Introduction

The present chapter provides an update to the findings contained in chapter 43 of the first *World Ocean Assessment* (United Nations, 2017b) on tropical and subtropical coral reefs. The update may contain overlaps or interactions with the content of other chapters in the present Assessment (e.g., chaps. 4–10, 13–15, and 25–27). Therefore, they will need to be read in conjunction with the present chapter.

Chapter 43 in the first Assessment contained comprehensive coverage of many aspects of the state of global coral reefs up to 2010, when coral reefs were estimated to cover 249,713–284,300 km^2 but were determined to have continually declined over the past 100 years (United Nations, 2017a). Tropical and subtropical coral reefs were identified as one of the most vulnerable ecosystems and, under a business-as-usual carbon emission scenario, it was projected that they could potentially be functionally extinct by 2050 (Intergovernmental Panel on Climate Change (IPCC), 2014). Anthropogenic-driven impacts exacerbated by population pressures were identified as the primary threat to reefs. They included climate change (e.g., ocean warming, ocean acidification and sea level rise); extractive activities (e.g., overfishing); pollution and sedimentation; and physical destruction. Other stressors identified and linked to the above stressors (in particular climate change) included coral diseases and crown-of-thorns starfish predation. The degree to which each stressor was affecting coral reefs was identified as varying considerably among different species and geographical regions.[1]

1 See A/66/298 and A/66/298/Corr.1.

2. Description of environmental changes between 2010 and 2020

The status of global coral reefs has not improved since they were described in the first Assessment. Recent global marine heatwave events (e.g., 2014–2017) (Eakin and others, 2019) have caused widespread mortality of corals through heat stress and associated bleaching, with recovery from such events uncertain (Leggat and others, 2019; Hughes and others, 2017a). Continued global declines in coral reef biodiversity (e.g., fishes) (Johnson and others, 2017), as well as regional changes in reef composition of coral species (e.g., to *Porites* dominance) (Moritz and others, 2018) have been reported since the first Assessment (see also sect. 5 below). There has been a 50–75 per cent decrease in coral cover globally over the past 30–40 years (Bruno and others, 2019). Few coral reef areas have not been affected (i.e., areas largely unaffected by direct human activities), with most occurring in areas afforded high protection status (Jones and others, 2018).

The steady degradation of coral reefs around the world continues to be closely linked to population growth and increased anthropogenic pressures, with an overlaid impact of climate change affecting even remote areas (e.g., Jarvis Island; Vargas-Ángel and others, 2019). The major threats include extractive activities, pollution (including run-off, chemicals), sedimentation, physical destruction and anthropogenic climate change. Although regional or local efforts will assist in mitigating decreases in coral cover, they are unlikely to offset the loss of coral caused by climate change (Bruno and others, 2019).

Globally, greenhouse gas levels have steadily increased (IPCC, 2018; see also chap. 5 of the present Assessment). Ocean warming hotspots (e.g., Australia, Brazil, India, Madagascar and South Africa) are appearing (Fordyce and others, 2019; Kerr and others, 2018; Popova and others, 2016), resulting in substantially altered ecosystems, for example, in Australia where coral has replaced kelp (Wernberg and others, 2016). Coral community changes, including delayed or reduced spawning (Birkeland, 2019) and coral reef areas now dominated by macroalgae (Johns and others, 2018) or cyanobacteria (de Bakker and others, 2017), are being seen globally (Hughes and others, 2018b). In association with the increased incidence of marine heatwaves (Smale and others, 2019), rates of thermal stress to corals have been steadily increasing (Lough and others, 2018). A 36-month global heatwave from 2014 to 2017 resulted in 75 per cent of the world's coral reefs experiencing bleaching and 30 per cent of coral reefs experiencing mortality as a result of bleaching (Babcock and others, 2019; Eakin and others, 2019). For some reefs, that was the first recorded bleaching event (e.g., southern offshore Great Barrier Reef; Hughes and others, 2017). Globally, the frequency of bleaching events has increased such that recovery is uncertain (Hughes and others, 2018a).

There are many other factors that cumulatively affect the quality and quantity of coral reefs globally. Ocean pH has decreased steadily, with net carbonate loss occurring in coral reef frameworks (Albright and others, 2016; Kuffner and others, 2019; Steiner and others, 2018). The physical destruction of coral reefs is increasing as a result of major category 4 and category 5 storms (e.g., Atlantic Ocean; Murakami and others, 2014). Flooding impacts associated with storm events have been severe and repetitive (Butler and others, 2015) and recovery is regionally variable (Adjeroud and others, 2018; Holbrook and others, 2018). There has been an increase in coral diseases globally (Ruiz-Moreno and others, 2012), which has been linked to thermal stress (Anyamba and others, 2019; Randall and van Woesik, 2015).

There are some areas where coral reefs are thriving (e.g., "bright spots"; Cinner and others, 2016a; Flower Garden Reefs; National Oceanic and Atmospheric Administration (NOAA), 2020) and replacing other habitats (e.g., kelp forests). With increased warming and strengthened currents, coral reefs have expanded into higher latitudes, for example, in Japan over approximately the past 80 years (Yamano and others, 2011; Kumagai and others, 2018) and eastern Australia over approximately the past 20 years (Baird and others, 2012; Booth and Sear, 2018).

3. Description of economic and social consequences and/or other economic or social changes

About 79 States Members of the United Nations have coral reefs in their maritime areas. Coral reefs are important as a source of income and protein to millions of people through fishing activities; a major source of income through tourism; and a basis for sociocultural identity (Cinner and others, 2016b; Kittinger and others, 2012). The value of goods and services derived from coral reefs was estimated at $9.9 trillion in 2012 (Costanza and others, 2014). It is now estimated that up to 500 million people globally benefit from coral reef services (Bruno and others, 2019), including 6 million fishers who depend directly on coral reefs (Teh and others, 2013). The economic value from tourism, fishing and coastal development, for example, across countries in Mesoamerica and the Coral Triangle, are valued at $20.1 billion per year (United Nations Environment Programme (UNEP) and others, 2018). Globally, the value from tourism per ha of coral reef is estimated to be above $400,000, with some reefs valued up to $7 million per ha (Spalding and others, 2017).

Coral reefs are important for coastal protection (valued at $170,205 per ha per year) and can be exploited for rock and sand (valued at $22,000 per ha per year) (Costanza and others, 2014). The annual value of flood risk reduction provided by coral reefs in the United States is estimated at above 18,000 lives and $1.805 billion (Storlazzi and others, 2019). Globally, the extra costs associated with coral reef loss associated with increased flooding from a major storm is estimated at $272 billion (Beck and others, 2018). In the United States, coral reefs prevent indirect damage worth over $699 million annually in the economic activity of individuals and over $272 million in avoided business interruption (Storlazzi and others, 2019).

The value of coral reefs for health and well-being far exceeds traditional economic valuations (UNEP and others, 2018). However, coral reefs have a complicated interaction with society (Cinner and others, 2016b), which makes it difficult to assign a dollar value to health and well-being benefits. Progress has been made in understanding effective governance arrangements for coral reef conservation and the importance of sustainable use (Aswani and others, 2015; Turner and others, 2018), in particular where exploitation expands faster than governance arrangements (Eriksson and others, 2015). Conflicts occur between community-based management of coral reefs and that of national or international management frameworks. Increased local knowledge has supported community-level ownership and control of coral reefs and their management (e.g., Hawaii; Schemmel and others, 2016; Solomon Islands; Shaver and others, 2018).

Coral reefs contribute to the lives of millions of humans globally, and coral reef health affects the ability to achieve the Sustainable Development Goals of the 2030 Agenda for Sustainable

Development.[2] The attainment of any, or all, Goals may be compromised through loss of healthy coral reefs. In particular, they contribute to the attainment of Goals 1, 2, 3 and 12 through income and nutrition, as well as Goals 3, 6, 11, 12, 13 and 14 through their aesthetic and natural values, healthy environmental conditions and the development of medical products. Healthy reefs maintain the integrity of island and coastal land and water resources and infrastructure, thereby contributing to Goals 6, 9, 11 and 13, and they are linked to the ability of countries to attract and maintain skilled communities (Goals 3, 4 and 10).

4. Key region-specific changes and consequences

4.1. Mediterranean Sea

Coral reef areas of the Mediterranean are subtropical and temperate in nature and limited in scale, with temperate reefs considered in chapter 7 of the present Assessment. Some corals, for example, *Oculina patagonica,* are expanding in range and abundance with increased temperature and light availability (Serrano and others, 2018). Other corals are decreasing in coverage (e.g., *Cladocora caespitose*) (Chefaoui and others, 2017) through changes from coralligenous to algae-dominated reefs (e.g., *Womersleyella setacea* and *Caulerpa cylindracea)*(Gatti and others, 2015).

4.2. Atlantic Ocean, in particular the wider Caribbean

Overall, 43 per cent of reefs in the Caribbean are under high or very high threat from human activities (International Coral Reef Initiative (ICRI), 2018a). Primary threats to Caribbean reefs are the same as those affecting global reefs (Mumby and others, 2014), including the rapid expansion of coral diseases (van Woesik and Randall, 2017), such as the novel "stony coral tissue loss disease" (Alvarez-Filip and others, 2019). Invasive lionfish (*Pterois volitans*) are now considered a threat to native reef fish populations and overall reef biodiversity (Chagaris and others, 2017).

Marine heatwaves occurring across the Caribbean region since the 1970s and, in particular, in 2015 and 2016 (Banon and others, 2018), have reduced live coral cover from over 70 per cent to approximately 14 per cent (ICRI, 2018a), although there is substantial regional variation (Jackson and others, 2014; Cortés and others, 2018; Muniz-Castillo and others, 2019). Van Hooidonk and others (2014) predict that most coral reefs in the Caribbean will suffer annual bleaching by the period 2045–2050, and ocean acidification could result in carbonate saturation levels dropping below those required to sustain coral reef accretion by 2050. Perry and others (2013) report that 37 per cent of reefs in the wider Caribbean are eroding and only 26 per cent are accreting. Significant reductions in calcification rates and the density of structure have been reported for such species as *Orbicella faveolata* in the Seaflower Biosphere Reserve (Lizcano-Sandoval and others, 2019), with strong indications that reefs in Florida are, in general, eroding (Kuffner and others, 2019). Major tropical storms in the western central Atlantic (five category 5 and three category 4 hurricanes in 2017, 2018 and 2019) are believed to have caused massive damage to coral reefs across the region, although official data have not yet been published.

There has been progress with regard to standardized monitoring and reporting of coral reef

2 See General Assembly resolution 70/1.

health across the Caribbean (e.g., Global Coral Reef Monitoring Network (GCRMN)-Caribbean, 2016). Reef report cards indicate improvement in reef health across the Mesoamerican Barrier Reef system over the past 10 years (McField and others, 2018). There have been enhanced efforts to improve the conservation and sustainable use of coral reefs and coastal environments across the region (Caribbean Challenge Initiative (CCI), 2019) and efforts to restore herbivores (ICRI, 2013; Vallès and Oxenford, 2018). Reef restoration and rehabilitation efforts have expanded in the wider Caribbean and are benefiting from new research and improved nursery and outplanting technologies (Lirman and Schopmeyer, 2016, Baums and others, 2019).

4.3. Indian Ocean

Coral reef abundance has been stable across the Indian Ocean since 2010, with the exception of reefs near Mozambique, which have decreased substantially (Obura and others, 2017). Global marine heatwaves have caused widespread bleaching (e.g., Maldives; Cowburn and others, 2019; north-western Australia; Keesing and others, 2019) across the region. Over 65 per cent of reefs in the Indian Ocean have been identified as being at risk as a result of local threats, with 33 per cent considered to be at high or very high risk (ICRI, 2018b). Some 19 per cent of coral reefs are contained within marine protected areas. However, only 25 per cent of the total number of marine protected areas are considered effective (ICRI, 2018c), and many lack management plans (Obura and others, 2017).

Coral reefs in the Red Sea and the Persian Gulf show resilience to high temperatures and mortality from bleaching (Howells and others, 2016), although calcification rates appear to be declining (Steiner and others, 2018). Harmful fishing practices, including the use of poison and dynamite, are decreasing in occurrence in the Indian Ocean (Obura and others, 2017), with notable exceptions, such as in the United Republic of Tanzania (Chevallier, 2017). Other direct impacts to coral reefs are increasing, such as the damage caused by anchoring by fishers and tour boats (Obura and others, 2017), as well as outbreaks of crown-of-thorns starfish (Saponari and others, 2018; Keesing and others, 2019).

4.4. Pacific Ocean

Threats to coral reefs in the Pacific Ocean, including the biodiverse Coral Triangle region, are similar to those experienced globally (ICRI, 2018c). A 3 per cent decrease in coral cover was reported from 1999 to 2016 (ICRI, 2018c) and widespread bleaching events have occurred throughout the region since 2015 (Moritz and others, 2018; Hughes and others, 2019), the impacts of which are only now being reported (Gorospe and others, 2018).

Outbreaks of crown-of-thorns starfish continue to occur across the Pacific Ocean, causing coral decline, including, for example, in French Polynesia (Kayal and others, 2012), Mexico (Rodríguez-Vilalobos and Ayala-Bocos, 2018), Japan (Yasuda, 2018), Australia (MacNeil and others, 2017) and other Pacific islands (Moritz and others, 2018).

Biodiversity of coral species across the Pacific is changing as a result of disturbance, with certain coral species, such as *Porites* spp., becoming increasingly dominant, while *Pocillopora* spp. showed significant decline (Moritz and others, 2018). Japan has reported a poleward range expansion of corals to former seaweed habitats (Yamano and others, 2011; Kumagai and others, 2018). Coral reef cover around the north-western side of the island of Hawaii decreased from about 44 per cent to 31 per cent between 2002 and 2014, mainly owing to human use-related activities (e.g., animal production; land development, deforestation and urban sprawl; fishing and recreation) and climate-related heatwaves (Gove and others, 2016).

About 88 per cent of reefs are under threat from human activities, especially localized threats (ICRI, 2018c). About 13 per cent (8,960 km^2) of coral reef areas across the Pacific Ocean are contained in protected areas, of which 20 per cent have formal management plans implemented (Moritz and others, 2018). Of those reefs contained in marine protected areas in the South-East Asian region, only 30 per cent are identified as having effective management and compliance measures in place (ICRI, 2018c).

5. Outlook

Coral abundance and cover have decreased from 2010 to 2019 and are expected to continue decreasing in the coming decades (Graham and others, 2017). Bleaching events since 2015 have led to reduced production of larvae and reduced recruitment, which will lead to reduced or delayed recovery (Hughes and others, 2018b). Climate projections (see also chaps. 5 and 9) suggest that many of the world's coral reefs will experience annual bleaching associated with warming temperatures by mid-century (Hughes and others, 2018b). Greater erosion, sedimentation and nutrient flow associated with increasing storm intensity (Walsh and others, 2016; Vitousek and others, 2017); increased mortality associated with a reduction in oxygen (Nelson and Altieri, 2019; Altieri and others, 2017); and drowning of coral reefs as sea levels increase (Perry and others, 2018; Storlazzi and others, 2019) have all been projected to occur. Deepwater areas appear a less likely option for thermal refuge for corals because the thermal relief provided by the deeper waters only occurs at certain times of the year and only certain species are tolerant of deepwater environments (Frade and others, 2018). Future coral communities are likely to be dominated by fewer species, which are resistant to high temperatures and bleaching (Moritz and others, 2018; Birkeland, 2019). The effects of ocean acidification should become more apparent in the coming decades, with the continued dissolution of reefs (Eyre and others, 2018; Birkeland, 2019).

Coral reefs are expected to expand into higher latitudes as a result of warming oceans and warm currents (Wilson and others, 2016, 2018). Coral habitats or species ranges are, however, expected to be bound between higher latitudes, where the saturation state of aragonite decreases, and lower latitudes, where warmer water temperatures result in thermal stress (Matz and others, 2018; Yara and others, 2012).

Greenhouse gases and climate change are considered by many reef scientists to be the key risk to future coral reefs (e.g., Beyer and others, 2018; Rinkevich, 2019). To combat losses of coral reefs, coral reef restoration techniques are being improved (van Oppen and others, 2017) and widely used and are demonstrating some success (Bayraktarov and others, 2019; Rinkevich, 2019). Further investigations are taking place to understand the response of corals to climate change and to develop methods that help corals to adapt to future conditions (e.g., alterations to symbiotic microalgae diversity; Rinkevich, 2019). Modelling approaches that identify risk envelopes for coral reefs may assist in prioritizing efforts so that they are focused on the reefs that have the greatest resilience and likelihood of survival (Beyer and others, 2018). Innovative sustainable funding mechanisms for supporting marine resource conservation and sustainable use, especially of coral reefs, are finding traction under the blue economy (e.g., Deutz and others, 2018).

Overall, the continued loss of coral reefs that is expected to occur in the coming decades will also result in the erosion of the many socioeconomic benefits of healthy coral reefs.

6. Key remaining knowledge gaps

The key knowledge gaps reported in the first Assessment included understanding of the responses of corals and coral-dependent species (such as fishes) to climate change and the spatial extent of mesophotic coral reefs (reefs found in lower light conditions at depths of 30–150 m). Progress has been made in addressing them, however, and the specific knowledge gaps today are somewhat different, although they remain in the same broad areas. The gaps are: (a) the responses of reef communities to climate change; (b) the socioeconomic value of coral reefs; and (c) the distribution and ecology of mesophotic coral reefs.

A greater understanding of the responses of reef communities to climate change is still required, although progress is occurring with regard to coral adaptation to change (e.g., Dziedzic and others, 2019). There is still limited understanding of coral reef responses to ocean acidification (Morais and others, 2018), and the current understanding is confounded by inaccuracy in measurements of net reef growth and erosion. In particular, there is a lack of understanding of the impacts of increasing ocean temperatures on the life cycles of reef taxa, changes to neurosensory function and metabolism in a variety of key reef-associated taxa and the cumulative effects of climate change and other stressors, such as nutrient enrichment, increased sediment load and overfishing, on coral reef systems. There is currently limited understanding of the role played by coralline algae and microbial communities in reef ecology and health (Cornwall and others, 2019; Ricci and others, 2019), although there are clear seasonal links between the microbiome and macroalgae abundance (Glasl and others, 2020). There are also significant geographic gaps in our understanding of coral reefs and their responses to climate change and other stressors, such as reduced oceanic oxygen and emerging pollutants. Western South Atlantic coral reef communities and deeper water coral reef communities are poorly described (Loya and others, 2016; Morais and others, 2018). Further information is needed to identify the mechanisms of coral diseases and how they are transmitted, in particular their relationship with coral bleaching events and poor water quality.

For the socioeconomic value of coral reefs, robust evaluations of the economic value of the ecosystem services that reefs provide at the local and national levels are lacking, so their value is not properly accounted for in the cost-benefit analyses of development projects. There are further knowledge gaps regarding the effectiveness of management tools and efforts to improve reef resilience, including coral restoration (Boström-Einarsson and others, 2020).

Finally, there is emerging evidence that mesophotic coral reefs occur widely (e.g., Baker and others, 2016) beyond the Atlantic (Loya and others, 2016), and further information is required regarding their biodiversity and ecological function.

7. Key remaining capacity-building gaps

The first Assessment identified capacity-building gaps at the local, national and regional levels. Those gaps remain in most developing countries and, in particular, a lack of qualified technicians and researchers limits the monitoring and management of coral reefs and, therefore, the capacity to identify change over time and to respond to changes. There have been substantial improvements in the development of new technologies to monitor coral reef systems (Bayley and Mogg, 2019; Hedley and others, 2016), but there is limited local capability for the utilization and application of such technology (e.g., Díaz and others, 2015; Timpte and others, 2018). Greater capability for implementing adaptation strategies in response to climate change (Cinner and others, 2018) and building adaptive management capabilities (Hoegh-Guldberg, 2018), both locally and globally, is needed. Public awareness and capacity-building for the management and sustainable use of expanding corals is an emerging issue at higher latitudes. The capabilities for coral reef rehabilitation are currently limited, and capabilities for transplanting and farming corals (Kittinger and others, 2016; Van Oppen and others, 2017) and maintaining them need to be developed.

References

Adjeroud, Mehdi, and others (2018). Recovery of coral assemblages despite acute and recurrent disturbances on a South Central Pacific reef. *Scientific Reports*, vol. 8, No. 1, art. 9680.

Albright, Rebecca, and others (2016). Reversal of ocean acidification enhances net coral reef calcification. *Nature*, vol. 531, No. 7594, p. 362.

Altieri, Andrew H., and others (2017). Tropical dead zones and mass mortalities on coral reefs. *Proceedings of the National Academy of Sciences*, vol. 114, No. 14, pp. 3660–3665.

Alvarez-Filip, Lorenzo, and others (2019). A rapid spread of the Stony Coral Tissue Loss Disease outbreak in the Mexican Caribbean. *PeerJ Preprints*, vol. 7, e27893v1.

Anyamba, Assaf, and others (2019). Global Disease outbreaks Associated with the 2015–2016 El Niño event. *Scientific Reports*, vol. 9, No. 1, art. 1930.

Aswani, Shankar, and others (2015). Scientific frontiers in the management of coral reefs. *Frontiers in Marine Science*, vol. 2, art. 50.

Babcock, Russell C., and others (2019). Severe continental-scale impacts of climate change are happening now: Extreme climate events impact marine habitat forming communities along 45% of Australia's coast. *Frontiers in Marine Science*, vol. 6, art. 411.

Baird, A.H., and others (2012). Pole-ward range expansion of *Acropora* spp. along the east coast of Australia. *Coral Reefs*, vol. 31, No. 4, pp. 1063–1063.

Baker, E., and others (2016) *Mesophotic coral ecosystems–a lifeboat for coral reefs?* United Nations Environment Programme and GRID-Arendal.

Banon, Ysabel, and others (2018). Thermal Stress and Bleaching in Coral Reef Communities during the 2014–2016 Caribbean Bleaching Event. In *AGU Fall Meeting Abstracts*.

Baums, Iliana B., and others (2019). Considerations for maximizing the adaptive potential of restored coral populations in the western Atlantic. *Ecological Applications*, vol. 29, No. 8, e01978. 10.1002/eap.1978

Bayley, Daniel T.I., and Andrew O.M. Mogg (2019). New advances in benthic monitoring technology and methodology. In *World Seas: An Environmental Evaluation*, pp. 121–132. Elsevier.

Bayraktarov, Elisa, and others (2019). Motivations, success and cost of coral reef restoration. *Restoration Ecology*.

Beck, Michael W., and others (2018). The global flood protection savings provided by coral reefs. *Nature Communications*, vol. 9, No. 1, art. 2186.

Beyer, Hawthorne L., and others (2018). Risk-sensitive planning for conserving coral reefs under rapid climate change. *Conservation Letters*, vol. 11, e12587.

Birkeland, Charles (2019). Global status of coral reefs: in combination, disturbances and stressors become ratchets. In *World Seas: An Environmental Evaluation*, pp. 35–56. Elsevier.

Booth, David J., and John Sear (2018). Coral expansion in Sydney and associated coral-reef fishes. *Coral Reefs*, vol. 37, No. 4, pp. 995–995.

Boström-Einarsson, Lisa, and others (2020). Coral restoration – A systematic review of current methods, successes, failures and future directions. *PloS ONE*, vol. 15, e0226631.

Bruno, John F., and others (2019). Climate change, coral loss, and the curious case of the parrotfish paradigm: Why don't marine protected areas improve reef resilience? *Annual Review of Marine Science*, vol. 11, pp. 307–334.

Butler, I.R., and others (2015). The cumulative impacts of repeated heavy rainfall, flooding and altered water quality on the high-latitude coral reefs of Hervey Bay, Queensland, Australia. *Marine Pollution Bulletin*, vol. 96, Nos. 1–2, pp. 356–367.

Caribbean Challenge Initiative (CCI) (2019). *Factsheet and Overview: Caribbean Challenge Initiative*. www.caribbeanbiodiversityfund.org/pdf/CCI_Overview_factSheet_HighRes.pdf.

Chagaris, David, and others (2017) An ecosystem-based approach to evaluating impacts and management of invasive lionfish. *Fisheries*, vol. 42, No.8, pp. 421–431, https://doi.org/10.1080/03632415.2017.1340273.

Chefaoui, Rosa M., and others (2017). Environmental drivers of distribution and reef development of the Mediterranean coral *Cladocora caespitosa*. *Coral Reefs*, vol. 36, No. 4, pp. 1195–1209.

Chevallier, Romy (2017). Safeguarding Tanzania's Coral Reefs: The Case of Illegal Blast Fishing.

Cinner, Joshua E., and others (2016a). Bright Spots among the World's Coral Reefs. *Nature*, vol. 535, p. 416.

Cinner, Joshua E., and others (2016b). A Framework for Understanding Climate Change Impacts on Coral Reef Social–Ecological Systems. *Regional Environmental Change*, vol. 16, No. 4, pp. 1133–1146.

Cinner, Joshua E., and others (2018). Building Adaptive Capacity to Climate Change in Tropical Coastal Communities. *Nature Climate Change*, vol. 8, No. 2, p. 117.

Cornwall, Christopher Edward, and others (2019). Impacts of ocean warming on coralline algae: knowledge gaps and key recommendations for future research. *Frontiers in Marine Science*, vol. 6, art. 186.

Cortés, Jorge, and others (2018). The CARICOMP Network of Caribbean Marine Laboratories (1985–2007): History, Key Findings and Lessons Learned. *Frontiers in Marine Science*, vol. 5, art. 519.

Costanza, Robert, and others (2014). Changes in the global value of ecosystem services. *Global Environmental Change*, vol. 26, pp. 152–158.

Cowburn, Benjamin, and others (2019). Evidence of coral bleaching avoidance, resistance and recovery in the Maldives during the 2016 mass-bleaching event. *Marine Ecology Progress Series*, vol. 626, pp. 53–67.

De Bakker, Didier M., and others (2017). 40 Years of benthic community change on the Caribbean reefs of Curaçao and Bonaire: the rise of slimy cyanobacterial mats. *Coral Reefs*, vol. 36, No. 2, pp. 355–367.

Deutz, Andrew, and others (2018). *Innovative Finance for Resilient Coasts and Communities. A Briefing Paper Prepared by The Nature Conservancy and the United Nations Development Programme for Environment and Climate Change Canada*. www.nature.org/content/dam/tnc/nature/en/documents/Innovative_Finance_Resilient_Coasts_and_Communities.pdf.

Díaz, Sandra, and others (2015). The IPBES Conceptual Framework—connecting nature and people. *Current Opinion in Environmental Sustainability*, vol. 14, pp. 1–16.

Dziedzic, Katherine E., and others (2019) Heritable variation in bleaching responses and its functional genomic basis in reef-building corals (*Orbicella faveolata*). *Molecular Ecology*, vol. 28, No. 9, pp. 2238–2253.

Eakin, C. Mark, and others (2019). The 2014–2017 global-scale coral bleaching event: insights and impacts. *Coral Reefs*, vol. 38, pp. 539–545.

Eriksson, Hampus and others (2015). Contagious exploitation of marine resources. *Frontiers in Ecology and the Environment*, vol. 13, No. 8, pp. 435–440.

Eyre, Bradley D., and others (2018). Coral reefs will transition to net dissolving before end of century. *Science*, vol. 359, No. 6378, pp. 908–911.

Fordyce, Alexander John, and others (2019). Marine heatwave hotspots in coral reef environments: physical drivers, ecophysiological outcomes and impact upon structural complexity. *Frontiers in Marine Science*, vol. 6, p. 498.

Frade, Pedro R., and others (2018). Deep reefs of the Great Barrier Reef offer limited thermal refuge during mass coral bleaching. *Nature Communications*, vol. 9, No. 1, pp. 3447.

Gatti, Giulia, and others (2015). Ecological change, sliding baselines and the importance of historical data: lessons from combing observational and quantitative data on a temperate reef over 70 years. *PloS One*, vol. 10, No. 2, e0118581.

Global Coral Reef Monitoring Network (GCRMN)-Caribbean (2016). *GCRMN-Caribbean Guidelines for Coral Reef Biophysical Monitoring*. UNEP(DEPI)/CAR WG.38/INF.17.

Glasl, B., and others (2020). Comparative genome-centric analysis reveals seasonal variation in the function of coral reef microbiomes. *ISME Journal*, vol. 14, pp. 1435–1450.

Gorospe, Kelvin D., and others (2018). Local biomass baselines and the recovery potential for Hawaiian coral reef fish communities. *Frontiers in Marine Science*, vol. 5, art. 162.

Gove, Jamison M., and others (2016). West Hawai'i integrated ecosystem assessment: ecosystem trends and status report. Pacific Islands Fisheries Science Centre.

Graham, Nicholas A.J., and others (2017). Human disruption of coral reef trophic structure. *Current Biology*, vol. 27, No. 2, pp. 231–236.

Hedley, John D., and others (2016). Remote sensing of coral reefs for monitoring and management: a review. *Remote Sensing*, vol. 8, No. 2, art. 118.

Hoegh-Guldberg, Ove, and others (2018). Securing a long-term future for coral reefs. *Trends in Ecology & Evolution*.

Holbrook, Sally J., and others (2018). Recruitment drives spatial variation in recovery rates of resilient coral reefs. *Scientific Reports*, vol. 8, No. 1, art. 7338.

Howells, Emily J., and others (2016). Host adaptation and unexpected symbiont partners enable reef-building corals to tolerate extreme temperatures. *Global Change Biology*, vol. 22, pp. 2702–2714. https://doi.org/10.1111/gcb.13250.

Hughes, Terry P., and others (2017a). Coral reefs in the Anthropocene. *Nature*, vol. 546, No. 7656, p. 82.

Hughes, Terry P., and others (2017b). Global warming and recurrent mass bleaching of corals. *Nature*, vol. 543, No. 7645, p. 373.

Hughes, Terry P., and others (2018a). Global warming transforms coral reef assemblages. *Nature*, vol. 556, No. 7702, p. 492.

Hughes, Terry P., and others (2018b). Spatial and temporal patterns of mass bleaching of corals in the Anthropocene. *Science*, vol. 359, No. 6371, pp. 80–83.

Hughes, Terry P., and others (2019). Ecological memory modifies the cumulative impact of recurrent climate extremes. *Nature Climate Change*, vol. 9, pp. 40–43. https://doi.org/10.1038/s41558-018-0351-2.

International Coral Reef Initiative (ICRI) (2013). *Recommendation on Addressing the Decline in Coral Reef Health throughout the Wider Caribbean: The Taking of Parrotfish and Similar Herbivores.* Adopted on 17 October 2013, at the 28th ICRI General Meeting (Belize City).

________ (2018a). *Caribbean Fact Sheet - Communicating the Economic and Social Importance of Coral Reefs for Caribbean Countries.* International Coral Reef Initiative.

International Coral Reef Initiative (ICRI) (2018b). *Communicating the Economic and Social Importance of Coral Reefs for Indian Ocean Countries.* International Coral Reef Initiative.

________ (2018c). *South Asia Factsheet - Communicating the Economic and Social Importance of Coral Reefs for South East Asian Countries.* International Coral Reef Initiative.

Intergovernmental Panel on Climate Change (IPCC) (2014). *Climate Change 2014 – Impacts, Adaptation and Vulnerability: Part A: Global and Sectoral Aspects: Working Group II Contribution to the IPCC Fifth Assessment Report* vol. 1. C.B Field, and others, eds. Cambridge: Cambridge University Press. https://doi.org/10.1017/CBO9781107415379.

________ (2018). Summary for policymakers. In *Global Warming of 1.5°C. An IPCC Special Report on the Impacts of Global Warming of 1.5°C above Pre-Industrial Levels and Related Global Greenhouse Gas Emission Pathways, in the Context of Strengthening the Global Response to the Threat of Climate Change, Sustainable Development, and Efforts to Eradicate Poverty.* V. Masson-Delmotte, and others, eds., p. 24.

Jackson, Jeremy, and others (2014). *Status and Trends of Caribbean Coral Reefs: 1970–2012.* Gland, Switzerland: Global Coral Reef Monitoring Network, ICUN.

Johns, Kerryn A., and others (2018). Macroalgal feedbacks and substrate properties maintain a coral reef regime shift. *Ecosphere*, vol. 9, No. 7, art. e02349.

Johnson, Christopher N., and others (2017). Biodiversity losses and conservation responses in the Anthropocene. *Science*, vol. 356, No. 6335, pp. 270–275.

Jones, Kendall R., and others (2018). The location and protection status of Earth's diminishing marine wilderness. *Current Biology*, vol. 28, No. 15, pp. 2506–2512.

Kayal, Mohsen, and others (2012). Predator Crown-of-Thorns Starfish (*Acanthaster planci*) Outbreak, Mass Mortality of Corals, and Cascading Effects on Reef Fish and Benthic Communities. *PLOS ONE*, vol. 7, No. 10, pp. 1–9. https://doi.org/10.1371/journal.pone.0047363.

Keesing, John K. and others (2019). Two time losers: selective feeding by crown-of-thorns starfish on corals most affected by successive coral-bleaching episodes on western Australian coral reefs. *Marine Biology*, vol. 166, No. 72. https://doi.org/10.1007/s00227-019-3515-3.

Kerr, Rodrigo, and others (2018). Northern Antarctic Peninsula: a marine climate hotspot of rapid changes on ecosystems and ocean dynamics. *Deep-Sea Research Part II Topical Studies in Oceanography*, vol. 149, pp. 4–9.

Kittinger, John, and others (2012). Human dimensions of coral reef social-ecological systems. *Ecology and Society*, vol. 17, No. 4. https://doi.org/10.5751/ES-05115-170417.

Kittinger, John, and others (2016). Restoring ecosystems, restoring community: socioeconomic and cultural dimensions of a community-based coral reef restoration project. *Regional Environmental Change*, vol. 16, No. 2, pp. 301–313.

Kuffner, Ilsa B., and others (2019). Improving estimates of coral reef construction and erosion with in situ measurements. *Limnology and Oceanography*. https://doi.org/10.1002/lno.11184.

Kumagai, Naoki H., and others (2018). Ocean currents and herbivory drive macroalgae-to-coral community shift under climate warming. *Proceedings of the National Academy of Sciences*, vol. 115, No. 36, pp. 8990–8995.

Leggat, William P., and others (2019). Rapid coral decay is associated with marine heatwave mortality events on reefs. *Current Biology*, vol. 29, No. 16, pp. 2723–2730. https://doi.org/10.1016/j.cub.2019.06.077.

Lirman, D. and S. Schopmeyer (2016). Ecological solutions to reef degradation: optimizing coral reef restoration in the Caribbean and Western Atlantic. *PeerJ* 4: e2597; https://doi.org/10.7717/peerj.2597.

Lizcano-Sandoval, Luis David, and others (2019). Climate change and Atlantic Multidecadal Oscillation as drivers of recent declines in coral growth rates in the Southwestern Caribbean. *Frontiers in Marine Science*, vol. 6, art. 38.

Lough, J.M., and others (2018). Increasing thermal stress for tropical coral reefs: 1871–2017. *Scientific Reports*, vol. 8, No. 1, art. 6079.

Loya, Yossi, and others (2016). *Theme Section on Mesophotic Coral Ecosystems: Advances in Knowledge and Future Perspectives. Coral Reefs*, vol. 35, pp. 1–9.

MacNeil, M., and others (2017). Age and growth of an outbreaking *Acanthaster* cf. *solaris* population within the Great Barrier Reef. *Diversity*, vol. 9, No. 1, art. 18.

Matz, Mikhail V., and others (2018) Potential and limits for rapid genetic adaptation to warming in a Great Barrier Reef coral. *PLOS Genetics*, 14(4), e1007220. https://doi.org/10.1371/journal.pgen.1007220.

Mcfield, Melanie, and others (2018). *2018 Mesoamerican Reef Report Card*. Healthy Reefs Initiative. https://doi.org/10.13140/RG.2.2.19679.36005.

Morais, Juliano, and others (2018). Research gaps of coral ecology in a changing world. *Marine Environmental Research*, vol. 140, pp. 243–250.

Moritz, Charlotte, and others, eds. (2018). Status and Trends of Coral Reefs of the Pacific, Global Coral Reef Monitoring Network.

Mumby, Peter J., and others (2014). Towards Reef Resilience and Sustainable Livelihoods: A handbook for Caribbean coral reef managers.

Muniz-Castillo, Aaron Israel, and others (2019). Three decades of heat stress exposure in Caribbean coral reefs: a new regional delineation to enhance conservation. *Scientific Reports*, vol. 9, art. 11013. https://doi.org/10.1038/s41598-019-47307-0.

Murakami, Hiroyuki, and others (2014). Contributing factors to the recent high level of accumulated cyclone energy (ACE) and power dissipation index (PDI) in the North Atlantic. *Journal of Climate*, vol. 27, No. 8, pp. 3023–3034.

Nelson, Hannah R., and Andrew H. Altieri (2019). Oxygen: the universal currency on coral reefs. *Coral Reefs*, vol. 38, No. 2, pp. 177–198.

National Oceanic and Atmospheric Administration (NOAA) (2020). *Coral reef condition: A status report for the Flower Garden Banks*. NOAA Coral Reef Conservation Program. Silver Spring, Maryland, United States.

Obura, David, and others (2017). *Coral Reef Status Report for the Western Indian Ocean*. Global Coral Reef Monitoring Network (GCRMN)/International Coral Reef Initiative (ICRI), p. 144.

Perry, Chris T., and others (2013). Caribbean-wide decline in carbonate production threatens coral reef growth. *Nature Communications*, vol. 4, No. 1, art. 1402.

Perry, Chris T., and others (2018). Loss of coral reef growth capacity to track future increases in sea level. *Nature*, vol. 558, No. 7710, pp. 396–400.

Popova, Ekaterina, and others (2016). From global to regional and back again: common climate stressors of marine ecosystems relevant for adaptation across five ocean warming hotspots. *Global Change Biology*, vol. 22, No. 6, pp. 2038–2053.

Randall, Carly J., and Robert van Woesik (2015). Contemporary white-band disease in Caribbean corals driven by climate change. *Nature Climate Change*, vol. 5, No. 4, p. 375.

Ricci, Francesco, and others (2019). Beneath the surface: community assembly and functions of the coral skeleton microbiome. *EcoEvoRxiv*. https://doi.org/10.32942/osf.io/9yjw8.

Rinkevich, Baruch (2019). The active reef restoration toolbox is a vehicle for coral resilience and adaptation in a changing world. *Journal of Marine Science and Engineering*, vol. 7, No. 7, art. 201.

Rodríguez-Vilalobos, J.C., and A. Ayala-Bocos (2018). Coral colonies in the eastern tropical Pacific: predation by *Acanthaster* cf. *solaris*. *Pacific Conservation Biology*, vol. 24, No. 4, pp. 419–420.

Ruiz-Moreno, Diego, and others (2012). Global coral disease prevalence associated with sea temperature anomalies and local factors. *Diseases of Aquatic Organisms*, vol. 100, No. 3, pp. 249–261.

Saponari, Luca, and others (2018). Monitoring and assessing a 2-year outbreak of the corallivorous seastar *Acanthaster planci* in Ari Atoll, Republic of Maldives. *Environmental Monitoring and Assessment*, vol. 190, No. 6, art. 344.

Schemmel, Eva, and others (2016). The codevelopment of coastal fisheries monitoring methods to support local management. *Ecology and Society*, vol. 21, No. 4.

Serrano, Eduard, and others (2018). Demographics of the zooxanthellate coral *Oculina patagonica* along the Mediterranean Iberian coast in relation to environmental parameters. *Science of The Total Environment*, vol. 634, pp. 1580–1592.

Shaver, Elizabeth C., and others (2018). Local management actions can increase coral resilience to thermally-induced bleaching. *Nature Ecology & Evolution*, vol. 2, No. 7, p. 1075.

Smale, Dan A., and others (2019). Marine heatwaves threaten global biodiversity and the provision of ecosystem services. *Nature Climate Change*, vol. 9, No. 4, p. 306.

Spalding, Mark, and others (2017). Mapping the global value and distribution of coral reef tourism. *Marine Policy*, vol. 82, pp. 104–113.

Steiner, Zvi, and others (2018). Water chemistry reveals a significant decline in coral calcification rates in the southern Red Sea. *Nature Communications*, vol. 9, No. 1, art. 3615.

Storlazzi, Curt D., and others (2019). Rigorously valuing the role of US coral reefs in coastal hazard risk reduction. US Geological Survey.

Teh, Louise S.L., and others (2013). A global estimate of the number of coral reef fishers. *PLoS One*, vol. 8, No. 6, e65397.

Timpte, Malte, and others (2018). Engaging diverse experts in a global environmental assessment: participation in the first work programme of IPBES and opportunities for improvement. *Innovation: The European Journal of Social Science Research*, vol. 31, No. sup1, pp. S15–S37.

Turner, Rachel A., and others (2018). Social fit of coral reef governance varies among individuals. *Conservation Letters*, vol. 11, No .3, e12422.

United Nations (2017a). Chapter 43: Tropical and sub-tropical coral reefs. In *The First Global Integrated Marine Assessment: World Ocean Assessment I*. Cambridge: Cambridge University Press.

________ (2017b). *The First Global Integrated Marine Assessment: World Ocean Assessment I*. Cambridge: Cambridge University Press.

United Nations Environment Programme (UNEP) and others (2018). *Plastics and Shallow Water Coral Reefs. Synthesis of the Science for Policy-Makers*. Nairobi: UNEP.

Vallès, Henri, and Hazel A. Oxenford (2018). Simple family-level parrotfish indicators are robust to survey method. *Ecological Indicators*, vol. 85, pp. 244–252.

Van Hooidonk, Ruben, and others (2014). Opposite latitudinal gradients in projected ocean acidification and bleaching impacts on coral reefs. *Global Change Biology*, vol. 20, No.1, pp. 103–112.

Van Oppen, Madeleine J.H., and others (2017). Shifting paradigms in restoration of the world's coral reefs. *Global Change Biology*, vol. 23, No. 9, pp. 3437–3448.

Van Woesik, Robert, and Carly J. Randall (2017). Coral disease hotspots in the Caribbean. *Ecosphere*, vol. 8, No. 5, art. e01814.

Vargas-Ángel, Bernardo, and others (2019). El Niño-associated catastrophic coral mortality at Jarvis Island, central Equatorial Pacific. *Coral Reefs* vol. 38, pp. 731–741.

Vitousek, Sean, and others (2017). Doubling of coastal flooding frequency within decades due to sea-level rise. *Scientific Reports*, vol. 7, No. 1, art. 1399.

Walsh, Kevin J.E., and others (2016). Tropical cyclones and climate change. *Wiley Interdisciplinary Reviews: Climate Change*, vol. 7, No. 1, pp. 65–89.

Wernberg, Thomas, and others (2016). Climate-driven regime shift of a temperate marine ecosystem. *Science*, vol. 353, No. 6295, pp. 169–172.

Wilson, Laura J., and others (2016). Climate-driven changes to ocean circulation and their inferred impacts on marine dispersal patterns. *Global Ecology and Biogeography*, vol. 25, No. 8, pp. 923–939.

Wilson, Shaun K., and others (2018). Climatic forcing and larval dispersal capabilities shape the replenishment of fishes and their habitat-forming biota on a tropical coral reef. *Ecology and Evolution*, vol. 8, No. 3, pp. 1918–1928.

Yamano, Hiroya, and others (2011). Rapid poleward range expansion of tropical reef corals in response to rising sea surface temperatures. *Geophysical Research Letters*, vol. 38, No. 4.

Yara, Yumiko, and others (2012). Ocean acidification limits temperature-induced poleward expansion of coral habitats around Japan. *Biogeosciences*, vol. 9, No. 12, pp. 4955–4968.

Yasuda, Nina (2018). Distribution Expansion and Historical Population Outbreak Patterns of Crown-of-Thorns Starfish, *Acanthaster planci* sensu lato, in Japan from 1912 to 2015. In *Coral Reef Studies of Japan*, pp. 125–148. Springer, Singapore.

Chapter 7E
Cold water corals

Contributors: Erik Cordes (convener), Malcolm R. Clark, Karen Evans (subchapter lead member), Sebastian Hennige and Georgios Kazanidis.

Keynote points

- Cold water coral and sponge ecosystems are common features along continental margins and mid-ocean ridges and on seamounts worldwide, providing habitats for numerous species and contributing to carbon sequestration.
- Basic knowledge of cold water coral biology and distribution is still limited to the few areas of the deep ocean that have been explored.
- Threats from fishing, offshore oil drilling, deep-sea mining, waste deposition and climate change continue. Some efforts to curb demersal trawling and establish marine protected areas have been effective. However, because of their slow-growing and long-lived nature, recovery from anthropogenic impacts can take decades to centuries.
- Cold water corals are highly sensitive to elevated temperatures and deoxygenation, but recent work suggests they are relatively resilient to ocean acidification, in particular when nutritional resources are plentiful.
- Future projected declines in cold water coral abundance will reduce the habitat available to commercially significant species, reduce carbon sequestration in deep waters, eliminate potential genetic resources and have an effect on numerous Sustainable Development Goals, in particular Goal 14, but also Goals 2, 10, and 12.[1]

1. Introduction and summary of the first *World Ocean Assessment*

Cold water corals (CWC) occur globally (figure I), forming important habitats that support a high diversity and biomass of associated organisms. Following the framework established in the first *World Ocean Assessment* (United Nations, 2017), the present chapter focuses on corals found below 200 m. CWC are found where hard substrata are available on continental margins, mid-ocean ridges (chap. 7O) and seamounts (chap. 7L) worldwide. Those habitats are components of the slopes of volcanic islands (included in chap. 7C), submarine canyons (chap. 7J) and fjords, as well as seamounts and pinnacles (chap. 7L), and ridges and plateaux (chap. 7O). Coral habitats may be found on the periphery of cold seeps and extinct hydrothermal vents (chap. 7P). CWC, sponges and associated species also interact directly with the overlying open ocean (chap. 7N) through benthic-pelagic coupling. Surface-derived productivity forms the energetic basis of the vast majority of the deep-sea food web, and CWC systems recycle nutrients that can fuel surface productivity via upwelling and the transport of nutrients by diel vertical migrators.

Modelling of global habitats predicts that major cold water framework-building scleractinian corals (figure II) are likely most abundant in:

(a) hard substrate areas necessary for settlement of coral larvae;

(b) waters supersaturated with aragonite;

(c) depths shallower than 1,500 m;

(d) water masses containing dissolved oxygen concentrations of more than 4 ml/l;

(e) waters that have a salinity range between 34 and 37 practical salinity units; and

1 See General Assembly resolution 70/1.

(f) temperatures between 5°C and 10°C (Davies and Guinotte, 2011).

However, CWC of a variety of taxa occupy wider niches in the deep ocean (Quattrini and others, 2013, 2017). Food is supplied to CWC through rapid downwelling (Davies and others, 2009), geostrophic currents, internal waves, tides, Taylor columns (eddies causing upwelling and enhancing food concentration; White and others, 2005), intermediate and bottom nepheloid layers (Mienis and others, 2007), and diel vertical migrators (Maier and others, 2019).

CWC ecosystems provide essential services for human communities and well-being (see also sect. 3). Demonstrated services include the discovery of novel marine genetic resources (chap. 23), carbon sequestration and significant aesthetic value (see Thurber and others, 2014, for a review). Most directly, CWC provide a habitat that acts as shelter and/or nursery grounds for commercially exploited or exploitable fish stocks (Baillon and others, 2012; Quattrini and others, 2012; Roberts and others, 2009). The Food and Agriculture Organization of the United Nations (FAO, 2009) recognizes taxa of CWC as indicators of vulnerable marine ecosystems and the General Assembly, in its resolutions 59/25 and 61/105, called for conservation measures to protect vulnerable marine ecosystems from anthropogenic impacts. The wealth of niches combined with high food availability make some CWC reefs "hotspots" of biodiversity and biomass, including hundreds of other sessile and mobile species (Cordes and others, 2008; Henry and Roberts, 2007), as well as carbon and nutrient cycling (Cathalot and others, 2015; van Oevelen and others, 2009).

Figure I
Map of global distribution of deep-sea corals, including subclass Octocorallia (gorgonian sea fans, soft corals) and order Scleractinia (stony corals)

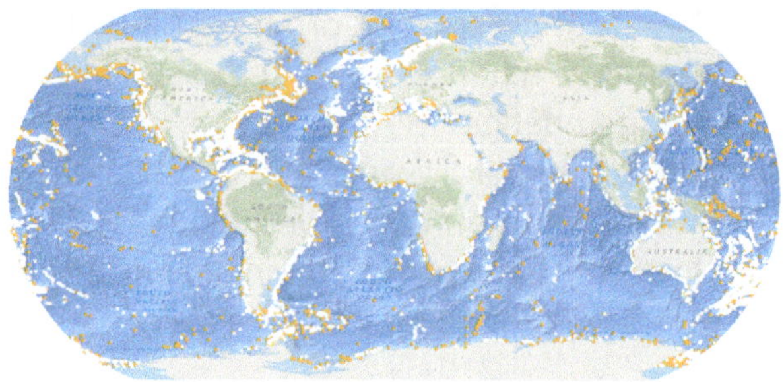

Octocorallia   Scleratinia

Sources: Coral occurrence records from Freiwald and others, 2017, available at https://data.unep-wcmc.org/datasets/3; Ocean Biodiversity Information System, available at https://obis.org; and National Oceanic and Atmospheric Administration Deep-Sea Coral and Sponge database, available at www.ncei.noaa.gov/maps/deep-sea-corals/mapSites.htm. Map courtesy of Dr. Jay Lunden.

Note: The underlying base map was created in ArcGIS Pro v.2.3, using data compiled from multiple sources and data providers, including the General Bathymetric Chart of the Oceans GEBCO_08 Grid, the National Oceanic and Atmospheric Administration, National Geographic, Garmin, Here, Geonames.org and Esri, as well as several additional contributors.

2. Description of environmental changes between 2010 and 2020

As sessile organisms with very slow growth rates, CWC and the frameworks they create (both living and dead) are extremely vulnerable to direct and indirect impacts from bottom trawling, oil and gas exploration, and deep-sea mining. While the current stressors on CWC are not significantly different from those identified in the first Assessment, the distribution of pressures and their magnitude has changed. As fishing activities (chap. 15) continue to operate in deeper waters and the search for oil and gas reserves (chap. 19) moves further offshore, those activities affect CWC gardens and reefs more frequently. Further, the accidental release of hydrocarbons (chap. 11) associated with oil and gas extraction can have drastic effects on those habitats, as highlighted by the 2010 Deepwater Horizon oil spill in the Gulf of Mexico (Fisher and others, 2014; White and others, 2012a). In the vicinity of the oil spill, where over 50 per cent of individual octocoral colonies were affected by the spill, their health and colony size has continued to decline (Hsing and others, 2013). Some recovery has been observed in coral colonies with less than 50 per cent coverage by oil and dispersant, although some branch loss has been observed (Hsing and others, 2013).

Figure II
Common cold-water scleractinian corals

Sources: A and B, courtesy of the Deep Search programme, United States Bureau of Ocean Energy Management, United States Geological Survey and National Oceanic and Atmospheric Administration; copyright WHOI; C, courtesy of Malcolm Clark, New Zealand National Institute of Water and Atmospheric Research; D, courtesy of Erik Cordes and the Schmidt Ocean Institute.

Notes: A, *Lophelia pertusa* and B, *Madrepora oculata* and a neptheld octocoral from the Atlantic coast, United States; C, *Solenosmilia variabilis* from the Pacific coast, New Zealand; D, *Enallopsamia profunda* from the Phoenix Islands in the central Pacific.

Figure III
Representative cold water octocoral gardens

Sources: A, courtesy of Erik Cordes and the Schmidt Ocean Institute; B, courtesy of Erik Cordes and the Research on Cold Seeps and How They Influence the Sea (ROC HITS) programme, United States National Science Foundation, and the Schmidt Ocean Institute; C, courtesy of Malcolm Clark, New Zealand National Institute of Water and Atmospheric Research; D, courtesy of J. Murray Roberts and the Changing Ocean Expedition, 2012.

Notes: A, high diversity of octocorals (primarily primnoids and plexaurids) and antipatharian corals from the Phoenix Islands in the central Pacific; B, yellow and purple *Paramuricea* spp. octocoral sea fan colonies with orange chirostylid crab associates and *Euplectella* spp. hexactinellid (glass) sponges to the left of the image; C, *Paragorgia* spp. octocorals, stylasterid hydrocorals and hexactinellid sponges from the Macquarie Ridge, New Zealand; D, a large *Leiopathes glaberrima* colony in the North-East Atlantic.

The impacts of fishing activities on CWC are well recognized, with bottom trawling, in particular, having strong direct physical (e.g., breaking or dislodging colonies) as well as secondary sedimentation (e.g., smothering individuals or colonies) effects (see review by Clark and others, 2016). Surveys of coral populations on seamount features off Australia and New Zealand show little sign of recovery 15 years after trawling was stopped, but where other unfished seamounts have been protected, they have dense coral populations (Clark and others, 2019; Williams and others, 2010). Recent work on the Hawaiian-Emperor seamount chain suggests there may be some regrowth of CWC between 300 m and 600 m deep 30–40 years after fishing was stopped (Baco and others, 2019). In general, the estimated extent of deep-sea trawling has decreased in recent decades (e.g., Victorero and others, 2018).

Predicting the response of CWC populations to anthropogenic disturbance requires knowledge of their current distribution and resilience. Recent predictive habitat models have led to new CWC discoveries in a productive, iterative process (i.e., Georgian and others, 2020) and to the development of new modelling techniques (Robert and others, 2016; Diesing and Thorsnes, 2018). Recent discoveries include populations of habitat-forming octocorals (figure III) on the Antarctic continental shelf (Ambroso and others, 2017), scleractinian coral reefs in low pH waters in the North Pacific (Baco and others, 2017; Gómez and others, 2018) and thousands of CWC mounds surviving under low dissolved oxygen along the Moroccan Atlantic continental margin (Wienberg and others, 2018).

An understanding of the reproductive factors influencing CWC distribution is also required to determine their recolonization potential. Recent studies indicate that *Lophelia pertusa* larvae are planktotrophic and reside in the upper water column for up to five weeks before settlement (Larsson and others, 2014; Strömberg and Larsson, 2017). Similarities in the genetic structure of the cup coral *Desmophyllum dianthus* (Miller and Gunasekera, 2017) over large areas suggest wide connectivity (Holland and others, 2019). Conversely, variable genetic structure in the reef-building stony coral *Solenosmilia variabilis* suggests that asexual reproduction and localized recruitment may be prevalent (Miller and Gunasekera, 2017). Zeng and others (2017) report genetic differentiation among three CWC species off New Zealand, primarily determined by regional and local currents (Dueñas and others, 2016; Holland and others, 2019). The limited number of population genetic studies of deepwater octocorals indicate that gene flow among populations is restricted to those residing at similar depths, with water masses creating barriers to larval dispersal and genetic exchange (Baco and Shank, 2005, Quattrini and others, 2015).

The increased use of long-term observatories, identified as a major capacity-building gap in the first Assessment, has highlighted the influence of seabed heterogeneity (Pierdomenico and others, 2017), hydrodynamics (Mienis and others, 2019) and in situ growth dynamics (Lartaud and others, 2017) on the spatial extent and morphology of CWC habitats at local scales (De Clippele and others, 2018), and CWC and sponge fauna composition at regional scales (van Soest and de Voogd, 2015; Radice and others, 2016). Modelling of interactions between tidal currents and CWC mounds suggests an enhanced downwelling of surface food particles that promotes the proliferation of benthic communities (Cyr and others, 2016; Soetaert and others, 2016). Data from the Lofoten-Vesterålen Cabled Ocean Observatory[2] have led to the identification of turbulent mixing in winter and spring along with vertically migrating zooplankton in the stratified waters of warmer months as the food supply mechanisms to CWC on the Norwegian continental shelf. They have also emphasized the benefits of sustained ocean observatories (Van Engeland and others, 2019).

Climate change remains a persistent and pervasive threat to CWC through global ocean warming, ocean acidification, deoxygenation, reduced food supply (figure IV) and the cumulative effects of those stressors (Hebbeln and others, 2019; Sweetman and others, 2017; Wienberg and Titschack, 2017). Currently, many CWC that occur at shallower depths appear to be near the limits of their temperature tolerance (Georgian and others, 2016b; Morato and others, 2020). That may be particularly significant in regions where the ocean temperature is

2 Available at http://love.statoil.com.

changing the most rapidly (Levin and Le Bris, 2015; see also chap. 5 of the present Assessment). Expanding oxygen minimum zones are perhaps more immediate threats to living corals (Fink and others, 2012; Lunden and others, 2014; Tamborrino and others, 2019).

At the deeper ends of their distribution, corals appear to be limited by aragonite and calcite saturation states. However, there are now several reports of scleractinian corals surviving and growing below the aragonite saturation horizon (Baco and others, 2017; Gómez and others, 2018) and octocorals persisting at depths near or below the calcite saturation horizon (Quattrini and others, 2017). In undersaturated waters, coral colonies can continue to calcify in laboratory experiments, although variable responses have been noted from different species and populations (i.e., Gammon and others, 2018; Georgian and others, 2016a). The differences in the response to ocean acidification by CWC suggest distinct interspecific sensitivity to environmental changes and the importance of food supply and energy allocation in the nature of the response (Kurmann and others, 2017; Glazier and others, 2020). Reduced tissue coverage of coral skeletons may make corals more vulnerable to acidification, since reduced pH can elevate the chemical dissolution (Hennige and others, 2015) and even bioerosion rate (Schönberg and others, 2017) of dead coral skeleton, which makes up much of the standing coral reef structure.

Figure IV
Modelled environmental changes at the deep (>200 m) sea floor (in the year 2100 relative to the present day)

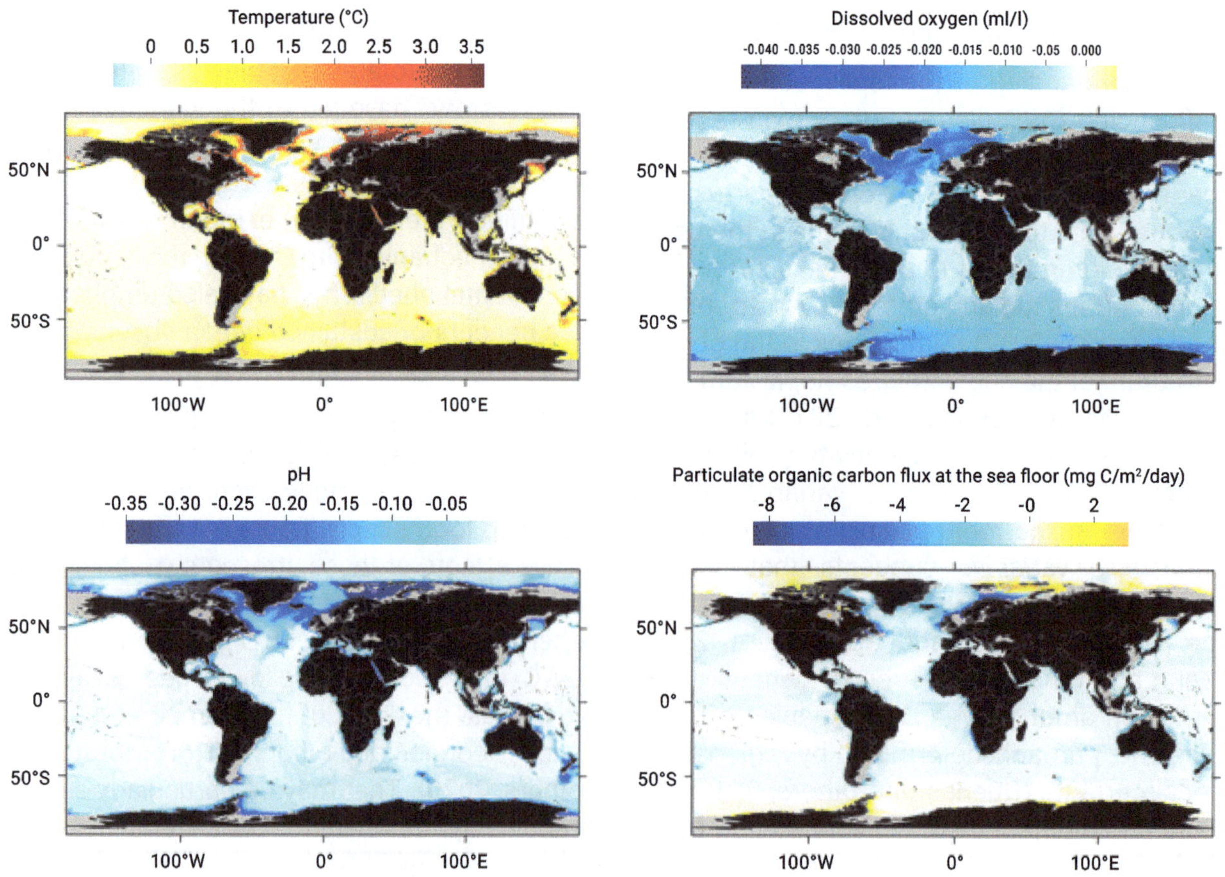

Source: Sweetman and others, 2017.

Laboratory experiments have demonstrated that CWC tolerance to high temperature, low pH and low dissolved oxygen improves when there is an abundance of nutritional resources. For example, regular food pulses to the hermatypic (reef-building) scleractinian *Lophelia pertusa* were highlighted as important for maintaining its metabolic rate (Georgian and others, 2016a; Maier and others, 2019). Recent studies have revealed that *L. pertusa* hosts a versatile microbiome, likely shaped by nutritional status or environmental conditions, while *Madrepora oculata*, another reef-forming CWC, has a more stable and consistent microbiome regardless of the underlying conditions (Meistertzheim and others, 2016). However, the degree to which that variability in the microbial community confers metabolic plasticity in coral species in their natural environment has not been determined. Experimental work has unravelled how sponges, key contributors to CWC ecosystems, can proliferate under food-limited conditions (Kazanidis and Witte, 2016; Kazanidis and others, 2018).

Emerging threats include marine debris, with microplastics (see also chap. 12) observed in corals at all depths in the ocean (Taylor and others, 2016), and physical disturbance associated with deep-sea mining (see also chap. 18), in particular on seamounts and in the vicinity of active and extinct hydrothermal deposits. There is the potential for direct restoration efforts in deep-sea coral communities to accelerate recovery from disturbance, although there have only been a limited number of pilot studies to examine such techniques (e.g., Boch and others, 2019).

3. Economic and social consequences

CWC overlap with human economic and social interests increasingly as the range of human activities in the deep sea expands. Changes in CWC distribution through removal or differential survivorship, along with changes in health and metabolism, will affect numerous Sustainable Development Goals. Beyond the obvious and direct changes of relevance to Goal 14, CWC and sponges are of increasing interest as genetic resources for the development of pharmaceutical products (Molinski and others, 2009; Rocha and others, 2011), which would be affected by the loss of species and habitat. CWC and sponge habitats actively sequester carbon through both feeding and carbonate precipitation in the deep ocean (Kahn and others, 2015; Soetaert and others, 2016). While there are few empirical rate estimates of those processes, carbon sequestration by CWC could contribute to mitigating global climate change (Goal 13), including ocean acidification (target 14.3).

CWC also provide habitat for numerous fishes, including many significant fisheries species, such as broadbill swordfish (*Xiphias gladius*), orange roughy (*Hoplostethus atlanticus*) and fishes of the snapper-grouper complex (Ross and Quattrini, 2009; Morato and others, 2020). CWC ecosystems also recycle nutrients at depth that are brought back to the surface through upwelling to fuel shallow-water productivity (White and others, 2012b; Soetaert and others, 2016). Therefore, changes in their distribution and ecosystem function will affect Goal 2 and associated sustained food security, as well as the use of marine resources (Goal 14). Changes in the availability of nutritional or genetic resources derived from CWC would disproportionately affect the economic benefits to small island developing States and least developed countries, thereby affecting the achievement of Goal 10.

Recent conservation efforts have resulted in the protection of CWC and sponge ecosystems, including bans on fishing activities by the

European Union and around many seamounts in the North and South Pacific, as well as the establishment of protected areas such as the Northeast Canyons and Seamounts Marine National Monument, the Phoenix Islands Protected Area, and the Pacific Remote Islands National Marine Monument.

4. Key region-specific changes and consequences

While human activities are affecting CWC globally, there is regional heterogeneity to the degree of those impacts. For example, deep-sea oil and gas activities are more widespread in the Gulf of Mexico (Cordes and others, 2016), leading to greater potential issues in that region. Deepwater oil and gas exploration is expanding in the Caribbean, in the South Atlantic (off the coasts of Brazil, Namibia and South Africa) and in the Indian Ocean (off Mozambique and South Africa). Those expansions and their potential associated impacts are occurring in regions in which the capacity to conduct and review deep-sea environmental assessments is lower, and future efforts should be focused on augmenting that capacity (see also sect. 7). Deep-sea bottom trawling tends to be focused in only some regions of the world (e.g., South-West Pacific, Indian Ocean). Although the amount of the sea floor that is affected and the overall effort have declined in recent decades, the limited distribution of such fisheries concentrates associated impacts to individual CWC reefs and gardens on the pinnacles and seamounts that are targeted.

Regional variability in the impacts of climate change will result in region-specific effects on CWC. For example, areas of upwelling (e.g., North Pacific) have an aragonite and calcite saturation depth that is relatively shallow. In those regions, where scleractinian corals are living close to saturation horizons, species are thus at greater potential risk to ocean acidification over relatively short timescales (i.e., Gómez and others, 2018). Projected changes in ocean circulation under climate change scenarios include a slowing down of the Atlantic meridional overturning circulation (Bryden and others, 2005; Thornalley and others, 2018), which is expected to have an impact on temperature, salinity and food supply to corals in the North Atlantic Ocean.

Emerging impacts associated with microplastics and deep-sea mining also vary regionally. The impacts associated with microplastics are expected to be higher in regions with marine canyons, as the structures facilitate the trapping and "channelling" of submerged matter (Fabri and others, 2019; Pham and others, 2014). Cobalt-rich ferromanganese crusts, the mining of which is driven by developing battery technologies, occur on seamounts and guyots. In recent years, exploration licences have been issued by the International Seabed Authority for areas in the Northwest Pacific Ocean and South Atlantic Ocean that host seamounts with stony corals and octocorals. Leases for the mining of polymetallic nodules in the Clarion-Clipperton Fracture Zone have been issued and mining of sea floor massive sulphides may soon begin off Papua New Guinea. Both regions contain CWC, including long-lived species of antipatharian black corals (Boschen and others, 2013; Molodtsova and Opresko, 2017), making recovery times from those types of removal activities extensive.

5. Outlook

Current trends suggest that human activities and the effects of global ocean change will continue to increase in deep waters. The responses of CWC could be through range shifts, alterations of metabolism and physiology or local and potentially widespread reductions in genetic diversity and even extinctions of species. Any and all of those responses would have an effect on the distribution and magnitude of the ecosystem services provided by CWC. Achievement of the Sustainable Development Goal target 14.5, the conservation of 10 per cent of coastal and marine areas, would significantly improve the outlook for CWC.

6. Key remaining knowledge gaps

At the most basic level, much of the ocean floor remains unmapped, although the Seabed 2030 project has made great advances and modern multibeam bathymetric surveys now cover approximately 20 per cent of the sea floor (General Bathymetric Chart of the Oceans (GEBCO) Compilation Group, 2020). Owing to the remoteness of the deep sea, current knowledge regarding CWC and the structures they form, as well as the variability in key environmental drivers, is still very limited. There is a clear need to monitor environmental variables (e.g., temperature, dissolved oxygen and pH), in particular in areas near the edge of coral species niches, such as the periphery of large oxygen minimum zones, near the aragonite saturation horizon or in basins where temperature is already high (i.e., the deep Mediterranean), as well as where CWC ecosystems are threatened by the cumulative stressors of human activities.

The resilience of CWC to changes in oceanographic conditions remains a major information gap. In particular, research on the impacts of deoxygenation is lacking compared with studies of ocean acidification (Levin and Le Bris, 2015). The long-term energetic costs associated with coral acclimation, or the potential for adaptation to any and all ocean change stressors, and combinations thereof, remain to be determined for most species. The dead coral framework has been largely understudied and understanding of bioerosion processes and impacts of ocean acidification is limited.

Basic biological information is still lacking for many coral species and, similarly, the applicability of using other species as "proxies" is uncertain. The majority of experimental studies conducted to date are on the "model organism" *Lophelia pertusa*. Application of experimental studies to other CWC species of a variety of taxonomic groups (i.e., octocorals, antipatharians) and to other deepwater groups, such as sponges, is necessary to test for the universality of the conclusions based on the model organism. Reproductive and age-and-growth studies are receiving more attention among researchers (e.g., Larcom and others, 2014) and the increasing use of "seascape genetics" (e.g., Miller and Gunasekera, 2017) can help managers to adopt more integrated broadscale management options. Nevertheless, improved marker development is required for genetics to support future coral connectivity research and/or molecular-based taxonomy (Quattrini and others, 2017).

Advances in modelling approaches, such as species distribution and habitat suitability modelling (Robert and others, 2016); the use of emerging technologies, such as machine learning (Osterloff and others, 2016); cross-sectoral collaboration (Murray and others, 2018); and appropriate data archiving in online databases will improve data availability and reduce processing time, which can improve assessments of the status of CWC and associated structures. Overcoming challenges associated with the limited standardization of studies through the development and use of standardized protocols for video acquisition and analysis will improve the comparability of data and thus upscaling from local to regional spatial scales (Davies and others, 2017; Girard and Fisher, 2018).

Although baseline surveys are often required prior to human industrial activity (Cordes and others, 2016), the Deepwater Horizon oil spill highlighted a lack of local information on CWC and the deep sea, in general. Baseline assessments of the status of those ecosystems have only recently been established, with the earliest baselines from the 1980s, and many CWC habitats continue to be discovered even in relatively well-explored regions. Furthermore, when surveys are conducted, the information generated is often proprietary and not made available publicly, which then limits the transfer of baseline information and its incorporation into further investigations and broader modelling efforts. In addition, surveys may be designed simply to look for hazards and not to characterize the environment or document the fauna and, therefore, may lack the potential to improve understanding of CWC habitats. However, the requirement for documentation of habitats associated with industrial activities is increasing. For example, detailed baseline data collection is a requirement for contractors undertaking exploration for deep-sea minerals in the Area[3] under the auspices of the International Seabed Authority, which may provide the means to address some of the current knowledge gaps associated with that deepwater habitat.

7. Key remaining capacity-building gaps

Gathering the data necessary to evaluate the status and trends of deep-sea ecosystems is difficult, time-consuming and expensive. There is an urgent need to develop the capacity to carry out such surveys, in particular in developing countries. Fundamental to such studies is the proper taxonomy of the species investigated, which is required to correctly assess population sizes and distributions, and to properly attribute impacts. For example, the iconic deep-sea coral species, *Lophelia pertusa*, is currently listed as *Desmophyllum pertusum* in the World Register of Marine Species database, based on evidence presented in Addamo and others (2016). However, the renaming remains controversial, as there are a large number of populations around the world for which there are few genetic or genomic data, and the genus *Desmophyllum* otherwise consists of only solitary species. Proper identification of CWC is itself a capacity gap, with a decline in the numbers of properly trained taxonomists in recent years, in particular for octocorals.

Access to and the proper expertise to use the tools required for studying CWC habitats (e.g., multibeam echo sounders, manned and unmanned deepwater vehicles) represent a major gap for many of the areas where CWC are abundant and their distribution overlaps with proposed industrial activities. The necessary tools and training to collect the appropriate baseline data allowing for an evaluation of the impacts of industrial activity on CWC need to be made available to, and ideally be located in, those countries in which the activities are taking place. Furthermore, where impacts have already occurred, there is little capacity anywhere in the world for deep-sea coral restoration. The development of effective techniques is a critical capacity gap that should be a major focus of future work and will become increasingly important in the future. Although the capacity-building gap is most obvious in developing States, the deep sea is so remote and unexplored that numerous capacity and information gaps remain in developed States.

---

3 The Area is the seabed and ocean floor and subsoil thereof, beyond the limits of national jurisdiction (United Nations Convention on the Law of the Sea, article 1).

References

Addamo, Anna Maria, and others. (2016). Merging scleractinian genera: the overwhelming genetic similarity between solitary *Desmophyllum* and colonial *Lophelia*. *BMC Evolutionary Biology*, vol. 16, No. 1, art. 108.

Ambroso, Stefano, and others (2017). Pristine populations of habitat-forming gorgonian species on the Antarctic continental shelf. *Scientific Reports*, vol. 7, No. 1, art. 12251.

Baco, Amy R., and Tim M. Shank (2005). Population genetic structure of the Hawaiian precious coral *Corallium lauuense* (Octocorallia: Coralliidae) using microsatellites. In *Cold-water corals and ecosystems*. Springer, Berlin, Heidelberg.

Baco, Amy R., and others (2017). Defying dissolution: discovery of deep-sea scleractinian coral reefs in the North Pacific. *Scientific Reports*, vol. 7, No. 1, art. 5436.

Baco, Amy R., and others (2019). Amid fields of rubble, scars, and lost gear, signs of recovery observed on seamounts on 30-to 40-year time scales. *Science Advances*, vol. 5, No. 8, eaaw4513.

Baillon, Sandrine, and others (2012). Deep cold-water corals as nurseries for fish larvae. *Frontiers in Ecology and the Environment*, vol. 10, No. 7, pp. 351–356.

Boch, Charles A., and others (2019). Coral translocation as a method to restore impacted deep-sea coral communities. *Frontiers in Marine Science*, vol. 6, art. 540.

Boschen, Rachel E., and others (2013) Mining of deep-sea seafloor massive sulfides: a review of the deposits, their benthic communities, impacts from mining, regulatory frameworks and management strategies. *Ocean & Coastal Management*, vol. 84, pp. 54–67.

Bryden, Harry L., and others (2005). Slowing of the Atlantic meridional overturning circulation at 25 N. *Nature*, vol. 438, No. 7068, pp. 655–657.

Cathalot, Cécile, and others (2015). Cold-water coral reefs and adjacent sponge grounds: Hotspots of benthic respiration and organic carbon cycling in the deep sea. *Frontiers in Marine Science*, vol. 2, art. 37.

Clark, Malcolm Ross, and others (2016). The impacts of deep-sea fisheries on benthic communities: a review. *ICES Journal of Marine Science*, vol. 73, No. suppl. 1, pp. i51–i69.

Clark, Malcolm Ross, and others (2019). Little evidence of benthic community resilience to bottom trawling on seamounts after 15 years. *Frontiers in Marine Science*, vol. 6, art. 63.

Cordes, Erik E., and others (2008). Coral communities of the deep Gulf of Mexico. *Deep Sea Research Part I: Oceanographic Research Papers*, vol. 55, No. 6, pp. 777–787.

Cordes, Erik E., and others (2016). Environmental impacts of the deep-water oil and gas industry: a review to guide management strategies. *Frontiers in Environmental Science*, vol. 4, art. 58. https://doi.org/10.3389/fenvs.2016.00058.

Cyr, Frédéric, and others (2016). On the influence of cold-water coral mound size on flow hydrodynamics, and vice versa. *Geophysical Research Letters*, vol. 43, No. 2, pp. 775–783.

Davies, Andrew J., and John M. Guinotte (2011). Global habitat suitability for framework-forming cold-water corals. *PloS One*, vol. 6, No. 4, e18483.

Davies, Andrew J., and others (2009). Downwelling and deep-water bottom currents as food supply mechanisms to the cold-water coral *Lophelia pertusa* (Scleractinia) at the Mingulay Reef Complex. *Limnology and Oceanography*, vol. 54, No. 2, pp. 620–629.

Davies, J.S., and others (2017). A new classification scheme of European cold-water coral habitats: implications for ecosystem-based management of the deep sea. *Deep Sea Research Part II: Topical Studies in Oceanography*, vol. 145, pp. 102–109.

De Clippele, L.H., and others (2018). The effect of local hydrodynamics on the spatial extent and morphology of cold-water coral habitats at Tisler Reef, Norway. *Coral Reefs*, vol. 37, No. 1, pp. 253–266.

Diesing, Markus, and Terje Thorsnes (2018). *Mapping of cold-water coral carbonate mounds based on geomorphometric features: an object-based approach. Geosciences*, vol. 8, No. 2, art. 34.

Dueñas, Luisa F., and others (2016). The Antarctic Circumpolar Current as a diversification trigger for deep-sea octocorals. *BMC Evolutionary Biology*, vol. 16, No. 1, art. 2.

Fabri, Marie-Claire, and others (2019). Evaluating the ecological status of cold-water coral habitats using non-invasive methods: An example from Cassidaigne canyon, northwestern Mediterranean Sea. *Progress in Oceanography*, vol. 178, art. 102172.

Fink, Hiske G., and others (2012). Oxygen control on Holocene cold-water coral development in the eastern Mediterranean Sea. *Deep Sea Research Part I: Oceanographic Research Papers*, vol. 62, pp. 89–96.

Fisher, Charles R., and others (2014). Footprint of *Deepwater Horizon* blowout impact to deep-water coral communities. *Proceedings of the National Academy of Sciences*, vol. 111, No. 32, pp. 11744–11749.

Food and Agriculture Organization of the United Nations (FAO) (2009). International Guidelines for the Management of Deep-Sea Fisheries in the High Seas. Rome.

Freiwald, A., and others (2017). Global distribution of cold-water corals (version 5.0). Fifth update to the-dataset in Freiwald and others (2004) by UNEP-WCMC, in collaboration with Andre Freiwald and John Guinotte. Cambridge (UK): United Nations Environment Programme World Conservation-Monitoring Centre.

Gammon, Malindi J., and others (2018). The physiological response of the deep-sea coral *Solenosmilia variabilis* to ocean acidification. *PeerJ*, vol. 6, e5236.

General Bathymetric Chart of the Oceans (GEBCO) Compilation Group (2020). GEBCO 2020 Grid (https://doi.org/10.5285/a29c5465-b138-234d-e053-6c86abc040b9).

Georgian, Samuel E., and others (2016a). Biogeographic variability in the physiological response of the cold-water coral *Lophelia pertusa* to ocean acidification. *Marine Ecology*, vol. 37, No. 6, pp. 1345–1359. https://doi.org/10.1111/maec.12373.

Georgian, Samuel E., and others (2016b). Oceanographic patterns and carbonate chemistry in the vicinity of cold-water coral reefs in the Gulf of Mexico: Implications for resilience in a changing ocean. *Limnology and Oceanography*, vol. 61, No. 2, pp. 648–665.

Georgian, Samuel E., and others (2020). Habitat suitability modelling to predict the spatial distribution of cold-water coral communities affected by the Deepwater Horizon oil spill. *Journal of Biogeography*.

Girard, Fanny, and Charles R. Fisher (2018). Long-term impact of the *Deepwater Horizon* oil spill on deep-sea corals detected after seven years of monitoring. *Biological Conservation*, vol. 225, pp. 117–127.

Glazier Amanda, and others (2020) Regulation of ion transport and energy metabolism enables certain coral genotypes to maintain calcification under experimental ocean acidification. *Molecular Ecology*, vol. 29, pp. 1657–1673.

Gómez, Carlos E., and others (2018). Growth and feeding of deep-sea coral *Lophelia pertusa* from the California margin under simulated ocean acidification conditions. *PeerJ*, vol. 6, e5671.

Hebbeln, Dierk, and others (2019). The fate of cold-water corals in a changing world: a geological perspective. *Frontiers in Marine Science*, vol. 6, art. 119.

Hennige, S.J., and others (2015). Hidden impacts of ocean acidification to live and dead coral framework. *Proceedings of the Royal Society B: Biological Sciences*, vol. 282, No. 1813, 20150990.

Henry, Lea-Anne, and J. Murray Roberts (2007). Biodiversity and ecological composition of macrobenthos on cold-water coral mounds and adjacent off-mound habitat in the bathyal Porcupine Seabight, NE Atlantic. *Deep Sea Research Part I: Oceanographic Research Papers*, vol. 54, No. 4, pp. 654–672.

Holland, L.P., and others (2019). Genetic connectivity of deep-sea corals in the New Zealand region. *New Zealand Aquatic Environment & Biodiversity Report, Wellington.*

Hsing, Pen-Yuan, and others (2013). Evidence of lasting impact of the *Deepwater Horizon* oil spill on a deep Gulf of Mexico coral community. *Elementa: Science of the Anthropocene*, vol. 1.

Kahn, Amanda S., and others (2015). Benthic grazing and carbon sequestration by deep-water glass sponge reefs. *Limnology and Oceanography*, vol. 60, No. 1, pp. 78–88.

Kazanidis, Georgios, and others (2018). Unravelling the versatile feeding and metabolic strategies of the cold-water ecosystem engineer *Spongosorites coralliophaga* (Stephens, 1915). *Deep Sea Research Part I: Oceanographic Research Papers*, vol. 141, pp. 71–82.

Kazanidis, Georgios, and Ursula F.M. Witte (2016). The trophic structure of *Spongosorites coralliophaga*-coral rubble communities at two northeast Atlantic cold water coral reefs. *Marine Biology Research*, vol. 12, No. 9, pp. 932–947.

Kurmann, Melissa, and others (2017). Intra-specific variation reveals potential for adaptation to ocean acidification in a cold-water coral from the Gulf of Mexico. *Frontiers in Marine Science*, vol. 4, art. 111.

Larcom, Elizabeth A., and others (2014). Growth rates, densities, and distribution of *Lophelia pertusa* on artificial structures in the Gulf of Mexico. *Deep Sea Research Part I: Oceanographic Research Papers*, vol. 85, pp. 101–109.

Larsson, Ann I., and others (2014). Embryogenesis and larval biology of the cold-water coral *Lophelia pertusa*. *PLoS One*, vol. 9, No. 7, e102222.

Lartaud, Frank, and others (2017). Growth patterns in long-lived coral species. In *Marine Animal Forests: The Ecology of Benthic Biodiversity Hotspots*, Sergio Rossi and others, eds. Springer International Publishing.

Levin, Lisa A., and Nadine Le Bris (2015). The deep ocean under climate change. *Science*, vol. 350, No. 6262, pp. 766–768.

Lunden, Jay J., and others (2014). Acute survivorship of the deep-sea coral *Lophelia pertusa* from the Gulf of Mexico under acidification, warming, and deoxygenation. *Frontiers in Marine Science*, vol. 1, art. 78.

Maier, Sandra R., and others (2019). Survival under conditions of variable food availability: Resource utilization and storage in the cold-water coral *Lophelia pertusa*. *Limnology and Oceanography.*

Meistertzheim, Anne-Leila, and others (2016). Patterns of bacteria-host associations suggest different ecological strategies between two reef building cold-water coral species. *Deep Sea Research Part I: Oceanographic Research Papers*, vol. 114, pp. 12–22.

Mienis, F., and others (2007). Hydrodynamic controls on cold-water coral growth and carbonate-mound development at the SW and SE Rockall Trough Margin, NE Atlantic Ocean. *Deep Sea Research Part I: Oceanographic Research Papers*, vol. 54, No. 9, pp. 1655–1674.

Mienis, F., and others (2019). Experimental assessment of the effects of coldwater coral patches on water flow. *Marine Ecology Progress Series*, vol. 609, pp. 101–117.

Miller, Karen J., and Rasanthi M. Gunasekera (2017). A comparison of genetic connectivity in two deep sea corals to examine whether seamounts are isolated islands or stepping stones for dispersal. *Scientific Reports*, vol. 7, art. 46103.

Molinski, Tadeusz, and others (2009). Drug development from marine natural products. *Nature Reviews Drug Discovery*, vol. 8, No. 1, pp. 69–85.

Molodtsova, Tina N., and Dennis M. Opresko (2017). Black corals (Anthozoa: Antipatharia) of the Clarion-Clipperton Fracture Zone. *Marine Biodiversity*, vol. 47, No. 2, pp. 349–365.

Morato, Telmo, and others (2020). Climate-induced changes in the suitable habitat of cold-water corals and commercially important deep-sea fishes in the North Atlantic. *Global Change Biology*, vol. 26, No. 4, pp. 2181–2202.

Murray, Fiona, and others (2018). Data challenges and opportunities for environmental management of North Sea oil and gas decommissioning in an era of blue growth. *Marine Policy*, vol. 97, pp. 130–138.

Osterloff, Jonas, and others (2016). A computer vision approach for monitoring the spatial and temporal shrimp distribution at the LoVe observatory. *Methods in Oceanography*, vol. 15, pp. 114–128.

Pham, Christopher K., and others (2014). Marine litter distribution and density in European seas, from the shelves to deep basins. *PloS One*, vol. 9, No. 4.

Pierdomenico, Martina, and others (2017). Megabenthic assemblages at the Hudson Canyon head (NW Atlantic margin): Habitat-faunal relationships. *Progress in Oceanography*, vol. 157, pp. 12–26.

Quattrini, Andrea M., and others (2012). Megafaunal-habitat associations at a deep-sea coral mound off North Carolina, USA. *Marine Biology*, vol. 159, No. 5, pp. 1079–1094.

Quattrini, Andrea M., and others (2013). Niche divergence by deep-sea octocorals in the genus *Callogorgia* across the continental slope of the Gulf of Mexico. *Molecular Ecology*, vol. 22, No. 15, pp. 4123–4140.

Quattrini, Andrea M., and others (2015). Testing the depth-differentiation hypothesis in a deepwater octocoral. *Proceedings of the Royal Society B: Biological Sciences*, vol. 282, No. 1807, 20150008.

Quattrini, Andrea M., and others (2017). Environmental filtering and neutral processes shape octocoral community assembly in the deep sea. *Oecologia*, vol. 183, No. 1, pp. 221–236.

Radice, Veronica Z., and others (2016). Vertical water mass structure in the North Atlantic influences the bathymetric distribution of species in the deep-sea coral genus *Paramuricea*. *Deep Sea Research Part I: Oceanographic Research Papers*, vol. 116, pp. 253–263.

Robert, Katleen, and others (2016). Improving predictive mapping of deep-water habitats: Considering multiple model outputs and ensemble techniques. *Deep Sea Research Part I: Oceanographic Research Papers*, vol. 113, pp. 80–89.

Roberts, J.M., and others (2009). Mingulay reef complex: an interdisciplinary study of cold-water coral habitat, hydrography and biodiversity. *Marine Ecology Progress Series*, vol. 397, pp. 139–151.

Rocha, Joana, and others (2011). Cnidarians as a source of new marine bioactive compounds—an overview of the last decade and future steps for bioprospecting. *Marine Drugs*, vol. 9, No. 10, pp. 1860–1886.

Ross, Steve W., and Andrea M. Quattrini (2009). Deep-sea reef fish assemblage patterns on the Blake Plateau (Western North Atlantic Ocean). *Marine Ecology-–an Evolutionary Perspective*, vol. 30, No. 1, pp. 74–92.

Schönberg, Christine H.L., and others (2017). Bioerosion: the other ocean acidification problem. *ICES Journal of Marine Science*, vol. 74, No. 4, pp. 895–925.

Soetaert, Karline, and others (2016). Ecosystem engineering creates a direct nutritional link between 600-m deep cold-water coral mounds and surface productivity. *Scientific Reports*, vol. 6, art. 35057.

Strömberg, Susanna M., and Ann I. Larsson (2017). Larval behavior and longevity in the cold-water coral *Lophelia pertusa* indicate potential for long distance dispersal. *Frontiers in Marine Science*, vol. 4, art. 411.

Sweetman, Andrew K., and others (2017). Major impacts of climate change on deep-sea benthic ecosystems. *Elementa: Science of the Anthropocene*, vol. 5, art. 4.

Tamborrino, Leonardo, and others (2019) Mid-Holocene extinction of cold-water corals on the Namibian shelf steered by the Benguela oxygen minimum zone. *Geology*, vol. 47, No. 12, pp. 1185–1188.

Taylor, M.L., and others (2016). Plastic microfibre ingestion by deep-sea organisms. *Scientific Reports*, vol. 6, art. 33997.

Thornalley, David J.R., and others (2018). Anomalously weak Labrador Sea convection and Atlantic overturning during the past 150 years. *Nature*, vol. 556, No. 7700, pp. 227–230.

Thurber, Andrew R., and others (2014). Ecosystem function and services provided by the deep sea. *Biogeosciences*, vol. 11, No. 14, pp. 3941–3963.

United Nations (2017). *The First Global Integrated Marine Assessment: World Ocean Assessment I*. Cambridge: Cambridge University Press.

Van Engeland, Tom, and others (2019). Cabled ocean observatory data reveal food supply mechanisms to a cold-water coral reef. *Progress in Oceanography*, vol. 172, pp. 51–64.

Van Oevelen, Dick, and others (2009). The cold-water coral community as hotspot of carbon cycling on continental margins: A food-web analysis from Rockall Bank (northeast Atlantic). *Limnology and Oceanography*, vol. 54, No. 6, pp. 1829–1844.

Van Soest, R.W.M., and N.J. de Voogd (2015). Sponge species composition of north-east Atlantic cold-water coral reefs compared in a bathyal to inshore gradient. *Journal of the Marine Biological Association of the United Kingdom*, vol. 95, No. 7, pp. 1461–1474.

Victorero, Lisette, and others (2018). Out of sight, but within reach: A global history of bottom-trawled deep-sea fisheries from > 400 m depth. *Frontiers in Marine Science*, vol. 5, No. 98.

White, Helen K., and others (2012a). Impact of the *Deepwater Horizon* oil spill on a deep-water coral community in the Gulf of Mexico. *Proceedings of the National Academy of Sciences*, vol. 109, No. 50, pp. 20303–20308.

White, Martin, and others (2005). Deep-water coral development as a function of hydrodynamics and surface productivity around the submarine banks of the Rockall Trough, NE Atlantic. In *Cold-Water Corals and Ecosystems*, pp. 503–514. Springer.

White, Martin and others (2012b). Cold-water coral ecosystem (Tisler Reef, Norwegian Shelf) may be a hotspot for carbon cycling. *Marine Ecology Progress Series*, vol. 465, pp. 11–23.

Wienberg, Claudia, and Jürgen Titschack (2017). Framework-forming scleractinian cold-water corals through space and time: a late Quaternary North Atlantic perspective. *Marine Animal Forests: The Ecology of Benthic Biodiversity Hotspots*, pp. 699–732.

Wienberg, Claudia, and others (2018). The giant Mauritanian cold-water mound province: Oxygen control on coral mound formation. *Quaternary Science Reviews*, vol. 185, pp. 135–152.

Williams, Alan, and others (2010). Seamount megabenthic assemblages fail to recover from trawling impacts. *Marine Ecology*, vol. 31, pp. 183–199.

Zeng, Cong, and others (2017). Population genetic structure and connectivity of deep-sea stony corals (Order Scleractinia) in the New Zealand region: Implications for the conservation and management of vulnerable marine ecosystems. *Evolutionary Applications*, vol. 10, No. 10, pp. 1040–1054.

Chapter 7F
Estuaries and deltas

Contributors: Colin D. Woodroffe (convener), Bing Qiao, Ronaldo Christofoletti, Dana E. Hunt, Pablo Muniz and Moriaki Yasuhara.

Keynote points

- Human populations, fishing, shipping, engineering activities, including upstream river dams, and recreation and tourism exert pressures on the resources and health of estuaries and deltas.
- Interactions between multiple stressors on estuary and delta habitats are not fully understood.
- A key gap remains in identifying measurable indices of ecosystem health and human well-being across diverse estuarine and deltaic systems.

1. Introduction

Estuaries and deltas, where major rivers meet the sea, are highly productive systems supporting diverse biota that are structured by temporally variable gradients in salinity, nutrients and other factors. The variability reflects both natural (e.g., precipitation, tides) and anthropogenic (e.g., development, contaminant loading) drivers. While they are often heavily populated and perturbed systems in their natural state, estuaries and deltas typically maintain biodiversity within a variety of ecosystems, many of which are the subject of other chapters in the present Assessment, such as mangroves (chap. 7H), salt marshes (chap. 7I), seagrass meadows (chap. 7G), sand and mud substrates (chap. 7B) and an often extensive intertidal zone (chap. 7A). The mouths of rivers are locations where fresh water mixes with the ocean, thus they are the receiving bodies for land-derived nutrients, sediments and pollutants (chaps. 10–13) and they often host invasive species, in particular from ballast water (chap. 22). Estuaries and deltas are valuable for their intrinsic biota and the commercial and subsistence fisheries they support (chap. 15), as well as the tourism and recreation activities they attract. Their total economic value was estimated at over $6.1 trillion in 2014, as outlined in the first *World Ocean Assessment* (United Nations, 2017).

Economically important, urbanized and industrialized, most estuaries and deltas have been transformed by human interventions. They are increasingly affected by the impacts of global change, including sea level rise, changes in precipitation and related natural hazards, such as cyclones and storm surges (Renaud and others, 2013). Most megacities are coastal, with the attendant heavy industry, urbanization and recreation activities that can harm such areas (Todd and others, 2019). The first Assessment contained a preliminary global, integrated assessment of the condition of estuaries and deltas. Based on limited data, few waterbodies were qualitatively assessed to be in very good condition, whereas the condition of 62 per cent of them was considered poor or very poor, and the quality of most was in decline.

The present chapter contains an update to the first Assessment, with an emphasis on the fact that estuaries and deltas provide unique habitats for many organisms, both of marine and coastal origins, as well as recreation, food provisioning and water sources for humans. Such environments are affected by short-term, event-driven disturbances, such as storms, and longer-term trends, such as climate change (Doney and others, 2012; Harris and others, 2018), which are often interconnected (e.g., storms that result in nuisance flooding, which is exacerbated by sea level rise). Advances have been made in observing systems, such as satellites, global observing networks and buoys, which are designed to capture rapid changes in environmental conditions. However, the capacity to monitor, model or interpret those observations is still inadequately developed for the optimal management of estuary and delta environments.

2. Documented changes in the state of estuaries and deltas

2.1. Environmental changes between 2010 and 2020

2.1.1. Water and sediments

There has generally been a continuing decline in the delivery of both water and sediments by rivers as a result of anthropogenic activities in catchments across the world, such as changing land management practices and the construction of dams (Li and others, 2018; Day and others, 2019; Dunn and others, 2019); however, the melting of terrestrial ice and permafrost may also increase freshwater inputs to high-latitude estuaries (chap. 3). Reduced sediment input accelerates the loss of coastal wetlands caused by erosion and affects soft-sediment deposit and suspension feeders, which is augmented by sand extraction (Anthony and others, 2015), whereas high levels of sedimentation can shade primary producers, such as seagrasses, and smother benthic organisms. Urbanization increases peak flow and decreases base flow into estuaries, resulting in potentially harmful variations in salinity and threatening intertidal ecosystems (Freeman and others, 2019).

2.1.2. Eutrophication

Nutrient loading (largely of nitrogen and phosphorous) remains a serious problem in estuaries, owing to their proximity to large cities and ever-increasing agriculture, silviculture and aquaculture (Pesce and others, 2018; Todd and others, 2019), along with domestic wastewater, fertilizers and animal wastes, which result in bottom-water hypoxia (Yasuhara and others, 2017; Breitburg and others, 2018a, 2018b). Eutrophication can also lead to blooms of cyanobacteria, dinoflagellates and sometimes macroalgae (Teichberg and others, 2010), including harmful algal blooms. The situation is stabilized or partly recovering in developed countries (e.g., Chesapeake Bay, United States; Osaka Bay, Japan) owing to improvements in sanitation and reduced nutrient loading (Lefcheck and others, 2018) but is rapidly getting worse along the coasts of populous Asian countries, owing to poor sanitation, elevated nutrient flux and further population growth (Boesch, 2019). Bottom-water hypoxia can lead to fish kills that affect local economies (Breitburg and others, 2018a, 2018b; Yasuhara and others, 2019).

2.1.3. Global change

Global change is already affecting estuaries and deltas. Poleward extension of fish and crustacean ranges has been observed (Hallett and others, 2017; Pecl and others, 2017). More frequent storms and weather extremes affect salinity and sedimentation (Prandle and Lane, 2015; Day and Rybczyk, 2019). Future increased temperatures could lead to localized extinctions and intensify microbial pathogen concentrations and public health risks (Robins and others, 2016). Higher sea levels will be compounded by riverine flooding, resulting in more extensive inundation of coastal areas (Moftakhari and others, 2015, 2017; Kumbier and others, 2018; Ikeuchi and others, 2017; Nichols and others, 2019). The annual cost of the flooding of coastal cities could be in the order of $60 billion–$63 billion by 2050 (Hallegatte and others, 2013), and it is projected that 1.46 per cent of the world population will be displaced by permanent flooding by 2200 (Desmet and others, 2018). Flooding may lead to significant habitat losses owing to coastal squeeze, where fixed infrastructure impedes landward migration of intertidal ecosystems (Doody, 2013; Phan and others, 2015).

2.1.4. Delta subsidence

Anthropogenic stresses are having a particular impact on deltas owing to high rates of relative sea level rise and socioeconomic vulnerability (Tessler and others, 2015; Hiatt and others, 2019). The impact of rising sea levels is exacerbated by subsidence in large deltas (megadeltas) owing to human activities, primarily groundwater extraction (Syvitski and others, 2009; Erban and others, 2014; Auerbach and others, 2015; Brown and Nicholls, 2015; Schmidt, 2015; Minderhoud and others, 2017, 2019; Wright and Wu, 2019). Protective infrastructure may be able to limit present-day threats; however, engineering solutions may not be feasible in densely populated or lower-income countries (Tessler and others, 2016).

2.1.5. Invasive species

Many estuaries and deltas host large ports and have serious invasive species issues related to ballast water discharge from ships (Astudillo and others, 2014; Shalovenkov, 2019). Invasive species can directly influence the decline of resources and the health of estuaries and deltas, affecting their ecology and balance, posing significant dangers to the biodiversity of both systems. The rate of invasive species introduction has been accelerating, reflecting increased shipping (Seebens and others, 2017). Overall, numbers of invasive species are approximately 30 times greater in high-income countries than in low-income countries, owing to trade and population, and the capacity to detect such invasions (Seebens and others, 2018). The observed rate of introductions has been getting slower in European seas, including the Mediterranean (Korpinen and others, 2019).

2.1.6. Degradation and restoration of ecosystem services

Estuaries and deltas provide essential provisioning, regulating, supporting and cultural ecosystem services (chap. 44 of the first Assessment). The systems provide recreation through such activities as boating, swimming, wildlife watching and fishing (Whitfield, 2017). Some organisms perform important roles as foundational members and create, modify and maintain habitats. Oysters, for example, form reefs that construct habitat, reduce erosion and improve water quality. However, in degraded estuaries, oysters are affected by overfishing, sediment loads and disease, as well as increased ocean acidity (Janis and others, 2016; Day and Rybczyk, 2019). The loss of seagrasses, salt marshes and mangroves, as well as water quality degradation (Reynolds and others, 2016; Schmidt and others, 2017), lead to a decline in juvenile fish diversity and abundance (Whitfield, 2017). Restoration efforts have been successful in relatively few estuaries but can also be integrated into natural shoreline protection strategies (Bilkovic and others, 2016; Ducrotoy and others, 2019).

2.2. Factors associated with the changes: drivers, pressures, impacts and response

Many human activities have degraded the health and productivity of estuaries and deltas, ranging from direct impacts, such as development that destroys habitat, to longer-term indirect impacts caused by global climate change (Cavallaro and others, 2018). There are increasing pressures from human habitation, intrusive coastal infrastructure, recreation, fisheries (finfish and shellfish), land reclamation and the filling of wetlands (Sengupta and others, 2018), resulting in environmental degradation and the loss of sensitive marine organisms (Buttigieg and others, 2018), which have led to increasing efforts to protect ecosystems for their intrinsic worth, for human health and for sustainable resource use. Additional human pressures, such as the development of large container ports with deep-draft shipping, also modify estuarine environments through dredging and the use of dredge spoils to nourish beaches or modify shorelines (Intergovernmental Panel on Climate Change (IPCC), 2019).

It is still difficult to predict the intensity and scale of drivers and pressures or the response of biological communities and ecosystem functions. Temperature, rainfall anomalies and sea level rise cause substantial impacts on estuarine ecosystems over both the short and the long term (Elliott and Whitfield, 2011; McLeod and others, 2011; Condie and others, 2012; Turra and others, 2013; Bernardino and others, 2015, 2016). Both long-term averages and shorter-term exceedances of physiological ranges will affect the metabolism, growth and reproduction of estuarine biota, which, combined with local eutrophication, may lead to acute oxygen depletion and the mass mortality of organisms (Gillanders and others, 2011). On longer timescales, ecological pressures from fishing activities are affecting fish populations and ecosystems (Muniz and others, 2019). For example, in Río de la Plata, fishing effort for artisanal and industrial fleets has remained constant or even declined slightly, but catches for the two most important species have reached their lowest values in the past 35 years (Gianelli and Defeo, 2017; García-Alonso and others, 2019).

Although many human activities have negative consequences for the health of individual estuaries and deltas, recent efforts have been made to restore the productivity of coastal waters, notably by developing nutrient and pollutant management plans, restoring ecosystems and keystone species, and protecting estuaries and deltas in parks and marine protected areas (Lefcheck and others, 2018; Boesch, 2019). In some places, such as the United States and Hong Kong, China, oyster reefs have been restored, so they now protect shorelines and filter the water column (Morris and others, 2019). In other locations, seagrasses, salt marshes or mangroves may serve similar purposes in protecting the coastline from storms and sea level rise, as well as providing critical habitat for juvenile fish and other biota.

3. Consequences of the changes for human communities, economies and well-being

Estuaries and deltas have socioeconomic and cultural importance, as they provide goods and services, including fishing resources and ecosystem processes. There are local traditional communities that rely on those resources for their livelihood, including subsistence fishing and income from tourist activities. Therefore, to understand changes and manage their impacts on estuaries and deltas, an integrated consideration of environmental, biological, cultural, economic and anthropologic issues is essential.

The World Health Organization has advocated the One Health concept to integrate the human-animal-ecosystem interface because it has been recognized that changes in any of those elements will affect the others. Declines in estuarine health owing to increased pollutants or invasive species can pose a direct threat to human health. The level of impact on humans depends on socioecological factors. Whereas urban populations may suffer from reduced storm protection and from the consumption of contaminated fish, local indigenous communities may also suffer from the loss of cultural values, sanitation issues and social inequality. Indigenous populations and local coastal communities have developed traditional knowledge and skills relevant to the conservation, sustainable use and management of estuaries (Breitburg and others, 2018b). Changes in estuaries owing to urbanization can lead to the loss of identity and cultural practices in communities that depend on such resources for their livelihoods.

There is now greater awareness regarding ecosystem services in estuaries and conflicts that have arisen owing to changes in the ecosystems (Nicholls and others, 2018). Science can be a powerful tool at the interface with policy to inform decision-making at local, regional and national levels and to integrate it into global goals, such as the 2030 Agenda for Sustainable Development[1] (Dietz, 2013; Howarth and Painter, 2016). The integration of public participation, including of indigenous peoples and local communities, with scientific analyses can lead to effective scientific communication, socialization and decision-making. Improved communication among stakeholders can enable an effective transfer of knowledge and adaptive management, for example, with social scientists helping to build trust among actors (Fischhoff, 2013). Citizen science, an innovative area with benefits for environmental and social sciences, could link traditional and scientific knowledge and help to develop integrated management in estuaries by including indigenous populations and local communities in scientific studies. Ecosystem complexity and connections with other habitats make joint management and collaboration between governments and local communities essential for maintaining coastal biodiversity and ecosystem functions (Teixeira and others, 2013; Brondizio and others, 2016).

Changes in estuarine and deltaic environments, ecosystem services and socioeconomic dynamics have implications for achieving the Sustainable Development Goals of the 2030 Agenda. For example, socioecological conflicts in estuaries, mainly related to indigenous peoples and local communities, are linked to aims related to poverty (Goal 1), gender equality (Goal 5), sanitation (Goal 6), resilient cities (Goal 11) and safe seafood resources (Goal 14). If it is possible to reverse impacts through positive actions consistent with the 2030 Agenda, a series of benefits for society could be achieved in a short time. The conservation of estuaries and their biodiversity and cultural diversity is particularly relevant to Sustainable Development Goal targets 14.2 and 14.5, which are related to promoting the protection and conservation of coastal resources (Neumann and others, 2017), and could also provide other services, such as increased ecotourism. The promotion of human engagement with nature strengthens efforts towards nature conservation in associated ecosystems. To achieve that goal, it is valuable to adopt an innovative approach, together with decision makers and society, to supporting the adaptive management, conservation and sustainable use of estuaries that will benefit human well-being for future generations (Szabo and others, 2015).

4. Key region-specific changes and consequences

Estuaries and deltas are widespread around the world, but there is no global inventory, and the category encompasses a range of geomorphological types. It was suggested in the first Assessment that there may be about 4,500 estuaries in total. However, a gridded global digital elevation model gives a more recent estimate of more than 53,000 estuaries (McSweeney and others, 2017). There are an estimated 1,200 intermittently closed partially saline lakes and lagoons, in particular along the swell-dominated coasts of southern Africa and eastern Australia. They will experience a different set of responses to climate change than estuaries that are continually open to the sea, including altered opening regimes, increased flooding and saltwater intrusion into surface water and groundwater (Carrasco and others, 2016). Adopting a similar approach, a recent study has suggested that there are about

1 See General Assembly resolution 70/1.

11,000 deltas worldwide; of those, 25 per cent have undergone a net land gain over recent decades as a consequence of deforestation-induced increases in fluvial sediment supply, whereas damming has resulted in reduced sediment and land loss in approximately 1,000 delta systems (Nienhuis and others, 2020).

The first Assessment contained a preliminary assessment of the condition of selected estuaries, with a classification by continent. There remain inadequate data to improve on that evaluation or to consider estuaries and deltas following the region-specific framework of the present Assessment. Several recent compilations do provide data for several regions that were previously poorly documented. For example, little information was available on Arctic estuaries, which are likely to become increasingly important as global warming opens up access to shipping in those regions (Kosyan, 2017). Regional compilations have provided more information for the southern hemisphere, including a focus on estuaries on the east coast of Africa and the western Indian Ocean (Diop and others, 2016), and a review of Brazilian estuaries (Lana and Bernardino, 2018). Relatively little information was previously available on the numerous estuaries and deltas along the 18,000-km coastline of China, which contains many large megacities, such as Shanghai and Guangzhou, that are very susceptible to coastal hazards from sea level rise and storm surges (Yin and others, 2012; Kuang and others, 2014; Chen and others, 2018). Those urbanized megadeltas are home to millions of people and often contain rich biodiversity, which faces threats, including from eutrophication, pollution, coastal modification and invasive species (Lai and others, 2016).

5. Outlook

Based on trends over recent decades, coastal zone populations will increase, with ongoing urbanization focused on estuaries and deltas. Those human stresses will be the principal pressures that continue to affect the biodiversity and habitat health of such coastal ecosystems. Climate change will exacerbate the stresses – increased frequency of storms appears likely and sea level rise is anticipated to accelerate, in particular in the case of large deltas that are experiencing subsidence. Good governance has the potential to maintain or improve the status of ecosystems, although protection of low-lying metropolises will require upgraded engineering infrastructure.

Estuarine and deltaic sustainability can be considered in terms of functional processes using geomorphic, ecological or economic perspectives (Mahoney and Bishop, 2018). Changes can lead to either enhanced or diminished sustainability, but most changes have been detrimental (Day and others, 2016). Ecosystem consequences that can be anticipated include the alteration of food webs owing to the loss of keystone, top predator or ecosystem engineer species; habitat losses owing to sea level rise and land reclamation; and the poleward migration of marine species to adapt to climate change. A reduction in wetlands through coastal squeeze and aquaculture activities is already apparent in many estuaries and deltas. Further increases in invasive species can be expected, although considerable progress has been made in identifying, setting priorities for and eradicating invasive organisms.

It is more difficult to predict the socioeconomic consequences of continued change in the system. However, greater population pressures and expanding urbanization around estuaries and on deltas are likely to mean more dredging to maintain navigability, the silting of channels and the erosion of shorelines, as well as losses of wetlands, with reduced access to recreation, fishing grounds and clean water. The desire to protect extensive residential, industrial and agricultural areas against storm surges and sea level rise will require huge investments in engineering solutions, and the risks of failure of such infrastructure appear catastrophic. In many areas, it will eventually become necessary to move inland. Even where the pressures of expanding human populations can be contained, significant investments will be required to restore critical habitats. Assessments of changing ecosystem services and implications for human well-being would benefit from improved monitoring and investment in scientific research. Integrated coastal planning is necessary for sustainable use and to extend conservation beyond protected areas, which may require broader strategies for funding, for example from public sources, and multisectoral cooperation. Coastal management may need to include new standards for building and construction, eco-labelling, innovative economic instruments for financing conservation and payments for ecosystem services, such as blue carbon sequestration.

6. Key remaining knowledge and capacity-building gaps

There are major challenges in managing land use in estuaries and deltas so that future generations can also enjoy the aesthetic, cultural and sustaining services that they provide (Elliott and others, 2019). Existing models lack sufficient spatial and temporal resolution to simulate future extreme events (Haigh and others, 2016; Robins and others, 2018), including compound flooding from both fluvial and oceanic sources. Such floods result in environmental degradation, including wetland erosion and eutrophication, and expose people to harmful waterborne pathogens (Yin and others, 2018). Relatively little is known about the long-term effects of rapid human interventions in deltas. The characterization of socioeconomic tipping points will have to be improved to avoid unacceptable changes. More evidence is needed to target coastal wetland conservation in those areas where it can be most beneficial or might alleviate the need for engineered protection works (Van Coppenolle and others, 2018; Van Coppenolle and Temmerman, 2019). The future resilience of megadeltas, and megacities within them, will depend on advances in resource and emergency strategies and investments in flood protection through engineered and living shorelines. Modelling, engineering and natural sciences need to be integrated with social science and public outreach (Bonebrake and others, 2018). Innovative technologies and nature-based solutions are already helping to reduce vulnerability to coastal hazards but collaborative science is needed so that people living in estuarine and deltaic locations have environmental information, reliable short-term and long-term predictions and appropriate observations to validate models, thus improving data-driven approaches to coastal resilience (Nichols and others, 2019).

References

Anthony, Edward J., and others (2015). Linking rapid erosion of the Mekong River delta to human activities. *Scientific Reports*, vol. 5, art. 14745.

Astudillo, Juan-Carlos, and others (2014). Status of six non-native marine species in the coastal environment of Hong Kong, 30 years after their first record. *BioInvasions Records*, vol. 3, No. 3, pp. 123–37. https://doi.org/10.3391/bir.2014.3.3.01.

Auerbach, L.W., and others (2015). Flood risk of natural and embanked landscapes on the Ganges–Brahmaputra tidal delta plain. *Nature Climate Change*, vol. 5, No. 2, p. 153.

Bernardino, Angelo Fraga, and others (2015). Predicting ecological changes on benthic estuarine assemblages through decadal climate trends along Brazilian Marine Ecoregions. *Estuarine, Coastal and Shelf Science*, vol. 166, pp. 74–82.

Bernardino, Angelo Fraga, and others (2016). Benthic estuarine communities in Brazil: moving forward to long term studies to assess climate change impacts. *Brazilian Journal of Oceanography*, vol. 64, No. SPE2, pp. 81–96.

Bilkovic, Donna, and others (2016). The role of living shorelines as estuarine habitat conservation strategies. *Coastal Management*, vol. 44, No. 3, pp. 161–174.

Boesch, Donald F. (2019). Barriers and bridges in abating coastal eutrophication. *Frontiers in Marine Science*, vol. 6, art. 123. https://doi.org/10.3389/fmars.2019.00123.

Bonebrake, Timothy C., and others (2018). Managing consequences of climate-driven species redistribution requires integration of ecology, conservation and social science. *Biological Reviews*, vol. 93, No. 1, pp. 284–305.

Breitburg, Denise and others (2018a). Declining oxygen in the global ocean and coastal waters. *Science*, vol. 359, No. 6371, eaam7240.

Breitburg, Denise and others (2018b). *The Ocean Is Losing Its Breath: Declining Oxygen in the World's Ocean and Coastal Waters; Summary for Policy Makers*. IOC/2018/TS/137 REV. Paris.

Brondizio, Eduardo S., and others (2016). Catalyzing action towards the sustainability of deltas. *Current Opinion in Environmental Sustainability*, vol. 19, pp. 182–94. https://doi.org/10.1016/j.cosust.2016.05.001.

Brown, S., and R.J. Nicholls (2015). Subsidence and human influences in mega deltas: the case of the Ganges–Brahmaputra–Meghna. *Science of the Total Environment*, vol. 527, pp. 362–374.

Buttigieg, Pier Luigi, and others (2018). Marine microbes in 4D—using time series observation to assess the dynamics of the ocean microbiome and its links to ocean health. *Current Opinion in Microbiology*, vol. 43, pp. 169–185.

Carrasco, A. Rita, and others (2016). Coastal lagoons and rising sea level: a review. *Earth-Science Reviews*, vol. 154, pp. 356–368.

Cavallaro, N., and others (2018). *USGCRP, 2018: Second State of the Carbon Cycle Report (SOCCR2): A Sustained Assessment Report*. Washington, D.C.: U.S. Global Change Research Program.

Chen, Shihong, and others (2018). Assessment of tropical cyclone disaster loss in Guangdong Province based on combined model. *Geomatics, Natural Hazards and Risk*, vol. 9, No. 1, pp. 431–441.

Condie, Scott A., and others (2012). Modelling ecological change over half a century in a subtropical estuary: impacts of climate change, land-use, urbanization and freshwater extraction. *Marine Ecology Progress Series*, vol. 457, pp. 43–66.

Day, John W., and John M. Rybczyk (2019). Chapter 36 – Global Change Impacts on the Future of Coastal Systems: Perverse Interactions Among Climate Change, Ecosystem Degradation, Energy Scarcity, and Population. In *Coasts and Estuaries*, Eric Wolanski and others, eds., pp. 621–39. Elsevier. https://doi.org/10.1016/B978-0-12-814003-1.00036-8.

Day, John W., and others (2016). Approaches to defining deltaic sustainability in the 21st century. *Estuarine, Coastal and Shelf Science*, vol. 183, pp. 275–291.

Day, John W., and others (2019). Chapter 9 – Delta Winners and Losers in the Anthropocene. In *Coasts and Estuaries*, Eric Wolanski, and others, eds., pp. 149–65. Elsevier. https://doi.org/10.1016/B978-0-12-814003-1.00009-5.

Desmet, Klaus, and others (2018). Evaluating the Economic Cost of Coastal Flooding. Working Paper 24918. National Bureau of Economic Research. https://doi.org/10.3386/w24918.

Dietz, Thomas (2013). Bringing values and deliberation to science communication. *Proceedings of the National Academy of Sciences*, vol. 110, No. Supplement 3, pp. 14081–14087.

Diop, Salif, and others (2016). *Estuaries: A Lifeline of Ecosystem Services in the Western Indian Ocean*. Springer.

Doney, Scott C., and others (2012). Climate change impacts on marine ecosystems. *Annual Review of Marine Science*, vol. 4, No. 1, pp. 11–37. https://doi.org/10.1146/annurev-marine-041911-111611.

Doody, J. Patrick (2013). Coastal squeeze and managed realignment in southeast England, does it tell us anything about the future? *Ocean & Coastal Management*, vol. 79, pp. 34–41.

Ducrotoy, J-P., and others (2019). Temperate estuaries: their ecology under future environmental changes. In *Coasts and Estuaries*, pp. 577–594. Elsevier.

Dunn, Frances E., and others (2019). Projections of declining fluvial sediment delivery to major deltas worldwide in response to climate change and anthropogenic stress. *Environmental Research Letters*, vol. 14, No. 8, 084034.

Elliott, Michael, and Alan K. Whitfield (2011). Challenging paradigms in estuarine ecology and management. *Estuarine, Coastal and Shelf Science*, vol. 94, No. 4, pp. 306–314.

Elliott, Michael, and others (2019). A Synthesis: What Is the Future for Coasts, Estuaries, Deltas and Other Transitional Habitats in 2050 and Beyond? In *Coasts and Estuaries*, pp. 1–28. Elsevier.

Erban, Laura E., and others (2014). Groundwater extraction, land subsidence, and sea level rise in the Mekong Delta, Vietnam. *Environmental Research Letters*, vol. 9, No. 8, 084010.

Fischhoff, Baruch (2013). The sciences of science communication. *Proceedings of the National Academy of Sciences*, vol. 110, No. Supplement 3, pp. 14033–14039.

Freeman, Lauren, and others (2019). Impacts of urbanization and development on estuarine ecosystems and water quality. *Estuaries and Coasts*, vol. 42, pp. 1821–1838.

García-Alonso, Javier, and others (2019). Río de la Plata: A Neotropical Estuarine System. In *Coasts and Estuaries*, pp. 45–56. Elsevier.

Gianelli, Ignacio, and Omar Defeo (2017). Uruguayan fisheries under an increasingly globalized scenario: long-term landings and bioeconomic trends. *Fisheries Research*, vol. 190, pp. 53–60.

Gillanders, Bronwyn M., and others (2011). Potential effects of climate change on Australian estuaries and fish utilising estuaries: a review. *Marine and Freshwater Research*, vol. 62, No. 9, pp. 1115–1131.

Haigh, Ivan D., and others (2016). Spatial and temporal analysis of extreme sea level and storm surge events around the coastline of the UK. *Scientific Data*, vol. 3, art. 160107.

Hallegatte, Stephane, and others (2013). Future flood losses in major coastal cities. *Nature Climate Change*, vol. 3, No. 9, p. 802.

Hallett, Chris S., and others (2017). Observed and predicted impacts of climate change on the estuaries of south-western Australia, a Mediterranean climate region. *Regional Environmental Change*, vol. 18, pp. 1357–73.

Harris, Rebecca M.B., and others (2018). Biological responses to the press and pulse of climate trends and extreme events. *Nature Climate Change*, vol. 8, No. 7, p. 579.

Hiatt, Matthew, and others (2019). Drivers and impacts of water level fluctuations in the Mississippi River delta: Implications for delta restoration. *Estuarine, Coastal and Shelf Science*, vol. 224, pp. 117–137.

Howarth, Candice, and James Painter (2016). Exploring the science-policy interface on climate change: the role of the IPCC in informing local decision-making in the UK. *Palgrave Communications*, vol. 2, No. 1, art. 16058.

Ikeuchi, Hiroaki, and others (2017). Compound simulation of fluvial floods and storm surges in a global coupled river-coast flood model: Model development and its application to 2007 Cyclone Sidr in Bangladesh. *Journal of Advances in Modeling Earth Systems*, vol. 9, No. 4, pp. 1847–1862.

Intergovernmental Panel on Climate Change (IPCC) (2019). *IPCC Special Report on the Ocean and Cryosphere in a Changing Climate.*

Janis, Samuel, and others (2016). Billion oyster project: linking public school teaching and learning to ecological restoration of New York Harbor using innovative applications of environmental and digital technologies. *International Journal of Digital Content Technology and Its Applications*, vol. 10, No. 1.

Korpinen, Samuli, and others (2019) *Multiple pressures and their combined effects in Europe's seas.* ETC/ICM Technical Report 4/2019: European Topic Centre on Inland, Coastal and Marine waters.

Kosyan, Ruben (2017). *The Diversity of Russian Estuaries and Lagoons Exposed to Human Influence.* Springer.

Kuang, Cuiping, and others (2014). Numerical assessment of the impacts of potential future sea level rise on hydrodynamics of the Yangtze River Estuary, China. *Journal of Coastal Research*, vol. 30, No. 3, pp. 586–597.

Kumbier, Kristian, and others (2018). Investigating compound flooding in an estuary using hydrodynamic modelling: a case study from the Shoalhaven River, Australia.

Lai, Racliffe W.S., and others (2016). Hong Kong's marine environments: History, challenges and opportunities. *Regional Studies in Marine Science*, vol. 8, pp. 259–273.

Lana, Paulo da Cunha, and Angelo F. Bernardino (2018). *Brazilian Estuaries: A Benthic Perspective.* 1st ed. Brazilian Marine Biodiversity. Springer International Publishing.

Lefcheck, Jonathan S., and others (2018). Long-term nutrient reductions lead to the unprecedented recovery of a temperate coastal region. *Proceedings of the National Academy of Sciences*, vol. 115, No. 14, pp. 3658–3662.

Li, Tong, and others (2018). Driving forces and their contribution to the recent decrease in sediment flux to ocean of major rivers in China. *Science of the Total Environment*, vol. 634, pp. 534–541.

Mahoney, Peter C., and Melanie J. Bishop (2018). Are geomorphological typologies for estuaries also useful for classifying their ecosystems? *Aquatic Conservation: Marine and Freshwater Ecosystems*, vol. 28, No. 5, pp. 1200–1208.

Mcleod, Elizabeth, and others (2011). A blueprint for blue carbon: toward an improved understanding of the role of vegetated coastal habitats in sequestering CO_2. *Frontiers in Ecology and the Environment*, vol. 9, No. 10, pp. 552–560.

McSweeney, S.L., and others (2017). Intermittently closed/open lakes and lagoons: their global distribution and boundary conditions. *Geomorphology*, vol. 292, pp. 142–152.

Minderhoud, P.S.J., and others (2017). Impacts of 25 years of groundwater extraction on subsidence in the Mekong delta, Vietnam. *Environmental Research Letters*, vol. 12, No. 6, 064006.

Minderhoud, P.S.J., and others (2019). Mekong delta much lower than previously assumed in sea level rise impact assessments. *Nature Communications*, vol. 10, No. 1, art. 3847. https://doi.org/10.1038/s41467-019-11602-1.

Moftakhari, Hamed R., and others (2015). Increased nuisance flooding along the coasts of the United States due to sea level rise: past and future. *Geophysical Research Letters*, vol. 42, No. 22, pp. 9846–9852.

Moftakhari, Hamed R., and others (2017). Cumulative hazard: The case of nuisance flooding. *Earth's Future*, vol. 5, No. 2, pp. 214–223.

Morris, Rebecca L., and others (2019). Design options, implementation issues and evaluating success of ecologically engineered shorelines.

Muniz, Pablo, and others (2019). Río de la Plata: Uruguay. In *World Seas: An Environmental Evaluation*, pp. 703–724. Elsevier.

Neumann, Barbara, and others (2017). Strong sustainability in coastal areas: a conceptual interpretation of SDG 14. *Sustainability Science*, vol. 12, No. 6, pp. 1019–1035.

Nicholls, Robert J., and others (2018). Erratum to: Ecosystem Services for Well-Being in Deltas: Integrated Assessment for Policy Analysis. In *Ecosystem Services for Well-Being in Deltas*, pp. E1–E1. Springer.

Nichols, Charles Reid, and others (2019). Collaborative science to enhance coastal resilience and adaptation. *Frontiers in Marine Science*, vol. 6, art. 404.

Nienhuis, J. H., and others (2020). Global-scale human impact on delta morphology has led to net land area gain. *Nature*, vol. 577, No. 7791, pp. 514–18. https://doi.org/10.1038/s41586-019-1905-9.

Pecl, Gretta T., and others (2017). Biodiversity redistribution under climate change: impacts on ecosystems and human well-being. *Science*, vol. 355, No. 6332, eaai9214.

Pesce, M., and others (2018). Modelling climate change impacts on nutrients and primary production in coastal waters. *Science of the Total Environment*, vol. 628, pp. 919–937.

Phan, Linh K., and others (2015). Coastal mangrove squeeze in the Mekong Delta. *Journal of Coastal Research*, vol. 31, No. 2, pp. 233–243. https://doi.org/10.2112/JCOASTRES-D-14-00049.1.

Prandle, David, and Andrew Lane (2015). Sensitivity of estuaries to sea level rise: vulnerability indices. *Estuarine, Coastal and Shelf Science*, vol. 160, pp. 60–68.

Renaud, Fabrice G., and others (2013). Tipping from the Holocene to the Anthropocene: How threatened are major world deltas? *Current Opinion in Environmental Sustainability*, vol. 5, No. 6, pp. 644–654.

Reynolds, Laura, and others (2016) Ecosystem services returned through seagrass restoration. *Restoration Ecology*, vol. 24, No. 5, pp. 583–588.

Robins, Peter E., and others (2016). Impact of climate change on UK estuaries: A review of past trends and potential projections. *Estuarine, Coastal and Shelf Science*, vol. 169, pp. 119–135.

Robins, Peter E., and others (2018). Improving estuary models by reducing uncertainties associated with river flows. *Estuarine, Coastal and Shelf Science*, vol. 207, pp. 63–73.

Schmidt, Allison L., and others (2017). Regional-scale differences in eutrophication effects on eelgrass-associated (*Zostera marina*) macrofauna. *Estuaries and Coasts*, vol. 40, No. 4, pp. 1096–1112.

Schmidt, Charles W. (2015). *Delta Subsidence: An Imminent Threat to Coastal Populations*. NLM-Export.

Seebens, Hanno, and others (2017). No saturation in the accumulation of alien species worldwide. *Nature Communications*, vol. 8, art. 14435.

Seebens, Hanno, and others (2018). Global rise in emerging alien species results from increased accessibility of new source pools. *Proceedings of the National Academy of Sciences*, vol. 115, No. 10, pp. E2264–E2273.

Sengupta, Dhritiraj, and others (2018). Building beyond land: an overview of coastal land reclamation in 16 global megacities. *Applied Geography*, vol. 90, pp. 229–238.

Shalovenkov, Nickolai (2019). Alien Species Invasion: Case Study of the Black Sea. In *Coasts and Estuaries*, pp. 547–568. Elsevier.

Syvitski, James, and others (2009) Sinking deltas due to human activities. *Nature Geoscience*, vol. 2, pp. 681–686.

Szabo, Sylvia, and others (2015). Sustainable development goals offer new opportunities for tropical delta regions. *Environment: Science and Policy for Sustainable Development*, vol. 57, No. 4, pp. 16–23.

Teichberg, Mirta, and others (2010). Eutrophication and macroalgal blooms in temperate and tropical coastal waters: nutrient enrichment experiments with *Ulva* spp. *Global Change Biology*, vol. 16, No. 9, pp. 2624–37. https://doi.org/10.1111/j.1365-2486.2009.02108.x.

Teixeira, João Batista, and others (2013). Traditional ecological knowledge and the mapping of benthic marine habitats. *Journal of Environmental Management*, vol. 115, pp. 241–250.

Tessler, Z.D., and others (2015). Profiling risk and sustainability in coastal deltas of the world. *Science*, vol. 349, No. 6248, pp. 638–643. https://doi.org/10.1126/science.aab3574.

Tessler, Z.D., and others (2016). A global empirical typology of anthropogenic drivers of environmental change in deltas. *Sustainability Science*, vol. 11, No. 4, pp. 525–537.

Todd, Peter A., and others (2019). Towards an urban marine ecology: characterizing the drivers, patterns and processes of marine ecosystems in coastal cities. *Oikos*.

Turra, Alexander, and others (2013). Global environmental changes: setting priorities for Latin American coastal habitats. *Global Change Biology*, vol. 19, No. 7, pp. 1965–1969.

United Nations (2017). *The First Global Integrated Marine Assessment: World Ocean Assessment I*. Cambridge: Cambridge University Press.

Van Coppenolle, Rebecca, and others (2018). Contribution of mangroves and salt marshes to nature-based mitigation of coastal flood risks in major deltas of the world. *Estuaries and Coasts*, vol. 41, No. 6, pp. 1699–1711.

Van Coppenolle, Rebecca, and Stijn Temmerman (2019). A global exploration of tidal wetland creation for nature-based flood risk mitigation in coastal cities. *Estuarine, Coastal and Shelf Science*, vol. 226, art. 106262.

Whitfield, Alan K. (2017). The role of seagrass meadows, mangrove forests, salt marshes and reed beds as nursery areas and food sources for fishes in estuaries. *Reviews in Fish Biology and Fisheries*, vol. 27, No. 1, pp. 75–110.

Wright, Lynn Donelson, and Wei Wu (2019). Pearl River Delta and Guangzhou (Canton) China. In *Tomorrow's Coasts: Complex and Impermanent*, pp. 193–205. Springer.

Yasuhara, Moriaki, and others (2017). Combining marine macroecology and palaeoecology in understanding biodiversity: microfossils as a model. *Biological Reviews*, vol. 92, No. 1, pp. 199–215.

Yasuhara, Moriaki, and others (2019). Palaeo-records of histories of deoxygenation and its ecosystem impact. Ocean deoxygenation: Everyone's problem. IUCN.

Yin, Jie, and others (2012). National assessment of coastal vulnerability to sea level rise for the Chinese coast. *Journal of Coastal Conservation*, vol. 16, No. 1, pp. 123–133.

Yin, Jiabo, and others (2018). Large increase in global storm runoff extremes driven by climate and anthropogenic changes. *Nature Communications*, vol. 9, No. 1, art. 4389.

Chapter 7G
Seagrass meadows

Contributors: Hugh Kirkman (convener), Elizabeth Sinclair and Paul Lavery.

Keynote points

- Seagrass meadows continue to decline at alarming rates, in particular where they are in conflict with human activities.
- Marine ecosystems are reconfigured as a result of climate-driven changes in species distributions, with some species projected to be functionally extinct by 2100.
- Blue carbon sequestration will play a role in mitigating climate change impacts.
- Successful long-term solutions around conservation and restoration will require a balance between social, economic and environmental drivers.

1. Introduction

Seagrasses are marine flowering plants that inhabit coastal waters. The most recent global assessment found that seagrasses had been disappearing at a rate of 110 km^2 per year since 1980 and that 29 per cent of the known areal extent had disappeared since seagrass areas were initially recorded in 1879 (Waycott and others, 2009). Those rates of decline accelerated from a median of 0.9 per cent per year before 1940 to 7 per cent per year since 1990. Seagrass loss rates are comparable to those reported for mangroves, coral reefs and tropical rainforests, placing seagrass meadows among the most threatened ecosystems on Earth. Declining seagrass health is the result of shifting environmental conditions attributable largely to coastal development, land reclamation, deforestation, seaweed and fish farming, overfishing and garbage dumping. Four main factors are responsible for seagrass degradation: poor water quality; physical disturbance; the degradation of food webs; and grazing.

Globally, seagrass meadows have continued to decline since the first World Ocean Assessment (United Nations, 2017; see also Unsworth and others, 2019). Effective management strategies need to be implemented to reverse the loss and enhance the fundamental role of seagrass meadows in coastal ocean habitats. A multifaceted and interdisciplinary perspective is needed to achieve global conservation of seagrass meadows (see box below). Long-term solutions will require a balance of social, economic and environmental drivers. Good examples are emerging, in particular where traditional owners and custodians are being consulted and enabled to resume broader custodial roles. Indigenous people have a culture that relates to the land and sea in a holistic way that also includes connections to powerful and significant places (National Oceans Office, 2002).

Global environmental changes have an effect on communities living near the sea or that are dependent on maritime resources. The loss and decline of meadows means that fishing grounds and nursery areas will continue to be lost and coastal areas will become less desirable for tourism, erosion control, marine education, fisheries nursery areas, clean water and clarity. The impacts of storm surges, erosion and inundation are being experienced in coastal communities around the globe. The science is clear that the protection and restoration of natural habitats can be a cost-effective complement to built infrastructure approaches for protecting communities (Ruckelshaus and others, 2016). Climate-driven range shifts in seagrasses (and many benthic marine species) are difficult to quantify owing to a lack of long-term data sets across complete species distributions. Range shifting with changing climates may lead to often unforeseen and dramatic consequences that may ultimately lead to biodiversity loss and the homogenization of communities (Brustolin and others, 2019). The combination of poleward migration of habitat-modifying species as a result of climate change (H. Kirkman, personal communication) and fishing pressure can result in reduced ecosystem resilience (e.g., Ling and Keane, 2018).

Specific means of recovery after seagrass loss are particularly difficult to quantify. O'Brien and others (2018) demonstrated four different scenarios of degradation and recovery for seagrass ecosystems. Recovery can be rapid, once conditions improve, but seagrass absence at landscape scales may persist for many decades. O'Brien and others (2018) proposed a framework modelling resilience.

Challenges and possible solutions to reduce seagrass meadow loss

Challenge 1: Societal recognition of seagrass importance. Explain what seagrasses are and their importance in coastal systems.

Challenge 2: Up-to-date information on status and condition. Explain and record the status and condition of many seagrass meadows.

Challenge 3: Identify threatening activities at local scales to target management actions accordingly. Suggest which threatening local activities can be managed

Challenge 4: Balancing the needs of people and planet. Expand the understanding of interactions between the socioeconomic and ecological elements of seagrass systems.

Challenge 5: Generating scientific research to support conservation actions. Use current high-profile seagrass research (food security and blue carbon) to engage wider research fields.

Challenge 6: Conservation action in an era of climate change. Use indicators that provide an early warning of seagrass climate change impacts. Innovate restoration techniques.

Source: Adapted from Unsworth and others, 2019.

2. Socioeconomic consequences

Coastal development is contributing to ongoing declines of marine and estuarine ecosystems globally. The increase in tourism and industrial and urban development in coastal areas will require more careful and sustainable management. Many socioecological systems rely on healthy seagrass meadows to support a multitude of important ecosystem services (Cullen-Unsworth and others, 2014). The loss of seagrass in one small area or estuary may seem negligible but, for an entire coastline, the cumulative effect of many disturbances is important to consider.

Seagrass grows in many estuarine areas, where development and ports often require extensive maintenance dredging. The physical removal of seagrass during dredging is minor, because the removal is only in the dredged channel, compared with the damage caused by shading from the dredge plume and the smothering of plants by sediment disturbed during the dredging. Risk mitigation opportunities have been assessed to reduce the impact on seagrass (Wu and others, 2018). Proposed development and other anthropogenic changes should be treated using the precautionary approach. Offsets or new plantings should be considered when seagrass meadows are to be lost or decline in area. However, offsets should not be encouraged, as they are rarely equal to lost habitat. Furthermore, in many areas, there may not be suitable areas to replant seagrass, leading to a net loss of seagrass meadow, and restoration success is not guaranteed.

Economic valuations of seagrass systems have not been widespread (Constanza and others, 2014), with most estimates under-valuing them. More derivative-based models linking ecological structure and function to all associated economic services will be essential for accurate estimations of their dollar value (Dewsbury and others, 2016).

Coral reefs (chap. 7D) and mangroves (chap. 7H) act synergistically in the production of economically useful food through the connectivity of nutrients and biota. Marine health is also enhanced by seagrass associated with the natural filtering of pathogens. Lamb and others (2017) found that, when seagrass meadows were present, there was a 50 per cent reduction in the relative abundance of potential bacterial pathogens of humans and marine organisms. Field surveys of more than 8,000 reef-building corals located adjacent to seagrass meadows showed twofold reductions in disease levels compared with corals at paired sites without adjacent seagrass meadows.

Seagrass meadows have an important cultural service; they play an unusual central role in preserving valuable submerged archaeological and historical heritage (Krause-Jensen and others, 2016). The age and growth rates of rhizomes can be determined by establishing the age of amphora, a realm of seagrass service that has been greatly overlooked.

3. Region-specific changes

There is a lack of data for many areas, or data are not being collected globally.

The lack of data is best illustrated with examples from islands throughout East Asian seas. South-East Asia and Australia have the highest diversity of seagrass species and habitat types but basic information on seagrass habitats is still lacking. Fortes and others (2018) pointed out that the known distribution, extent, species diversity, research and knowledge gaps of seagrasses in South-East Asia are not well presented. They estimated that biogeographic region of the marine ecoregions of the world as approximately 36,700 km^2, but that is likely an underestimate, as some ecoregions were not well represented and updated information was lacking.

The shallow East Asian island biotopes have extensive coral reefs, bordered with seagrass and mangroves. The major seagrass meadows in India grow along the south-eastern coast in the Gulf of Mannar between India and Sri Lanka and Palk Bay and in the lagoons of islands of Lakshadweep in the Arabian Sea to Andaman and Nicobar Islands in the Bay of Bengal. The Gulf is a national marine park, it covers an approximate area of 10,500 km^2 and it comprises 21 islands located parallel to the coastline. Management, mapping and monitoring are required to sustain the valuable nursery area for the local people who are using the meadows as their fishing grounds. Awareness of the uses of seagrass and about the important ecosystem should be created among the local people. Seagrasses play a key part in supporting Indonesia as the world's second largest producer of seafood. The perilous state of its seagrasses will compromise their resilience to climate change and result in the loss of their high ecosystem service value (Unsworth and others, 2018).

Some 800 islands off Myeik, Myanmar, have coral reefs and mangroves, with seagrass on their lee side. The valuable seagrass habitats have not been mapped but are probably over-exploited for fishes and crustaceans (H. Kirkman, personal communication).

Arias-Ortiz and others (2018) reported damage to 36 per cent of seagrass meadows in Shark Bay, Western Australia, following a marine heatwave in the period 2010–2011. Shark Bay has the largest carbon stock reported for a seagrass ecosystem, containing up to 1.3 per cent of the total carbon stored within the top metre of seagrass sediments worldwide.

Large areas of seagrass along the Queensland coast of eastern Australia have been lost as a result of cyclones and the associated increased turbidity and pollutant run-off. Such cyclones are predicted to become more intense as a result of climate change. *Posidonia oceanica* meadows in the Mediterranean are also declining at alarming rates owing to climate change and human activities (Telesca and others, 2015).

There is also a lack of coordination concerning which types of data are collected and how the data are collected, which results in data sets that cannot be analysed between regions.

Globally, not only in Australia and Asia, there are losses of seagrass. In the Caribbean, there are major changes with the invasion of *Halophila stipulacea*. Also, the Mediterranean is experiencing tropicalization (Hyndes and others, 2016).

4. Outlook

General awareness of seagrass and the important ecosystem services offered by it is improving globally. Conceptual and mathematical models are helping managers to reach a science-based management process for seagrass. Seagrass services can be organized into a drivers, pressures, state, impact and response framework, which has been adopted in some seagrass ecosystems around the world, including the European Union, for State of Environment reporting (Kelble and others, 2013). The goal of such a framework is to reach a science-based consensus in defining characteristics and fundamental regulating processes of seagrass ecosystems that is sustainable and capable of providing diverse ecosystem services. It is necessary to consider regional, social, political, cultural, economic and public health factors, in a research and management context, with ecological variables, to achieve that goal.

Monitoring, evaluation and reporting to managers are essential in determining whether management action, recovery or deterioration is taking place. A tiered monitoring approach that is designed by both scientists and managers should be introduced wherever management of seagrass is required. Neckles and others (2012) provided a conceptual model for such an approach. Zimmerman and others (2015) used a mathematical model to predict the effects of ocean warming, acidification and water quality on eelgrass in the Chesapeake Bay.

5. Key remaining knowledge gaps

An assessment of gaps in seagrass knowledge for Australian seagrasses identified deficiencies covering many research fields, including taxonomy and systematics, physiology, population biology, sediment biogeochemistry and microbiology, ecosystem function, faunal habitats, threats, rehabilitation and restoration, mapping and monitoring, and management tools (York and others, 2017). Those knowledge gaps apply globally and need to be addressed if the systems are to be effectively managed and capable of providing diverse ecosystem services. It is necessary to consider regional, social, political, cultural, economic and public health factors, in a research and management context, with ecological variables, to achieve those goals. Research programmes that address the interactions of multiple stressors simultaneously should be considered. The systems are extremely complex and it is difficult to accurately predict all the downstream impacts of multiple-stressor experiments. Progress in filling the knowledge gaps will rely on technological advances in remote sensing, genomics, microsensors, computer modelling and statistical analyses. Interdisciplinary approaches will continue to broaden the understanding of the complex interactions among seagrasses and their environment. Ecosystem services for seagrass meadows, with a focus on fishing grounds, and the damaging activities that threaten their existence should encourage expert opinion to elicit potential solutions to prevent further loss. It is expected that restoration will be increasingly necessary to mitigate seagrass disturbance (Statton and others, 2018).

There are also knowledge gaps in sociocultural-economic-themed research, despite growing awareness of the importance of seagrass-human relationships. Suggested solutions include more education in communities that do not understand the usefulness of seagrass, for example, on the active removal of seagrass in Maldives and other tourist destinations[1] and excessive build-up of beach wrack as a result of building inappropriate beach infrastructure, such as groynes or marinas.

6. Key remaining capacity-building gaps

6.1. Mapping

The detail and definition of seagrass distribution maps need to be improved. The most current global map of seagrass distribution was made nearly 10 years ago. It should be updated with further mapping of seagrass meadows and associated habitats. Mapping of seagrass meadows is being carried out in many different parts of the world, yet there are no centres for incorporating products into a global map and collecting metadata. The map would serve to indicate updated loss or gain in seagrass areas, seagrass diversity and the need for further information. Satellite technologies are also improving and are being used more frequently in seagrass mapping efforts.

Seagrass meadows are dynamic systems that add to the challenges of mapping and monitoring, especially as some species are ephemeral (colonizing or opportunistic) and/or prone to storm damage. Duffy and others (2019) envisioned an ecosystem map with a marine biodiversity observation network linking biodiversity to environmental geophysical variables.

There is currently no repository where information can be shared and changes can be tracked at ecologically relevant scales. That would require seagrass ecologists and managers to define a suite of metrics and sampling methodologies that result in data that can be shared and compared across regions. Detailed mapping can be done with drones or multi-sideband sonar. Hamana and Komatsu

1 Available at www.maldivesresilientreefs.com/campaigns/seagrass.

(2016) used a narrow multibeam sonar system to detect seagrass meadows and estimate their relative abundance. Gumusay and others (2019) reviewed the literature on seagrass mapping, monitoring and detection applications using acoustic systems. High-quality aerial imagery databases can monitor local to regional change in seagrass cover (e.g., Evans and others, 2018). However, none of those options are species-specific, so ground truthing is required. Water clarity is also an issue after storms, during dredging and as a result of land-based pollution and eutrophication. Eutrophication occurs when high levels of nutrients enter the sea and are taken up by macroalgae, which grow on seagrass leaves and can smother the seagrass and prevent photosynthesis.

6.2. Carbon sequestration

Seagrass ecosystems have greater capacity to sequester carbon than terrestrial ecosystems (Macreadie and others, 2019). Together with tidal marshes and mangroves, seagrasses contribute approximately 50 per cent of the total carbon sequestered in marine sediments (blue carbon), despite occupying only 0.2 per cent of the ocean area, and their organic carbon sequestration rates exceed those of terrestrial forests per unit area by one to two orders of magnitude. Seagrass meadows sequester carbon dioxide through photosynthesis and store large quantities in the plants but also, more importantly, in the sediment below (Mcleod and others, 2011; Fourqurean and others, 2012). Sediment sequestration is highly variable across species. Large seagrasses, such as *Posidonia* spp., form mattes several metres deep that can remain stored for millenniums (Mateo and others, 1997). Mazarrasa and others (2015) estimated the mean particulate inorganic carbon accumulation rate in seagrass sediments as 126.3 ± 31.05 g per m^2 per year. Based on the global extent of seagrass meadows (177,000 to 600,000 km^2), the ecosystems globally store between 11 and 39 picograms of particulate inorganic carbon in the top metre of sediment and accumulate between 22 and 75 teragrams of particulate inorganic carbon per year. Unfortunately, the dense meadows are under continuing threat (see earlier estimates of global loss rate), and with the loss comes the emission of the CO_2 stored within the meadows. For example, the ongoing loss of seagrass in Australia has been estimated to emit up to 3 million tons of CO_2 to the atmosphere every year, and increasing annual CO_2 emissions from land use change by 12–21 per cent (Serrano and others, 2019). Furthermore, damage and decline in seagrass meadows will lower the level of ongoing sequestration of carbon dioxide.

The potential exists to avoid the emissions of greenhouse gases, and enhance their sequestration, through the conservation and restoration of vegetated coastal ecosystems, which have been recognized by many nations as a means of achieving their policy objectives in relation to greenhouse gas abatement (Martin and others, 2016). However, for that potential to be realized, a number of key information gaps and policy issues need to be addressed. A recent overview by Macreadie and others (2019) provides a comprehensive road map for the coming decades on future research in blue carbon science.

6.3. Climate change

Changing climate and ocean conditions are affecting valuable marine resources and the communities that depend on them. Climate change has an effect on seagrass meadows over a number of themes. Extreme climatic events were predicted to become more frequent and severe (Intergovernmental Panel on Climate Change (IPCC), 2013), causing rapid ecosystem change, the scale of which is likely to be greater than that caused by a gradually changing climate (Wernberg and others, 2016). Climate vulnerability assessments with respect to key marine species and habitats should be conducted to better understand which species are most vulnerable and may require innovative management strategies.[2] Ecosystem reconfigurations arising from climate-driven changes in species distributions are expected to have profound ecological, social and economic implications, as temperate ecosystems are replaced by tropical species (Vergés and others, 2014). Range shifts can also be caused by invasive species, which alter the ecosystem services delivered. For example, the Caribbean has become infested with *Halophila stipulacea* over the past 20 years, and research shows dramatic shifts in fish assemblages as well as impact to turtles. Non-native *Zostera japonica* has also altered the estuarine communities that it colonizes (Vergés and others, 2014). It is predicted that climate change may lead to fewer but more intensive storms in many places (Gera and others, 2014) and, hence, local disturbance and loss of water quality for weeks to months at a time. That will disturb seagrass meadows and cause management and policy changes. Although loss of seagrass calcareous epiphytes may be beneficial through acidity (refer to chap. 5 for details on ocean acidification), under high carbon dioxide, nutrients and temperature, Alsterberg and others (2013) predict that non-calcareous epiphytes, such as filamentous algae and diatoms, will increase. That may lead to shifts in the epiphyte community structure from less palatable calcareous to more palatable algae. In addition, decreased production of grazing deterrent phenolics by seagrasses under high carbon dioxide (Arnold and others, 2012) may increase the palatability of seagrass leaves for a number of invertebrate and fish grazers. Positive effects of increased carbon dioxide on seagrass physiology may help to ameliorate negative effects of other environmental stressors known to affect seagrass growth and survival, although the combination of increasing temperatures and reduced light will likely negate increases in available carbon (Collier and others, 2018).

There may also be range shifting in some species, namely the replacement of temperate seagrass species with tropical species and changes in community structure. There is limited evidence of range shifting in seagrasses, at present, although current projections for warming range edges, such as the Mediterranean Sea and Shark Bay, Australia, suggest functional extinction of their large temperate seagrass species by 2100 with accelerating warming (Hyndes and others, 2016).

6.4. Restoration and recovery

Recovery and mitigation of disturbance regimes have typically been the first line of defence (and cheaper), but ecological restoration or intervention is becoming increasingly necessary in a rapidly changing environment and is potentially a more effective management strategy where seagrass habitat has already been lost or is heavily degraded (Statton and others, 2018). Restoration success has improved but is still extremely limited. Restoration and recovery of seagrass meadows is an important activity, with some significant advances in restoration successes being documented (e.g., Orth and others, 2017; Wendländer and others, 2019). Although restoration will be increasingly necessary to mitigate seagrass disturbance, it is critical to emphasize that restoration is never 100 per cent successful and is often far less than 50 per cent successful. Furthermore, restored and degraded beds rarely provide the same level of ecosystem services. Thus, relying on restoration will lead to continued global loss of seagrasses. The protection of existing meadows from further wholesale destruction will be a far better use of resources.

2 See www.fisheries.noaa.gov/national/climate/climate-vulnerability-assessments.

References

Alsterberg, Christian, and others (2013). Consumers mediate the effects of experimental ocean acidification and warming on primary producers. *Proceedings of the National Academy of Sciences*, vol. 110, No. 21, pp. 8603–8608.

Arias-Ortiz, Ariane, and others (2018). A marine heatwave drives massive losses from the world's largest seagrass carbon stocks. *Nature Climate Change*, vol. 8, No. 4, p. 338.

Arnold, Thomas, and others (2012). Ocean acidification and the loss of phenolic substances in marine plants. *PLoS One*, vol. 7, No. 4, e35107.

Brustolin, Marco Colossi, and others (2019). Future ocean climate homogenizes communities across habitats through diversity loss and rise of generalist species. *Global Change Biology*, vol. 25, No. 10, pp. 3539–3548.

Collier, Catherine J., and others (2018). Losing a winner: thermal stress and local pressures outweigh the positive effects of ocean acidification for tropical seagrasses. *New Phytologist*, vol. 219, No. 3, pp. 1005–1017.

Constanza, Robert, and others (2014) Changes in the global value of ecosystem services. *Global Environmental Change*, vol. 26, pp. 152–158.

Cullen-Unsworth, L.C., and others (2014). Seagrass meadows globally as a coupled social-ecological system: Implications for human wellbeing. *Marine Pollution Bulletin*, vol. 83, pp. 387–397. http://dx.doi.org/10.1016/j.marpolbul.2013.06.001.

Dewsbury, Bryan M., and others (2016). A review of seagrass economic valuations: Gaps and progress in valuation approaches. *Ecosystem Services*, vol. 18, pp. 68–77.

Duffy, J. Emmett, and others (2019). Toward a coordinated global observing system for seagrasses and marine macroalgae. *Frontiers in Marine Science*, vol. 6, art. 317.

Evans, Suzanna M., and others (2018). Seagrass on the brink: decline of threatened seagrass *Posidonia australis* continues following protection. *PloS One*, vol. 13, No. 4.

Fortes, Miguel D., and others (2018). Seagrass in Southeast Asia: a review of status and knowledge gaps, and a road map for conservation. *Botanica Marina*, vol. 61, No. 3, pp. 269–288.

Fourqurean, James W., and others (2012). Seagrass ecosystems as a globally significant carbon stock. *Nature Geoscience*, vol. 5, No. 7, p. 505.

Gera, Alessandro, and others (2014). The effect of a centenary storm on the long-lived seagrass *Posidonia oceanica. Limnology Oceanography*, vol. 59, pp. 1910–1918.

Gumusay, Mustafa Umit, and others (2019). A review of seagrass detection, mapping and monitoring applications using acoustic systems. *European Journal of Remote Sensing*, vol. 52, No. 1, pp. 1–29.

Hamana, Masahiro, and Teruhisa Komatsu (2016). Real-time classification of seagrass meadows on flat bottom with bathymetric data measured by a narrow multibeam sonar system. *Remote Sensing*, vol. 8, No. 2, art. 96.

Hyndes, Glenn A., and others (2016). Accelerating tropicalization and the transformation of temperate seagrass meadows. *Bioscience*, vol. 66, No. 11, pp. 938–948.

Intergovernmental Panel on Climate Change (IPCC) (2013). Climate Change 2013: The Physical Science Basis. Contribution of Working Group I to the Fifth Assessment Report of the Intergovernmental Panel on Climate, T.F. Stocker and others, eds. Cambridge and New York: Cambridge University Press.

Kelble, Christopher R., and others (2013). The EBM-DPSER Conceptual Model: Integrating Ecosystem Services into the DPSIR Framework. *PLOS ONE*, vol. 8, No. 8, pp. 1–12. https://doi.org/10.1371/journal.pone.0070766.

Krause-Jensen, Dorte, and others (2016). Seagrass sedimentary deposits as security vaults and time capsules of the human past. *Ambio*, vol. 48, No. 4, pp. 325–335.

Lamb, Joleah B., and others (2017). Seagrass ecosystems reduce exposure to bacterial pathogens of humans, fishes, and invertebrates. *Science*, vol. 355, No. 6326, pp. 731–733.

Ling, Scott D., and John P. Keane (2018). Resurvey of the Longspined Sea Urchin (*Centrostephanus rodgersii*) and associated barren reef in Tasmania. Institute for Marine and Antarctic Studies Report. University of Tasmania, Hobart, 52 p.

Macreadie, Peter I., and others (2019). The future of Blue Carbon science. *Nature Communications*, vol. 10, No. 1, pp. 1–13.

Martin, A., and others (2016). Blue Carbon - Nationally Determined Contributions Inventory. Appendix to: Coastal blue carbon ecosystems. Opportunities for Nationally Determined Contributions. Published by GRID-Arendal, Norway. ISBN: 978-82-7701-161-5.

Mateo, Miguel A., and others (1997). Dynamics of millenary organic deposits resulting from the growth of the Mediterranean seagrass *Posidonia oceanica*. *Estuarine, Coastal and Shelf Science*, vol. 44, No. 1, pp. 103–110.

Mazarrasa, I., and others (2015). Seagrass meadows as a globally significant carbonate reservoir. *Biogeosciences*, vol. 12, No. 16, pp. 4993–5003. https://doi.org/10.5194/bg-12-4993-2015.

Mcleod, Elizabeth, and others (2011). A blueprint for blue carbon: toward an improved understanding of the role of vegetated coastal habitats in sequestering CO_2. *Frontiers in Ecology and the Environment*, vol. 9, No. 10, pp. 552–560.

National Oceans Office (2002). *Sea Country: An Indigenous Perspective: The South-East Regional Marine Plan*. Assessment Reports.

Neckles, Hilary A., and others (2012). Integrating scales of seagrass monitoring to meet conservation needs. *Estuaries and Coasts*, vol. 35, No. 1, pp. 23–46. https://doi.org/10.1007/s12237-011-9410-x.

O'Brien, Katherine R., and others (2018). Seagrass ecosystem trajectory depends on the relative timescales of resistance, recovery and disturbance. *Marine Pollution Bulletin*, vol. 134, pp. 166–176.

Orth, Robert J., and others (2017). Submersed aquatic vegetation in Chesapeake Bay: sentinel species in a changing world. *Bioscience*, vol. 67, No. 8, pp. 698–712.

Ruckelshaus, Mary H., and others (2016). Evaluating the benefits of green infrastructure for coastal areas: location, location, location. *Coastal Management*, vol. 44, No. 5, pp. 504–16. https://doi.org/10.1080/08920753.2016.1208882.

Serrano, Oscar, and others (2019). Australian vegetated coastal ecosystems as global hotspots for climate change mitigation. *Nature Communications*, vol. 10, No. 1, pp. 1–10.

Statton, John, and others (2018). Decline and restoration ecology of Australian seagrasses. In *Seagrasses of Australia*, pp. 665–704. Springer.

Telesca, Luca, and others (2015). Seagrass meadows (*Posidonia oceanica*) distribution and trajectories of change. *Scientific Reports*, vol. 5, art. 12505.

United Nations (2017). *The First Global Integrated Marine Assessment: World Ocean Assessment I*. Cambridge: Cambridge University Press.

Unsworth, Richard K.F., and others (2018). Indonesia's globally significant seagrass meadows are under widespread threat. *Science of the Total Environment*, vol. 634, pp. 279–286.

Unsworth, Richard, K.F., and others (2019). Global challenges for seagrass conservation. *Ambio*, vol. 48, No. 8, pp. 801–815.

Vergés, Adriana, and others (2014). The tropicalization of temperate marine ecosystems: climate-mediated changes in herbivory and community phase shifts. *Proceedings of the Royal Society B: Biological Sciences*, vol. 281, No. 1789, p20140846.

Waycott, Michelle, and others (2009). Accelerating loss of seagrasses across the globe threatens coastal ecosystems. *Proceedings of the National Academy of Sciences*, vol. 106, No. 30, pp. 12377–12381.

Wendländer, Nele Svenja, and others (2019). Assessing methods for restoring seagrass (*Zostera muelleri*) in Australia's subtropical waters. *Marine and Freshwater Research*.

Wernberg, Thomas, and others (2016). Climate-driven regime shift of a temperate marine ecosystem. *Science*, vol. 353, No. 6295, pp. 169–172.

Wu, Paul Pao-Yen, and others (2018). Managing seagrass resilience under cumulative dredging affecting light: Predicting risk using dynamic Bayesian networks. *Journal of Applied Ecology*, vol. 55, No. 3, pp. 1339–1350.

York, Paul H., and others (2017). Identifying knowledge gaps in seagrass research and management: an Australian perspective. *Marine Environmental Research*, vol. 127, pp. 163–172.

Zimmerman, Richard C., and others. (2015) Predicting effects of ocean warming, acidification, and water quality on Chesapeake region eelgrass. *Limnology and Oceanography*, vol. 60, No. 5, pp.1781–1804.

Chapter 7H
Mangroves

Contributors: José Souto Rosa Filho (convener), Nahid Abdel Rahim Osman and Colin D. Woodroffe.

Keynote points

- Despite their ecological and socioeconomic importance, especially as carbon sinks, mangrove forest areas have been decreasing annually.
- Although deforestation continues in most areas, afforestation and the replanting of mangroves on all continents have partially decreased the speed of mangrove area loss from about 2 per cent per year to less than 0.4 per cent per year.
- Increasing human population density and unplanned development in the coastal zone are the main threats to mangrove forests.
- Global climatic change, such as rising sea levels and temperature, is causing expansion of mangroves towards the poles and landward into salt marsh in some areas.
- Local actions and international agreements have helped mangrove conservation, but bureaucracy and the lack of commitment on the part of local, state, and national governments, and local communities have limited their success.

1. Introduction

Mangroves occur in 118 countries, occupy estuaries and shores in tropical and subtropical regions (Tomlinson, 2016), and are home to a recorded 73 species and hybrids, with the highest diversity and the largest extent in Asia (Spalding, 2010). Mangroves are key ecosystems at the interface between sea and land, and they influence, and are affected by, most human activities (United Nations, 2017b; Feller and others, 2017).

Mangrove forests, although they account for only 0.7 per cent of the total tropical forests in the world, provide seafood, firewood and timber, as well as such services as shoreline protection, carbon export and sequestration, and bioremediation of wastes, as noted in chapter 48 of the first *World Ocean Assessment* (United Nations, 2017a). In addition to goods and services, mangroves also provide cultural ecosystem services that are an important part of the lives, livelihoods and cultural identity of local communities and primary stakeholders (Mitra, 2020a).

Despite their ecological, socioeconomic and cultural value, mangroves are among the most threatened ecosystems on the planet. Mangroves are being destroyed at rates three to five times greater than average rates of forest loss, and over a quarter of original mangrove cover has already disappeared (United Nations Environment Programme (UNEP), 2014; Richards and Friess, 2016). Although mangroves continue to be degraded and have been lost from specific regions, conservation initiatives, rehabilitation efforts, natural regeneration and climate change-related expansion have resulted in gains in other places (Feller and others, 2017).

Considerable changes in the extent of mangrove area have been observed in response to natural environmental drivers in areas far from direct human impacts (Rioja-Nieto and others, 2017; Lucas and others, 2018). Worldwide mangrove cover had been estimated at 181,000 km^2 by Spalding and others (1997), but that value was revised downward to 137,760 km^2 by Giri and others (2011), which was similar to estimates provided by the Global Mangrove Watch (Bunting and others, 2018).

Increasing human population density in the coastal zone is the main threat to mangrove forests, specifically unplanned urban development, aquaculture, conversion to agricultural use, such as rice farming, and overexploitation of timber (United Nations, 2017b; Ferreira and Lacerda, 2016; Thomas and others, 2018; Romañach and others, 2018). However, globally, mangrove loss rates dropped from about 2 per cent per year to less than 0.4 per cent per year between the late twentieth and early twenty-first centuries (Friess and others, 2019b, 2020).

In the past decade, the quality and availability of global-scale mangrove distribution data have been improving (Hamilton and Casey, 2016; Ferreira and Lacerda, 2016; Thomas and others, 2018; Romañach and others, 2018; Saintilan and others, 2019; Lucas and others, 2020) and many national initiatives (public and/or public-private partnerships) have been launched to try to better understand mangrove changes (Schaeffer-Novelli and others, 2016). One of the main approaches to mangrove assessment has been the use of satellite data (Giri and others, 2011; Li and others, 2013; Duncan and others, 2017; Jayakumar, 2019; Lymburner and others, 2019).

Recently, the advent of cloud computing platforms, such as Google Earth Engine (Gorelick and others, 2017) and Amazon Web Services (Chen and others, 2017; Lucas and others, 2020), which combine several PB of orbital and geospatial data with statistical analysis resources, have enabled more reliable estimates of local, regional and global mangrove cover and changes over successive years (Diniz and others, 2019).

The present chapter builds on the observations in chapter 48 of the first Assessment. Mangroves are marine plants, which are covered in chapter 6G of the present Assessment, are often found in estuaries and deltas, which are discussed in chapter 7F, and share many characteristics with salt marshes (chap. 7I) and seagrass meadows (chap. 7G).

2. Documented change in state of mangroves between 2010 and 2020

The area occupied by mangroves has been declining annually worldwide. The status of mangrove forests varies by country and region (Romañach and others, 2018). Although deforestation continues in most areas, afforestation and the replanting of mangroves in some areas have decreased the speed of forest losses (Li and others, 2013; Cavanaugh and others, 2014; Ferreira and Lacerda, 2016; Friess and others, 2019b, 2020). A 12 per cent increase in mangrove area was recorded by Almahasheer and others (2016) in the Red Sea. In New Zealand, mangroves have expanded rapidly during the past 50–80 years, as a result of accelerated estuarine infilling and vertical accretion of tidal flats (Horstman and others, 2018).

Mangrove loss owing to human activities has been recorded across all regions where mangroves occur. The main driver of mangrove destruction is increased human density in the coastal zone (Branoff, 2017; Saifullah, 2017; Romañach and others, 2018). The most frequent anthropogenic activity in mangroves has been the conversion of mangrove areas to aquaculture or agricultural use (Thomas and others, 2018; see chap. 16 of the present Assessment). Other factors associated with mangrove losses are logging, erosion and sedimentation (see chap. 13), salt production (Feller and others, 2017) and cattle grazing (Ferreira and Lacerda, 2016; Thomas and others, 2018).

Mangroves are unpopular in some communities because they are often perceived as "taking over" or being of "low value" (e.g., mangroves "spoil" a beautiful sandy beach by making it muddy). Destruction of mangroves in New Zealand is mostly by local communities (with or without consent) owing to a lack of community or agency awareness of mangrove values, or fear of loss of support by elected members of the government or community.

Global climate change has also been associated with changes in mangrove distribution (see chaps. 4 and 9), such as poleward and landward expansion (Cavanaugh and others, 2014; Saintilan and others, 2019), except where hard infrastructure prevents retreat, resulting in wetland contraction, called coastal squeeze (Leo and others, 2019). Extreme climate events can increase mangrove mortality caused by extreme droughts (Sippo and others, 2018), and an increase in carbon dioxide and nitrogen enrichment can augment the growth of other vegetation, thus suppressing the growth of mangrove seedlings (McKee and Rooth, 2008; Zhang and others, 2012).

Extensive dieback of mangroves was recorded along 1,000 km of the southern Gulf of Carpentaria in Australia in 2015 and 2016. The area is sparsely inhabited, and the event appears to have been associated with an unusually lengthy period of severe drought conditions, unprecedented high temperatures and a temporary drop in sea level (Duke and others, 2017). Similar dieback also occurred at that time in other parts of northern Australia (Asbridge and others, 2019).

2.1. Impacts of the change on, and interactions with, other components of the marine system

2.1.1. Carbon sequestration in mangroves

Mangroves are well known for their ability to accumulate high amounts of carbon (Tomlinson, 2016; Donato and others, 2011, 2012; Estrada and Soares, 2017; Kauffman and others, 2018; Lagomasino and others, 2019) and, in fact, they sequester four times more carbon than rainforests (Rovai and others, 2018; Twilley and others, 2018). Recent assessments have suggested that global mangrove biomass ranges from 1.91 to 2.83 petagrams (Pg) (Hutchison and others, 2014a; Tang and others, 2018), whereas global mangrove carbon stock has been estimated at 5.03 Pg by Simard and others (2018). At the global scale, the average above-ground biomass density has been estimated at 1.46 megagrams per km^2 (Tang and others, 2018). Rovai and others (2018) predicted a total global budget of 2.26 Pg carbon in mangrove soils. Globally, mangroves stored 4.19 Pg of carbon in 2012, with Brazil, Indonesia, Malaysia and Papua New Guinea accounting for more than 50 per cent of the global stock (Hamilton and Friess, 2018).

2.1.2. Loss of biodiversity

Mangroves are among the most productive ecosystems in the world (Alongi, 2008) and produce large amounts of litter (fallen leaves, branches and other debris) that are used by diverse fauna. Mangrove trees also offer hard substrate (aerial roots, trunks, branches, pneumatophores and leaves) for a myriad of invertebrates and plants (Hogarth, 2015; Rosa Filho and others, 2018). In addition to the substantial contribution they make to marine ecosystems, mangroves are used by more than 400 terrestrial mammal, amphibian and reptile species around the world (Rog and others, 2016), as the mangroves provide them with a refuge from anthropogenic disturbance. The global decline in mangrove habitats has had a negative impact on biodiversity, with cascading effects on the natural ecosystem functioning of other associated estuarine and coastal ecosystems, thus affecting at least three critical ecosystem services: the number of viable fishing activities (33 per cent decline); the provision of nursery habitats (69 per cent decline); and detoxification services provided by wetlands (63 per cent decline) (Worm and others, 2006; Barbier and others, 2011).

2.1.3. Impacts on populations of invertebrates and fishes in adjacent habitats

High primary productivity and habitat complexity in mangrove forests make them important areas for larvae and juveniles of invertebrates and fishes (Saenger and others, 2012; Lee and others, 2014). Some species of crustaceans and fishes that live in rivers and offshore ocean water or coral reefs need mangroves for breeding and/or juvenile growth (Sheaves and others, 2012; Bertini and others, 2014; Hogarth, 2015). In addition to ecological importance, some mangrove crustaceans and molluscs have high economic and cultural value in several countries (Abdullah and others, 2016; Beitl, 2018; Figueira and others, 2020). In recent years, considerable progress has been made in the bioprospecting of mangrove-derived microbes (Mitra, 2020b) and the microphytobenthic primary production in mangrove forests (Kwon and others, 2020).

2.1.4. Reduction in coastal protection

Mangroves can directly attenuate waves, thus enhancing drag against wave energy, because of the density of their trunks and root systems, such as pneumatophores or prop roots. Their complex root systems, which are important for sediment stabilization, also reduce both storm surges and tsunami wave impacts further inland (Marois and Mitsch, 2015; Sheng and Zou, 2017). The protective mangrove fringe in Viet Nam has been lost in recent decades, initially through defoliation by herbicides and, subsequently, by conversion to aquaculture and coastal development (Phan and others, 2015; Thinh and Hens, 2017; Truong and others, 2017; Fagherazzi and others, 2017; Veettil and others, 2019).

2.1.5. Displacement of salt marshes

The poleward and landward expansion of mangroves as a result of increasing temperatures and sea level rise has been at the expense of salt marshes and has already been recorded in several regions (Record and others, 2013; Saintilan and others, 2014, 2019; Kelleway and others, 2016; Hickey and others, 2017; Feller and others, 2017; Osland and others, 2017). Mangroves continue to expand into adjacent salt marsh areas where the two wetland types occur together (Yando and others, 2016; Pérez and others, 2017), which can result in greater carbon storage, as well as changes in associated fauna (Smee and others, 2017), although in some cases it remains unclear whether that is a consequence of climate change or other anthropogenic causes (Boon, 2017).

3. Consequences of the changes for human communities, economies and well-being

Mangroves play an important cultural and socioeconomic role for communities around the tropics (Walters and others, 2008; UNEP, 2014). Ecosystem services provided globally by mangroves and tidal marshes have been valued at \$19.4 per km^2 per year (Costanza and others, 2014). The consequences of mangrove forest destruction are related mainly but not exclusively to the loss of biodiversity and related effects on the capture of fish, molluscs and invertebrates, and to reduced coastal protection, which affects coastal constructions and adjacent marine habitats (Bertini and others, 2014; Hogarth, 2015; Sheng and Zou, 2017).

A reduction in mangrove area will lead to a decrease in the primary production of mangroves and the biodiversity and abundance of associated fauna and, consequently, will affect coastal and offshore fishing activities. A global meta-analysis designed to formally and statistically test the relationship between mangroves and catches of marine fishing resources showed that mangroves had a strong effect on catches in a variety of mangrove settings around the world (Carrasquilla-Henau and Juanes, 2017). It has been estimated that each mangrove hectare produces $0.2–$12,305 worth of fishes and $17.5–$3,412 worth of mixed species (Hutchison and others, 2014b).

Studies have shown that mangroves are able to protect coastal areas and habitats (coral reefs and/or seagrass meadows) against waves, cyclones, tsunamis and flooding (Marois and Mitsch, 2015; Sheng and Zou, 2017; Veettil and others, 2019). The fragmentation of mangroves will significantly reduce their role in coastal protection (Lee and others, 2019). A fully established mangrove fringe can reduce wave energy by 20 per cent per 100 m of mangroves (Mazda and others, 2006). Mangroves also have a significant effect on the extent of inundation and damage caused by coastal flooding. It has been estimated that, if all mangroves were lost, 18 million more people would be flooded every year, on average, an increase of almost 40 per cent, and the annual damage to property would increase by 16 per cent and $82 billion (Reguero and others, 2018).

The effects of the predicted expansion of mangroves are still poorly understood. Modelling studies have demonstrated that climate change will cause range shifts in species, expanding the geographical distribution of some species and enhancing the number of species present in some areas (Record and others, 2013; Saintilan and others, 2014, 2019; Simard and others, 2018). The changes may cause an increase in primary production and the habitat complexity of coastal areas that are presently mangrove-free, which, in turn, may favour local biodiversity, fishing activities and coastal protection (Lee and others, 2014).

Local studies of mangrove carbon sequestration often contrast with regional models of likely response (Hayes and others, 2017; Sasmito and others, 2020). Over short timescales, the greater contribution that expanding mangroves make to blue carbon storage may be difficult to detect (Rogers and others, 2019a) but, over longer timescales, sea level rise appears to augment below-ground carbon sequestration in coastal wetlands (Krauss and others, 2017; Rogers and others, 2019b). The management of mangrove areas is not always based on research outcomes; site managers often identify anthropogenic disturbances as key threats while, in contrast, the bulk of research focuses on natural disturbances, including climate change and sea level rise (Canty and others, 2018). Although sea level rise has been regarded as a major threat to mangrove shorelines (Lovelock and others, 2015), it is becoming apparent that rapid sedimentation beneath mangroves can, at least partially, offset the impacts (Woodroffe and others, 2016; Schuerch and others, 2018).

Mangroves, with their strong link to coastal fishing activities, are particularly relevant to Sustainable Development Goal 14, life below water.[1] Mangroves may also contribute to the achievement of other Goals, including Goal 2, zero hunger, and Goal 13, climate action, through the provisioning of ecosystem services, such as fishing grounds, and carbon sequestration and storage (Friess and others, 2019b). Several other Goals should benefit local communities that derive direct and indirect livelihoods from mangroves, including Goal 1, no poverty, Goal 11, sustainable cities and communities, and Goal 15, life on land.

1 See General Assembly resolution 70/1.

4. Key region-specific changes and consequences

As pointed out in previous sections, the area occupied by mangroves has been decreasing globally, which has serious economic, ecological and social implications (Lee and others, 2014; Branoff, 2017; Saifullah, 2017; Romañach and others, 2018; Mitra, 2020a). Mangroves have been threatened worldwide, in large part owing to anthropogenic impacts, including logging, conversion to aquaculture and agricultural use, urbanization, pollution and climate change (UNEP, 2014; Ward and others, 2016; Thomas and others, 2018).

In the seminal work that warned of "a world without mangroves", Duke and others (2007) predicted that, if nothing was done, the world could be deprived of mangroves and their ecosystem services by the end of the twenty-first century. However, since then, huge efforts, including local action and international agreements, have been put into mangrove rehabilitation and creation (Feller and others, 2017). Globally, between the late twentieth and early twenty-first century, mangrove loss rates dropped greatly (Friess and others, 2019b). Those results have prompted speculation that mangrove conservation just may have shifted from a pessimistic to a more optimistic trajectory (Friess and others, 2020).

Despite recent mangrove conservation successes, it is still too soon to assume that there is a general decrease in mangrove loss, as such progress in not evenly distributed worldwide. In some countries in South-East Asia, mangroves have been destroyed at rates of between 0.70 and 0.41 per cent per year (Friess and others, 2019b). New deforestation frontiers are also beginning to emerge in regions that have not previously experienced significant mangrove loss, in particular in South-East Asia and West Africa (Friess and others, 2020).

Various national and international conservation policy instruments have helped to reduce the loss of or increase mangrove area in some countries (Ferreira and Lacerda, 2016; Friess and others, 2019a, 2020). In Brazil, for example, 75 per cent of mangroves have remained unchanged for two decades or more; 10 per cent have remained stable for between one and two decades; and 15 per cent have remained stable for 10 or fewer years (Diniz and others, 2019). An analysis of multi-temporal Landsat data (1972, 2000 and 2013) showed that the area covered by mangroves in the Red Sea had increased by about 0.29 per cent per year, with a total expansion of 12 per cent over the 41 years from 1972 to 2013 (Almahasheer and others, 2016).

In some regions, in addition to human-assisted rehabilitation, natural increases of mangroves have been recorded. More than 15 per cent of mangroves deforested in South-East Asia between 2000 and 2012 reverted to mangrove area (Friess and others, 2019b), in part through natural colonization. Episodic mangrove colonization has also increased mangrove area on the north coast of South America (Gardel and others, 2011) and on the Firth of Thames coast in New Zealand (Swales and others, 2015).

Climate change, mainly increases in temperature, decreases in low temperature and freeze events, and changes in water availability (Saintilan and others, 2014; Cavanaugh and others, 2014) have favoured the poleward expansion of mangroves, as already recorded in Australia, China, Mexico (Pacific coast), Peru, South Africa and the United States (Atlantic coast) (Saintilan and others, 2014, 2019; Cavanaugh and others, 2014; Osland and others, 2017; Smee and others, 2017). Although increases in mangrove area at their range limit are unlikely to drastically increase global mangrove area, they can make substantial contributions to mangrove extent in those locations (Friess and others, 2019a).

5. Outlook

Based on the information presently available, it is possible to predict the continued reduction of mangrove area, in view of the continuity of human actions that have led to the loss of mangrove forests in most areas around the world (Friess and others, 2020). In areas where reforestation initiatives and management and conservation measures are being implemented, a reduction in the speed of mangrove destruction may occur.

As a result of further mangrove destruction, a reduction in productivity in estuarine areas is expected, which will have a cascading effect. Accordingly, loss of biodiversity is likely to continue in coastal areas, including in mangroves, seagrass meadows and coral reefs. In view of the role mangroves play as nursery grounds for invertebrates and fishes, a reduction in fish stocks in coastal and even offshore catches of fish, crustaceans and molluscs can be expected (El-Regal and Ibrahim, 2014). That may lead to further increases in economic losses resulting from the absence of mangroves.

The total or partial (loss of structure) destruction of mangroves, whether it be natural, for example following subsidence related to earthquakes (Albert and others, 2017), or anthropogenic, will diminish their coastal protection function. The loss of mangroves is likely to increase the damage resulting from cyclones (Cavanaugh and others, 2014; Asbridge and others, 2018; Montgomery and others, 2019; Zhang and others, 2020), tsunamis and floods (Asbridge and others, 2016; Menéndez and others, 2020), thus raising costs for the reconstruction and maintenance of coastal facilities, as well as increasing the threat to the lives of populations living near the coast.

6. Key remaining knowledge and capacity-building gaps

A range of new methodologies has been developed to study mangrove forests, including the use of terrestrial, airborne and satellite sensors (Kamal and Phinn, 2011; Koedsin and Vaiphasa, 2013; Zhu and others, 2015; Mackenzie and others, 2016; Olagoke and others, 2016; Duncan and others, 2017; Owers and others, 2018; Warfield and Leon, 2019; Wang and others, 2020). However, there remains a lack of reliable surveys on the status of mangroves at the global and regional scales, and of standardization of methods for assessing mangroves. Although there has been a recent drive to address those knowledge gaps, especially in South America and South-East Asia, further research is required to enable researchers to determine the processes that influence both vulnerability and resilience to climate change (Ward and others, 2016). The gap is even more prominent in the poorest developing countries.

There are few detailed studies (distinct taxonomic groups, large temporal and spatial scales) demonstrating the interconnectivity, and mechanisms involved, between mangrove forests and the biodiversity of adjacent coastal environments (Saunders and others, 2014). Further research needs to be done on the sustainability and interrelationships of the habitats (marshes, seagrass and coral reefs) and between mangroves and catches of marine fishing resources (coastal and oceanic) in order to increase the capabilities of coastal managers and to empower local communities to more effectively conserve such resources.

A gap remains with respect to capacity-building in the restoration of degraded mangroves and abandoned aquaculture ponds within former mangrove areas (Paul and others, 2017; Worthington and Spalding, 2018; Van Bijsterveldt and others, 2020). Restoration of mangrove areas needs to be more widely adopted; in some cases, restoration efforts have owed more to economic incentives than to conservation objectives (Aheto and others, 2016). Wider rehabilitation could lead to much healthier mangroves throughout the tropics.

Global distribution of mangroves (blue shading), showing diversity as numbers of specific taxa (species and nominal hybrids)

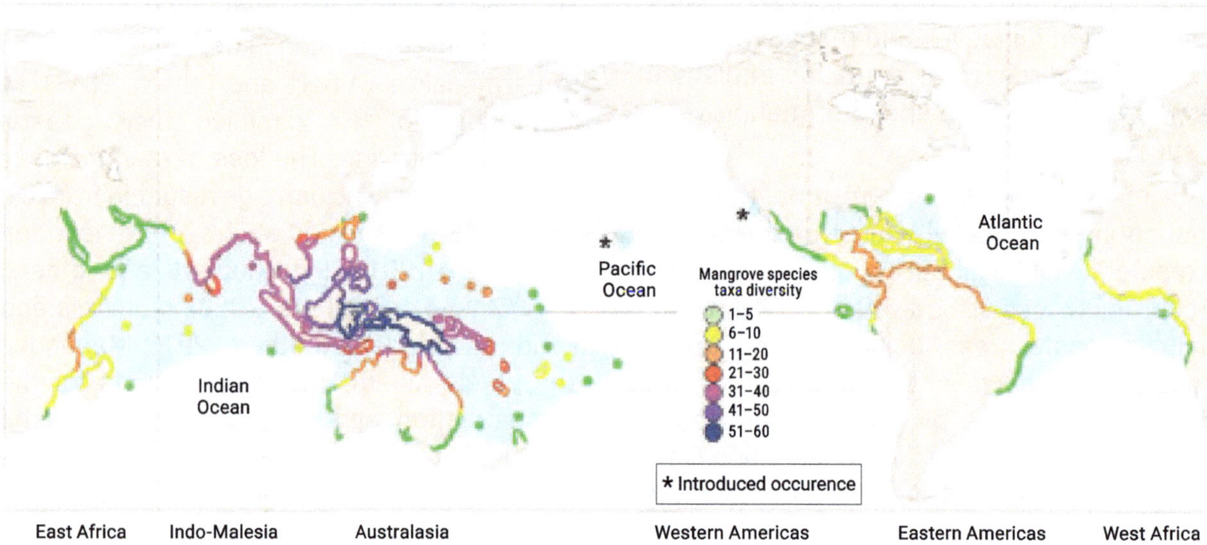

Source: UNEP, 2014.

References

Abdullah, Abu Nasar, and others (2016). Economic dependence on mangrove forest resources for livelihoods in the Sundarbans, Bangladesh. *Forests Policy and Economics*, vol. 64, pp. 15–24.

Aheto, Denis Worlanyo, and others (2016). Community-based mangrove forest management: implications for local livelihoods and coastal resource conservation along the Volta estuary catchment area of Ghana. *Ocean & Coastal Management*, vol. 127, pp. 43–54.

Albert, Simon, and others (2017). Winners and losers as mangrove, coral and seagrass ecosystems respond to sea-level rise in Solomon Islands. *Environmental Research Letters*, vol. 12, No. 9, 094009.

Almahasheer, Hanan, and others (2016). Decadal stability of Red Sea mangroves. *Estuarine, Coastal and Shelf Science*, vol. 169, pp. 164–172.

Alongi, Daniel M. (2008). Mangrove forests: resilience, protection from tsunamis, and responses to global climate change. *Estuarine, Coastal and Shelf Science*, vol. 76, No. 1, pp. 1–13.

Asbridge, Emma, and others (2016). Mangrove response to environmental change in Australia's Gulf of Carpentaria. *Ecology and Evolution*, vol. 6, No. 11, pp. 3523–3539.

Asbridge, Emma, and others (2018). The extent of mangrove change and potential for recovery following severe Tropical Cyclone Yasi, Hinchinbrook Island, Queensland, Australia. *Ecology and Evolution*, vol. 8, No. 21, pp. 10416–10434.

Asbridge, Emma, and others (2019). Assessing the distribution and drivers of mangrove dieback in Kakadu National Park, northern Australia. *Estuarine, Coastal and Shelf Science*, vol. 228, art. 106353.

Barbier, Edward B., and others (2011). The value of estuarine and coastal ecosystem services. *Ecological Monographs*, vol. 81, No. 2, pp. 169–193.

Beitl, Christine (2018). Rights-based Approaches in Ecuador's Fishery for Mangrove Cockles. In FAO Case Study for Tenure and User Rights in Fisheries (2018). *Proceedings*.

Bertini, Giovana, and others (2014). A test of large-scale reproductive migration in females of the amphidromous shrimp *Macrobrachium acanthurus* (Caridea: Palaemonidae) from south-eastern Brazil. *Marine and Freshwater Research*, vol. 65, No. 1, pp. 81–93.

Boon, Paul I. (2017). Are mangroves in Victoria (south-eastern Australia) already responding to climate change? *Marine and Freshwater Research*, vol. 68, No. 12, pp. 2366–2374.

Branoff, Benjamin L. (2017). Quantifying the influence of urban land use on mangrove biology and ecology: a meta-analysis. *Global Ecology and Biogeography*, vol. 26, No. 11, pp. 1339–1356.

Bunting, Pete, and others (2018). The global mangrove watch – a new 2010 global baseline of mangrove extent. *Remote Sensing*, vol. 10, No. 10, art. 1669.

Canty, Steven W.J., and others (2018). Dichotomy of mangrove management: a review of research and policy in the Mesoamerican reef region. *Ocean & Coastal Management*, vol. 157, pp. 40–49.

Carrasquilla-Henao, Mauricio, and Francis Juanes (2017). Mangroves enhance local fisheries catches: a global meta-analysis. *Fish and Fisheries*, vol. 18, No. 1, pp. 79–93.

Cavanaugh, Kyle C., and others (2014). Poleward expansion of mangroves is a threshold response to decreased frequency of extreme cold events. *Proceedings of the National Academy of Sciences*, vol. 111, No. 2, pp. 723–727.

Chen, Xiuhong, and others (2017). Running climate model on a commercial cloud computing environment: a case study using Community Earth System Model (CESM) on Amazon AWS. *Computers & Geosciences*, vol. 98, pp. 21–25.

Costanza, Robert, and others (2014). Changes in the global value of ecosystem services. *Global Environmental Change*, vol. 26, pp. 152–158.

Diniz, Cesar, and others (2019). Brazilian Mangrove Status: Three Decades of Satellite Data Analysis. *Remote Sensing*, vol. 11, No. 7, art. 808.

Donato, Daniel C., and others (2011). Mangroves among the most carbon-rich forests in the tropics. *Nature Geoscience*, vol. 4, No. 5, pp. 293–297.

Donato, Daniel C., and others (2012). Whole-island carbon stocks in the tropical pacific: implications for mangrove conservation and upland restoration. *Journal of Environmental Management*, vol. 97, pp. 89–96. https://doi.org/10.1016/j.jenvman.2011.12.004.

Duke, Norman C., and others (2007). A world without mangroves? *Science*, vol. 317, pp. 41–42.

Duke, Norman C., and others (2017). Large-scale dieback of mangroves in Australia's Gulf of Carpentaria: a severe ecosystem response, coincidental with an unusually extreme weather event. *Marine and Freshwater Research*, vol. 68, No. 10, pp. 1816–1829.

Duncan, Clare, and others (2017). Satellite remote sensing to monitor mangrove forest resilience and resistance to sea level rise. *Methods in Ecology and Evolution*, vol. 9, No. 8, pp. 1837–52. https://doi.org/10.1111/2041-210X.12923.

El-Regal, Mohamed A. Abu, and Nesreen K. Ibrahim (2014). Role of mangroves as a nursery ground for juvenile reef fishes in the southern Egyptian Red Sea. *The Egyptian Journal of Aquatic Research*, vol. 40, No. 1, pp. 71–78.

Estrada, Gustavo C.D., and Mario L.G. Soares (2017). Global patterns of aboveground carbon stock and sequestration in mangroves. *Anais da Academia Brasileira de Ciências*, vol. 89, No. 2, pp. 973–989.

Fagherazzi, Sergio, and others (2017). Buried alive or washed away: the challenging life of mangroves in the Mekong Delta. *Oceanography*, vol. 30, No. 3, pp. 48–59.

Feller, Ilka C., and others (2017). The state of the world's mangroves in the 21st century under climate change. *Hydrobiologia*, vol. 803, No. 1, pp. 1–12.

Ferreira, Alexander Cesar, and Luiz Drude Lacerda (2016). Degradation and conservation of Brazilian mangroves, status and perspectives. *Ocean & Coastal Management*, vol. 125, pp. 38–46.

Friess, Daniel A., and others (2019a). SDG 14: Life below Water – Impacts on Mangroves. In *Sustainable Development Goals: Their Impacts on Forests and People*, P. Katila and others, eds., pp. 445–481. Cambridge: Cambridge University Press.

Friess, Daniel A., and others (2019b). The state of the world's mangrove forests: past, present, and future. *Annual Review of Environment and Resources*, vol. 44, pp. 89–115.

Friess, Daniel A., and others (2020). Mangroves give cause for conservation optimism, for now. *Current Biology*, vol. 30, No. 4, pp. R153–R154.

Gardel, Antonie, and others (2011). Wave-formed mud bars: their morphodynamics and role in opportunistic mangrove colonization. *Journal of Coastal Research*, Special issue 64: Proceedings of the 11th International Coastal Symposium, pp. 384–387.

Giri, Chandra, and others (2011). Status and distribution of mangrove forests of the world using earth observation satellite data. *Global Ecology and Biogeography*, vol. 20, No. 1, pp. 154–159.

Gorelick, Noel, and others (2017). Google Earth Engine: Planetary-scale geospatial analysis for everyone. *Remote Sensing of Environment*, vol. 202, pp. 18–27.

Hamilton, Stuart E., and Daniel Casey (2016). Creation of a high spatio-temporal resolution global database of continuous mangrove forest cover for the 21st century (CGMFC-21). *Global Ecology and Biogeography*, vol. 25, No. 6, pp. 729–738.

Hamilton, Stuart E., and Daniel A. Friess (2018). Global carbon stocks and potential emissions due to mangrove deforestation from 2000 to 2012. *Nature Climate Change*, vol. 8, No. 3, pp. 240–244.

Hayes, Matthew A., and others (2017). Dynamics of sediment carbon stocks across intertidal wetland habitats of Moreton Bay, Australia. *Global Change Biology*, vol. 23, No. 10, pp. 4222–4234.

Hickey, Sharyn M., and others (2017). Is climate change shifting the poleward limit of mangroves? *Estuaries and Coasts*, vol. 40, No. 5, pp. 1215–1226.

Hogarth, Peter J. (2015). *The Biology of Mangroves and Seagrasses*. 3rd ed. Oxford University Press.

Horstman, Erik M., and others (2018). The dynamics of expanding mangroves in *New Zealand. Threats to Mangrove Forests: Hazards, Vulnerability, and Management*, C. Makowski and C.W. Finkl, eds., pp. 23–52. Springer.

Hutchison, James, and others (2014a). Predicting global patterns in mangrove forest biomass. *Conservation Letters*, vol. 7, No. 3, pp. 233–240.

Hutchison, James, and others (2014b). *The Role of Mangroves in Fisheries Enhancement*. The Nature Conservancy and Wetlands International.

Jayakumar, K. (2019). Chapter 15 – Managing Mangrove Forests Using Open Source-Based WebGIS. In *Coastal Management*, R.R. Krishnamurthy, and others, eds., pp. 301–21. Academic Press. https://doi.org/10.1016/B978-0-12-810473-6.00016-9.

Kamal, Muhammad, and Stuart Phinn (2011). Hyperspectral data for mangrove species mapping: a comparison of pixel-based and object-based approach. *Remote Sensing*, vol. 3, No. 10, pp. 2222–2242.

Kauffman, J. Boone, and others (2018). Carbon stocks of mangroves and salt marshes of the Amazon region, Brazil. *Biology Letters*, vol. 14, No. 9, 20180208. https://doi.org/10.1098/rsbl.2018.0208.

Kelleway, Jeffrey J., and others (2016). Seventy years of continuous encroachment substantially increases 'blue carbon' capacity as mangroves replace intertidal salt marshes. *Global Change Biology*, vol. 22, No. 3, pp. 1097–1109.

Kwon, Bong-Oh, and others (2020). Spatiotemporal variability in microphytobenthic primary production across bare intertidal flat, saltmarsh, and mangrove forest of Asia and Australia. *Marine Pollution Bulletin*, vol. 151, p. 110707

Koedsin, Werapong, and Chaichoke Vaiphasa (2013). Discrimination of tropical mangroves at the species level with EO-1 Hyperion data. *Remote Sensing*, vol. 5, No. 7, pp. 3562–3582.

Krauss, Ken W., and others (2017). Created mangrove wetlands store belowground carbon and surface elevation change enables them to adjust to sea-level rise. *Scientific Reports*, vol. 7, No. 1, pp. 1–11.

Lagomasino, David, and others (2019). Measuring mangrove carbon loss and gain in deltas. *Environmental Research Letters*, vol. 14, No. 2, 025002.

Lee, Shing Yip, and others (2014). Ecological role and services of tropical mangrove ecosystems: a reassessment. *Global Ecology and Biogeography*, vol. 23, No. 7, pp. 726–43. https://doi.org/10.1111/geb.12155.

Lee, Shing Yip, and others (2019). Better restoration policies are needed to conserve mangrove ecosystems. *Nature Ecology & Evolution*, vol. 3, No. 6, pp. 870–872.

Leo, Kelly L., and others (2019). Coastal habitat squeeze: a review of adaptation solutions for saltmarsh, mangrove and beach habitats. *Ocean & Coastal Management*, vol. 175, pp. 180–190.

Li, Mingshi S., and others (2013). Change and fragmentation trends of Zhanjiang mangrove forests in southern China using multi-temporal Landsat imagery (1977–2010). *Estuarine, Coastal and Shelf Science*, vol. 130, pp. 111–120.

Lovelock, Catherine E., and others (2015). The vulnerability of Indo-Pacific mangrove forests to sea-level rise. *Nature*, vol. 526, No. 7574, pp. 559–563.

Lucas, Richard and others (2018). Historical perspectives on the mangroves of Kakadu National Park. *Marine and Freshwater Research*, vol. 69, No. 7, pp. 1047–1063.

Lucas, Richard and others (2020). Structural characterisation of mangrove forests achieved through combining multiple sources of remote sensing data. *Remote Sensing of Environment*, vol. 237, p. 111543.

Lymburner, Leo, and others (2019). Mapping the multi-decadal mangrove dynamics of the Australian coastline. *Remote Sensing of Environment*, vol. 238, 111185.

MacKenzie, Richard A., and others (2016). Sedimentation and belowground carbon accumulation rates in mangrove forests that differ in diversity and land use: a tale of two mangroves. *Wetlands Ecology and Management*, vol. 24, No. 2, pp. 245–261.

Marois, Darryl E., and William J. Mitsch (2015). Coastal protection from tsunamis and cyclones provided by mangrove wetlands – a review. *International Journal of Biodiversity Science, Ecosystem Services & Management*, vol. 11, No. 1, pp. 71–83.

Mazda, Yoshihiro, and others (2006). Wave reduction in a mangrove forest dominated by *Sonneratia* sp. *Wetlands Ecology and Management*, vol. 14, No. 4, pp. 365–378.

McKee, Karen, and Jill E. Rooth (2008). Where temperate meets tropical: multi-factorial effects of elevated CO_2, nitrogen enrichment, and competition on a mangrove-salt marsh community. *Global Change Biology*, vol. 14, No. 5, pp. 971–984.

Menéndez, Pelayo, and others (2020). The global flood protection benefits of mangroves. *Scientific Reports*, vol. 10, No. 1, pp. 1–11.

Mitra, Abhijit (2020a). Mangroves: A Natural Ecosystem of Cultural and Religious Convergence. In *Mangrove Forests in India*, pp. 337–352. Cham, Switzerland: Springer.

________ (2020b). Ecosystem services of mangroves: an overview. In *Mangrove Forests in India*, pp. 1–32. Cham, Switzerland: Springer.

Montgomery, John M., and others (2019). Attenuation of storm surges by coastal mangroves. *Geophysical Research Letters*, vol. 46, No. 5, pp. 2680–2689.

Olagoke, Adewole, and others (2016). Extended biomass allometric equations for large mangrove trees from terrestrial LiDAR data. *Tree*, vol. 30, No. 3, pp. 935–947.

Osland, Michael J. (2017). Mangrove expansion and contraction at a poleward range limit: climate extremes and land-ocean temperature gradients. *Ecology*, vol. 98, No. 1, pp. 125–137.

Owers, Christopher J., and others (2018). Terrestrial laser scanning to quantify above-ground biomass of structurally complex coastal wetland vegetation. *Estuarine, Coastal and Shelf Science*, vol. 204, pp. 164–176.

Paul, Ashis K., and others (2017). Mangrove degradation in the Sundarbans. In *Coastal Wetlands: Alteration and Remediation*, pp. 357–392. Springer.

Pérez, Alexander, and others (2017). Changes in organic carbon accumulation driven by mangrove expansion and deforestation in a New Zealand estuary. *Estuarine, Coastal and Shelf Science*, vol. 192, pp. 108–116.

Phan, Linh K., and others (2015). Coastal mangrove squeeze in the Mekong Delta. *Journal of Coastal Research*, vol. 31, No. 2, pp. 233–243.

Record, S., and others (2013). Projecting global mangrove species and community distributions under climate change. *Ecosphere*, vol. 4, No. 3, art. 34. https://doi.org/10.1890/ES12-00296.1.

Richards, Daniel R., and Daniel A. Friess (2016). Rates and drivers of mangrove deforestation in Southeast Asia, 2000–2012. *Proceedings of the National Academy of Sciences*, vol. 113, No. 2, pp. 344–349.

Rioja-Nieto, Rodolfo, and others (2017). Environmental drivers of decadal change of a mangrove forest in the North coast of the Yucatan peninsula, Mexico. *Journal of Coastal Conservation*, vol. 21, No. 1, pp. 167–175.

Rog, Stefanie M., and others (2016). More than marine: revealing the critical importance of mangrove ecosystems for terrestrial vertebrates. *Diversity and Distributions*, vol. 23, No. 2, pp. 221–230.

Rogers, Kerrylee, and others (2019a). Mangrove dynamics and blue carbon sequestration. *Biology Letters*, vol. 15, No. 3, 20180471.

Rogers, Kerrylee, and others (2019b). Wetland carbon storage controlled by millennial-scale variation in relative sea-level rise. *Nature*, vol. 567, No. 7746, pp. 91–95.

Romañach, Stephanie S., and others (2018). Conservation and restoration of mangroves: global status, perspectives, and prognosis. *Ocean & Coastal Management*, vol. 154, pp. 72–82. https://doi.org/10.1016/j.ocecoaman.2018.01.009.

Rosa Filho, José Souto, and others (2018). Benthic Estuarine Assemblages of the Brazilian North Coast (Amazonia Ecoregion). In *Brazilian Estuaries: A Benthic Perspective*, Paulo da Cunha Lana and Angelo Fraga Bernardino, eds., pp. 39–74. Cham, Switzerland: Springer International Publishing. https://doi.org/10.1007/978-3-319-77779-5_2.

Rovai, André S., and others (2018). Global controls on carbon storage in mangrove soils. *Nature Climate Change*, vol. 8, No. 6, pp. 534–538.

Saenger, Peter, and others (2012). *A Review of Mangrove and Seagrass Ecosystems and Their Linkage to Fisheries and Fisheries Management*. Bangkok: FAO Regional Office for Asia and the Pacific.

Saifullah, S.M. (2017). The effect of global warming (climate change) on mangroves of Indus Delta with relevance to other prevailing anthropogenic stresses a critical review. European Academic Research, vol. 5, pp. 2110–2138.

Saintilan, Neil, and others (2014). Mangrove expansion and salt marsh decline at mangrove poleward limits. *Global Change Biology*, vol. 20, No. 1, pp. 147–157.

Saintilan, Neil, and others (2019). Climate change impacts on the coastal wetlands of Australia. *Wetlands*, vol. 39, No. 6, pp. 1145–1154.

Sasmito, Sigit D., and others (2020). Mangrove blue carbon stocks and dynamics are controlled by hydrogeomorphic settings and land-use change. *Global Change Biology*, vol. 26, No. 5, pp. 3028–3039.

Saunders, Megan I., and others (2014). Interdependency of tropical marine ecosystems in response to climate change. *Nature Climate Change*, vol. 4, No. 8, pp. 724–729.

Schaeffer-Novelli, Yara, and others (2016). Climate changes in mangrove forests and salt marshes. *Brazilian Journal of Oceanography*, vol. 64, No. Spe2, pp. 37–52.

Sheaves, Marcus, and others (2012). Importance of estuarine mangroves to juvenile banana prawns. *Estuarine, Coastal and Shelf Science*, vol. 114, pp. 208–219.

Sheng, Y. Peter, and Ruizhi Zou (2017). Assessing the role of mangrove forest in reducing coastal inundation during major hurricanes. *Hydrobiologia*, vol. 803, No. 1, pp. 87–103.

Simard, Marc, and others (2018). Mangrove canopy height globally related to precipitation, temperature and cyclone frequency. *Nature Geoscience*, vol. 12, No. 1, pp. 40–45.

Sippo, James Z., and others (2018). Mangrove mortality in a changing climate: an overview. *Estuarine, Coastal and Shelf Science*, vol. 215, pp. 241–249.

Smee, Delbert L., and others (2017). Mangrove expansion into salt marshes alters associated faunal communities. *Estuarine, Coastal and Shelf Science*, vol. 187, pp. 306–313.

Spalding, Mark (2010). *World Atlas of Mangroves*. Routledge.

Spalding, M.D., and others (1997). *World Mangrove Alias*. Okinawa, Japan: The International Society for Mangrove Ecosystems.

Swales A., and others (2015). Mangrove-forest evolution in a sediment-rich estuarine system: opportunists or agents of geomorphic change? *Earth Surface Processes and Landforms*, vol. 40, No. 1, pp. 1672–1687.

Tang, Wenwu, and others (2018). Big geospatial data analytics for global mangrove biomass and carbon estimation. *Sustainability*, vol. 10, No. 2, art. 472.

Thinh, Nguyen An, and Luc Hens (2017). A Digital Shoreline Analysis System (DSAS) applied on mangrove shoreline changes along the Giao Thuy coastal area (Nam Dinh, Vietnam) during 2005–2014. *Vietnam Journal of Earth Sciences*, vol. 39, No. 1, pp. 87–96.

Thomas, Nathan, and others (2018). Mapping mangrove extent and change: a globally applicable approach. *Remote Sensing*, vol. 10, No. 9, art. 1466.

Tomlinson, P. Barry (2016). *The Botany of Mangroves*. 2nd ed. Cambridge University Press.

Truong, Son Hong, and others (2017). Estuarine mangrove squeeze in the Mekong Delta, Vietnam. *Journal of Coastal Research*, vol. 33, No. 4, pp. 747–763.

Twilley, Robert R., and others (2018). Coastal morphology explains global blue carbon distributions. *Frontiers in Ecology and the Environment*, vol. 16, No. 9, pp. 503–508.

United Nations (2017a). Chapter 48: Mangroves. In *The First Global Integrated Marine Assessment: World Ocean Assessment I*. Cambridge: Cambridge University Press.

________ (2017b). *The First Global Integrated Marine Assessment: World Ocean Assessment I*. Cambridge: Cambridge University Press.

United Nations Environment Programme (UNEP) (2014). The Importance of Mangroves to People: A Call to Action. Hanneke van Lavieren and others, eds. Cambridge: United Nations Environment Programme World Conservation Monitoring Centre.

Van Bijsterveldt, Celine E.J., and others (2020). How to restore mangroves for greenbelt creation along eroding coasts with abandoned aquaculture ponds. *Estuarine, Coastal and Shelf Science*, vol. 235, art. 106576. https://doi.org/10.1016/j.ecss.2019.106576.

Walters, Bradley B., and others (2008). Ethnobiology, socio-economics and management of mangrove forests: a review. *Aquatic Botany*, vol. 89, No. 2, pp. 220–236.

Wang, Dezhi, and others (2020). Estimating aboveground biomass of the mangrove forests on northeast Hainan Island in China using an upscaling method from field plots, UAV-LiDAR data and Sentinel-2 imagery. *International Journal of Applied Earth Observation and Geoinformation*, vol. 85, 101986.

Ward, Raymond D., and others (2016). Impacts of climate change on mangrove ecosystems: a region by region overview. *Ecosystem Health and Sustainability*, vol. 2, No. 4, art. e01211.

Warfield, Angus D., and Javier X. Leon (2019). Estimating Mangrove Forest Volume Using Terrestrial Laser Scanning and UAV-Derived Structure-from-Motion. *Drones*, vol. 3, No. 2, art. 32.

Woodroffe, Colin D., and others (2016). Mangrove sedimentation and response to relative sea-level rise. *Annual Review of Marine Science*, vol. 8, pp. 243–266.

Worm, Boris, and others (2006). Impacts of biodiversity loss on ocean ecosystem services. *Science*, vol. 314, No. 5800, pp. 787–790.

Worthington, Thomas, and Mark Spalding (2018). *Mangrove Restoration Potential: A Global Map Highlighting a Critical Opportunity.*

Yando, Erik S., and others (2016). Salt marsh-mangrove ecotones: using structural gradients to investigate the effects of woody plant encroachment on plant-soil interactions and ecosystem carbon pools. *Journal of Ecology*, vol. 104, No. 4, pp. 1020–1031.

Zhang, Caiyun, and others (2020). Modelling risk of mangroves to tropical cyclones: a case study of Hurricane Irma. *Estuarine, Coastal and Shelf Science*, vol. 224, pp. 108–116.

Zhang, Yihui, and others (2012). Interactions between mangroves and exotic *Spartina* in an anthropogenically disturbed estuary in southern China. *Ecology*, vol. 93, No. 3, pp. 588–597.

Zhu, Yuanhui, and others (2015). Retrieval of mangrove aboveground biomass at the individual species level with worldview-2 images. *Remote Sensing*, vol. 7, No. 9, pp. 12192–12214.

Chapter 7I
Salt marshes

Contributors: Judith S. Weis (convener), Luis M. Pinheiro and Katherine E.A. Segarra.

Keynote points

- Salt marshes, as defined in the first *World Ocean Assessment* (United Nations, 2017) are intertidal, coastal systems that are regularly flooded with salt or brackish water and dominated by salt-tolerant plants adapted to regular or occasional immersion by tides.
- Salt marshes serve as nesting, nursery and feeding grounds for numerous species of birds, fishes, molluscs and crustaceans, including some commercially important species.
- Salt marshes are very effective "blue carbon" sinks, as they sequester CO_2 owing to their high levels of primary production and low rates of decomposition, but they can also produce greenhouse gas emissions.
- Salt marsh area is declining globally despite protective measures in many areas.
- Sea level rise poses the biggest threat, and marshes must either increase their elevation to keep pace with rising seas or move inland. Since the first Assessment, that has changed from a future issue to a present reality. If coastal development or restrictions on sediment supply and delivery make either adjustment difficult, salt marshes are converted to mudflats and open water.
- Many marshes worldwide are already showing signs of waterlogging, which indicates that they are not elevating rapidly enough.
- Some evidence suggests that marshes with certain invasive plants may be better able to keep up with sea level rise.

1. Introduction

Salt marshes occur on every continent except Antarctica (Mcowen and others, 2017; see figure below). They are more prevalent in temperate climates than in subtropical and tropical regions where mangrove forests occur. Salt marshes are highly productive ecosystems that provide critical ecosystem services, such as coastal protection, erosion prevention, nutrient cycling, habitat for various fish and bird species and carbon sequestration (Barbier and others, 2011).

The global extent of salt marshes is shrinking, primarily owing to sea level rise and more frequent and intense coastal storms (Cahoon, 2006; Duarte and others, 2013). Global historical coverage has declined 25–50 per cent since 1980 (Crooks and others, 2011; Duarte and others, 2008) owing to many factors, including filling for agriculture and development. Many remaining wetland ecosystems are showing evidence of eutrophication, waterlogging and disease (Short and others, 2016). Impacts on salt marshes will largely depend on the relative rate of sea level rise and other factors, such as those mentioned above (Adam, 2002). Salt marshes are losing ground to mangroves at the low-latitude end of their distribution owing to warming (Saintilan and others, 2014). The expansion of mangroves into salt marshes may result in increased storm protection and carbon storage (Doughty and others, 2016) but may also result in a decline in habitat for some animals.

Salt marshes are key coastal ecosystems that provide ecosystem services to humans, such as food and protection from storm surges and waves, by attenuating flooding (owing to wave height reduction) and through protection from pollutants and "blue carbon" sequestration. According to Macreadie and others (2013), the global carbon burial in salt marshes (up to 87.2 ± 9.6 teragrams (Tg) of carbon (C)/year, based on preliminary assessments) appears to exceed that of tropical rainforests (53 Tg C/year), although they occupy a much smaller area (0.1–2 per cent). However, greenhouse gas emissions can also be significant at some sites, in particular in wetlands with lower salinities and high organic matter content (Bartlett and others, 1987; Poffenbarger and others, 2011; Huertas and others, 2019). As concerns methane emissions, bubble-mediated fluxes (ebullition) can also play an important role in shallow areas with significant tidal variations, promoting gas release at low tide, owing to the lower water pressure (Duarte and others, 2007; Baulch and others, 2011; Call and others, 2015; Huertas and others, 2019). The atmospheric fluxes of greenhouse gases, mainly methane (CH_4) and nitrous oxide (N_2O), have greater global warming potential than CO_2 (Duarte and others, 2007; Roughan and others, 2018). Anthropogenic stressors, including nutrient pollution and salinity changes, may increase greenhouse gas fluxes in the future (Chmura and others, 2016; Yin and others, 2015; Roughan and others, 2018; Doroski and others, 2019).

Global distribution of salt marshes

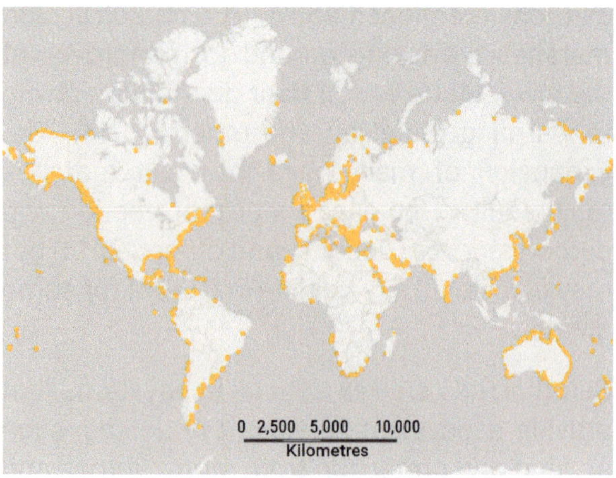

Source: Data from Mcowen and others, 2017.
Note: Created using ArcGIS Version 10.4.

The first Assessment stated that the key threats to salt marshes were land reclamation, coastal development, dredging, sea level rise and eutrophication, and sea level rise was identified as the largest climate-related threat to salt marshes. According to the United States National Oceanic and Atmospheric Administration (2019), the global mean water level in the ocean rose by 3.6 mm per year from 2006 to 2015, which was 2.5 times the average rate per year throughout most of the twentieth century. By the end of the century, global mean sea level is likely to rise at least 0.3 m above 2000 levels, even if greenhouse gas emissions follow a relatively low pathway in the coming decades. Nicholls and others (1999) predict that sea level rise of 1 m will eliminate 46 per cent of the world's coastal wetlands. Sea level rise varies regionally and differences from the predicted mean global range could exceed ±30 per cent (Oppenheimer and others, in press). Salt marshes may be able to migrate inland or increase their elevation in response to sea level rise, but that varies with local conditions, including subsidence in some areas. Subsidence is caused primarily by groundwater withdrawal but also by glacial isostatic readjustment, soil compaction and settling of fill (Eggleston and others, 2013). Managed realignment may be precluded by coastal development, which limits the inland area available for relocation. Such "coastal squeeze" occurs when rising sea levels advance the low-water mark, while high water is fixed by shoreline structures (Doody, 2004). A meta-analysis by Kirwan and others (2016) indicated that marshes are generally building at rates similar to or exceeding historical sea level rise and that process-based models predict survival under a wide range of future sea level scenarios. They argue that marsh vulnerability tends to be overstated because assessment methods often fail to consider feedback processes that accelerate soil building with sea level rise and the potential for marshes to migrate inland, a phenomenon affirmed by a recent global analysis by Rogers and others (2019). Those enhanced accretion rates are accompanied by enhanced rates of carbon burial, suggesting a reduction in the rate of climate change (McTigue and others, 2019). Sea level rise may also enhance rates of carbon burial. Scheider and others (2018) found that historical marsh loss in the Chesapeake Bay, United States, was compensated for by the conversion of upland to marshes. Schuerch and others (2018) stressed the importance of upland space ("accommodation space") for marsh migration, which requires the absence of human built infrastructure. If coastal squeeze were not an issue, marsh migration inland would be much easier in most places but could be restricted by steep slopes.

Since the first Assessment, the loss of marshes to rising sea levels has become an issue for the present, rather than the future. The use of recent surface elevation tables and assessment tools to examine the rate of marsh accretion as compared with sea level rise has provided data documenting the loss. Remote sensing techniques, such as light detection and ranging and aerial photography, also show the extent of losses and can be used periodically to monitor rates of change.

2. Description of the environmental changes between 2010 and 2020

At the global level, the extent of salt marshes is shrinking. However, as the rate of sea level rise is not identical at all sites, there are regional differences. Crosby and others (2016) synthesized available data and found that the rate of local sea level rise outpaced salt marsh accretion rates in many sites in Europe and the United States, indicating that under even the most optimistic Intergovernmental Panel on Climate Change emissions scenario, 60 per cent of the marshes studied will accrete less than the rate of sea level rise by 2100. The observed worldwide increase in storm surges over the past few years affects the water level and salinity in tidal salt marshes which, in turn, may affect greenhouse gas emissions (CO_2, CH_4 and N_2O) from those sites (Capooci and others, 2019).

While sea level rise is the major driver of wetland loss, eutrophication can also contribute (Deegan and others, 2012). Eutrophication increases above-ground biomass, decreases root biomass and increases microbial decomposition, resulting in plant instability, which causes creek-bank collapse, with areas of marsh converted to unvegetated mud (see also chap. 10). Overfishing of some fish species has led to increased populations of the herbivorous marsh crab (*Sesarma reticulatum*), whose consumption of marsh grasses has caused marsh dieback in some areas (Bertness and others, 2014; see also chap. 15 of the present Assessment). The death of rhizomes, from waterlogging and drought, is responsible for marsh dieback in other areas (Elmer and others, 2013). Marshes with inadequate sediment supply are most vulnerable to sea level rise (see chap. 13).

The loss of coastal marshes has an impact on other components of the marine system. With reduced marshes, there is likely to be less overall productivity in associated estuarine systems.

3. Consequences of the changes for human communities, economies and well-being

Narayan and others (2017) found that tidal wetlands in New Jersey, United States, prevented flood damage amounting to $625 million in the wake of Hurricane Sandy. They estimated a reduction of 16 per cent in annual flood losses attributable to salt marshes, with greater reductions at lower elevation. Coastal wetlands have the capacity to reduce property damage and prevent costs associated with storm surges (Rezaie and others, 2020). With less marsh area and more intense and frequent coastal storms and extreme events, there will be less protection for human communities, greater storm damage and reduced resilience. Fish stocks may also decline with the reduction of habitat for juvenile and larval stages (see chap. 15 regarding commercial fisheries), contributing to disruptions in income and food security for seafood-reliant communities.

Loss and degradation of salt marshes and the reduction of the ecosystem services and protection they provide will affect the achievement of the Sustainable Development Goals,[1] in particular Goals 1, 2 and 8, through reduced ecosystem services and food supply, Goal 11, owing to reduced protection of coastal areas from extreme events, Goal 13, through their potential to sequester blue carbon, but also their potential for emitting greenhouse gases, and Goals 14 and 15, owing to impacts on ecosystems. The loss of tidal marshes will also have socioeconomic consequences. Reduced numbers and types of goods provided would most likely lead to reduced yields of fisheries, less sequestration of pollutants, less carbon storage and storm abatement, and increased nitrogen and methane emissions to the atmosphere. Effects on human health may also result from increased pollutants in salt marsh animals used for human consumption, as well as poorer water quality if salt marshes do not remove pathogens and pollutants from wastewater.

4. Key region-specific changes and consequences

Salt marshes occupy a considerable area with variable vegetation along European coasts. Protection of much of the salt marsh area is increasing within the Natura 2000 network (European Commission, 2007). In terms of vegetation, North Atlantic salt marshes are mainly colonized by *Salicornia* spp. and other annuals, as well as *Spartina* swards (*Spartinion maritimae*) (Bortolus and others, 2019), whereas Mediterranean species higher up on the shore are usually more resistant to desiccation. In general, salt marshes along the Mediterranean experience minimal tidal differences and are considered microtidal, whereas those on Atlantic shores generally experience significant tidal variations. In general, areas with greater tidal amplitude will have more severe effects from sea level rise (Devlin and others, 2017).

Salt marshes in South Africa include many rarely-flooded supratidal marshes that support halophytic communities (Adams and others, 2016). Owing to wave action and high sediment availability, over 90 per cent of the estuaries have restricted inlets, with most closing temporarily when a sandbar forms (Cooper, 2001). Sea level rise, increased storms and wave

1 See General Assembly resolution 70/1.

height, and changed river discharge will influence inundation patterns, salinity gradients and sediment biogeochemistry (Van Niekerk, 2018). If there is available land, salt marshes will migrate inland (Tabot and Adams, 2013; Veldkornet and others, 2015). Sea level rise will produce more open conditions, in particular if the mouth of the estuary is sheltered from wave action and little sediment is available (Van Niekerk, 2018). However, drought and reduced freshwater inflow will result in mouth closure, flooding and dieback of salt marsh plants.

Salt marshes in China are dominated by native *Phragmites australis* or *Spartina alterniflora*, the latter of which was introduced from England and North America (Gu and others, 2018; Wan and others, 2009). The areal extent of salt marshes declined by about 59 per cent between the 1980s and the 2010s, largely owing to land reclamation (Gu and others, 2018; Tian and others, 2016). To combat that trend, China has implemented policy measures to restore and conserve salt marshes, such as establishing protected areas, ecological redlining and strictly regulating reclamation. Such management strategies are recent and, therefore, their effectiveness is yet to be proved (Bai and others, 2018).

5. Outlook

In order to persist, salt marshes must either elevate at a rate equal to sea level rise, which requires obtaining enough new sediment, or migrate inland, which requires undeveloped land immediately inland of the marsh and an appropriate slope. A continued loss of marshes, with the concurrent loss of ecosystem services and biodiversity, is anticipated in many areas. A decrease in their spatial extent will reduce ecosystem service provision. The loss of salt marshes not only reduces their capacity to act as carbon sinks, but the related degradation and disturbance also contribute to the release of carbon back into the atmosphere in the form of CO_2 (Pendleton and others, 2012) and the emission of other greenhouse gases, such as N_2O and CH_4.

Peteet and others (2018) found that urban development greatly reduced inputs of mineral sediment, but organic matter allowed vertical accumulation to outpace sea level for a time. However, reduced mineral content caused structural weakness and edge failure, and they concluded that marsh survival would require mineral sediment addition. Borchert and others (2018) further showed that migration corridors were particularly important in urbanized estuaries with coastal development where there is no space for wetlands to move inland and adapt to sea level rise.

A meta-analysis by Davidson and others (2018) found that certain invasive plants enabled biomass and carbon storage potential to increase by over 100 per cent. Because plants such as the invasive *Phragmites australis* grow larger and faster, the ecosystem can store more carbon, and that plant also promotes increased marsh elevation. Rooth and Stevenson (2000) found greater rates of litter production and of mineral and organic sediment trapping in *P. australis*. Therefore, that species may provide a strategy to combat sea level rise, albeit with a reduced diversity of plants in the marsh and some faunal changes. That information has not yet altered restoration policies and projects in which the plant is removed. The invasion of *Spartina alterniflora* in salt marshes in China (Zhang and others, 2004; Zuo and others, 2012) and South America (Bortolus and others, 2015) has created new vegetated areas, thus reducing the degree of wetland loss.

Since marshes are one of the most highly productive ecosystems on the planet and are home to many endangered species, their loss will have significant repercussions for overall productivity, biodiversity and ecosystem services. Significant impacts are expected from the loss of their nursery function for juvenile fishes and invertebrates. Marsh loss will also affect birds, as marshes are important breeding, foraging, overwintering and migration stop-off points (United Nations Environment Programme (UNEP), 2019).

Indices of resilience developed for tidal salt marshes by Raposa and others (2016) suggest that Pacific marshes are likely to be more resilient than their Atlantic counterparts, largely owing to differences in the percentage of vegetation currently located below mean high water. Such indices provide a way to evaluate resilience, inform management and prioritize areas of marsh restoration.

Reduced marsh habitat would have socioeconomic consequences, including reductions in fish stocks, storm protection, carbon and pollution sequestration, and water quality.

6. Key remaining knowledge gaps

While there has been some study of carbon dioxide fluxes in marshes (Forbrich and Giblin, 2015; Wei and others, 2020), more needs to be known about greenhouse gas fluxes at the sediment-water and water-atmosphere interfaces in salt marshes. The influence of tides on methane escape from the sediment-water interface has been documented (Duarte and others, 2007; Poffenbarger and others, 2011; Baulch and others, 2011; Call and others, 2015; Segarra and others, 2013; Huertas and others, 2019) but the quantification of greenhouse gas fluxes from such systems is still largely unknown. The same applies to N_2O emissions, for which experimental studies have shown the influence of nutrient input (Bulseco and others, 2019) but, again, there is no sustained long-term measurement and quantification of the fluxes involved. The future role of salt marshes in the global carbon and greenhouse gas budgets is largely unknown owing to the ongoing changes in their extent and hydrographic, nutrient and salinity (Poffenbarger and others, 2011) regimes.

Another gap relates to how to increase salt marsh resilience to sea level rise. The best techniques for leaving some *Phragmites australis* in place and how well that can accelerate marsh elevation are not yet known. Maintaining migration corridors for marshes to migrate inland is important in many areas and needs more investigation and political will. One possible way of increasing marsh elevation is "thin layer deposition", that is, spraying sediments from tidal creeks onto the marsh surface (Ford and others, 1999). Additional actions may include the artificial supply of dredging materials to reach a sufficiently high accretion rate to allow the tidal flat to adapt to sea level rise (Mendelssohn and Kuhn, 2003). The long-term effectiveness and how often the procedures will have to be done is unknown. When marshes are eroding at the edge, "living shorelines", in the form of oyster reefs, "reef balls" or rocks, can be placed at the edge to prevent further erosion (Bilkovic and others, 2017). Living shorelines have enhanced the resilience of marshes to hurricanes more than either hard edges or natural marshes (Smith and others, 2016). Another approach is to create floating marshes (Streb and others, 2019). However, since such approaches are relatively new, their continued effectiveness in the face of sea level rise remains to be seen.

7. Key remaining capacity-building gaps

As of 2020, 985 coastal wetlands are designated Ramsar Convention sites.[2] The sites, covering nearly 75 million ha, are recognized for their significant value to humanity and are to be managed in such a way as to maintain their ecological character and to promote wise use.[3] The number of Ramsar sites has steadily increased since the Convention entered into force in 1975. However, since global wetland extent has decreased markedly over that time, the efficacy of Ramsar policy implementation is questionable (Finlayson, 2012). According to scientists around the world, immediate action is urgently needed to transition to more sustainable practices (Ripple and others, 2017) and to reduce the loss of critical natural habitats that provide ecosystem services, such as wetlands and salt marshes (Finlayson, 2019; Finlayson and others, 2019).

Some countries lack adequate expertise or resources to study and rehabilitate salt marshes. In the United States, the State of Louisiana has one of the most comprehensive coastal restoration plans in North America and the largest investment in marsh creation ($17.1 billion), and it will use dredged material and sediment diversion to build and maintain coastal land. Such investments underlie the high monetary and organizational costs of major restoration projects. In many areas of the world, such intervention would be beyond the capacity of individual States. Also, some countries have the capacity but have not yet prioritized the conservation of coastal wetlands. It will take large investments of time and resources on a global level, with large-scale governmental awareness, agreements and common commitments, to reverse the deleterious trends observed in salt marshes today. Together, sea level rise and human development threaten marsh extent globally. Conservation and restoration efforts must acknowledge that, with accelerating sea level rise, marsh habitat is a moving target. Undeveloped coastal lands could convert to marsh in the next century; however, human development is the main obstacle to that occurring.

References

Adams, J.B., and others (2016). Distribution of macrophyte species and habitats in South African estuaries. *South African Journal of Botany*, vol. 107, pp. 5–11.

Adam, P. (2002). Saltmarshes in a time of change. *Environmental Conservation*, vol. 29, No. 1, pp. 39–61.

Bai, Y., and others (2018). Developing China's ecological redline policy using ecosystem services assessments for land use planning. *Nature Communications*, vol. 9, art. 3034 . https://doi.org/10.1038/s41467-018-05306-1.

Barbier, E.B., and others (2011). The value of estuarine and coastal ecosystem services. *Ecological Monographs*, vol. 81, No. 2, pp. 169–193.

Bartlett, K., and others (1987). Methane emissions along a salt marsh salinity gradient. *Biogeochemistry*, vol. 4, No. 3, pp. 183–202.

2 Convention on Wetlands of International Importance especially as Waterfowl Habitat (United Nations, *Treaty Series*, vol. 996, No. 14583).

3 List of sites available at https://rsis.ramsar.org.

Baulch, H.M., and others (2011). Diffusive and ebullitive transport of methane and nitrous oxide from streams: are bubble-mediated fluxes important? *Journal of Geophysical Research*, vol. 116, G04028, https://doi.org/10.1029/2011JG001656.

Bertness, M.D., and others (2014). Experimental predator removal causes rapid salt marsh die-off. *Ecology Letters*, vol. 17, No. 7, pp. 830–835.

Bilkovic, D.M., and others (2017). *Living Shorelines: The Science and Management of Nature-Based Coastal Protection*. CRC Press.

Borchert, S.M., and others (2018). Coastal wetland adaptation to sea level rise: Quantifying potential for landward migration and coastal squeeze. *Journal of Applied Ecology*, vol. 55, No. 6, pp. 2876–2887.

Bortolus A., and others (2015). Reimagining South American coasts: unveiling the hidden invasion history of an iconic ecological engineer. *Diversity and Distributions*, vol. 21, pp. 1267–1283.

Bortolus A., and others (2019). Supporting *Spartina:* interdisciplinary perspective shows *Spartina* as a distinct solid genus. *Ecology*, vol. 100, No. 11, e02863. https://doi.org/10.1002/ecy.2863.

Bulseco, A.N., and others (2019). Nitrate addition stimulates microbial decomposition of organic matter in salt marsh sediments. *Global Change Biology*, vol. 25, No. 10, pp. 3224–3241.

Cahoon, D.R. (2006). A review of major storm impacts on coastal wetland elevations. *Estuaries and Coasts*, vol. 29, No. 6, pp. 889–898.

Call, M., and others (2015). Spatial and temporal variability of carbon dioxide and methane fluxes over semi-diurnal and spring-neap-spring timescales in a mangrove creek. *Geochimica et Cosmochimica Acta*, vol. 150, pp. 211–225. https://doi.org/10.1016/j.gca.2014.11.023.

Capooci, M., and others (2019). Experimental influence of storm-surge salinity on soil greenhouse gas emissions from a tidal salt marsh. *Science of the Total Environment*, vol. 686, pp. 1164–1172.

Chmura, G.L., and others (2016). Greenhouse gas fluxes from salt marshes exposed to chronic nutrient enrichment. *PloS One*, vol. 11, No. 2, e0149937.

Cooper, J.A.G. (2001). Geomorphological variability among microtidal estuaries from the wave-dominated South African coast. *Geomorphology*, vol. 40, Nos. 1–2, pp. 99–122.

Crooks, S., and others (2011). Mitigating Climate Change through Restoration and Management of Coastal Wetlands and Near-Shore Marine Ecosystems: Challenges and Opportunities. *Environment Department Papers; Marine Ecosystem Series*, No. 121.

Crosby, S., and others (2016). Salt marsh persistence is threatened by predicted sea-level rise. *Estuarine, Coastal and Shelf Science*, vol. 181, pp. 93–99.

Davidson, I.C., and others (2018). Differential effects of biological invasions on coastal blue carbon: a global review and meta-analysis. *Global Change Biology*, vol. 24, No. 11, pp. 5218–5230.

Deegan, L.A., and others (2012). Coastal eutrophication as a driver of salt marsh loss. *Nature*, vol. 490, No. 7420, pp. 388–392.

Devlin, A., and others (2017). Coupling of sea level and tidal range changes, with implications for future water levels. *Scientific Reports*, vol. 7, art. 17021. https://doi.org/10.1038/s41598-017-17056-z.

Doody, J.P. (2004). "Coastal squeeze"—an historical perspective. *Journal of Coastal Conservation*, vol. 10, No. 1, pp. 129–138.

Doroski, A.A., and others (2019). Greenhouse gas fluxes from coastal wetlands at the intersection of urban pollution and saltwater intrusion: a soil core experiment. *Soil Biology and Biochemistry*, vol. 131, pp. 44–53.

Doughty, C.L., and others (2016). Mangrove range expansion rapidly increases coastal wetland carbon storage. *Estuaries and Coasts*, vol. 39, No. 2, pp. 385–396.

Duarte, C.M., and others (2008). The charisma of coastal ecosystems: addressing the imbalance. *Estuaries and Coasts*, vol. 31, No. 2, pp. 233–238.

Duarte, C.M., and others (2013). The role of coastal plant communities for climate change mitigation and adaptation. *Nature Climate Change*, vol. 3, No. 11, pp. 961–968.

Duarte, H., and others (2007). High-resolution seismic imaging of gas accumulations and seepage in the sediments of the Ria de Aveiro barrier lagoon (Portugal). *Geo-Marine Letters*, vol. 27, Nos. 2–4, pp. 115–126.

Eggleston, J., and others, 2013, Land subsidence and relative sea-level rise in the southern Chesapeake Bay region: U.S. Geological Survey Circular 1392. http://dx.doi.org/10.3133/cir1392.

Elmer, W.H., and others (2013). Sudden vegetation dieback in Atlantic and Gulf Coast salt marshes. *Plant Diseases*, vol. 97, No. 4, pp. 436-445.

European Commission (2007). *The Interpretation Manual of European Union Habitats–EUR27.* European Commission DG Environment, Brussels.

Finlayson, C.M. (2012). Forty years of wetland conservation and wise use. *Aquatic Conservation: Marine and Freshwater Ecosystems*, vol. 22, No. 2, pp. 139–143.

________ (2019). Addressing the decline in wetland biodiversity. *The Ecological Citizen*, vol. 2, pp. 139–40.

Finlayson, C.M., and others (2019). The second warning to humanity–providing a context for wetland management and policy. *Wetlands*, vol. 39, No. 1, pp. 1–5.

Forbrich I., and A. Giblin (2015) Marsh-atmosphere CO_2 exchange in a New England salt marsh. *JGR Biosciences*, vol. 20, No. 9, pp. 1825–1838.

Ford, M.A., and others (1999). Restoring marsh elevation in a rapidly subsiding salt marsh by thin-layer deposition of dredged material. *Ecological Engineering*, vol. 12, Nos. 3–4, pp. 189–205.

Gu, J., and others (2018). Losses of salt marsh in China: trends, threats and management. *Estuarine, Coastal and Shelf Science*, vol. 214, pp. 98–109.

Huertas, I., and others (2019) Methane emissions from the salt marshes of Doñana Wetlands: spatio-temporal variability and controlling factors. *Frontiers in Ecology and Evolution*. https://doi.org/10.3389/fevo.2019.00032.

Kirwan, M.L., and others (2016). Overestimation of marsh vulnerability to sea level rise. *Nature Climate Change*, vol. 6, No. 3, pp. 253–260.

Macreadie, P.I., and others (2013). Loss of 'Blue Carbon' from Coastal Salt Marshes Following Habitat Disturbance. *PLoS One*, vol. 8, No. 7, e69244. https://doi.org/10.1371/journal.pone.0069244.

Mcowen, C.J., and others (2017). A global map of saltmarshes. *Biodiversity Data Journal*, No. 5, e11764. Paper: https://doi.org/10.3897/BDJ.5.e11764; Data URL: http://data.unep-wcmc.org/datasets/43 (v.6)

McTigue, N., and others (2019) Sea level rise explains changing carbon accumulation rates in a salt marsh over the past two millennia. *Journal of Geophysical Research: Biogeosciences*, vol. 124, No. 10, pp. 2945–2957.

Mendelssohn, I.A., and N.L. Kuhn (2003). Sediment subsidy: effects on soil-plant responses in a rapidly submerging coastal salt marsh. *Ecological Engineering*, vol. 21, Nos. 2–3, pp. 115–128.

Narayan, and others (2017). The value of coastal wetlands for flood damage reduction in the northeastern USA. *Scientific Reports*, vol. 7, No. 1, pp. 1–12.

National Oceanic and Atmospheric Administration (NOAA) (2019). Climate Change: Global Sea Level (www.climate.gov/news-features/understanding-climate/climate-change-global-sea-level).

Nicholls, R.J., and others (1999). Increasing flood risk and wetland losses due to global sea-level rise: regional and global analyses. *Global Environmental Change*, vol. 9, pp. S69–S87.

Oppenheimer, M., and others (in press). Sea Level Rise and Implications for Low-Lying Islands, Coasts and Communities. In *IPCC Special Report on the Ocean and Cryosphere in a Changing Climate*, H-O. Pörtner and others, eds.

Pendleton, L., and others (2012). Estimating global "blue carbon" emissions from conversion and degradation of vegetated coastal ecosystems. *PloS One*, vol. 7, No. 9.

Peteet, D.M., and others (2018). Sediment starvation destroys New York City marshes' resistance to sea level rise. *Proceedings of the National Academy of Sciences*, vol. 115, No. 41, pp. 10281–10286.

Poffenbarger, H., and others (2011). Salinity influence on methane emissions from tidal marshes. *Wetlands*, vol. 31, pp. 831–842. https://doi.org/10.1007/s13157-011-0197-0.

Raposa, K.B., and others (2016). Assessing tidal marsh resilience to sea-level rise at broad geographic scales with multi-metric indices. *Biological Conservation*, vol. 204, pp. 263–275.

Rezaie, A., and others (2020). Valuing natural habitats for enhancing coastal resilience: wetlands reduce property damage from storm surge and sea level rise. *PLoS One*, vol. 15, No. 1, pp. 1–17. https://doi.org/10.1371/journal.pone.0226275.

Ripple, William J., and others (2017). World scientists' warning to humanity: A second notice. *BioScience*, vol. 67, No. 12, pp. 1026–1028.

Rogers, K., and others (2019). Impacts and adaptation options for estuarine vegetation in a large city. *Landscape and Urban Planning*, vol. 182, pp. 1–11.

Rooth, J., and J.C. Stevenson (2000). Sediment deposition patterns in *Phragmites australis* communities: Implications for coastal areas threatened by rising sea-level. *Wetlands Ecology and Management*, vol. 8, Nos. 2–3, pp. 173–183.

Roughan, B.L., and others (2018). Nitrous oxide emissions could reduce the blue carbon value of marshes on eutrophic estuaries. *Environmental Research Letters*, vol. 13, No. 4, 044034.

Saintilan, N., and others (2014). Mangrove expansion and salt marsh decline at mangrove poleward limits. *Global Change Biology*, vol. 20, No. 1, pp. 147–157.

Scheider N., and others (2018). Massive upland to wetland conversion compensated for historical marsh loss in Chesapeake Bay, USA. *Estuaries and Coasts*, vol. 41, pp. 940–951.

Schuerch, M., and others (2018) Future response of global coastal wetlands to sea level rise. *Nature*, vol. 561, pp. 231–234.

Segarra, K., and others (2013). Seasonal variations of methane fluxes from an unvegetated tidal freshwater mudflat (Hammersmith Creek, GA). *Biogeochemistry*, vol. 115, No. 1, pp. 349–61. https://doi.org/10.1007/s10533-013-9840-6.

Short, F.T., and others (2016). Impacts of climate change on submerged and emergent wetland plants. *Aquatic Botany*, vol. 135, pp. 3–17.

Smith, C.S., and others (2016). Living shorelines enhanced the resilience of saltmarshes to Hurricane Matthew (2016). *Ecological Applications*, vol. 28, No. 4, pp. 871–877.

Streb, C., and others (2019). Adapting floating wetland design to advance performance in urban waterfronts. *Wetland Science and Practice*, vol. 36, No. 2, pp. 106–113.

Tabot, P.T., and J.B. Adams (2013). Ecophysiology of salt marsh plants and predicted responses to climate change in South Africa. *Ocean & Coastal Management*, vol. 80, pp. 89–99.

Tian, B., and others (2016). Drivers, trends, and potential impacts of long-term coastal reclamation in China from 1985 to 2010. *Estuarine, Coastal and Shelf Science*, vol. 170, pp. 83–90.

United Nations (2017). *The First Global Integrated Marine Assessment: World Ocean Assessment I*. Cambridge: Cambridge University Press.

United Nations Environment Programme (UNEP) (2019). Biodiversity A-Z. 2019. www.biodiversitya-z.org.

Van Niekerk, L. (2018). Approaches to detecting and assessing patterns, processes and responses to change in South African estuaries. PhD thesis. Port Elizabeth, South Africa: Nelson Mandela University.

Veldkornet, D.A., and others (2015). Where do you draw the line? Determining the transition thresholds between estuarine salt marshes and terrestrial vegetation. *South African Journal of Botany*, vol. 101, pp. 153–159.

Wan, S.W., and others (2009). The positive and negative effects of exotic *Spartina alterniflora* in China. *Ecological Engineering*, vol. 35, pp. 444–452.

Wei, S., and others (2020). Effect of tidal flooding on ecosystem CO_2 and CH4 fluxes in a salt marsh in the Yellow River Delta. *Estuarine Coastal and Shelf Science*, vol. 232, 106512.

Yin, S., and others (2015). *Spartina alterniflora* invasions impact CH4 and N2O fluxes from a salt marsh in eastern China. *Ecological Engineering*, vol. 81, pp. 192–199.

Zhang, R., and others (2004). Formation of *Spartina alterniflora* salt marshes on the coast of Jiangsu Province, China. *Ecological Engineering*, vol. 23, pp. 95–10.

Zuo, P., and others (2012). Distribution of *Spartina* spp. along China's coast. *Ecological Engineering*, vol. 40, pp. 160–166.

Chapter 7J **Continental slopes and submarine canyons**

Contributors: Lisa A. Levin (convener), Peter Auster, Malcolm R. Clark, Jason M. Hall-Spencer, Russell Hopcroft, Jeroen Ingels, Anna Metaxas, Bhavani E. Narayanaswamy, Joshua T. Tuhumwire (subchapter lead member) and Moriaki Yasuhara.

Keynote points

- Continental slopes represent 5.2 per cent of the ocean, with over one fifth of the slope comprised of submarine canyons; they are critical transition areas between the continental shelf and the deep sea and are important for carbon burial and as habitats for species of ecological and economic importance.
- Strong vertical hydrographic gradients, complex geomorphic features and fluid fluxes from the sea floor make canyon and slope faunal communities highly heterogeneous.
- Hundreds of newly discovered methane seep, coral and sponge habitats enhance biodiversity and host novel interactions with surrounding sediments.
- Canyons can be hotspots of biological activity but their communities do not always differ from those on adjacent slopes, which are also highly productive; slope and basin sediments can be an archive of historical information about climate effects on biodiversity.
- Naturally occurring oxygen minimum zones reveal that biodiversity is highly sensitive to oxygenation; expansion of low oxygen zones will reduce biodiversity; projected declines in pH and food supply are likely to affect cold water coral ecosystems.
- Owing to their proximity to shore, slopes and canyons are subject to expansion of deepwater oil and gas activities, offshore energy installations, bottom fisheries and, potentially, mineral mining activities, as well as to increasing contamination, including litter and mine tailings from land.
- Exploration has accelerated the discovery of new ecosystem functions and services, including novel productivity and carbon transfer mechanisms, nursery grounds, and contaminant and waste transfer. However, most canyons and slope areas remain largely unexplored, with major questions about species ranges, ecological connectivity, benthopelagic linkages, sensitivity to climate and direct disturbance remaining unanswered, in particular in the southern hemisphere and along African and South American margins.
- Better integration of climate science, connectivity research, conservation biology and resource management, combined with increased taxonomic and geographic expertise, will improve the distribution of knowledge, technology, analytical tools and methodologies required to advance global understanding and promote sustainability of slope and canyon ecosystems.

1. Introduction

The continental slope represents a deepening of the sea floor out from the shelf edge (about 200 m depth) to the upper limit of the continental rise, where steepness decreases. It covers a total of 19.6 million km^2, representing 5.2 per cent of the ocean (table 1; Harris and others, 2014). That environment was discussed briefly in chapter 36F (United Nations, 2017a) of the first *World Ocean Assessment* (United Nations, 2017c), as a component of deep-sea margins. The continental slope is typically cut by steep-walled canyons (see United Nations, 2017b), with as many as 9,477 known canyons covering nearly 4.4 million km^2 (table 1) and many more that are undiscovered. The slope also encompasses other geomorphic and geochemical features, such as basins, banks, scarps, seamounts and methane seeps (see figure below). Slopes and canyons are major transition areas between shallow and deep waters, transporting (and transforming) sediments, organic matter, water, organisms, contaminants and debris (Puig and others, 2014; Leduc and others, 2018). Continental slopes can be highly productive, accounting for extensive carbon burial and nutrient recycling, and are therefore important in societal well-being (Levin and Sibuet, 2012).

Table 1
Cover and number of slopes and canyons in the world ocean

Feature	Total ocean	Arctic Ocean	Indian Ocean	Mediter-ranean Sea	North Atlantic	North Pacific	South Atlantic	South Pacific	Southern Ocean
Area of slope (km^2)	19 606 260	913 590	4 189 700	906 590	3 436 150	4 752 240	1 591 830	3 201 000	615 170
Percentage of total slope area	100.00	7.03	5.88	30.00	7.68	5.80	3.94	3.67	3.03
Total area of canyons (km^2)	4 393 650	359 650	760 420	163 040	738 430	816 580	291 290	694 790	569 440
Number of canyons	9 477	404	1 590	817	1 548	2 085	453	2 009	571
Percentage of slope that is canyon	22.4	16.1	11.2	13.8	10.4	10.2	8.9	10.2	15.1

Source: Harris and others, 2014.

The strong (usually vertical) gradients in temperature, oxygenation, CO_2, hydrodynamics, particulate fluxes and sediment transport that characterize slopes and canyons shape their biological communities (see figure below). Fluxes of particulate organic carbon and large organic falls (dead marine mammal and fish carcasses, wood and algae) from surface and shelf waters, and geochemical fluxes (methane, sulphide and hydrogen) from within the seabed create significant heterogeneity of energy sources for slope and canyon ecosystems. Additional environmental heterogeneity on slopes at scales from 10 m to hundreds of km derives from variation in sedimentary sources, oceanographic conditions, dynamic geologic processes and frameworks created by habitat-forming species (Kelly and others, 2010).

Slopes and canyons are strongly influenced by climate change and, because they are the closest deep-sea environments to human populations, they experience greater pressures from human activity than other deep-sea systems. Major anthropogenic influences include physical and biological disturbance from fishing, shipping, intentional and unintentional disposal of waste from land (e.g., mine tailings disposal, litter and contaminants), organic matter loading (e.g., sewage, nutrients and industrial inputs), oil and gas activities and potential mineral mining. Climate-related changes leading to warming, oxygen loss and changes in storm intensity and frequency superimpose their effects on those other human disturbances.

Habitat features on continental slopes and canyons, with effects of changing conditions

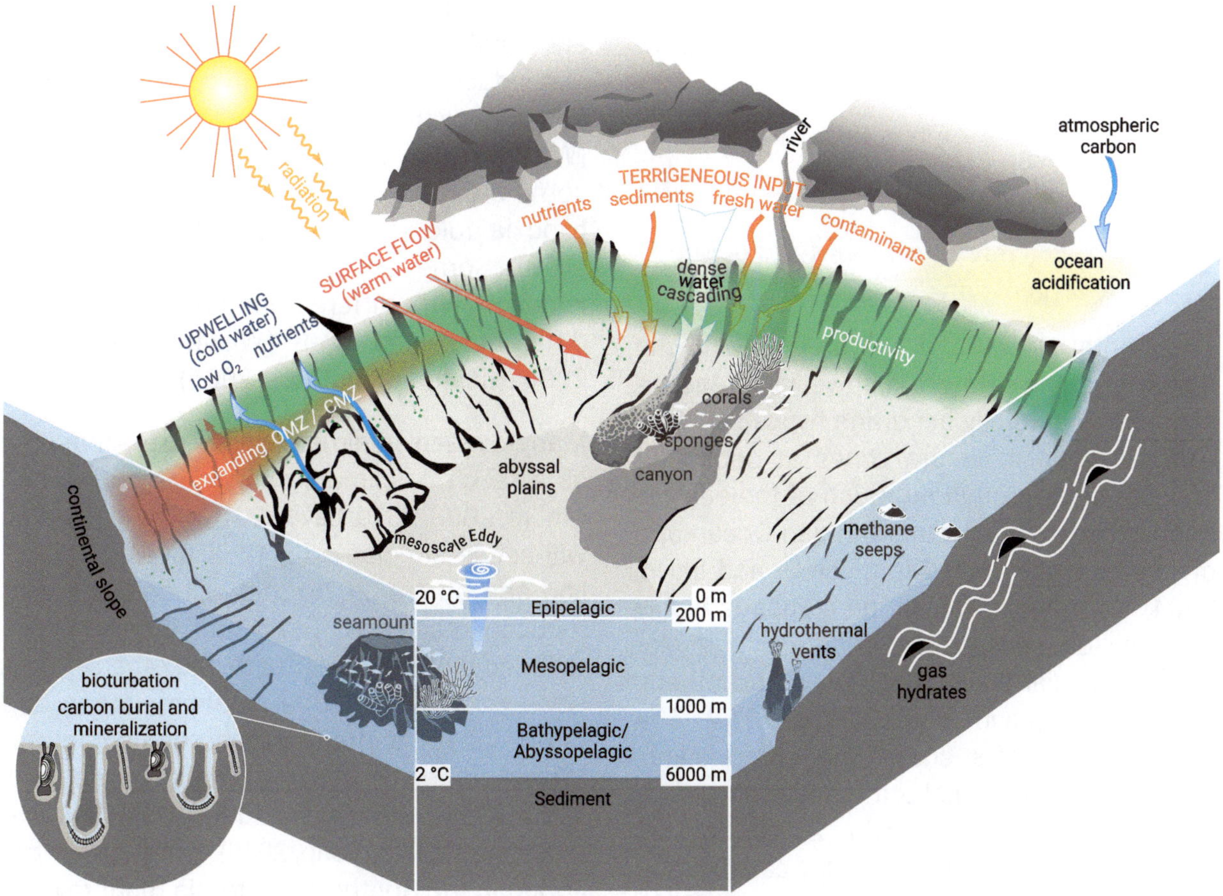

Source: Modified from Levin and Sibuet, 2012, available at https://doi.org/10.1146/annurev-marine-120709-142714, with assistance from Andres Algeria.

Abbreviations: CMZ, carbon minimum zone ; OMZ, oxygen minimum zone.

2. Developments in understanding of slopes and canyons

2.1. Slope biodiversity

Continental margin fauna exhibits strong depth zonation, with major changes in composition at the shelf-slope transition (300–500 m depths), along the upper slope (1,000 m depth) and between 2,000 and 3,000 m (Carney, 2005). Water mass layering across continental slopes contributes to depth zonation and to high beta diversity in macrobenthos (e.g., Narayanaswamy and others, 2010), meiobenthos (Danovaro and others, 2009; Bianchelli and others, 2010), fishes (e.g., Priede and others, 2010) and megafauna (e.g., Hunter and others, 2011), and to depressed alpha diversity within low-oxygen water masses (Sellanes and others, 2010; Gooday and others, 2010). Faunal diversity is generally highest at mid-slope depths (1,500–2,500 m) (Rex and Rowe, 1983; Rex and Etter, 2010; Menot and others, 2010), linked to heterogeneity of sediments, productivity and water flow (Levin and others, 2001). Faunal density maximums on slopes (e.g., crustaceans, brittle stars, sponges) routinely occur at the edges of low-oxygen water masses beneath upwelling areas, on topographic highs and in canyons (Levin, 2003; De Leo and others, 2010; Domke and others, 2017), where increased particle fluxes enhance food supply.

2.2. Changing environmental conditions on slopes and canyons

2.2.1. Oxygen minimum zones and ocean deoxygenation

Continental slopes and canyons beneath the highly productive waters in upwelling areas are exposed to naturally occurring hypoxic waters called oxygen minimum zones at depths of 100–1,200 m throughout much of the eastern Pacific Ocean, the northern Indian Ocean and off the coast of West Africa (Helly and Levin, 2004). Oxygen availability on slopes is highly dynamic on seasonal, inter-annual and glacial-interglacial timescales (Levin and others, 2015a; Huang and others, 2018, 2019). The low-oxygen waters shape the composition, diversity and functional attributes of the water column and benthos (Levin, 2003), as well as individual body size, growth, calcification and reproduction (Sato and others, 2018), and cause strong zonation of megafauna and macrofauna across oxygen gradients of oxygen minimum zones on slopes (Wishner and others, 1995; Levin, 2003; Gooday and others, 2009, 2010; Hunter and others, 2011; Levin and Gallo, 2019) and canyons (De Leo and others, 2012; Domke and others, 2017). Macrofaunal invertebrate diversity (Levin and Gage, 1998; Sperling and others, 2016) and fish diversity (Gallo and others, 2020) drop dramatically at oxygen concentrations below 7 µMol O_2, and patterns of fish abundance, catch and effort can vary with oxygen availability (Bertrand and others, 2011; Keller and others, 2015; Salvatteci and others, 2019; De Leo and others, 2017).

Planktonic communities within oxygen minimum zones also respond strongly to oxygenation in space and time (Ekau and others, 2010; Gilly and others, 2013; Seibel and others, 2016; Tutasi and Escribano, 2020), exhibiting distinct zonation, edge effects, density maximums and sharp diversity thresholds (Wishner and others, 2008, 2013). Very small oxygen gradients can induce large changes in community composition over relatively small spatial scales (Wishner and others, 2008, 2018). However, some crustacean plankters exhibit surprising hypoxia tolerance (Seibel and others, 2016, 2018). Some copepods exhibit ontogenetic zonation, with different life stages thriving at different oxygen levels (Wishner and others, 2000; Hidalgo and others, 2005).

Many oxygen minimum zones are expanding as a result of climate change (Stramma and others, 2008, 2010; Levin, 2018). The expansion will cause anticipated loss of diversity (Sperling and others, 2016) that will coincide with reduced resilience (Levin and others, 2013), reduced bioturbation (Smith and others, 2000; Levin and others, 2009; Schimmelmann and others, 2016), shifts from macrofaunal to protozoan carbon processing (Woulds and others, 2007, 2009) and altered food webs (Sperling and others, 2013; Gallo, 2018). Long-term monitoring in the Southern California Bight reveals declines in abundances of mesopelagic fish larvae (Koslow and others, 2011) and shoaling of the daytime upper and lower boundaries of the deep scattering layer (fish and large invertebrates) (Netburn and Koslow, 2015) in conjunction with shoaling hypoxia over the past 25 years. The vision of squid, octopus and crab larvae in southern California can be impaired by reduced oxygen availability (McCormick and others, 2019). In the south-eastern Pacific, the oxygen minimum zone is highly sensitive to climate variability, with high seasonal to inter-annual oxygen variability generated by the El Niño Southern Oscillation, fluctuations in the Peru-Chile Undercurrent, the Equatorial Undercurrent and eddies (Czeschel and others, 2015; Pizarro-Koch and others, 2018; Espinoza-Morriberón and others, 2019). Multiple proxies suggest that recent oxygenation (recorded since 1999) has occurred, associated with the deepening of the oxygen minimum zone (Graco and others, 2017; Cardich and others, 2019).

2.2.2. Ocean acidification

Slopes and canyons are increasingly vulnerable to ocean acidification. Under Representative Concentration Pathway (RCP) 8.5 (a business-as-usual climate scenario), the average pH is expected to decline by 0.14 units on slopes and 0.11 units in canyons by 2100 (table 2; Food and Agriculture Organization of the United Nations (FAO), 2019; Bindoff and others, 2019). The North Atlantic Ocean is particularly vulnerable because deepwater formation propagates surface-derived changes in carbonate chemistry to the ocean interior, and the western boundary current advects them further away; RCP 8.5 projects pH declines of 0.3 units for 14 per cent of the slope below 500 m and 15 per cent of canyons by 2100 (Gehlen and others, 2014). Taxa living where low oxygen and high CO_2 occur naturally may be less vulnerable to the impacts of ocean acidification, but ocean basins differ. In the Indian Ocean, macrofaunal biodiversity is more influenced by elevated CO_2 than lowered oxygen, whereas the reverse is true in the East Pacific Ocean, where low oxygen better explains biodiversity trends (Taylor and others, 2014; Sperling and others, 2016; Sato and others, 2018). It is necessary to strengthen monitoring systems, such as the Global Ocean Acidification Observing Network, to evaluate changes in the seawater carbonate system on slopes and canyons.

Table 2

Projected climate changes given as mean (minimum, maximum) at the deep sea floor for continental slopes, canyons and cold water corals mapped from 200 m to 2,500 m under RCP 8.5 and 2.6 from the present to the period 2081–2100 using three Earth-system models

	Temperature (°C)	pH	Dissolved oxygen (µMol/kg)	POC flux
	RCP 2.6	RCP 2.6	RCP 2.6	RCP 2.6
Continental slopes	+0.30 (-0.44, +2.30)	-0.06 (-0.19, -0.02)	-3.1 (-49.3, +61.7)	-0.39 (-16.0, +3.9)
Canyons	+0.31 (-0.27, +1.76)	-0.05 (-0.13, +0.01)	-3.54 (-44.66, +29.30)	-0.33 (-10.53, +3.53)
Cold water corals	+4.3 (-0.29, +1.85)	-0.07 (-0.13, 0.0)	-3.5 (-25.6, +24.7)	-0.7 (-10.5, +3.4)
	RCP 8.5	RCP 8.5	RCP 8.5	RCP 8.5
Continental slopes	+0.75 (-8.4, +4.4)	-0.14 (-0.02, -0.44)	-10.2 (-67.8, +53.82)	-0.66 (-33.33, +10.3)
Canyons	+0.19 (-0.03, +1.14)	-0.11 (-0.35, +0.02)	-0.80 (-28.76, +10.07)	-0.80 (-28.76, +10.07)
Cold water corals	+0.96 (-0.42, +3.84)	-0.15 (-0.39, +0.001)	-10.6 (-59.2, +11.1)	-1.69 (-20.1, +4.6)

Source: Adapted from Table 5.5 from Bindoff, N.L., and others, 2019.

Abbreviations: POC, particulate organic carbon; RCP, Representative Concentration Pathway.

2.2.3. Food supply

Food supply to slope and canyon ecosystems derives largely from the flux of organic matter from ocean surface waters. Slopes and canyons are projected to experience reductions in particulate organic carbon flux by the period 2081–2100 under a range of emissions scenarios (except in the Southern Ocean and the Arctic Ocean) with concomitant reductions in benthic biomass (Jones and others, 2014; Yool and others, 2017; Intergovernmental Panel on Climate Change (IPCC), 2019); however, the declines in particulate organic carbon flux are projected to be 30–50 per cent less under RCP 2.6 than under RCP 8.5 (IPCC, 2019; table 2). The overall contributions of chemosynthetic production to slope and canyon food webs has yet to be quantified, but future warming-induced dissociation of methane from buried gas hydrates (Biastoch and others, 2011) could increase the contributions.

2.3. Continental slopes as a unique palaeoecology archive

The continental slope serves as a unique setting, critical to understanding historical deep-sea biodiversity dynamics. Areas experiencing very high sedimentation rates allow for the reconstruction of past oceanographic conditions and biodiversity response at decadal-centennial timescales for the past tens of thousands of years based on ostracods (Yasuhara and Cronin, 2008; Yasuhara and others, 2017; Yasuhara, 2018). For example, in the North Atlantic Ocean, abrupt temperature change has affected deep-sea benthic abundance and biodiversity based on palaeoecological records from the past 20,000 years (Yasuhara and others, 2008, 2014, 2016; Yasuhara and Danovaro, 2016).

2.4. Habitat heterogeneity

2.4.1. Slope-canyon comparisons

Canyons are considered to be a key source of heterogeneity and biodiversity. The prokaryotic and eukaryotic microbial plankton communities appear similar in Mediterranean canyon and slope settings (Celussi and others, 2018; Diociaiuti and others, 2019), although more viruses and viral infections were documented within the Bisagno Canyon than on the adjacent slope (Corinaldesi and others, 2019a, 2019b). A strong link between mesoscale processes, and the cascading of dense water, in particular, can influence the biogeochemistry (Chiggiato and others, 2016), microbes (Luna and others, 2016), organic matter deposition, microbial production and viral activity (Rastelli and others, 2018) of canyons, and may be particularly critical for supporting deepwater coral habitat (Taviani and others, 2019). Recent comparisons suggest no significant differences in biomass, density or composition in foraminifera (Di Bella and others, 2019), metazoan meiofauna (Bianchelli and others, 2010; Bianchelli and Danovaro, 2019; Carugati and others, 2019) and macrofauna (Harriague and others, 2019) between slopes and canyons of the Mediterranean. In contrast, higher densities of deposit feeders (sipunculids and holothurian species) and meiofauna occur within New Zealand canyons (700–1,500 m depth) rather than on the adjacent slope (Rowden and others, 2016; Rosli and others, 2016), possibly owing to differences in topographic complexity and higher organic matter availability (Leduc and others, 2014, 2016; Rowden and others, 2016). High heterogeneity also promotes enhanced diversity locally and regionally within the Mediterranean canyons (Gambi and others, 2019; Bianchelli and Danovaro, 2019; Carugati and others, 2019) and the Northeast Atlantic (Ingels and Vanreusel, 2013; Ingels and others, 2011), with high species turnover between canyons (Harriague and others, 2019). The presence of deepwater ahermatypic corals (scleractinian and octocorals) at bathyal depths enhances the density and influences the composition and diversity of adjacent sediment communities in the Gulf of Mexico (Demopoulos and others, 2014, 2016; Bourque and Demopoulos, 2018), with different corals associated with different infaunal communities, possibly mediated by habitat differences. The presence of deepwater corals decouples normal depth-density and diversity patterns in the region (Wei and others, 2010). Disruption of coral habitats will therefore likely affect nearby slope infauna (Bourque and Demopoulos, 2018). Overall, variability between habitats in benthic species composition and abundance by region (Bowden and others, 2016; Leduc and others, 2016) can limit the ability to draw general conclusions on the differences between canyons and slopes. Zooplankton and, in particular, krill are often observed in higher abundances over the shelf break and slope (Lu and others, 2003; Lowe and others, 2018). A variety of processes can lead to such aggregations of zooplankton and fish (Genin, 2004), and it appears that canyons may also contribute greatly to krill hotspots (Santora and others, 2018), although their impact on smaller zooplankton is poorly documented.

2.4.2. Geomorphic heterogeneity

Finer-scale geomorphological aspects of slopes and canyons, including water depth, sediment type, acoustic backscatter, wave exposure and seabed rugosity, can be used to identify slopes and canyons as habitats and predict benthic communities in the absence of sampling (Harris and Baker, 2020; Kenchington and others, 2014; Pierdomenico and others, 2015, 2019; Fanelli and others, 2018; Huang, Zhi, and others, 2018). Geomorphic features, shaped by erosion, sediment transport, deposition and tectonic instability (Lastras and others, 2008), as well as by biology (Marsh and others, 2018; Lo Iacono and others, 2019), can now be mapped effectively using remotely operated vehicles and autonomous underwater vehicles (Huvenne and others, 2018). Geomorphic landscape features may underpin spatial planning, marine protected area design, research planning and economic resource assessment (Harris and Baker, 2020; Ismail and others, 2015; Hogg and others, 2016), emphasizing the relevance of ongoing efforts to map the entire deep-sea floor (Mayer and others, 2018). The interaction of large bathymetric or tectonic features with bottom currents can lead to the exposure or deposition of mineral hard grounds, crusts and nodules, including those formed of ferromanganese and phosphorites (Muiños and others, 2013), and can cause slope instability (Teixeira and others, 2019).

2.4.3. Geochemical heterogeneity

The biodiversity of slope and canyon ecosystems is influenced by sea floor seepage of methane and other hydrocarbon-rich fluids (Levin, 2005; Egger and others, 2018). Methane seeps host distinct megafaunal communities and are dominated by chemoautotrophic fauna (see chap. 7P). The recent advent of acoustic bubble plume detection methods has revealed the ubiquitous nature and high abundance of seeps (Riedel and others, 2018; Skarke and others, 2014). Ocean warming and altered circulation, which may promote degassing, may already be increasing the number of slope seepage sites (Phrampus and Hornbach, 2012; Johnson and others, 2015). New explorations reveal seep influence on the background slope and canyon communities (Levin and others, 2016a) through the provision of chemosynthetic food sources (Seabrook and others, 2019; Rathburn and others 2009; Goffredi and others, 2020), nursery habitat (Treude and others, 2011; Sen and others, 2019) and hardground (carbonate) as substrate (Levin and others, 2015b, 2017), and by stimulating water column production (D'souza and others, 2016).

2.5. Connectivity of populations

Fragmented populations, communities and ecosystems can remain viable or recover from disturbance through ecological connectivity, defined as the exchange of individuals, species or resources. On continental slopes and canyons, heterogeneously distributed hard substrate supports deepwater corals and sponges that are vulnerable to disturbance from fishing and exhibit life-history characteristics that are not conducive to population resilience or recovery, such as longevity, slow growth and recruitment (Reed and others, 2007; Huvenne and others, 2016; Bennecke and Metaxas, 2017). Understanding spatial variation in reproductive potential (Fountain and others, 2019) and the use of hydrodynamic models to assess patterns of connectivity can assist in developing effective conservation strategies (Kool and others, 2013, 2015; Metaxas and others, 2019). Recent genetic studies have advanced the understanding of dispersal distances and source-sink dynamics, which vary among cold water coral and sponge species in slope environments at the regional to geomorphic feature level (Zeng and others, 2017, 2019; Holland and others, 2019), and regional and local currents can act as routes or barriers for larval dispersal (Dueñas and others, 2016; Holland and others, 2019; Zeng and others, 2019).

3. Ecosystem services and benefits on slopes and in canyons

Ecosystem services provided by slopes and canyons include carbon sequestration and nutrient recycling, fisheries, biodiversity support and waste disposal, with emerging interest in mining of non-renewable resources (Fernandez-Arcaya and others, 2017).

3.1. Fisheries

Numerous deepwater fisheries rely on outer shelves and bathyal slopes, even within some oxygen minimum zones (Keller and others, 2015). Canyons serve as key feeding, spawning and recruitment grounds for economically valuable fishes (D'Onghia and others, 2015) and shellfish (Sardà and others, 2009). Fishes often, although not always (Ross and others, 2015), occur in greater abundance, are larger and have faster rates of maturity in canyons, as has been shown for sharks, conger, hake and common pandora (Sion and others, 2019). The discovery of close associations of some commercial fish and shellfish species with canyon and slope methane seeps (Sellanes and others, 2008; Bowden and others, 2013; Grupe and others, 2015; Seabrook and others, 2019) suggests the possible contribution of chemosynthetic ecosystems to continental margin fisheries (Levin and others, 2016a), and it led the United States Pacific Fishery Management Council to designate methane seeps as an essential fish habitat for Pacific coast groundfish (Pacific Coast Groundfish Fishery Management Plan, amendment 28).[1]

3.2. Supporting and regulating services

The exploration of slopes and canyons accelerates the ongoing discovery of new functions and services, such as the emerging role of demersal and deepwater fish on continental slopes in transferring carbon from the deep scattering layer to greater depths in the ocean (Trueman and others, 2014; Gallo, 2018; Vieira and others, 2019). Nursery support functions have been found on slopes off the coast of Costa Rica for octopus at 3,000 m and fish eggs attached within xenophyophores (giant protozoans) (Levin and Rouse, 2019), for elasmobranch egg cases associated with methane seeps on slopes off Chile and the Mediterranean (Treude

1 Available at www.pcouncil.org/groundfish/fishery-management-plan.

and others, 2011) and in gorgonian coral fields (Etnoyer and Warrenchuk, 2007). Physical processes within canyons contribute to the upwelling of nutrients to the shelf and the offshore transport of shelf productivity to deeper waters (Fernandez-Arcaya and others, 2017). Other canyon processes remove and bury contaminants and wastes and support biodiversity by providing refugia from fishing pressure.

3.3. Energy

Oil and gas exploitation has expanded to water depths of more than 3,000 m on continental slopes in the Gulf of Mexico, off the coasts of Angola and Brazil, and elsewhere (Merrie and others, 2014). Canyons accumulate organic matter and are increasingly targeted for hydrocarbon extraction. For example, 24 per cent of Australian canyons occur within oil and gas leases (Fernandez-Arcaya and others, 2017). For some countries, oil and gas represent an important source of income. However, environmental impacts result from exploration, routine operations and hydrocarbon spills (Cordes and others, 2016).

Although still in the early stages of development, offshore renewable energy in the form of wind infrastructure may eventually use floating structures over waters up to 1,000 m deep (Bosch and others, 2018).

3.4. Natural products

There is growing interest in prospecting for bioactive compounds the deep sea, although as of 2016, less than 3 per cent of known marine metabolites were derived from organisms in cold water (Soldatou and Baker, 2017). Bacteria and fungi from deep-sea sediments on continental slopes have been revealed to be a rich source of compounds with antibacterial, antifungal, anticancer and cytotoxicity properties (Skropeta and Wei, 2014). Invertebrates, in particular octocorals and demosponges, which are common in canyons, are also targets for biodiscovery (Winder and others, 2011; Leal and others, 2012; Blunt and others, 2013; Fernandez-Arcaya and others, 2017), as deep-sea sponge metabolites have antitumour properties (Wright and others, 2017).

4. Human impacts

A recent review identified four major categories of human impacts on canyons: bottom contact fisheries; oil and gas exploration and exploitation; contaminants, litter and mine tailings from land; and climate stressors (Fernandez-Arcaya and others, 2017). Those same activities affect continental slopes, as does the potential mining of minerals (e.g., sand, phosphorites) and gas hydrates.

Commercial fisheries on continental slopes and in canyons, as covered in the first Assessment, remain a major source of direct disturbance to deep-sea benthic communities (Pusceddu and others, 2014; Clark and others, 2016). Bottom trawling causes considerable modification of the sea floor, increasing suspended sediment concentration (Daly and others, 2018; Paradis and others, 2018a), changing sediment distribution and properties (Martín and others, 2014a, 2014b; Paradis and others, 2018b) and acting as a cumulative stressor in ecosystems under oxygen stress (De Leo and others, 2017; Levin and Gallo, 2019). Fishing activities produce litter and debris through lost lines, nets and pots (e.g., Pham and others, 2014; Maldonado and others, 2015; Quattrini and others, 2015; Vieira and others, 2015; Tubau and others, 2015; Woodall and others, 2015; Lastras and others, 2016; Cau and others, 2017; Giusti and others, 2019), which entangle or physically damage a variety of marine species, including cold water coral (Aymà and others, 2019). Invasive species can spread through attachment or associations with such debris and litter, which is a further concern.

Contaminants, sediments, detrital organic matter, plastics and other marine debris readily move from shelf waters into canyons (Salvadó and others, 2017, 2019; Tamburrino and others, 2019) and the deep sea (Puig and others, 2014; Leduc and others, 2018). Toxic metal (e.g., cadmium) accumulation in sediments promotes microbial metal tolerance (Papale and others, 2018). Plastic litter (e.g., wrappers, bags, bottles) is ubiquitous on the sea floor on the continental slope throughout the world, in particular under well-travelled ship routes (Gerigny and others, 2019; Mecho and others, 2020). Microplastics, which transport adsorbed persistent organic pollutants, have been found in animals sampled from deep slopes and canyons (Woodall and others, 2014; Taylor and others, 2016; Courtene-Jones and others, 2017, 2019). Disposal of mine tailings produced as fine particle waste after extracting metals from ore on land can introduce such metals as arsenic, cobalt, nickel, mercury, lead and zinc and processing wastes (sodium cyanide, lime) to slopes and canyons (Reichelt-Brushett, 2012; Ramirez-Llodra and others, 2015). In 2015 alone, seven countries piped tailings from 16 mines into the ocean (Vare and others, 2018). Tailings may cause faunal mortality through direct smothering or poisoning or through altered species interactions, as well as bioaccumulation.

The emergence of the potential mining of slope areas to exploit phosphate resources (off the coasts of Mexico, Namibia, New Zealand and South Africa) and sea floor massive sulphides in seamount/pinnacle or back-arc settings represent additional threats to slope environments (Levin and others, 2016b). Gas hydrates (frozen methane) buried on continental margins attract significant exploitation interest (Chong and others, 2016). Gas hydrate exploitation and release may generate environmental impacts on continental slopes and in canyons that resemble or exceed those documented for traditional oil and gas exploitation in deep water (Cordes and others, 2016; Olsen and others, 2016). The physical instability of slopes and canyons is an important issue to consider in the management of human activities in such habitats.

5. Key remaining knowledge gaps

Most canyons and slopes remain uninvestigated, in particular in the southern hemisphere and on the margins of developing countries. Half of all relevant publications focus on only 11 canyons globally (Matos and others, 2018). Several knowledge gaps were identified in the first Assessment, but they still exist, for the most part. They include the characterization of small-taxon biodiversity on hard substrates (such as in canyons) that are difficult to sample. Furthermore, species ranges, connectivity patterns and long-term trends in resilience and sensitivity to natural, climate and other anthropogenic disturbances remain poorly known for many slope environments around the world. The greatest climate-induced changes in the deep-sea environment are expected at bathyal depths coinciding with extensive areas that support productive fisheries or high biodiversity (Sweetman and others, 2017; see table 2).

Current conservation of canyon and slope ecosystems generally depends on the use of physiographic, geomorphological and oceanographic proxies and species-community inventories to locate vulnerable resources for planning and management (e.g., Van den Beld and others, 2017; Auster and others, 2020). Such knowledge allows the use of species distribution models (e.g., for deep-sea corals and sponges, as in Ross and others, 2019; Kinlan and others, 2020; Pearman and others, 2020; Morato and others, 2020) to inform policy about the geospatial extent of conservation targets and could guide such programmes as Natura 2000, which provides a mechanism to include deep-sea areas of canyons and slopes under protective management (Serrano and others, 2017; Van den Beld and others, 2017).

For the sustainable management of slope and canyon ecosystems and resource use, key scientific questions need to be addressed, including the major influences on the connectivity of populations and their ability to recover from disturbance; the roles of source-sink dynamics, niche specialization and species interactions in structuring diversity; whether extreme conditions (low oxygen and pH, high hydrogen sulphide, low carbonate saturation states) and highly developed mutualistic or facilitative relationships (e.g., symbioses, commensalism) change species assembly rules, adaptability or diversity-function relationships; and whether there are suitable indicator taxa or assemblages for ecosystem health that can be used as proxies (see Levin and Sibuet, 2012, supplemental appendix). Such information will inform the designation of ecologically important or vulnerable habitats, such as ecologically or biologically significant areas (Secretariat of the Convention on Biological Diversity, 2009) and vulnerable marine ecosystems (FAO, 2009). Other key management questions include: (a) the sustainability of deep-sea tailings disposal and its environmental impacts relative to those on land; (b) how to incorporate hundreds of newly discovered locations of seeps into the management of human activities; and (c) how benthopelagic coupling and carbon transfer carried out by demersal fishes (that feed on migrating plankton) will be affected by fishing and by climate-induced changes in surface production and phytoplankton composition, oxygenation and acidification.

Challenges in addressing those knowledge gaps result from undersampling associated with the remoteness, vastness and heterogeneity of the environments. Some 66 per cent of the continental slope seabed bathymetry from 200 to 1,000 m and 72 per cent from 1,000 to 3,000 m remain unmapped (Mayer and others, 2018). An even larger area of the sea floor has never been surveyed for biology, including significant portions of the African and South American margins. Often, pre-exploration assessments for the oil and gas industry provide the first characterizations of deep margins (Pabis and others, 2019). Time series (or continuous) observations are needed on slopes and canyons to characterize natural variability and response to climate change and to evaluate sensitivity to the impacts of human activities, which will require collaborations across sectors and jurisdictions (Evans and others, 2019; Garçon and others 2019; Levin and others, 2019; Vieira and others, 2019). Accelerated transfer of knowledge and technology as well as science infrastructure in developing countries can go a long way towards filling the gaps, as outlined in section 6 below.

6. Key remaining capacity-building gaps

For most of the deep ocean, a lack of taxonomic expertise is a major roadblock in advancing studies on biodiversity (Fontaine and others, 2012; Horton and others, 2017). Some researchers increasingly favour using DNA as an alternative tool to morphology-based taxonomy (Sinniger and others, 2016), while others argue that there is a need for naming species in order to support marine conservation and the development of ocean-based industry (Horton and others, 2017; Glover and others, 2018). Geographic bias in slope and canyon research, towards the exclusive economic zones of developed States bordering the North Atlantic Ocean and the North Pacific Ocean, and around Oceania, reflects the reality of access, financial resources and the interests of the industries involved in resource extraction. That leads to limited global understanding of biodiversity patterns and drivers and has consequences for the distribution of expertise, which, for slopes and canyons, is based in developed regions, as well as in China, India and, to a lesser extent, Brazil and Chile. It has also resulted in the uneven distribution of the technology, analytical tools and methodologies required to advance a global understanding of slope and canyon ecosystems.

Some solutions may be sought through the expanded engagement of developing country scientists in offshore observing programmes (e.g., the Array for Real-time Geostrophic Oceanography (Argo), the Global Ocean Ship-based Hydrographic Investigations Program (GO-SHIP), the International Ocean Sustained Interdisciplinary Timeseries Environment Observation System programme (OceanSITES)), observing networks (e.g., Global Ocean Acidification Observing Network, Global Ocean Oxygen Network) and scientific networks (Deep Ocean Observing Strategy, Deep Ocean Stewardship Initiative, International Network for Submarine Canyon Investigation and Scientific Exchange). That goal can be achieved, in part, through training courses, cruise opportunities, synthesis workshops or steering committee membership, but personal mentoring that provides scientific support and financial resources are critical elements. The United Nations Decade of Ocean Science for Sustainable Development (2021–2030) could provide the catalyst for bridging such capacity gaps.

Slopes and canyons represent a large source of deep-sea biodiversity in part owing to high geomorphic, geochemical and environmental heterogeneity. That biodiversity is still in a discovery phase and largely unprotected, but it is increasingly vulnerable to the confluence of changing climate and growing human extractive activity, contamination and waste disposal on continental margins. Improved ocean observation, biodiversity characterizations, taxonomic knowledge and technology transfer are needed, in particular in the southern hemisphere.

References

Auster, P.J., and others (2020). A Scientific Basis for Designation of the Northeast Canyons and Seamounts Marine National Monument. *Frontiers in Marine Science*, vol. 7, article 566, https://doi.org/10.3389/fmars.2020.00566.

Aymà, Anna, and others (2019). Occurrence of Living Cold-Water Corals at Large Depths Within Submarine Canyons of the Northwestern Mediterranean Sea. In *Mediterranean Cold-Water Corals: Past, Present and Future*, pp. 271–284. Springer.

Bennecke, Swaantje, and Anna Metaxas (2017). Effectiveness of a deep-water coral conservation area: evaluation of its boundaries and changes in octocoral communities over 13 years. *Deep Sea Research Part II: Topical Studies in Oceanography*, vol. 137, pp. 420–435.

Bertrand A., and others (2011) Oxygen: a fundamental property regulating pelagic ecosystem structure in the coastal southeastern tropical Pacific. *PLoS ONE*, vol. 6, No. 12, article e29558. https://doi.org/10.1371/journal.pone.0029558.

Bianchelli, S., and others (2010). Metazoan meiofauna in deep-sea canyons and adjacent open slopes: a large-scale comparison with focus on the rare taxa. *Deep Sea Research Part I: Oceanographic Research Papers*, vol. 57, No. 3, pp. 420–433.

Bianchelli, Silvia, and Roberto Danovaro (2019). Meiofaunal biodiversity in submarine canyons of the Mediterranean Sea: a meta-analysis. *Progress in Oceanography*, vol. 170, pp. 69–80.

Biastoch, A., and others. (2011). Rising Arctic Ocean temperatures cause gas hydrate destabilization and ocean acidification. *Geophysical Research Letters*, vol. 38, L08602, https://doi.org/10.1029/2011GL047222, 2011.

Bindoff, N.L., and others (2019). *Changing Ocean, Marine Ecosystems, and Dependent Communities*. IPCC Special Report on the Ocean and Cryosphere in a Changing Climate, H-O. Pörtner and others, eds.

Blunt, J., and others (2013). Natural product reports. *Natural Product Reports*, vol. 39, pp. 237–323. https://doi.org/10.1039/C2NP20112G.

Bosch, Jonathan, and others (2018). Temporally explicit and spatially resolved global offshore wind energy potentials. *Energy*, vol. 163, pp. 766–781.

Bourque, Jill R., and Amanda W.J. Demopoulos (2018). The influence of different deep-sea coral habitats on sediment macrofaunal community structure and function. *PeerJ*, vol. 6, e5276.

Bowden, David A., and others (2013). Cold seep epifaunal communities on the Hikurangi Margin, New Zealand: composition, succession, and vulnerability to human activities. *PLoS One*, vol. 8, No. 10, e76869.

Bowden, David A., and others (2016). Deep-sea seabed habitats: Do they support distinct mega-epifaunal communities that have different vulnerabilities to anthropogenic disturbance? *Deep Sea Research Part I: Oceanographic Research Papers*, vol. 107, pp. 31–47.

Cardich, J., and others (2019). Multidecadal changes in marine subsurface oxygenation off Central Peru during the last ca. 170 years. *Frontiers in Marine Science*, vol. 6, article 270, https://doi.org/10.3389/fmars.2019.00270.

Carney, Robert S. (2005). Zonation of deep biota on continental margins. In *Oceanography and Marine Biology*, pp. 221–288. CRC Press.

Carugati, L., and others (2019). Patterns and drivers of meiofaunal assemblages in the canyons Polcevera and Bisagno of the Ligurian Sea (NW Mediterranean Sea). *Progress in Oceanography*, vol. 175, pp. 81–91.

Cau, Alessandro, and others (2017). Submarine canyons along the upper Sardinian slope (Central Western Mediterranean) as repositories for derelict fishing gears. *Marine Pollution Bulletin*, vol. 123, Nos. 1–2, pp. 357–364.

Celussi, Mauro, and others (2018). Planktonic prokaryote and protist communities in a submarine canyon system in the Ligurian Sea (NW Mediterranean). *Progress in Oceanography*, vol. 168, pp. 210–221.

Chiggiato, Jacopo, and others (2016). Dense-water bottom currents in the Southern Adriatic Sea in spring 2012. *Marine Geology*, vol. 375, pp. 134–145.

Chong, Rong Zheng, and others (2016). Review of natural gas hydrates as an energy resource: Prospects and challenges. *Applied Energy*, vol. 162, pp. 1633–1652

Clark, Malcolm R., and others (2016). The impacts of deep-sea fisheries on benthic communities: a review. *ICES Journal of Marine Science*, vol. 73, No. suppl. 1, pp. i51–i69.

Cordes, Erik E., and others (2016). Environmental impacts of the deep-water oil and gas industry: a review to guide management strategies. *Frontiers in Environmental Science*, vol. 4, art. 58.

Corinaldesi, Cinzia, and others (2019a). High diversity of benthic bacterial and archaeal assemblages in deep-Mediterranean canyons and adjacent slopes. *Progress in Oceanography*, vol. 171, pp. 154–161.

Corinaldesi, Cinzia, and others (2019b). High rates of viral lysis stimulate prokaryotic turnover and C recycling in bathypelagic waters of a Ligurian canyon (Mediterranean Sea). *Progress in Oceanography*, vol. 171, pp. 70–75.

Courtene-Jones, Winnie, and others (2017). Microplastic pollution identified in deep-sea water and ingested by benthic invertebrates in the Rockall Trough, North Atlantic Ocean. *Environmental Pollution*, vol. 231, pp. 271–280.

Courtene-Jones, Winnie, and others (2019). Consistent microplastic ingestion by deep-sea invertebrates over the last four decades (1976–2015), a study from the North East Atlantic. *Environmental Pollution*, vol. 244, pp. 503–512.

Czeschel, Rena, and others (2015). Circulation, eddies, oxygen and nutrient changes in the eastern tropical South Pacific Ocean. *Ocean Science*, vol. 11, pp. 455–470, https://doi.org/10.5194/os-11-455-2015.

Daly, Eoghan, and others (2018). Bottom trawling at Whittard Canyon: evidence for seabed modification, trawl plumes and food source heterogeneity. *Progress in Oceanography.*

Danovaro, R., and others (2009). α-, β-, γ-, δ-and ε-diversity of deep-sea nematodes in canyons and open slopes of Northeast Atlantic and Mediterranean margins. *Marine Ecology Progress Series*, vol. 396, pp. 197–209.

De Leo, Fabio C., and others (2010). Submarine canyons: hotspots of benthic biomass and productivity in the deep sea. *Proceedings of the Royal Society B: Biological Sciences*, vol. 277, No. 1695, pp. 2783–2792.

De Leo, Fabio C., and others (2012). The effects of submarine canyons and the oxygen minimum zone on deep-sea fish assemblages off Hawai'i. *Deep Sea Research Part I: Oceanographic Research Papers*, vol. 64, pp. 54–70.

De Leo, Fabio C., and others (2017). Bottom trawling and oxygen minimum zone influences on continental slope benthic community structure off Vancouver Island (NE Pacific). *Deep Sea Research Part II: Topical Studies in Oceanography*, vol. 137, pp. 404–419.

Demopoulos, Amanda W.J., and others (2016). Impacts of the Deepwater Horizon oil spill on deep-sea coral-associated sediment communities. *Marine Ecology Progress Series*, vol. 561, pp. 51–68.

Demopoulos, Amanda W.J., and others (2014). Biodiversity and community composition of sediment macrofauna associated with deep-sea *Lophelia pertusa* habitats in the Gulf of Mexico. *Deep Sea Research Part I: Oceanographic Research Papers*, vol. 93, pp. 91–103.

Di Bella, L., and others (2019). Living foraminiferal assemblages in two submarine canyons (Polcevera and Bisagno) of the Ligurian basin (Mediterranean Sea). *Progress in Oceanography*, vol. 173, pp. 114–133.

Diociaiuti, Tommaso, and others (2019). Vertical distribution of microbial communities abundance and biomass in two NW Mediterranean Sea submarine canyons. *Progress in Oceanography*, vol. 175, pp. 14–23.

Domke, Lia, and others (2017). Influence of an oxygen minimum zone and macroalgal enrichment on benthic megafaunal community composition in a NE Pacific submarine canyon. *Marine Ecology*, vol. 38, No. 6, pe12481.

D'Onghia, Gianfranco, and others (2015). Exploring composition and behaviour of fish fauna by in situ observations in the Bari Canyon (Southern Adriatic Sea, Central Mediterranean). *Marine Ecology*, vol. 36, No. 3, pp. 541–556.

D'souza, N.A., and others (2016). Elevated surface chlorophyll associated with natural oil seeps in the Gulf of Mexico. *Nature Geoscience*, vol. 9, No. 3, p. 215.

Dueñas, Luisa F., and others (2016). The Antarctic Circumpolar Current as a diversification trigger for deep-sea octocorals. *BMC Evolutionary Biology*, vol. 16, No. 1, art. 2.

Egger, Matthias, and others (2018). Global diffusive fluxes of methane in marine sediments. *Nature Geoscience*, vol. 11, No. 6, p. 421.

Ekau, Werner, and others (2010). Impacts of hypoxia on the structure and processes in pelagic communities (zooplankton, macro-invertebrates and fish). *Biogeosciences*, vol. 7, No. 5, pp. 1669–1699.

Espinoza-Morriberón, D., and others. (2017). Impacts of El Niño events on the Peruvian upwelling system productivity. *Journal of Geophysical Research Oceans*, vol. 122, pp. 5423–5444, https://doi.org/10.1002/2016JC012439.

Etnoyer, Peter, and Jon Warrenchuk (2007). A catshark nursery in a deep gorgonian field in the Mississippi Canyon, Gulf of Mexico. *Bulletin of Marine Science*, vol. 81, No. 3, pp. 553–559.

Evans, Karen, and others (2019). The Global Integrated World Ocean Assessment: Linking Observations to Science and Policy Across Multiple Scales. *Frontiers in Marine Science*, vol. 6, art. 298.

Fanelli, Emanuela, and others (2018). Deep-sea mobile megafauna of Mediterranean submarine canyons and open slopes: analysis of spatial and bathymetric gradients. *Progress in Oceanography*, vol. 168, pp. 23–34.

Food and Agriculture Organization of the United Nations (FAO) (2009). *International Guidelines for the Management of Deep-Sea Fisheries in the High-Seas*. Rome.

________ (2019). *Deep-Ocean Climate Change Impacts on Habitat, Fish and Fisheries*. Fisheries and Aquaculture Technical Paper 638. Rome.

Fernandez-Arcaya, Ulla, and others (2017). Ecological role of submarine canyons and need for canyon conservation: a review. *Frontiers in Marine Science*, vol. 4, art. 5.

Fontaine, Benoît, and others (2012). 21 years of shelf life between discovery and description of new species. *Current Biology*, vol. 22, No. 22, pp. R943–R944.

Fountain, Christopher Tyler, and others (2019). Individual and Population Level Variation in the Reproductive Potential of Deep-Sea Corals From Different Regions Within the Gulf of Maine. *Frontiers in Marine Science*, vol. 6, art. 172.

Gallo, Natalya D. (2018). Influence of ocean deoxygenation on demersal fish communities: lessons from upwelling margins and oxygen minimum zones. PhD Thesis, UC San Diego.

Gallo, Natalya D., and others (2020). Dissolved oxygen and temperature best predict of deep-sea fish community structure in the Gulf of California with implications for climate change. In *Marine Ecology Progress Series*, vol. 637, pp.159–180.

Gambi, Cristina, and others (2019). Biodiversity and distribution of meiofauna in the Gioia, Petrace and Dohrn Canyons (Tyrrhenian Sea). *Progress in Oceanography*, vol. 171, pp. 162–174.

Garçon, Véronique, and others (2019). Multidisciplinary Observing in the World Ocean's Oxygen Minimum Zone regions: from climate to fish – the VOICE initiative. *Frontiers in Marine Science*, vol. 6, art. 722.

Gehlen, M., and others (2014). Projected pH reductions by 2100 might put deep North Atlantic biodiversity at risk. *Biogeosciences*, vol. 11, No. 23, pp. 6955–6967. https://doi.org/10.5194/bg-11-6955-2014.

Genin, Amatzia (2004). Bio-physical coupling in the formation of zooplankton and fish aggregations over abrupt topographies. *Journal of Marine Systems*, vol. 50, No. 1, pp. 3–20. https://doi.org/10.1016/j.jmarsys.2003.10.008.

Gerigny, O., and others (2019) Seafloor litter from the continental shelf and canyons in French Mediterranean water: distribution, typologies and trends. Marine Pollution Bulletin, vol. 146, pp. 653–666. https://doi.org/10.1016/j.marpolbul.2019.07.030.

Gilly, William F., and others (2013). Oceanographic and biological effects of shoaling of the oxygen minimum zone. *Annual Review of Marine Science*, vol. 5, pp. 393–420.

Giusti, M., and others (2019). Coral forests and Derelict Fishing Gears in submarine canyon systems of the Ligurian Sea. *Progress in Oceanography*, vol. 178, art. 102186.

Glover, Adrian G., and others (2018). Point of View: Managing a sustainable deep-sea 'blue economy' requires knowledge of what actually lives there. *ELife*, vol. 7, e41319.

Goffredi, S.K., and others (2020). Methanotrophic bacterial symbionts fuel dense populations of deep-sea feather duster worms (Sabellida, Annelida) and extend the spatial influence of methane seepage. *Science Advances*, vol. 6, No. 14.

Gooday, A.J., and others (2009). Faunal responses to oxygen gradients on the Pakistan margin: a comparison of foraminiferans, macrofauna and megafauna. *Deep Sea Research Part II: Topical Studies in Oceanography*, vol. 56, Nos. 6–7, pp. 488–502.

Gooday, A.J., and others (2010). Habitat heterogeneity and its influence on benthic biodiversity in oxygen minimum zones. *Marine Ecology*, vol. 31, No. 1, pp. 125–147.

Graco M., and others. (2017). The OMZ and nutrients features as a signature of interannual and low frequency variability off the Peruvian upwelling system. *Biogeosciences,* vol. 14, pp. 4601–4617. https://doi.org/10.5194/bg-14-4601-2017.

Grupe, Benjamin M., and others (2015). Methane seep ecosystem functions and services from a recently discovered southern California seep. *Marine Ecology*, vol. 36, pp. 91–108.

Harriague, Anabella Covazzi, and others (2019). Macrofaunal assemblages in canyon and adjacent slope of the NW and Central Mediterranean systems. *Progress in Oceanography*, vol. 171, pp. 38–48. Harris, Peter, and others (2014). Geomorphology of the oceans. *Marine Geology*, vol. 352. https://doi.org/10.1016/j.margeo.2014.01.011.

Harris, Peter T., and Elaine K. Baker (2020). GeoHab atlas of seafloor geomorphic features and benthic habitats–synthesis and lessons learned. In *Seafloor Geomorphology as Benthic Habitat*, pp. 969–990. Elsevier.

Helly, John J., and Lisa A. Levin (2004). Global distribution of naturally occurring marine hypoxia on continental margins. *Deep Sea Research Part I: Oceanographic Research Papers*, vol. 51, No. 9, pp. 1159–1168.

Hidalgo, Pamela, and others (2005). Ontogenetic vertical distribution and diel migration of the copepod *Eucalanus inermis* in the oxygen minimum zone off northern Chile (20–21° S). *Journal of Plankton Research*, vol. 27, pp. 519–529.

Hogg, Oliver T., and others (2016). Landscape mapping at sub-Antarctic South Georgia provides a protocol for underpinning large-scale marine protected areas. *Scientific Reports*, vol. 6, art. 33163.

Holland, L.P., and others (2019). *Genetic connectivity of deep-sea corals in the New Zealand region*. New Zealand Aquatic Environment & Biodiversity Report No. 245, Wellington.

Horton, Tammy, and others (2017). Improving nomenclatural consistency: a decade of experience in the World Register of Marine Species. *European Journal of Taxonomy*, No. 389.

Huang, Huai-Hsuan May, and others (2018). Benthic biotic response to climate changes over the last 700,000 years in a deep marginal sea: impacts of deoxygenation and the Mid-Brunhes Event. *Paleoceanography and Paleoclimatology*, vol. 33, No. 7, pp. 766–777.

Huang, Huai-Hsuan May, and others (2019). Deep-sea ostracod faunal dynamics in a marginal sea: biotic response to oxygen variability and mid-Pleistocene global changes. *Paleobiology*, vol. 45, No .1, pp. 85–97.

Huang, Zhi, and others (2018). A conceptual surrogacy framework to evaluate the habitat potential of submarine canyons. *Progress in Oceanography,* vol. 169, pp. 199–213. https://doi.org/10.1016/j.pocean.2017.11.007.

Hunter, William R., and others (2011). Epi-benthic megafaunal zonation across an oxygen minimum zone at the Indian continental margin. *Deep Sea Research Part I: Oceanographic Research Papers*, vol. 58, No. 6, pp. 699–710.

Huvenne, V.A.I., and others (2016). Effectiveness of a deep-sea cold-water coral Marine Protected Area, following eight years of fisheries closure. *Biological Conservation*, vol. 200, pp. 60–69.

Huvenne, V.A.I., and others (2018). ROVs and AUVs. In *Submarine Geomorphology*, pp. 93–108. Springer.

Ingels, Jeroen, and others (2009). Nematode diversity and its relation to the quantity and quality of sedimentary organic matter in the deep Nazaré Canyon, Western Iberian Margin. *Deep Sea Research Part I: Oceanographic Research Papers*, vol. 56, No. 9, pp. 1521–1539.

Ingels, Jeroen, and others (2011). Structural and functional diversity of Nematoda in relation with environmental variables in the Setúbal and Cascais canyons, Western Iberian Margin. *Deep Sea Research Part II: Topical Studies in Oceanography*, vol. 58, Nos. 23–24, pp. 2354–2368.

Ingels, Jeroen, and Ann Vanreusel (2013). The importance of different spatial scales in determining structural and functional characteristics of deep-sea infauna communities. *Biogeosciences*, vol. 10, No. 7, pp. 4547–4563.

Intergovernmental Panel on Climate Change (IPCC) (2019). Summary for Policymakers. In *IPCC Special Report on the Ocean and Cryosphere in a Changing Climate*, H-O. Pörtner and others, eds.

Ismail, Khaira, and others (2015). Objective automated classification technique for marine landscape mapping in submarine canyons. *Marine Geology*, vol. 362, pp. 17–32.

Johnson, H. Paul, and others (2015). Analysis of bubble plume distributions to evaluate methane hydrate decomposition on the continental slope. *Geochemistry, Geophysics, Geosystems*, vol. 16, No. 11, pp. 3825–3839.

Jones, Daniel O.B., and others (2014). Global reductions in seafloor biomass in response to climate change. *Global Change Biology*, vol. 20, No. 6, pp. 1861–1872.

Keller, Aimee A., and others (2015). Occurrence of demersal fishes in relation to near-bottom oxygen levels within the California Current large marine ecosystem. *Fisheries Oceanography*, vol. 24, No. 2, pp. 162–176.

Kelly, Noreen E., and others (2010). Biodiversity of the deep-sea continental margin bordering the Gulf of Maine (NW Atlantic): relationships among sub-regions and to shelf systems. *PloS One*, vol. 5, No. 11, e13832.

Kenchington, E.L., and others (2014). Limited depth zonation among bathyal epibenthic megafauna of the Gully submarine canyon, northwest Atlantic. *Deep Sea Research Part II: Topical Studies in Oceanography*, vol. 104, pp. 67–82.

Kinlan, B.P., and others (2020). Predictive modeling of suitable habitat for deep-sea corals offshore the Northeast United States. *Deep Sea Research Part I: Oceanographic Research Papers*, vol. 158,https://doi.org/10.1016/j.dsr.2020.103229.

Kool, Johnathan T., and others (2015). Simulated larval connectivity among Australia's southwest submarine canyons. *Marine Ecology Progress Series*, vol. 539, pp. 77–91.

Kool, Johnathan T., and others (2013). Population connectivity: recent advances and new perspectives. *Landscape Ecology*, vol. 28, No. 2, pp. 165–185.

Koslow, J. Anthony, and others (2011). Impact of declining intermediate-water oxygen on deepwater fishes in the California Current. *Marine Ecology Progress Series*, vol. 436, pp. 207–218.

Lastras, G., and others (2008). Geomorphology and sedimentary features in the Central Portuguese submarine canyons, Western Iberian margin. *Geomorphology*, vol. 103, No. 3, pp. 310–329.

Lastras, G., and others (2016). Cold-Water Corals and Anthropogenic Impacts in La Fonera Submarine Canyon Head, Northwestern Mediterranean Sea. *PLoS One*, vol. 11, No. 5, pp. 1–36. https://doi.org/10.1371/journal.pone.0155729.

Leal, M.C., and others (2012). Trends in the discovery of new marine natural products from invertebrates over the last two decades – where and what are we bioprospecting. *PLoS One*, vol. 7, No. 1, art. e30580, https://doi.org/10.1371/journal.pone.0030580.

Leduc, Daniel, and others (2014). Unusually high food availability in Kaikoura Canyon linked to distinct deep-sea nematode community. *Deep Sea Research Part II: Topical Studies in Oceanography*, vol. 104, pp. 310–318.

Leduc, Daniel, and others (2016). Limited differences among habitats in deep-sea macro-infaunal communities off New Zealand: implications for their vulnerability to anthropogenic disturbance. *Marine Ecology*, vol. 37, No. 4, pp. 845–866.

Leduc, Daniel, and others (2018). *Quantifying the Transfer of Terrestrial Organic Matter into Two Contrasting New Zealand Submarine Canyon Systems Using Bulk and Compound-Specific Stable Isotopes.* https://doi.org/10.13140/RG.2.2.24107.08482.

Levin, Lisa A. (2003). Oxygen minimum zone benthos: Adaptation and community response to hypoxia. *Oceanography and Marine Biology: An Annual Review,* vol. 41, pp. 1–45.

________(2005). Ecology of cold seep sediments: interactions of fauna with flow, chemistry and microbes. In *Oceanography and Marine Biology,* pp. 11–56. CRC Press.

________(2018). Manifestation, drivers, and emergence of open ocean deoxygenation. *Annual Review of Marine Science,* vol. 10, pp. 229–260.

Levin, Lisa A., and Greg Rouse (2019). Giant Protists (Xenophyophores) Function as Fish Nurseries. *Ecology.* https://doi.org/10.1002/ecy.2933.

Levin, Lisa A., and John D. Gage (1998). Relationships between oxygen, organic matter and the diversity of bathyal macrofauna. *Deep Sea Research Part II: Topical Studies in Oceanography,* vol. 45, No. 1–3, pp. 129–163.

Levin, Lisa A., and Myriam Sibuet (2012). Understanding continental margin biodiversity: a new imperative. *Annual Review of Marine Science,* vol. 4, pp. 79–112.

Levin, Lisa A., and Natalya D. Gallo (2019). Chapter 8.5: Continental margin benthic and demersal biota. In *Ocean Deoxygenation – Everyone's Problem: Causes, Impacts, Consequences and Solutions,* D. Laffoley and J.M. Baxter, eds. Gland: IUCN.

Levin, Lisa A., and others (2001). Environmental influences on regional deep-sea species diversity. *Annual Review of Ecology and Systematics,* vol. 32, No. 1, pp. 51–93.

Levin, Lisa A., and others (2009). Oxygen and organic matter thresholds for benthic faunal activity on the Pakistan margin oxygen minimum zone (700–1100 m). *Deep Sea Research Part II: Topical Studies in Oceanography,* vol. 56, Nos. 6–7, pp. 449–471.

Levin, Lisa A., and others (2013). Macrofaunal colonization across the Indian Margin oxygen minimum zone. *Biogeosciences,* vol. 10, pp. 7161–77.

Levin, Lisa A., and others (2015a). Biodiversity on the rocks: macrofauna inhabiting authigenic carbonate at Costa Rica methane seeps. *PLoS One,* vol. 10, No. 7, e0131080.

Levin, Lisa A., and others (2015b). Comparative biogeochemistry-ecosystem-human interactions on dynamic continental margins. *Journal of Marine Systems,* vol. 141, pp. 3–17.

Levin, Lisa A., and others (2016a). Defining "serious harm" to the marine environment in the context of deep-seabed mining. *Marine Policy,* vol. 74, pp. 245–259.

Levin, Lisa A., and others (2016b). Hydrothermal vents and methane seeps: rethinking the sphere of influence. *Frontiers in Marine Science,* vol. 3, art. 72.

Levin, Lisa A., and others (2017). Methane seepage effects on biodiversity and biological traits of macrofauna inhabiting authigenic carbonates. *Deep Sea Research Part II: Topical Studies in Oceanography,* vol. 137, pp. 26–41.

Levin, Lisa A., and others (2019). Global observing needs in the deep ocean. *Frontiers in Marine Science,* vol. 6, art. 241.https://doi.org/10.3389/fmars.2019.00241.

Lo Iacono, Claudio, and others (2019). 15 Habitat Mapping of Cold-Water Corals in the Mediterranean Sea. In *Mediterranean Cold-Water Corals: Past, Present and Future: Understanding the Deep-Sea Realms of Coral,* Covadonga Orejas and Carlos Jiménez, eds., pp. 157–171. Cham, Switzerland: Springer International Publishing. https://doi.org/10.1007/978-3-319-91608-8_15.

Lowe, Michael R., and others (2018). Drivers of euphausiid distribution and abundance in the Northeast U.S. Shelf Large Marine Ecosystem. *ICES Journal of Marine Science,* vol. 75, No. 4, pp. 1280–95. https://doi.org/10.1093/icesjms/fsx247.

Lu, Beiwei, and others (2003). Cross-shore separation of adult and juvenile euphausiids in a shelf-break alongshore current. *Progress in Oceanography*, vol. 57, No. 3, pp. 381–404. https://doi.org/10.1016/S0079-6611(03)00107-1.

Luna, Gian Marco, and others (2016). Dense water plumes modulate richness and productivity of deep sea microbes. *Environmental Microbiology*, vol. 18, No. 12, pp. 4537–4548.

Maldonado, Manuel, and others (2015). Aggregated clumps of lithistid sponges: a singular, reef-like bathyal habitat with relevant paleontological connections. *PloS One*, vol. 10, No. 5, e0125378.

Marsh, Leigh, and others (2018). Geomorphological evidence of large vertebrates interacting with the seafloor at abyssal depths in a region designated for deep-sea mining. *Royal Society Open Science*, vol. 5, No.8, 180286.

Martín, Jacobo, and others (2014a). Impact of bottom trawling on deep-sea sediment properties along the flanks of a submarine canyon. *PloS One*, vol. 9, No. 8, e104536.

Martín, Jacobo, and others (2014b). Trawling-induced daily sediment resuspension in the flank of a Mediterranean submarine canyon. *Deep Sea Research Part II: Topical Studies in Oceanography*, vol. 104, pp. 174–183.

Matos, F.L., and others (2018). Canyons pride and prejudice: Exploring the submarine canyon research landscape, a history of geographic and thematic bias. *Progress in Oceanography*, vol. 169, pp. 6–19.

Mayer, Larry, and others (2018). The Nippon Foundation—GEBCO seabed 2030 project: The quest to see the world's oceans completely mapped by 2030. *Geosciences*, vol. 8, No. 2, art. 63.

McCormick, Lillian R., and others (2019). Vision is highly sensitive to oxygen availability in marine invertebrate larvae. *Journal of Experimental Biology*, vol. 222, No. 10, jeb200899.

Mecho, A., and others (2020). Deep-sea litter in the Gulf of Cadiz (Northeastern Atlantic, Spain). *Marine Pollution Bulletin*, vol. 153, 110969. https://doi.org/10.1016/j.marpolbul.2020.110969.

Menot, Lenaick, and others (2010). New perceptions of continental margin biodiversity. *Life in the World's Oceans: Diversity, Distribution, and Abundance, Edited by: McIntyre, AD*79–103.

Merrie, Andrew, and others (2014). An ocean of surprises – trends in human use, unexpected dynamics and governance challenges in areas beyond national jurisdiction. *Global Environmental Change*, vol. 27, pp. 19–31.

Metaxas, Anna, and others (2019). Hydrodynamic connectivity of habitats of deep-water corals in Corsair Canyon, Northwest Atlantic: a case for cross-boundary conservation. *Frontiers in Marine Science*, vol. 6, art. 159.

Morato, Telmo, and others. (2020). Climate-induced changes in the suitable habitat of cold-water corals and commercially important deep-sea fishes in the North Atlantic. *Global Change Biology*, vol. 26, pp. 2181–2202. https://doi.org/10.1111/gcb.14996.

Muiños, Susana Bolhão, and others (2013). Deep-sea Fe-Mn Crusts from the Northeast Atlantic Ocean: Composition and Resource Considerations. *Marine Georesources & Geotechnology*, vol. 31, No. 1, pp. 40–70.https://doi.org/10.1080/1064119X.2012.661215.

Narayanaswamy, Bhavani E., and others (2010). Deep-water macrofaunal diversity in the Faroe-Shetland region (NE Atlantic): a margin subject to an unusual thermal regime. *Marine Ecology*, vol. 31, No. 1, pp. 237–246.

Netburn, Amanda N., and J. Anthony Koslow (2015). Dissolved oxygen as a constraint on daytime deep scattering layer depth in the southern California current ecosystem. *Deep Sea Research Part I: Oceanographic Research Papers*, vol. 104, pp. 149–158.

Olsen, B.R., and others (2016). Environmental challenges related to offshore mining and gas hydrate extraction. *Miljødirektoratet. Rapport M-532.*

Pabis, Krzysztof, and others (2019). Natural and anthropogenic factors influencing abundance of the benthic macrofauna along the shelf and slope of the Gulf of Guinea, a large marine ecosystem off West Africa. *Oceanologia*.

Papale, Maria, and others (2018). Heavy-metal resistant microorganisms in sediments from submarine canyons and the adjacent continental slope in the northeastern Ligurian margin (Western Mediterranean Sea). *Progress in Oceanography*, vol. 168, pp. 155–168.

Paradis, Sarah, and others (2018a). Enhancement of sedimentation rates in the Foix Canyon after the renewal of trawling fleets in the early XXIst century. *Deep Sea Research Part I: Oceanographic Research Papers*, vol. 132, pp. 51–59.

Paradis, Sarah, and others (2018b). Spatial distribution of sedimentation-rate increases in Blanes Canyon caused by technification of bottom trawling fleet. *Progress in Oceanography*, vol. 169, pp. 241–252.

Pearman, T.R.R., and others (2020). Improving the predictive capability of benthic species distribution models by incorporating oceanographic data – towards holistic ecological modelling of a submarine canyon. *Progress in Oceanography*, vol. 184, art. 102338.

Pham, Christopher K., and others (2014). Marine litter distribution and density in European seas, from the shelves to deep basins. *PloS One*, vol. 9, No. 4, e95839.

Phrampus, Benjamin J., and Matthew J. Hornbach (2012). Recent changes to the Gulf Stream causing widespread gas hydrate destabilization. *Nature*, vol. 490, No. 7421, p. 527.

Pierdomenico, M. and others (2019). Megafauna distribution along active submarine canyons of the central Mediterranean: relationships with environmental variables. *Progress in Oceanography*, vol. 171, pp. 49–69.

Pierdomenico, M. and others (2015). Sedimentary facies, geomorphic features and habitat distribution at the Hudson Canyon head from AUV multibeam data. *Deep Sea Research Part II: Topical Studies in Oceanography*, vol. 121, pp. 112–125.

Pizarro-Koch, M., and others (2018). Seasonal variability of the southern tip of the Oxygen Minimum Zone in the Eastern South Pacific (30°–38°S): A modeling study. *Journal of Geophysical Research: Oceans*, vol. 124, pp. 8574–8604. https://doi.org/10.1029/2019JC015201.

Priede, Imants G., and others (2010). Deep-sea demersal fish species richness in the Porcupine Seabight, NE Atlantic Ocean: global and regional patterns. *Marine Ecology*, vol. 31, No. 1, pp. 247–260.

Puig, Pere, and others (2014). Contemporary sediment-transport processes in submarine canyons. *Annual Review of Marine Science*, vol. 6, pp. 53–77.

Pusceddu, Antonio, and others (2014). Chronic and intensive bottom trawling impairs deep-sea biodiversity and ecosystem functioning. *Proceedings of the National Academy of Sciences*, vol. 111, No. 24, pp. 8861–8866.

Quattrini, Andrea M., and others (2015). Exploration of the canyon-incised continental margin of the northeastern United States reveals dynamic habitats and diverse communities. *PLoS One*, vol. 10, No. 10, e0139904.

Ramirez-Llodra, Eva, and others (2015). Submarine and deep-sea mine tailing placements: a review of current practices, environmental issues, natural analogs and knowledge gaps in Norway and internationally. *Marine Pollution Bulletin*, vol. 97, No. 1, pp. 13–35. https://doi.org/10.1016/j.marpolbul.2015.05.062.

Rastelli, Eugenio, and others (2018). Rapid response of benthic deep-sea microbes (viruses and prokaryotes) to an intense dense shelf water cascading event in a submarine canyon of the NW Mediterranean Sea. *Progress in Oceanography*, vol. 168, pp. 35–42.

Rathburn, A.E., and others (2009). Geological and biological heterogeneity of the Aleutian margin (1965–4822 m). *Progress in Oceanography*, vol. 80, Nos. 1–2, pp. 22–50.

Reed, John K., and others (2007). Impacts of bottom trawling on a deep-water Oculina coral ecosystem off Florida. *Bulletin of Marine Science*, vol. 81, No. 3, pp. 481–496.

Reichelt-Brushett, Amanda (2012). Risk assessment and ecotoxicology: limitations and recommendations for ocean disposal of mine waste in the coral triangle. *Oceanography*, vol. 25, No. 4, pp. 40–51.

Rex, Michael A., and Gilbert T. Rowe (1983). Geographic patterns of species diversity in the deep-sea benthos. In *The Sea*, pp. 453–472. New York: Wiley.

Rex, Michael A., and Ron J. Etter (2010). *Deep-Sea Biodiversity: Pattern and Scale*. Cambridge: Harvard University Press.

Riedel, Michael, and others (2018). Distributed natural gas venting offshore along the Cascadia margin. *Nature Communications*, vol. 9, No. 1, pp. 1–14.

Rosli, Norliana, and others (2016). Differences in meiofauna communities with sediment depth are greater than habitat effects on the New Zealand continental margin: implications for vulnerability to anthropogenic disturbance. *PeerJ*, vol. 4, e2154.

Ross, Rebecca E., and others (2019). Combining distribution and dispersal models to identify a particularly vulnerable marine ecosystem. *Frontiers in Marine Science*, vol. 6, art. 574. https://doi.org/10.3389/fmars.2019.00574.

Ross, Steve W., and others (2015). Demersal fish distribution and habitat use within and near Baltimore and Norfolk Canyons, US middle Atlantic slope. *Deep Sea Research Part I: Oceanographic Research Papers*, vol. 103, pp. 137–154.

Rowden, Ashley A., and others (2016). Habitat differences in deep-sea megafaunal communities off New Zealand: implications for vulnerability to anthropogenic disturbance and management. *Frontiers in Marine Science*, vol. 3, art. 241.

Salvadó, Joan A., and others (2017). Transfer of lipid molecules and polycyclic aromatic hydrocarbons to open marine waters by dense water cascading events. *Progress in Oceanography*, vol. 159, pp. 178–194.

Salvadó, Joan A., and others (2019). Influence of deep water formation by open-sea convection on the transport of low hydrophobicity organic pollutants in the NW Mediterranean Sea. *Science of the Total Environment*, vol. 647, pp. 597–605.

Salvatteci, Renato, and others (2019). Fish debris in sediments from the last 25 kyr in the Humboldt Current reveal the role of productivity and oxygen on small pelagic fishes. *Progress in Oceanography*, vol. 176, art. 102114. https://doi.org/10.1016/j.pocean.2019.05.006.

Santora, Jarrod A., and others (2018). Submarine canyons represent an essential habitat network for krill hotspots in a Large Marine Ecosystem. *Scientific Reports*, vol. 8, No. 1, art. 7579. https://doi.org/10.1038/s41598-018-25742-9.

Sardà, F., and others (2009). Relationship between environment and the occurrence of the deep-water rose shrimp *Aristeus antennatus* (Risso, 1816) in the Blanes submarine canyon (NW Mediterranean). *Progress in Oceanography*, vol. 82, No. 4, pp. 227–238.

Sato, Kirk N., and others (2018). Response of sea urchin fitness traits to environmental gradients across the southern California oxygen minimum zone. *Frontiers in Marine Science*, vol. 5, art. 258.

Schimmelmann, Arndt, and others (2016). Varves in marine sediments: a review. *Earth-Science Reviews*, vol. 159, pp. 215–246.

Seabrook, Sarah, and others (2019). Flipping for Food: The use of a methane seep by Tanner Crabs (*Chionoecetes tanneri*). *Frontiers in Marine Science*, vol. 6, art. 43.

Secretariat of the Convention on Biological Diversity (2009). COP 12 Decision *XII/22*. Marine and coastal biodiversity: ecologically or biologically significant marine areas (EBSAs). www.cbd.int/decision/cop/?id=13385.

Seibel, Brad A., and others (2016). Hypoxia tolerance and metabolic suppression in oxygen minimum zone euphausiids: implications for ocean deoxygenation and biogeochemical cycles. *Integrative and Comparative Biology*, vol. 56, No. 4, pp. 510–523.

Seibel, Brad A., and others (2018). Metabolic suppression in the pelagic crab, *Pleuroncodes planipes*, in oxygen minimum zones. *Comparative Biochemistry and Physiology Part B: Biochemistry and Molecular Biology*, vol. 224, pp. 88–97.

Sellanes, Javier, and others (2010). Diversity patterns along and across the Chilean margin: a continental slope encompassing oxygen gradients and methane seep benthic habitats. *Marine Ecology*, vol. 31, No. 1, pp. 111–124.

Sellanes, Javier, and others (2008). Megafauna community structure and trophic relationships at the recently discovered Concepción Methane Seep Area, Chile, 36 S. *ICES Journal of Marine Science*, vol. 65, No. 7, pp. 1102–1111.

Sen, Arunima, and others (2019). Atypical biological features of a new cold seep site on the Lofoten-Vesterålen continental margin (northern Norway). *Scientific Reports*, vol. 9, No. 1, art. 1762.

Serrano, A., and others (2017). Deep-sea benthic habitats modeling and mapping in a NE Atlantic seamount (Galicia Bank) *Deep-sea Research Part 1*, vol. 126, pp. 115–127

Sinniger, Frédéric, and others (2016). Worldwide analysis of sedimentary DNA reveals major gaps in taxonomic knowledge of deep-sea benthos. *Frontiers in Marine Science*, vol. 3, art. 92.

Sion, Letizia, and others (2019). Does the Bari Canyon (Central Mediterranean) influence the fish distribution and abundance? *Progress in Oceanography*, vol. 170, pp. 81–92.

Skarke, Adam, and others (2014). Widespread methane leakage from the sea floor on the northern US Atlantic margin. *Nature Geoscience*, vol. 7, No. 9, p. 657.

Skropeta, D., and L. Wei (2014). Recent advances in deep-sea natural products. *Natural Product Reports*, vol. 31, pp. 999–1025. https://doi.org/10.1039/C3NP70118B.

Smith, Craig R., and others (2000). Variations in bioturbation across the oxygen minimum zone in the northwest Arabian Sea. *Deep Sea Research Part II: Topical Studies in Oceanography*, vol. 47, Nos. 1–2, pp. 227–257.

Soldatou, Sylvia, and Bill J. Baker (2017). Cold-water marine natural products, 2006 to 2016. *Natural Product Reports*, vol. 34, pp. 585–626. https://doi.org/10.1039/C6NP00127K.

Sperling, Erik A., and others (2013). Oxygen, ecology, and the Cambrian radiation of animals. *Proceedings of the National Academy of Sciences*, vol. 110, No. 33, pp. 13446–13451.

Sperling, Erik A., and others (2016). Biodiversity response to natural gradients of multiple stressors on continental margins. *Proceedings of the Royal Society B: Biological Sciences*, vol. 283, No. 1829, 20160637.

Stramma, Lothar, and others (2008). Expanding oxygen-minimum zones in the tropical oceans. *Science*, vol. 320, No. 5876, pp. 655–658.

Stramma, Lothar, and others (2010). Ocean oxygen minima expansions and their biological impacts. *Deep Sea Research Part I: Oceanographic Research Papers*, vol. 57, No. 4, pp. 587–595.

Sweetman, Andrew K., and others (2017). Major impacts of climate change on deep-sea benthic ecosystems. *Elementa: Science of the Anthropocene*, vol. 5, No. 4.

Tamburrino, Stella, and others (2019). Pathways of inorganic and organic contaminants from land to deep sea: the case study of the Gulf of Cagliari (W Tyrrhenian Sea). *Science of the Total Environment*, vol. 647, pp. 334–341.

Taviani, Marco, and others (2019). U/Th dating records of cold-water coral colonization in submarine canyons and adjacent sectors of the southern Adriatic Sea since the Last Glacial Maximum. *Progress in Oceanography*, vol. 175.

Taylor, M.L., and others (2016). Plastic microfibre ingestion by deep-sea organisms. *Scientific Reports*, vol. 6, art. 33997.

Taylor, J.R., and others (2014). Physiological effects of environmental acidification in the deep-sea urchin *Strongylocentrotus fragilis*. *Biogeosciences*, vol. 11, No. 5, pp. 1413–1423.

Teixeira, Manuel, and others (2019). Interaction of along slope and downslope processes in the Alentejo Margin (SW Iberia) – Implications on slope stability. *Marine Geology*, vol. 410, pp. 88–108. https://doi.org/10.1016/j.margeo.2018.12.011.

Treude, Tina, and others (2011). Elasmobranch egg capsules associated with modern and ancient cold seeps: a nursery for marine deep-water predators. *Marine Ecology Progress Series*, vol. 437, pp. 175–181.

Trueman, C.N., and others (2014). Trophic interactions of fish communities at midwater depths enhance long-term carbon storage and benthic production on continental slopes. *Proceedings of the Royal Society B: Biological Sciences*, vol. 281, No. 1787, 20140669.

Tubau, Xavier, and others (2015). Marine litter on the floor of deep submarine canyons of the Northwestern Mediterranean Sea: the role of hydrodynamic processes. *Progress in Oceanography*, vol. 134, pp. 379–403.

Tutasi, Pritha, and Ruben Escribano (2020). Zooplankton diel vertical migration and downward C flux into the oxygen minimum zone in the highly productive upwelling region off northern Chile. *Biogeosciences*, vol. 17, pp. 455–473.

United Nations (2017a). Chapter 36F: Open ocean deep sea. In *The First Global Integrated Marine Assessment: World Ocean Assessment I*. Cambridge: Cambridge University Press.

________(2017b). Chapter 51: Biological communities on seamounts and other submarine features potentially threatened by disturbance. In *The First Global Integrated Marine Assessment: World Ocean Assessment I*. Cambridge: Cambridge University Press.

________(2017c). *The First Global Integrated Marine Assessment: World Ocean Assessment I*. Cambridge: Cambridge University Press.

Van den Beld, I.M., and others (2017). Cold-water coral habitats in submarine canyons of the Bay of Biscay. *Frontiers in Marine Science*, vol. 7, art. 118, https://doi.org/10.3389/fmars.2017.00118.

Vare, Lindsay L., and others (2018). Scientific considerations for the assessment and management of mine tailings disposal in the deep sea. *Frontiers in Marine Science*, vol. 5, art. 17.

Vieira, Rui P., and others (2015). Lost fishing gear and litter at Gorringe Bank (NE Atlantic). *Journal of Sea Research*, vol. 100, pp. 91–98.

Vieira, Rui P., and others (2019). Deep-water fisheries along the British Isles continental slopes: status, ecosystem effects and future perspectives. *Journal of Fish Biology*.

Wei, Chih-Lin, and others (2010). Bathymetric zonation of deep-sea macrofauna in relation to export of surface phytoplankton production. *Marine Ecology Progress Series*, vol. 399, pp. 1–14.

Winder, Priscilla L., and others (2011). Natural Products from the Lithistida: A Review of the Literature since 2000. *Marine Drugs*, vol. 9, pp. 2643–2682; https://doi.org/10.3390/md9122643.

Wishner, Karen F., and others (1995). Pelagic and benthic ecology of the lower interface of the Eastern Tropical Pacific oxygen minimum zone. *Deep Sea Research Part I: Oceanographic Research Papers*, vol. 42, No. 1, pp. 93–115.

Wishner, Karen F., and others (2008). Vertical zonation and distributions of calanoid copepods through the lower oxycline of the Arabian Sea oxygen minimum zone. *Progress in Oceanography*, vol. 78, No. 2, pp. 163–191.

Wishner, Karen F., and others (2013). Zooplankton in the eastern tropical north Pacific: boundary effects of oxygen minimum zone expansion. *Deep Sea Research Part I: Oceanographic Research Papers*, vol. 79, pp. 122–140.

Wishner, Karen F., and others (2018). Ocean deoxygenation and zooplankton: Very small oxygen differences matter. *Science Advances*, vol. 4, No. 12, eaau5180.

Wishner, Karen F., and others (2000). Living in suboxia: ecology of an Arabian Sea oxygen minimum zone copepod. *Limnology and Oceanography*, vol. 45, No. 7, pp. 1576–1593.

Woodall, Lucy C., and others (2014). The deep sea is a major sink for microplastic debris. *Royal Society Open Science*, vol. 1, No. 4, 140317.

Woodall, Lucy C., and others (2015). Deep-sea litter: a comparison of seamounts, banks and a ridge in the Atlantic and Indian Oceans reveals both environmental and anthropogenic factors impact accumulation and composition. *Frontiers in Marine Science*, vol. 2, art. 3.

Woulds, Clare, and others (2007). Oxygen as a control on sea floor biological communities and their roles in sedimentary carbon cycling. *Limnology and Oceanography*, vol. 52, No. 4, pp. 1698–1709.

Woulds, Clare, and others (2009). The short-term fate of organic carbon in marine sediments: comparing the Pakistan margin to other regions. *Deep Sea Research Part II: Topical Studies in Oceanography*, vol. 56, Nos. 6–7, pp. 393–402.

Wright, Amy E., and others (2017). Analogues of the potent antitumor compound Leiodermatolide from a deep-water sponge of the genus *Leiodermatium*. *Journal of Natural Products*, vol. 80, pp. 735–73, https://doi.org/10.1021/acs.jnatprod.6b01140.

Yasuhara, Moriaki (2018). Marine biodiversity in space and time: what tiny fossils tell. *Mètode Science Studies Journal-Annual Review*, No. 9.

Yasuhara, Moriaki, and others (2008). Abrupt climate change and collapse of deep-sea ecosystems. *Proceedings of the National Academy of Sciences*, vol. 105, No. 5, pp. 1556–1560.

Yasuhara, Moriaki, and others (2014). Response of deep-sea biodiversity to abrupt deglacial and Holocene climate changes in the North Atlantic Ocean. *Global Ecology and Biogeography*, vol. 23, No. 9, pp. 957–967.

Yasuhara, Moriaki, and others (2016). Biodiversity–ecosystem functioning relationships in long-term time series and palaeoecological records: deep sea as a test bed. *Philosophical Transactions of the Royal Society B: Biological Sciences*, vol. 371, No. 1694, 20150282.

Yasuhara, Moriaki, and others (2017). Combining marine macroecology and palaeoecology in understanding biodiversity: microfossils as a model. *Biological Reviews*, vol. 92, No. 1, pp. 199–215.

Yasuhara, Moriaki, and Thomas M. Cronin (2008). Climatic influences on deep-sea ostracode (Crustacea) diversity for the last three million years. *Ecology*, vol. 89, No. sp. 11, pp. S53–S65.

Yasuhara, Moriaki, and Roberto Danovaro (2016). Temperature impacts on deep-sea biodiversity. *Biological Reviews*, vol. 91, No. 2, pp. 275–287.

Yool, Andrew, and others (2017). Big in the benthos: Future change of seafloor community biomass in a global, body size-resolved model. *Global Change Biology*, vol. 23, No. 9, pp. 3554–3566.

Zeng, Cong and others (2017). Population genetic structure and connectivity of deep-sea stony corals (Order Scleractinia) in the New Zealand region: Implications for the conservation and management of vulnerable marine ecosystems. *Evolutionary Applications*, vol. 10, No. 10, pp. 1040–1054.

Zeng, Cong and others (2019). The use of spatially explicit genetic variation data from four deep-sea sponges to inform the protection of Vulnerable Marine Ecosystems. *Scientific Reports*, vol. 9, No. 1, art. 5482.

Chapter 7K
High-latitude ice

Contributors: Grant R. Bigg (convener), Maurizio Azzaro, Karen Evans (subchapter lead member), Huw Griffiths and Moriaki Yasuhara.

Keynote points

- High-latitude ice habitats are characterized by high, but geographically variable, declines in sea ice extent as a consequence of climate change.
- The loss of Arctic sea ice habitat and Antarctic ice shelves allows expansion of both pelagic and benthic species into the newly open water environments.
- In general, however, many ice-dependent species are decreasing in abundance and their spatial distributions may also be reducing, in particular in the Arctic.
- Decreasing sea ice extent in the Arctic provides increased opportunities for a range of human activities, including fishing, navigation and hydrocarbon exploration, with positive implications for several Sustainable Development Goals.[1]
- Many of those activities, however, will remain marginal for some time as a seasonally ice-free Arctic is not expected until later in the century.
- Decreasing sea ice will, however, reduce local community access to subsistence hunting opportunities.

1. Introduction

The present subchapter contains an update to chapter 46 of the first *World Ocean Assessment* (United Nations, 2017a). It also extends the coverage of high-latitude sea ice environments to include a discussion of habitats associated with icebergs and ice shelves. The subchapter overlaps with the high-latitude biodiversity aspects of many of the subchapters in chapter 6 of the present Assessment. However, in the present subchapter, the emphasis is on the use of marine ice habitats and interactions between organisms within those habitats. Furthermore, because high-latitude ice is intrinsically both a coastal and open ocean habitat, it interacts with several other habitats (e.g., benthic, open ocean and coastal-related habitats) that are covered in other subchapters of chapter 7 of the present Assessment.

The baseline state for the discussion of high-latitude ice habitats in the first Assessment (United Nations, 2017b) was one of massive and rapid change. That degree of change is, to some extent, intrinsic to the habitat itself, which experiences strong seasonal fluctuations between minimal ice coverage in high summer and maximal ice coverage in late winter. However, the mean sea ice habitat itself was altering dramatically, with ice extent, ice thickness and mean ice age all declining rapidly in the Arctic. In the Southern Ocean, change in the sea ice habitat was less notable, although several ice shelves on the Antarctic Peninsula had collapsed over previous decades (Vaughan and others, 2013). Those changes to habitats had concomitant responses in associated ecosystems (United Nations, 2017b). Iconic marine and terrestrial species that have adapted to the sea ice habitat, for example, polar bears, narwhals, seals and various seabirds, were found to be in decline both in abundance and geographic distribution. Sea ice algae were identified as playing a major role in the primary production of those habitats; the expansion of open ocean environments led to increased phytoplankton blooms. Both of those changes implied an altered base to the high-latitude food chain. In general, the expansion of open ocean environments was leading to a concomitant increase

1 See General Assembly resolution 70/1.

in the abundance and geographic distribution of open ocean species. In the Southern Ocean, it was uncertain whether changes in sea ice habitats were affecting keystone species and, in particular, krill populations.

While major advances in the understanding of marine biological polar sciences (Robinson, 2009; Stoddart, 2010) during the International Polar Year (2007–2008) provided novel information for the first Assessment, advances in knowledge available for the second *World Ocean Assessment* have been the result of a variety of more limited initiatives.

2. Description of the environmental changes between 2010 and 2020

The overriding environmental change in the high-latitude ice habitat since the first Assessment has been a continuation of past change (figure I; see also chap. 5 of the present Assessment). The greatest advances in knowledge, capacity and the establishment of trends are largely associated with national and international programmes, such as the Multidisciplinary Drifting Observatory for the Study of Arctic Climate (MOSAiC) and the Antarctic Circumnavigation Expedition, and long-standing international organizations, such as the multinational Arctic Council and the Commission for the Conservation of Antarctic Marine Living Resources. Annual, and regularly scheduled, summaries of Arctic change, including ice habitats, are issued by States – for example, the *Arctic Report Card* of the United States National Oceanic and Atmospheric Administration (Richter-Menge and others, 2019) and the *State of the Arctic Ocean Report 2019* of Fisheries and Oceans Canada (Niemi and others, 2019) – and by international committees – for example, the "State of Arctic science report" of the International Arctic Science Committee (2020) and the Scientific Committee for Antarctic Research (2020). More global summaries, again including ice habitat change, are issued through the American Meteorological Society (Blunden and Arndt, 2019). The Arctic Council has produced 25-year pan-Arctic summaries of changes in the cryosphere (Arctic Monitoring and Assessment Programme (AMAP), 2017) and biodiversity (Conservation of Arctic Flora and Fauna Programme (CAFF), 2017).

2.1. Sea ice habitats

In the Arctic, ongoing long-term declines in sea ice extent (see also chap. 5), both in summer and winter, have occurred. The summer Arctic sea ice extent has reached a new, reduced, mean position, although it may be temporary (Vaughan and others, 2013). The new minimum also applies to sea ice thickness, through the loss of significant amounts of multi-year ice after 2007, and the maintenance of the reduction in the years since then (Serreze and Meier, 2019). It is worth noting that, while there is general Arctic sea ice decline, the Pacific sector of the Arctic is losing its ice much more quickly than is the case for other sectors of the Arctic, including the Canadian Arctic archipelago (see figure X in chap. 5).

In the Southern Ocean, although there has been strong inter-annual variability, similar to that noted in the first Assessment, there has essentially been no long-term change in the sea ice extent for summer or winter (figure I; see also chap. 5). From 2017 to 2019, however, January (minimum) levels have been consistently below the levels registered since satellite records began in 1979, especially in the regular ice-covered zones of the Weddell Sea and the Amundsen Sea. That may be a consequence of recent oceanographic warming in the Southern Ocean (Meehl and others, 2019).

Figure I

Trends in summer and winter sea ice extent for both the northern and southern hemisphere polar regions since satellite imagery became available in 1979

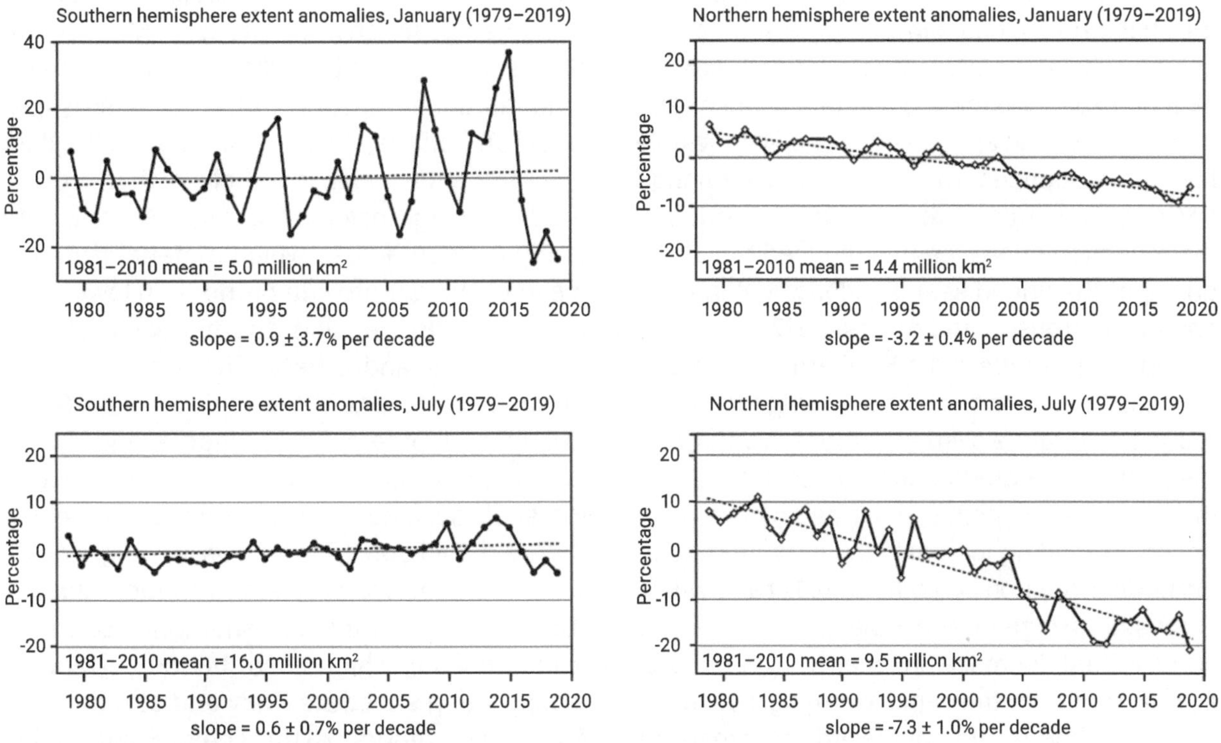

Sources: Fetterer and others, 2017; and the National Snow and Ice Data Center sea ice index, available at https://nsidc.org/data/seaice_index/compare_trends, which provides daily and monthly updates of Arctic and Antarctic sea ice extents and trends.

Note: The slope of the trend line in each panel is shown; the northern hemisphere trends are statistically significant at the 0.01 level, while the southern hemisphere trends are not significant.

The rapidly changing nature of the physical environment, combined with the relative inaccessibility of the polar oceans, means that studies have largely focused on climate change scenarios (see also chap. 5), especially at the base of the trophic system, rather than the identification of historical change. Limited studies of sea ice brine communities suggest no change as yet in relation to increased CO_2 concentrations or decreased pH (McMinn and others, 2017). However, phytoplankton productivity under the sea ice has been found to be unexpectedly high (Arrigo and others, 2012). Such changes may have positive impacts on benthic organisms and upper ocean organisms by increasing the food supply of particulate organic carbon to lower trophic levels (Oxtoby and others, 2017; Yasuhara and others, 2012; Xu and others, 2018). Diatoms from within the sea ice have been found to sustain under-ice production during winter in the northeast Chukchi Shelf (Koch and others, 2020).

The impact of the decreasing Arctic sea ice on populations of marine mammals and seabirds is species-specific and depends on the extent to which individual species rely on the sea ice habitat. While the ivory gull (*Pagophila eburnea*) has been identified as utilizing the Arctic's marginal ice zone and nearby open sea, Gilg and others (2016) found that approximately 80 per cent of seabird species were foraging in the increasingly rare high-concentration sea ice. That variable use of ice habitat may indicate adaptability under a changing climate.

Decreasing Arctic sea ice has led to general reductions of approximately 10 per cent in seabird numbers in the Bering Sea (Renner and others, 2016). There is some evidence that, as prey habitats are changing, species such as the beluga whale (*Delphinapterus leucas*) are exploiting expanded marine habitats (Hauser and others, 2018) and are generally showing flexible feeding responses to environmental change (O'Corry-Crowe and others, 2016). In contrast, the reduction in sea ice has reduced the abundance of the ringed seal (*Pusa hispida*) in Hudson Bay (Ferguson and others, 2017), and their distributional range in the Svalbard Archipelago has also contracted, which is leading to a major reduction in range overlap in those islands with the Arctic's top predator, the polar bear (*Ursus maritimus*). In response, polar bears have been observed feeding increasingly on ground-nesting birds (Hamilton and others, 2017) and whale carcasses (Pagano and others, 2020), with a concomitant increase in energy expenditure. In the Antarctic, the rapid warming has been shown to lead to the southward movement of krill (*Euphausia superba*) populations, with decreases in density, but increases in individual body length (Atkinson and others, 2019). Hückstädt and others (2020) suggest that that is likely to have negative consequences for species dependent on krill, such as the crabeater seal (*Lobodon carcinophaga*).

2.2. Ice shelf and iceberg habitats

The ice habitats of both ice shelves and icebergs extend up to hundreds of metres below the ocean surface, which means that their marine signatures are very different from those of sea ice, both in terms of their impact on the surrounding ocean and in the type of habitat that their subaerial and submarine surfaces provide. Ice shelves provide stable breeding platforms with direct access to the ocean where terminal thickness allows, and they have been utilized by species that are dependent on ice shelves for breeding – the emperor penguin (*Aptenodytes forsteri*), for example – for many years (Wienecke, 2012; Fretwell and others, 2014). The subaerial surfaces of ice shelves provide habitats for microbial mats, especially where aeolian or glacially entrained sediments are present (Mueller and others, 2006), thus providing a mechanism for long-distance transport of the organisms (Cefarelli and others, 2016). However, it is the dark environments under ice shelves that provide the surprisingly diverse habitats. Most of them are in the benthos, to which material from the ice shelves can provide nutrients (Hawes and others, 2018), leading to microbial activity (Vick-Majors and others, 2016) and a range of species present in the meiobenthos (Pawlowski and others, 2005; Ingole and Singh, 2010). Some organisms utilize the submarine ice shelf surface more directly. They include the bald rockcod (*Pagothenia borchgrevinki*), which forages for prey along the ice surface (Gutt, 2002), and the sea anemone (*Edwardsiella andrillae*), which uses the ice surface as a supporting substrate (Daly and others, 2013; Murray and others, 2016). The break-up of ice shelves in both the Arctic and Antarctic has led to the regional loss of that unique, dark environment, but significant biodiversity has spread into the regions newly exposed to surface inputs, leading to major carbon drawdown (Barnes and others, 2018).

Icebergs vary in size, from free-floating fractures from ice shelves, in particular but not exclusively in the Antarctic, to fragments of ice a few tens of metres in size broken off from the calving terminus of a tidewater glacier. As ecosystems, they therefore vary in their marine contribution greatly. At one extreme, they are effectively free-moving pieces of ice shelves, with the capacity for significant seabird nesting and feeding platforms in both the Antarctic (Ruhl and others, 2011; Joiris, 2018) and the Arctic. In the latter, both the ivory gull (Nachtsheim and others, 2016) and the kittiwake (*Rissa tridactyla*; Joiris, 2018) have been found in abundance on and near icebergs of various sizes. It has been speculated that past movement of giant icebergs in the Antarctic may have helped to facilitate the distribution of the Adélie penguin (*Pygoscelis adeliae*) through ice transport (Shepherd and others, 2005). Such large icebergs can also have negative impacts on ecosystems. If a giant iceberg grounds for long periods off an existing penguin colony, its presence, and the associated spread of fast ice, can block the passage of individual birds, preventing access to foraging grounds and leading to considerable chick mortality (Kooyman and others, 2007; Wilson and others, 2016). In addition, the grounding and scouring of bottom sediments by large icebergs is a physical disturbance and has a serious impact on benthic organisms (Kaiser and others, 2013; Yasuhara and others, 2007). In areas with frequent iceberg passage, such as extensive areas along the coastlines of Antarctica and Greenland (Bigg, 2015), as much as 30 per cent of the seabed may be disturbed in any one year, with up to two thirds of the benthic fauna in that area killed (Barnes, 2017). With an ecosystem recovery time of several years, the destruction could lead to significant loss in the short term in the ability of the area to act as a carbon store, in particular in shallow seas (Barnes and others, 2018).

The melting of icebergs allows for the input of nutrients and trace elements that are held in or on the ice into the water, creating a distinctive and productive local ecosystem (Smith and others, 2007; Smith and others, 2013). The melting process, with its associated upwelling of relatively fresh plumes, aids the input of nutrients into the surface waters (figure II), which can have chlorophyll concentrations that are 4–10 times above the background level. In association, near icebergs, there is an elevated bacterial population and a community composition that is different from that in the undisturbed water nearby (Kaufmann and others, 2011; Dinasquet and others, 2017). Further away, the combination of increased nutrients around the iceberg (Helly and others, 2011), as well as iron (Raiswell and others, 2008; De Jong and others, 2015) and silica (Hawkings and others, 2017) from the englacial debris released by the melting, leads to increases in phytoplankton levels (Vernet and others, 2011) and potential impacts on carbon sequestration (Cefarelli and others, 2016; Duprat and others, 2016).

The decay of ice shelves (e.g., Fettweis and others, 2017; Rignot and others, 2019) would be expected to lead to greater iceberg numbers; however, comprehensive, long-term iceberg number estimates both in the Arctic and Antarctic are lacking. Records of icebergs off Newfoundland (Bigg and others, 2014) and satellite-derived records of medium-small icebergs north of 66° S in the Southern Ocean (Tournadre and others, 2016) both report increasing numbers. The calving of giant icebergs (> 18 km in length) from ice shelves in the Antarctic, while very episodic in nature, also shows some evidence of recent increases in both number (figure III; see also the Antarctic Iceberg Tracking Database) and magnitude.

The likely increase in icebergs in both hemispheres has probably led to increased production and impact on coastal benthic ecosystems in recent years, but there is currently little evidence, with information on the impacts of iceberg flux largely derived from the Southern Ocean.

Figure II
Arctic/Antarctic ecosystem on and around an iceberg

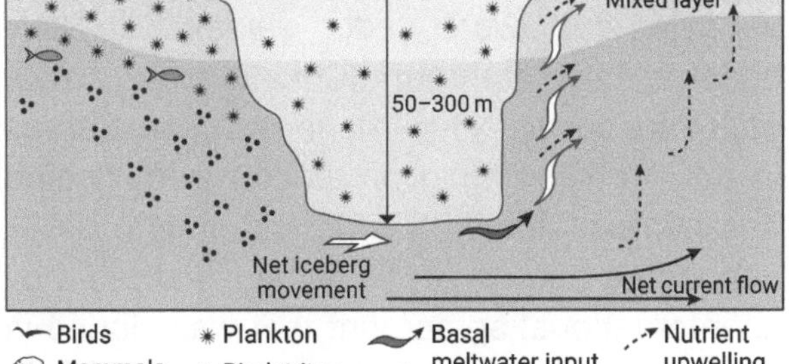

Source: Bigg, 2015. Reprinted with permission.

Figure III
Number of annual Antarctic giant iceberg calving events

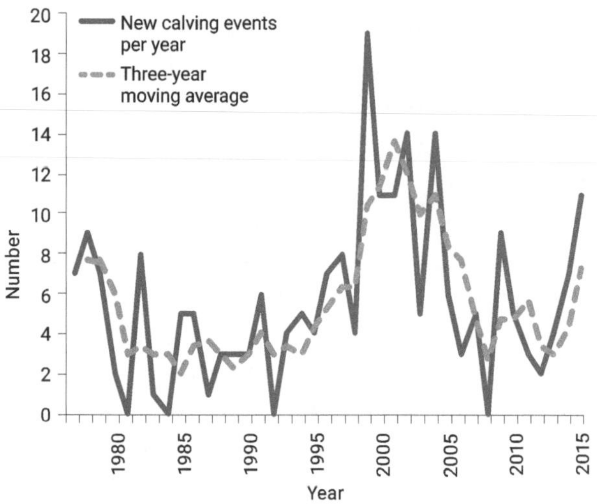

Sources: See Budge and Long, 2017; and the Antarctic Iceberg Tracking Database, available at www.scp.byu.edu/data/iceberg.
Note: Giant icebergs are greater than 18 km in one length dimension, but there is no consistent area/volume estimate available over the timescale.

3. Economic and social consequences

Historically, high-latitude ice habitats have experienced low levels of human activity, principally by indigenous inhabitants of the Arctic and its periphery. The continuing retreat of the habitat as a result of global warming, and the advance of human use of polar regions, is rapidly changing the relevance of the habitat for humanity, with associated economic and social consequences. While the decrease in sea ice increases opportunities for transoceanic shipping and the exploitation of sea floor hydrocarbon resources, the main driver for increased use of the Arctic so far is fishing (Eguíluz and others, 2016). More open ocean species can move north into now ice-free waters, increasing fishing opportunities, although the fish that rely on the sea ice habitat, such as polar cod (*Boreogadus saida*), will likely become less common (Christiansen, 2017). There are currently few marine protected areas in the Arctic offering protection from fishing or other exploitation (Harris and others, 2018), although a ban on Arctic fishing, instituted by an international agreement that was signed in October 2018, will limit the expansion of fishing activities in the Arctic for the next decade or more once 10 countries have ratified the agreement (European Commission, 2019). As of June 2020, however, only eight countries had done so. The initiative links directly to Sustainable Development Goal 14.

The direct impact of sea level rise from glacier melting and the associated freeing of once frozen coastlines in the Arctic is affecting, yet providing many opportunities for, communities and industries (Richter-Menge and others, 2019). Negative impacts include the loss of coastal ice roads, the elevation of flood levels, changes in nesting areas and alongshore coastal sediment transport, a reduction in subsistence hunting ranges, the release of previously trapped pollutants and even the loss of some coastal communities. Potential economic opportunities include the opening up of areas for ocean fishing activity, maritime transportation and new shipping routes and enhanced opportunities for renewable energy installations, as well as increasing opportunities for hydrocarbon exploitation. Those opportunities, however, have the potential to increase the risks associated with the activities, for example, habitat contamination from catastrophes such as oil spills (Cappello and others, 2014). It is worth noting that oil encased in sea ice does not readily degrade (Loftus and others, 2020).

Over time, as ice-free and therefore viable routes through the Arctic north of the Russian Federation (the Northern Sea Route) expanded, so too the number of vessels using the routes increased, with over 70 vessels sailing through the Northern Sea Route in 2013. However, the number, if not the tonnage, of vessels using the route has decreased in recent years, not exceeding 40 since 2014 (Northern Sea Route Information Office, 2019; Centre for High North Logistics Information Office, n.d.).

Oil and gas activities in the Arctic are variable. Canada has recently expanded a moratorium on issuing new drilling licences in its Arctic exclusive economic zone to prohibit all offshore oil and gas activities until the end of 2021 (Vigliotti, 2019). In the Arctic waters of the United States, the analogous drilling ban introduced in 2016 was removed in 2017 but restored in 2019. Its future remains subject to legal appeal (Gilmer, 2020). Western Arctic waters of the Russian Federation have seen limited drilling in recent years, but expansion is on hold for economic reasons and as a result of sanctions, although recent reports suggest that drilling may resume in 2020 or 2021 (Staalesen, 2019).

Most changes observed in the Arctic ice habitat have mixed consequences in terms of the Sustainable Development Goals, with hydrocarbon exploitation providing greater access to energy sources (Goal 7) and, with increased shipping, tourism and fishing enhancing local economic activity (Goal 8). However, those activities may work against creating a sustainable environment enriched by biodiversity (Goal 14) by causing further climate change and emissions (Goal 13), with associated pollution (Goals 12 and 14).

Some fishing grounds in the Antarctic, such as those for krill, occur in coastal waters in the South Atlantic and the Weddell Sea, where sea ice has shown signs of decrease. The broader implications of those decreases on the broader ecosystem and associated fishing grounds, however, are not yet clear. Hydrocarbon exploration has started on the plateau surrounding the Falkland Islands (Malvinas)[2] (MacAulay, 2015), although assessment of the associated environmental risks has only just begun and the area lies outside of the Antarctic governance system (Bigg and others, 2018). In the light of the importance of krill as a food source for a growing aquaculture industry, long-term management strategies for that species are beginning to be implemented in the area protected by the Commission for the Conservation of Antarctic Marine Living Resources (Commission for the Conservation of Antarctic Marine Living Resources (CCAMLR), 2019). Marine protected areas in some particular locations might help to address some of the management issues, and it will require more changes to be undertaken by the Antarctic Treaty System, in particular the Commission for the Conservation of Antarctic Marine Living Resources. A Ross Sea marine protected area was established in 2016, and other proposed marine protected areas, such as in the Weddell Sea and East Antarctica and near the Antarctic Peninsula, are being considered by members of the Commission.

2 A dispute exists between the Governments of Argentina and the United Kingdom of Great Britain and Northern Ireland concerning sovereignty over the Falkland Islands (Malvinas).

4. Outlook

The outlook for polar ice habitats remains very much as it was for the first Assessment. Arctic sea ice is expected to continue to retreat and thin, with the prospect of a seasonally ice-free Arctic very likely within the twenty-first century, although the timing of that key environmental event is still very uncertain (Serreze and Meier, 2019). Antarctic sea ice, while currently stable, is projected to decrease over the century (Naughton and others, 2018), mostly because of ocean warming. The latter is expected to affect Antarctic ice shelves by encouraging subsurface melting of up to 41–129 per cent by the end of the century (Naughten and others, 2018), with associated increases in iceberg calving. Continued warming in the Arctic is expected to result in increased melting of the Greenland Ice Sheet (Barry, 2017) and probably increased, if episodic, iceberg production.

The decrease in sea ice and ice shelves will continue to open up opportunities for the expansion of both pelagic and sea floor species, which will benefit from wider and improved feeding conditions (Christiansen, 2017), while threatening the viability of fish, in particular the polar cod (see Christiansen, 2017), and marine mammal populations for sea ice-dependent species (United Nations, 2017a). Many studies suggest that sea ice algae will become vulnerable to climate change, with reduced biodiversity and population declines (Hardge and others, 2017; Kiko and others, 2017). On the other hand, phytoplankton blooms may become more widespread, at least early in the summer before nutrient limitation occurs, under thinner, more lead-prone, snow-covered sea ice in the Arctic Ocean (Assmy and others, 2017; see also chap. 6A of the present Assessment). Such changes may have more wide-ranging impacts on carbon export, with seasonal sea ice zones switching to carbon sinks (Abelmann and others, 2015; Rapp and others, 2018). Decreasing sea ice may also reduce inputs of plastics to the Arctic Ocean, as sea ice currently contains, in orders of magnitude, more microplastics than the Arctic Ocean itself (see chap. 12 of the present Assessment; also Kanhai and others, 2020). In the Southern Ocean, where sea ice has demonstrated little long-term trend to date, it is known that individual-level specialization is lowest at sites where the inter-annual variability in sea ice is highest (McMullin and others, 2017), suggesting that there is scope for adaptation in a more variable future climate.

The opening up of the Arctic to navigation, fishing and exploitation of the sea floor and deeper resources will have major implications for high-latitude ice ecosystems (Harris and others, 2018) and for human populations, including indigenous peoples, that are reliant on high-latitude ice habitats. It will also have implications for achieving a number of Sustainable Development Goals. However, despite the first vessel sailing through the Northern Sea Route in August 2017 without being accompanied by an icebreaker (High North News, 2018), it is likely that cargo shipping will continue to need accompaniment, unless it is an "ice class" vessel, for the foreseeable future (Kiiski and others, 2018). As a result, Arctic routes will likely remain of secondary importance for some decades. Other factors limiting the use of such new shipping routes are the potential negative impact of increased shipping on Arctic marine mammals (Hauser and others, 2018), the unwanted facilitation of the transfer of non-indigenous species and the possible complex radiative feedback of ship exhaust fumes on the Arctic climate (Stephenson and others, 2018), with the latter potentially slowing the tendency for increases in ice-free periods.

5. Key remaining knowledge and capacity-building gaps

The inaccessibility of the high latitudes means that the ice habitat remains relatively poorly understood. Sea ice environments are currently the best studied of the marine ice habitats considered in the present subchapter, but, even for sea ice, a comprehensive food web study is yet to be conducted. Many food web studies have focused on just one aspect (Dickinson and others, 2016). In general, the understanding of the three-dimensional nature of ice habitats (Bluhm and others, 2018), the range and number of species within them and their spatial and temporal variability is still very limited (Christiansen, 2017). The lack of data extends also to the impact of the presence or absence of such habitats on the surrounding ocean and carbon sequestration (Barnes, 2017).

Similarly, the difficulty of access to ice shelves, marine areas near glaciers (Zappalà and others, 2017) and, in particular, the submarine environment beneath them, makes gaining new information about that ice habitat rare. Much analysis has been, and will remain, from remote sensing, with new satellite systems promising to revolutionize first-order knowledge of the habitats. It will be important to ensure ready and universal access to the new data produced by the observing platforms in order to address current knowledge and capacity gaps.

References

Abelmann, Andrea, and others (2015). The seasonal sea-ice zone in the glacial Southern Ocean as a carbon sink. *Nature Communications*, vol. 6, art. 8136.

Arctic Monitoring and Assessment Programme (AMAP) (2017). Snow, water, ice and permafrost in the Arctic. Summary for Policy-makers. https://swipa.amap.no.

Arrigo, Kevin R., and others (2012). Massive phytoplankton blooms under Arctic sea ice. *Science*, vol. 336, No. 6087, pp. 1408–1408.

Assmy, Philipp, and others (2017). Leads in Arctic pack ice enable early phytoplankton blooms below snow-covered sea ice. *Scientific Reports*, vol. 7, art. 40850.

Atkinson, Angus, and others (2019). Krill (*Euphausia superba*) distribution contracts southward during rapid regional warming. *Nature Climate Change*, vol. 9, No. 2, pp. 142–147.

Barnes, David K.A. (2017). Polar zoobenthos blue carbon storage increases with sea ice losses, because across-shelf growth gains from longer algal blooms outweigh ice scour mortality in the shallows. *Global Change Biology*, vol. 23, No. 12, pp. 5083–5091.

Barnes, David K.A., and others (2018). Icebergs, sea ice, blue carbon and Antarctic climate feedbacks. *Philosophical Transactions of the Royal Society A: Mathematical, Physical and Engineering Sciences*, vol. 376, No. 2122, 20170176.

Barry, Roger G. (2017). The Arctic cryosphere in the twenty-first century. *Geographical Review*, vol. 107, No. 1, pp. 69–88.

Bigg, Grant R. (2015). *Icebergs: Their Science and Links to Global Change*. Cambridge: Cambridge University Press.

Bigg, Grant R., and others (2018). A model for assessing iceberg hazard. *Natural Hazards*, vol. 92, No. 2, pp. 1113–1136.

Bigg, Grant R., and others (2014). A century of variation in the dependence of Greenland iceberg calving on ice sheet surface mass balance and regional climate change. *Proceedings of the Royal Society A: Mathematical, Physical and Engineering Sciences*, vol. 470, No. 2166, 20130662.

Bluhm, Bodil A., and others (2018). Sea ice meiofauna distribution on local to pan-Arctic scales. *Ecology and Evolution*, vol. 8, No. 4, pp. 2350–2364.

Blunden, Jessica, and Derek S. Arndt, eds. (2019). State of the Climate in 2018. *Bulletin of the American Meteorological Society*, vol. 100, No. 9, pp. Si–S305.

Budge, Jeffrey S., and David G. Long (2017). A comprehensive Database for Antarctic iceberg tracking using scatterometer data. *IEEE Journal of Selected Topics in Applied Earth Observations*, vol. 11, No. 2, https://doi.org/10.1109/JSTARS.2017.2784186.

Cappello, Simone, and others (2014). STRANgE, integrated physical–biological–mechanical system for recovery in of the "oil spill" in Antarctic environment. *Reviews in Environmental Science and Bio/Technology*, vol. 13, No. 4, pp. 369–375.

Cefarelli, Adrián O., and others (2016). Diatoms (Bacillariophyceae) associated with free-drifting Antarctic icebergs: taxonomy and distribution. *Polar Biology*, vol. 39, No.3 , pp. 443–459.

Centre for High North Logistics Information Office (n.d.). "Northern Sea Route transit statistics". Available at http://arctic-lio.com/category/statistics.

Christiansen, Jørgen S. (2017). No future for Euro-Arctic ocean fishes? *Marine Ecology Progress Series*, vol. 575, pp. 217–227.

Commission for the Conservation of Antarctic Marine Living Resources (CCAMLR) (2019). CCAMLR: Commission for the Conservation of Antarctic Marine Living Resources website. 2019. www.ccamlr.org.

Conservation of Arctic Flora and Fauna Programme (CAFF) (2017). State of the Arctic Marine Biodiversity Report. www.arcticbiodiversity.is/marine.

Daly, Marymegan, and others (2013). *Edwardsiella andrillae*, a new species of sea anemone from Antarctic Ice. *PloS One*, vol. 8, No. 12, e83476.

De Jong, J.T.M. and others (2015). Sources and fluxes of dissolved iron in the Bellingshausen Sea (West Antarctica): The importance of sea ice, icebergs and the continental margin. *Marine Chemistry*, vol. 177, pp. 518–535.

Dickinson, Iain, and others (2016). Microbes and the Arctic Ocean. In *Their World: A Diversity of Microbial Environments*, pp. 341–381.

Dinasquet, Julie, and others (2017). Mixing of water masses caused by a drifting iceberg affects bacterial activity, community composition and substrate utilization capability in the Southern Ocean. *Environmental Microbiology*, vol. 19, No. 6, pp. 2453–2467.

Duprat, Luis P.A.M., and others (2016). Enhanced Southern Ocean marine productivity due to fertilization by giant icebergs. *Nature Geoscience*, vol. 9, No. 3, p. 219.

Eguíluz, Victor M., and others (2016). A quantitative assessment of Arctic shipping in 2010–2014. *Scientific Reports*, vol. 6, art. 30682.

European Commission (2019). EU and Arctic partners enter historic agreement to prevent unregulated fishing in high seas. Fisheries – European Commission. 2019. https://ec.europa.eu/fisheries/eu-and-arctic-partners-enter-historic-agreement-prevent-unregulated-fishing-high-seas_en.

Ferguson, Steven H., and others (2017). Demographic, ecological, and physiological responses of ringed seals to an abrupt decline in sea ice availability. *PeerJ*, vol. 5, e2957.

Fetterer, Florence, and others (2017). *Sea Ice Index, Version 3*. Boulder, Colorado: NSIDC: National Snow and Ice Data Center. https://nsidc.org/data/G02135/versions/3.

Fettweis, Xavier, and others (2017). Reconstructions of the 1900–2015 Greenland ice sheet surface mass balance using the regional climate MAR model. *The Cryosphere*, vol. 11, pp. 1015–1033.

Fretwell, Peter T., and others (2014). Emperor penguins breeding on iceshelves. *PLoS One*, vol. 9, No.1, e85285.

Gilg, Olivier, and others (2016). Living on the edge of a shrinking habitat: the ivory gull, *Pagophila eburnea*, an endangered sea-ice specialist. *Biology Letters*, vol. 12, No. 11, 20160277.

Gilmer, Ellen M. (2020). Judges weight Trump's bid to reopen parts of Arctic to drilling. https://news.bloomberglaw.com/environment-and-energy/judges-weigh-trumps-bid-to-reopen-parts-of-arctic-to-drilling.

Gutt, Julian (2002). The Antarctic ice shelf: an extreme habitat for notothenioid fish. *Polar Biology*, vol. 25, No. 4, pp. 320–322.

Hamilton, Charmain D., and others (2017). An Arctic predator-prey system in flux: climate change impacts on coastal space use by polar bears and ringed seals. *Journal of Animal Ecology*, vol. 86, No. 5, pp. 1054–1064.

Hardge, Kristin, and others (2017). The importance of sea ice for exchange of habitat-specific protist communities in the Central Arctic Ocean. *Journal of Marine Systems*, vol. 165, pp. 124–138.

Harris, Peter T., and others (2018). Arctic marine conservation is not prepared for the coming melt. *ICES Journal of Marine Science*, vol. 75, No. 1, pp. 61–71.

Hauser, Donna D.W., and others (2018). Vulnerability of Arctic marine mammals to vessel traffic in the increasingly ice-free Northwest Passage and Northern Sea Route. *Proceedings of the National Academy of Sciences*, vol. 115, No. 29, pp. 7617–7622.

Hawes, I., and others (2018). The "Dirty Ice" of the McMurdo Ice Shelf: analogues for biological oases during the Cryogenian. *Geobiology*, vol. 16, No. 4, pp. 369–377.

Hawkings, Jon R., and others (2017). Ice sheets as a missing source of silica to the polar oceans. *Nature Communications*, vol. 8, art. 14198.

Helly, John J., and others (2011). Cooling, dilution and mixing of ocean water by free-drifting icebergs in the Weddell Sea. *Deep Sea Research Part II: Topical Studies in Oceanography*, vol. 58, No. 11–12, pp. 1346–1363.

High North News (2018). The Northern Sea Route is alive and well. www.highnorthnews.com/en/op-ed-northern-sea-route-alive-and-well.

Hückstädt, Luis A., and others (2020). Projected shifts in the foraging habitat of crabeater seals along the Antarctic Peninsula. *Nature Climate Change*, vol. 10, No. 5, pp. 472–477.

International Arctic Science Committee (2020). State of Arctic Science Report, 2020. International Arctic Science Committee. pp. 1–26.

Ingole, B.S., and Ravail Singh (2010). Biodiversity and community structure of freeliving marine nematodes from the Larsemann Ice Shelf, East Antarctica. *Current Science*, vol. 99, No. 10, pp. 1413–1419.

Joiris, Claude R. (2018). Hotspots of kittiwakes *Rissa tridactyla* on icebergs off southwest Greenland in autumn. *Polar Biology*, vol. 41, No. 11, pp. 2375–2378.

Kaiser, Stefanie, and others (2013). Patterns, processes and vulnerability of Southern Ocean benthos: a decadal leap in knowledge and understanding. *Marine Biology*, vol. 160, No. 9, pp. 2295–2317.

Kanhai, La Daana K., and others (2020). Microplastics in sea ice and seawater beneath ice floes. *Scientific Reports*, vol. 10, No. 11, art. 5004.

Kaufmann, Ronald S., and others (2011). Composition and structure of macrozooplankton and micronekton communities in the vicinity of free-drifting Antarctic icebergs. *Deep Sea Research Part II: Topical Studies in Oceanography*, vol. 58, Nos. 11–12, pp. 1469–1484.

Kiiski, Tuomas, and others (2018). Long-term dynamics of shipping and icebreaker capacity along the Northern Sea Route. *Maritime Economics & Logistics*, vol. 20, No .3, pp. 375–399.

Kiko, Rainer, and others (2017). Colonization of newly forming Arctic sea ice by meiofauna: a case study for the future Arctic? *Polar Biology*, vol. 40, No. 6, pp. 1277–1288.

Koch, Chelsea Wegner, and others (2020). Seasonal and latitudinal variations in sea ice algae deposition in the Northern Bering and Chukchi Seas determined by algal biomarkers. *PLoS One*, vol. 15, No. 4. https://doi.org/10.1371/journal.pone.0231178.

Kooyman, Gerald L., and others (2007). Effects of giant icebergs on two emperor penguin colonies in the Ross Sea, Antarctica. *Antarctic Science*, vol. 19, No. 1, pp. 31–38.

Loftus, Synnove, and others (2020). Biodegradation of weathered crude oil in seawater with frazil ice. *Marine Pollution Bulletin*, vol. 154. https://doi.org/10.1016/j.marpolbul.2020.111090.

MacAulay, F. (2015). Sea Lion Field discovery and appraisal: a turning point for the North Falkland Basin. *Petroleum Geoscience*, vol. 21, Nos. 2–3, pp. 111–124.

McMinn, Andrew, and others (2017). Effects of CO_2 concentration on a late summer surface sea ice community. *Marine Biology*, vol. 164, No. 4, art. 87.

McMullin, Rebecca M., and others (2017). Trophic position of Antarctic ice fishes reflects food web structure along a gradient in sea ice persistence. *Marine Ecology Progress Series*, vol. 564, pp. 87–98.

Meehl, Gerald A., and others (2019). Sustained ocean changes contributed to sudden Antarctic sea ice retreat in late 2016. *Nature Communications*, vol. 10, No. 1, art. 14.

Mueller, Derek R., and others (2006). Environmental gradients, fragmented habitats, and microbiota of a northern ice shelf cryoecosystem, Ellesmere Island, Canada. *Arctic, Antarctic, and Alpine Research*, vol. 38, No. 4, pp. 593–607.

Murray, Alison E., and others (2016). Microbiome composition and diversity of the ice-dwelling sea anemone, *Edwardsiella andrillae. Integrative and Comparative Biology*, vol. 56, No. 4, pp. 542–555.

Nachtsheim, Dominik A., and others (2016). A gravel-covered iceberg provides an offshore breeding site for ivory gulls *Pagophila eburnea* off Northeast Greenland. *Polar Biology*, vol. 39, No. 4, pp. 755–758.

Naughten, Kaitlin A., and others (2018). Future projections of Antarctic ice shelf melting based on CMIP5 scenarios. *Journal of Climate*, vol. 31, No. 13, pp. 5243–5261.

Niemi, Andrea, and others (2019). State of Canada's Arctic Seas. Canadian Technical Report of Fisheries and Aquatic Sciences 3344, pp. xv–189.

Northern Sea Route Information Office, 2019 (https://arctic-lio.com).

O'Corry-Crowe, Greg, and others (2016). Genetic profiling links changing sea-ice to shifting beluga whale migration patterns. *Biology Letters*, vol. 12, No. 11, 20160404.

Oxtoby, L.E., and others (2017). Resource partitioning between Pacific walruses and bearded seals in the Alaska Arctic and sub-Arctic. *Oecologia*, vol. 184, No. 2, pp. 385–398.

Pagano, Anthony M., and others (2020). The seasonal energetic landscape of an apex marine carnivore, the polar bear. *Ecology*, vol. 101, No. 3, e02959.

Pawlowski, Jan, and others (2005). Allogromiid foraminifera and gromiids from under the Ross Ice Shelf: morphological and molecular diversity. *Polar Biology*, vol. 28, No. 7, pp. 514–522.

Raiswell, Rob, and others (2008). Bioavailable iron in the Southern Ocean: the significance of the iceberg conveyor belt. *Geochemical Transactions*, vol. 9, No. 1, No. 7.

Rapp, Josephine Z., and others (2018). Effects of ice-algal aggregate export on the connectivity of bacterial communities in the central Arctic Ocean. *Frontiers in Microbiology*, vol. 9, art. 1035.

Renner, Martin, and others (2016). Timing of ice retreat alters seabird abundances and distributions in the southeast Bering Sea. *Biology Letters*, vol. 12, No. 9, 20160276.

Richter-Menge, Jackie, and others, eds. (2019). *Arctic Report Card*. https://arctic.noaa.gov/Report-Card.

Rignot, Eric, and others (2019). Four decades of Antarctic Ice Sheet mass balance from 1979–2017. *Proceedings of the National Academy of Sciences*, vol. 116, No. 4, pp. 1095–1103.

Robinson, Sharon A. (2009). Introduction: Climate change biology at the ends of the Earth-International Polar year special issue. *Global Change Biology*, vol. 15, No. 7, pp. 1615–1617.

Ruhl, Henry A., and others (2011). Seabird aggregation around free-drifting icebergs in the northwest Weddell and Scotia Seas. *Deep Sea Research Part II: Topical Studies in Oceanography*, vol. 58, Nos. 11–12, pp. 1497–1504.

Scientific Committee for Antarctic Research (2020). Scientific Committee for Antarctic Research website. www.scar.org.

Serreze, Mark C., and Walter N. Meier (2019). The Arctic's sea ice cover: trends, variability, predictability, and comparisons to the Antarctic. *Annals of the New York Academy of Sciences*, vol. 1436, No. 1, pp. 36–53.

Shepherd, L.D., and others (2005). Microevolution and mega-icebergs in the Antarctic. *Proceedings of the National Academy of Sciences*, vol. 102, No. 46, pp. 16717–16722.

Smith, Kenneth L., and others (2013). Icebergs as unique Lagrangian ecosystems in polar seas. *Annual Review of Marine Science*, vol. 5, pp. 269–287.

Smith, Kenneth L., and others (2007). Free-drifting icebergs: hot spots of chemical and biological enrichment in the Weddell Sea. *Science*, vol. 317, No. 5837, pp. 478–482.

Staalesen, Atle (2019). Russia's biggest oil company announces more offshore Arctic drilling. *Arctic Today*. www.arctictoday.com/russias-biggest-oil-company-announces-more-offshore-arctic- drilling.

Stephenson, Scott R., and others (2018). Climatic responses to future trans-Arctic shipping. *Geophysical Research Letters*, vol. 45, No. 18, pp. 9898–9908.

Stoddart, Michael (2010). Antarctic biology in the 21st century – advances in, and beyond the international polar year 2007–2008. *Polar Science*, vol. 4, No. 2, pp. 97–101.

Tournadre, J., and others (2016). Antarctic icebergs distributions 1992–2014. *Journal of Geophysical Research: Oceans*, vol. 121, No. 1, pp. 327–349.

United Nations (2017a). Chapter 46: High-latitude ice and the biodiversity dependent on it. In *The First Global Integrated Marine Assessment: World Ocean Assessment I*. Cambridge: Cambridge University Press.

________ (2017b). *The First Global Integrated Marine Assessment: World Ocean Assessment I*. Cambridge: Cambridge University Press.

Vaughan, David G., and others (2013). Observations: cryosphere. *Climate Change*, vol. 2103, pp. 317–382.

Vernet, M., and others (2011). Impacts on phytoplankton dynamics by free-drifting icebergs in the NW Weddell Sea. *Deep Sea Research Part II: Topical Studies in Oceanography*, vol. 58, No.11–12, pp. 1422–1435.

Vick-Majors, Trista J., and others (2016). Biogeochemistry and microbial diversity in the marine cavity beneath the McMurdo Ice Shelf, Antarctica. *Limnology and Oceanography*, vol. 61, No. 2, pp. 572–586.

Vigliotti, Marco (2019). Trudeau government expands moratorium on oil and gas work in Arctic waters. https://ipolitics.ca/2019/08/08/trudeau-government-expands-moratorium-on-oil-and-gas-work-in-arctic-waters.

Wienecke, Barbara (2012). Emperor penguins at the West Ice Shelf. *Polar Biology*, vol. 35, No. 9, pp. 1289–1296.

Wilson, Kerry-Jayne, and others (2016). The impact of the giant iceberg B09B on population size and breeding success of Adélie penguins in Commonwealth Bay, Antarctica. *Antarctic Science*, vol. 28, No. 3, pp. 187–193.

Xu, Zhiqiang, and others (2018). Inter-annual variation of the summer zooplankton community in the Chukchi Sea: spatial heterogeneity during a decade of rapid ice decline. *Polar Biology*, vol. 41, No. 9, pp. 1827–1843.

Yasuhara, Moriaki, and others (2007). Modern benthic ostracodes from Lutzow-Holm Bay, East Antarctica: paleoceanographic, paleobiogeographic, and evolutionary significance. *Micropaleontology*, vol. 53, No. 6, pp. 469–496.

Yasuhara, Moriaki, and others (2012). Patterns and controlling factors of species diversity in the Arctic Ocean. *Journal of Biogeography*, vol. 39, No. 11, pp. 2081–2088.

Zappalà, G., and others (2017). New Advanced Technology Devices for Operational Oceanography in Extreme Conditions. *International Journal of Sustainable Development and Planning*, vol. 12, No. 1, pp. 61–70.

Chapter 7L
Seamounts and pinnacles

Contributors: Malcolm R. Clark (convener), Angelo F. Bernardino, J. Murray Roberts, Bhavani E. Narayanaswamy, Paul Snelgrove and Joshua T. Tuhumwire (subchapter lead member).

Keynote points

- Seamounts and pinnacles are common topographic features of the global ocean.
- Sampling effort has increased in recent years but only a small percentage of seamounts has been sampled in detail.
- Limited sampling, combined with high environmental variability among seamounts, constrains biodiversity knowledge.
- Fishing, especially bottom trawling, constitutes the greatest current threat to seamount ecosystems but marine debris or litter, climate change and potential seabed mining are additional concerns. However, initiatives to protect seamounts are increasing.
- Recent time series research on deep seamounts shows limited, if any, recovery of stony coral communities over 15–20-year periods.

1. Introduction

Seamounts, knolls and pinnacles, collectively termed seamounts in the present subchapter, are submerged volcanoes, rising hundreds to thousands of metres above the sea floor. Estimates of their numbers vary, depending on the data source and algorithms used, from several tens of thousands of seamounts and over 100,000 knolls (Yesson and others, 2011; Harris and others, 2014) to extrapolations of more than 100,000 seamounts and 25 million knolls and pinnacles (Wessel and others, 2010). They collectively cover up to 20 per cent of the deep-sea floor (Yesson and others, 2011).

Three important characteristics distinguish seamounts from surrounding deep-sea habitat (Clark, 2009): their topography provides a range of depths for different communities; their typical hard rock surfaces contrast with the fine unconsolidated sediments that cover much of the sea floor; and their physical structure can alter local hydrography and currents, concentrating species and productivity. Such factors can result in richer and more diverse benthic communities than those on adjacent slope or abyssal plain habitats.

Depth-related environmental parameters strongly influence seamount species composition, together with sea floor type and character (Clark and others, 2010). Seamounts provide significant ecosystem services and often host aggregations of fishes, with substantial commercial activity. Annual landings of primary seamount species have fluctuated around 100,000 tons since the 1990s, dominated by armourheads, alfonsinos, oreosomatids and orange roughy (Clark and others, 2007; Watson and others, 2007).

Several "ecological paradigms" developed around seamounts, regarding them as unique environments and hotspots of biodiversity and endemicity. However, many seamounts are not highly isolated (Rowden and others, 2010a), and most do not have high levels of endemism. They share many species with other deep-sea habitats (Howell and others, 2010; Narayanaswamy and others, 2013), although variability in seamount topography and physical dynamics can induce high species turnover and distinct assemblage or abundance characteristics (Schlacher and others, 2014).

Chapter 51 of the first *World Ocean Assessment* (United Nations, 2017a) and parts of its chapters 34, 35 and 36F drew on multiple seamount reviews (e.g., Pitcher and others, 2007; Clark and others, 2010; Staudigel and others, 2010) and findings of the Census of Marine Life programme (e.g., Rowden and others, 2010a; Stocks and others, 2012; Clark and others, 2012). However, limited sampling constrains the understanding of seamounts. There are about 700 seamounts with data in the SeamountsOnline (Stocks, 2010) and Seamount Ecosystem Evaluation Framework databases (Kvile and others, 2014), but only about 300 have been surveyed in detail, with few in equatorial latitudes or with summits deeper than 2,000 m. Therefore, the structure, function and connectivity of seamount ecosystems remains largely unknown (Clark and others, 2012).

2. Description of changes in knowledge between 2010 and 2020

Over the past five years, several national or international research programmes have yielded considerable ecological information on seamount environments. They are briefly described in section 5, but key ecological results are synthesized below, drawing on a review of seamount ecology (Rogers, 2018).

While seamounts rarely exist in steady-state oceanographic flows (Lavelle and Mohn, 2010), there can be circulation entrainment and homogenization of the water column (Meredith and others, 2015). Internal wave formation can upwell nutrients, enhancing primary productivity in summit areas (Turnewitsch and others, 2016; Read and Pollard, 2017). However, elevated primary productivity may be transient (e.g., Lemos and others, 2018), rarely increasing zooplankton abundances. Zooplankton communities over seamounts reflect the surrounding oceanic waters (e.g., Carmo and others, 2013; Denda and Christiansen, 2014; Denda and others, 2017), but blockage of migrating zooplankton and micronekton by seamount topography can enhance predation by fishes and shrimps (e.g., Nishida and others, 2016; Preciado and others, 2017; Letessier and others, 2017).

Seamounts with shallow summits can support macroalgae, and macroalgal forests have recently been recorded in the North Atlantic (Ramos and others, 2016; Stefanoudis and others, 2019) and the North-East Pacific (Du Preez and others, 2016). Extensive rhodolith beds have been discovered on seamounts and ridges in the South-West Atlantic (Meirelles and others, 2015), the North-West Atlantic (Stefanoudis and others, 2019) and the South-West Pacific (Clark and others, 2017). Such algae can be important in the carbonate budgets of mesophotic ecosystems. Pereira-Filho and others (2012) estimated that carbonate production from the rhodoliths of four seamounts of the Vitoria-Trinidade seamount chain off Brazil was 1.5 x 10^{-3} gigatons per year.

Knowledge of seamount communities has improved through the modelling of species distributions related to physico-chemical conditions. The modelling, especially for deep-sea corals, which can be abundant on seamounts (e.g., Rowden and others, 2010b; Tracey and others, 2011), suggests that key environmental variables include calcite and/or aragonite saturation depths, topographic aspect, temperature, salinity, oxygen levels and particulate organic carbon (e.g., Davies and Guinotte, 2011; Yesson and others, 2012, 2017; Anderson and others, 2016a). Nevertheless, models can perform poorly, depending on the resolution of environmental data (Anderson and others, 2016b; Rowden and others, 2017). New data will also influence such models, as seen by the discovery of stony coral reef on seamounts in the North-West Pacific Ocean, in poor aragonite saturation conditions (Baco and others, 2017). Environmental parameters may also influence species replacement and species richness differently (Victorero and others, 2018).

Connectivity mechanisms between seamounts have been a focus of recent research. Seamounts can act as "stepping stones" across large regions, but there is no consistent pattern (Rowden and others, 2010a). The reef-building stony coral (*Solenosmilia variabilis*) and the cup coral (*Desmophyllum dianthus*) have similar widespread distributions across the southern hemisphere, but similar genetic structure for *D. dianthus* over large areas (thousands of km) contrasts with variation in *S. variabilis* between individual seamounts in close proximity (tens of km) (Miller and Gunasekera, 2017). The latter mechanism of "self-recruiting" populations also occurs in bivalve molluscs (Beeston and others, 2018). Spatial patterns of connectivity vary among species (e.g., Zeng and others, 2017) and even within the same genus (Pante and others, 2015). Currents can provide routes or barriers for larval dispersal (Dueñas and others, 2016; Holland and others, 2019).

The research highlights variability in environmental factors and faunal communities between seamounts, making it impossible to generalize about the ecology of seamounts, and emphasizes the importance of sampling across a wide range of physical and geographical seamount characteristics (Clark and others, 2012).

3. Description of economic and social changes

Artisanal fisheries date back to the 1500s and, even today, small-scale fisheries close to oceanic islands are important for employment, with estimated catches (mainly tuna) of between 150,000 and 250,000 tons per year (Da Silva and Pinho, 2007). Deep-sea demersal fisheries for such species as alfonsino and orange roughy have generally declined since the mid-1990s (Clark and others, 2007; Watson and others, 2007; Pitcher and others, 2010) to current levels below 100,000 tons per year. In addition to fishes, small invertebrate fisheries on seamounts have targeted lobsters in the South Atlantic Ocean and the southern Indian Ocean and deep-sea red crab in the North-East Atlantic Ocean (Rogers, 2018).

Seamounts host ferromanganese crusts that contain cobalt, nickel and rare earth elements that have commercial potential (Hein and others, 2013). Five exploration contracts for the crusts exist with the International Seabed Authority: four cover seamounts in the North-West Pacific, and the fifth, the Rio Grande Rise off Brazil.[1] There is currently no mining for deep-sea minerals, but mining operations could significantly impact seamount ecosystems (e.g., Levin and others, 2016; Miller and others, 2018). Hence, regulations are being developed by the Authority to balance potential exploitation with environmental conservation.

Litter and plastic debris represent a growing concern. Lost fishing lines, nets and pots (e.g., Maldonado and others, 2015; Vieira and others, 2015; Woodall and others, 2015) entangle or physically damage seamount-associated species. Microplastics have been found in animals sampled from southern Indian Ocean seamounts, as well as in the sediment (Woodall and others, 2014; Taylor and others, 2016). There is also concern that invasive species can spread through such debris and litter.

Recovery from fishing or potential mining, along with a return to economic or social value, could be very slow. Seamount fisheries can remove much of the benthic fauna, leading to declines in biodiversity and abundance

1 Available at www.isa.org.jm/deep-seabed-minerals-contractors.

(Clark and others, 2015). Seamounts in the Hawaiian-Emperor seamount chain still yield sporadic, small catches (e.g., Bensch and others, 2008) and several small orange roughy fisheries off New Zealand and Tasmania have reopened (Food and Agriculture Organization of the United Nations (FAO), 2018). Benthic habitats, however, may require decades to recover. Time series surveys off New Zealand show few signs of change in stony coral communities 15 years after closure to trawling (Clark and others, 2019), although anemones and small corals may have increased on some seamounts off Tasmania (Clark and others, 2010). A recent survey in the North Pacific shows some potential recovery from trawling in the 1970s (Baco and others, 2019), but Japanese research on a seamount previously fished for precious corals indicates no sign of recovery (Bruckner, 2014).

4. Key region-specific research in recent years

4.1. Arctic Ocean

Little work has been undertaken on seamounts in Arctic waters. However, in 2017, high levels of sponge density and diversity were found on the Schultz Massif seamount, possibly associated with the occurrence of warmer, oxygen-rich and food-rich currents (Jones and others, 2018).

4.2. North Atlantic Ocean

Recent work has focused on seamounts in the North-East Atlantic. Data from the Anton Dohrn seamount have revealed 13 biotopes, 10 of which met the criteria for vulnerable marine ecosystems (Davies and others, 2015). The Hebrides Terrace seamount was surveyed for the first time in 2012, and cold water coral habitats (Henry and others, 2014) and a spawning site for the deep-water skate (*Bathyraja richardsoni*) (Henry and others, 2016) were identified. The European Union Atlas programme has completed surveys of the Bowditch seamount (Bermuda) and the Formigas seamount (Azores) and, on Tropic seamount, it reported extensive areas of sponge (*Poliopogon amadou*), octocorals, *Solenosmilia variabilis* coral reefs, xenophyophores and crinoid fields (Ramiro-Sánchez and others, 2019). Tropic seamount has extensive areas of ferromanganese crusts of potential mining interest (Murton and others, 2017).

4.3. South Atlantic Ocean

Recent work associated with mineral and oil and gas exploration has advanced physico-chemical descriptions of the Vitoria-Trindade seamount chain and the Rio Grande Rise (Bernardino and Sumida, 2017; Montserrat and others, 2019). Seamounts share species pools with nearby continental slopes, but there are structurally distinct substrates across regions (O'Hara and others, 2010; Bernardino and others, 2016; Almada and Bernardino, 2017), which suggests a high diversity of benthic and pelagic fauna (Perez and others, 2018).

Surveys by the United Kingdom have also been carried out on seamounts off the Saint Helena, Ascension and Tristan da Cunha islands.[2]

4.4. Indian Ocean

Seamounts in the Indian Ocean region remain poorly studied, although several seamounts on the South-West Indian Ridge and the Madagascar Ridge have been surveyed in recent years (Rogers, 2016). The surveys found distinct

2 See www.bas.ac.uk/project/protecting-marine-ecosystems-in-the-south-atlantic.

microbial and phytoplankton communities across the ridge (Djurhuus and others, 2017; Sonnekus and others, 2017), high cephalopod diversity (Laptikhovsky and others, 2017) and high benthic faunal diversity between seamounts.

4.5. North Pacific Ocean

The United States has been active in the northern and central Pacific Ocean, with campaigns in 2015, 2016 and 2017 that included mapping and remotely operated vehicle dives on seamounts and ridges of the Papahānaumokuākea Marine National Monument (including the Hawaiian-Emperor seamount chain) around a number of United States islands in the central Pacific and extended down to Samoa, Tokelau and the Cook Islands. There were 18 dives on seamounts of the Musician seamounts chain. Most seamounts had only one remotely operated vehicle dive but revealed diverse and abundant benthic deep-sea coral and sponge communities (Kennedy and others, 2019). Further work with remotely operated vehicle dives, coring and trawls on four seamounts in the Gulf of Alaska was carried out in 2019.

Chinese researchers have carried out a number of surveys of seamounts in the north-western Pacific, including the Caroline, Yap and Magellan seamounts.

Increased survey work is also being carried out by deep-sea minerals contractors with exploration licences for cobalt crusts in the north-western Pacific. Contractors (COMRA (China), KIOST (Republic of Korea), Russian Federation, JOGMEC (Japan)) sampled 11 seamounts in 2017 and 2018 and found many new species among benthic communities of sponges, corals and echinoderms (e.g., Wang and others, 2016; Dong and others, 2017). Some sampling of abyssal hills and seamounts in the Clarion-Clipperton Fracture Zone has also occurred.

In the Canadian North Pacific, Fisheries and Oceans Canada has conducted a baseline characterization of several seamounts. In 2018, an autonomous monitoring array was deployed to collect environmental data on Dellwood seamount, along with hydrophones to detect whale presence. Thirty potential long-term monitoring sites have been established with photographic surveys. A science plan is being developed for SGaan Kinghlas-Bowie seamount and other Pacific seamounts.

4.6. South Pacific Ocean

Research has focused on investigating the potential recovery of benthic communities from bottom trawling. Further time series surveys were completed off New Zealand in 2015 (Clark and others, 2019) and off Tasmania in 2018.[3] The original stony coral reefs show little recovery in the 15–20 years since trawling stopped on some seamounts.

Recent seamount surveys have occurred off the Galapagos Islands (Ecuador) (2015, 2016) down to 3,000 m, as well as in the Nazca Desventuradas Marine Park (Chile) (2016) and Australia (Nanson and others, 2018).

4.7. Southern Ocean

A number of seamounts and ridges have been sampled in Antarctic waters in recent years. New Zealand surveys included "Long Ridge" (part of the Pacific-Antarctic Ridge) in 2018, and the Scott seamount chain in 2019.

3 See https://ecos.csiro.au/deep-sea-life.

5. Outlook

Significant research has occurred in the past decade. Planned future international initiatives by the Global Seamounts Project,[4] the new InterRidge Working Group on Seamounts and Islands Associated with Mid-Ocean Ridges, and initiatives to survey more seamounts under the United Nations Decade of Ocean Science for Sustainable Development can complement ongoing national research directed at local biodiversity or fisheries impacts issues. In 2017, the European Union, Brazil and South Africa signed the Belém Statement, a pan-Atlantic framework funding iAtlantic (2019–2023),[5] through which several seamount surveys will take place in the Atlantic.

The emergence of potential deep-sea mining on seamounts is a threat, but the International Seabed Authority requires considerable baseline environmental data as a condition of exploration in areas beyond national jurisdiction, which will increase seamount knowledge in several regions over the next 10 years.

Climate change effects over the next decade are difficult to predict but reviews by Rogers (2015) and Sweetman and others (2017) suggest that climate change remains a major threat to seamount communities, with rising temperatures, declining oxygen concentrations and the shallowing of the aragonite saturation horizon. Some seamount fauna, such as cold water corals, are vulnerable to changes in water mass characteristics (e.g., Guinotte and others, 2006; Matos and others, 2017; Hebbeln and others, 2019), and the greatest changes at bathyal depths (Sweetman and others, 2017) include seamounts that support productive fisheries or high biodiversity. However, because seamounts cover a wider depth range, they may be less susceptible to ocean acidification changes than the surrounding sea floor and act as temporary refugia (Tittensor and others, 2010).

Globally, seamount protection may increase, adding to national conservation efforts in the North Atlantic, the South-West Pacific and the North-East Pacific (Morato and others, 2010) and more recent closures to fishing activities off the west coasts of Canada and Chile, in the North Atlantic (Natura 2000) and throughout Hawaiian waters. Tourist funding may also increase future seamount conservation (Ison and others, 2021). Some seamount areas in the eastern Pacific have been protected from potential seabed mining by a regional environmental management plan adopted by the International Seabed Authority. Regional fisheries management organizations will likely identify increased numbers of seamounts as vulnerable marine ecosystems (FAO, 2009), and numerous seamounts are classified as ecologically or biologically significant areas (Secretariat of the Convention on Biological Diversity, 2009).

6. Key remaining knowledge gaps

Knowledge gaps identified in the first Assessment (United Nations, 2017b) largely remain: a limited number of seamounts sampled to date (some progress); predictive habitat suitability models developed, but not tested (some progress); pelagic components of seamount ecosystems poorly known, especially deep bathyal (still largely true); multiple stressors

[4] Available at https://osf.io/xtg5c.
[5] Available at www.iatlantic.eu.

need evaluation, including habitat disturbance, pollutants, climate change, acidification and deoxygenation, and need to be considered together (remains a major gap); and a limited understanding of the efficacy of closed areas to date (some progress).

Several global and national seamount data sets exist, including Seamount Catalog (mainly geological);[6] SeamountsOnline (biological); Seamount Ecosystem Evaluation Framework (ecological); for New Zealand (Rowden and others, 2008); Azores (Morato and others, 2008), and the western South Pacific (Allain and others, 2008). However, they have not been extensively updated as to which seamounts have been sampled since the first Assessment. An updated register of seamount surveys and sampling effort is urgently needed.

Given that so few seamounts have been surveyed globally, major gaps in the scientific understanding of biodiversity scales and patterns on seamounts and resilience to climate change and human activities remain (Clark and others, 2012). The collection of such baseline data requires a suite of multiple tools, including remote sensing, direct sampling and visual surveys (see Clark and others, 2016). Accurate and consistent taxonomic identification of seamount fauna is a problem common to many deep-sea habitats. The need to improve consistency in sampling across multiple disciplines inspired development of the General Ocean Survey and Sampling Iterative Protocol (Woodall and others, 2018). New techniques, such as artificial intelligence for species identification, eDNA, seascape genetic approaches and deep Argo floats, will likely be used widely and are necessary to improve knowledge of seamount environments. More time series studies are needed to address the long-term resilience and recovery potential of affected seamount communities and to inform their future management.

7. Key remaining capacity-building gaps

Seamount environments can play an important role in the deep-sea ecosystem. Deep-sea surveys must include seamounts to improve the understanding of their ecosystem structure and function as a prerequisite to human activities. However, the capacity gap in scientific capabilities and the remote and unexplored nature of deep-sea habitats, in general, indicate significant capacity and information gaps even in well-developed, industrialized countries. Across all countries, there are seamounts that overlap in distribution with current or proposed industrial activities, such as fisheries in the western Pacific, offshore oil and gas development in the Caribbean and Africa, and potential deep-sea mining in seamount and ridge areas of the North-West and South-West Pacific Ocean and the Indian Ocean. Considerable collaboration and cooperation between developing and developed nations is needed to build scientific and management capacity.

6 Available at https://earthref.org/SC.

References

Allain, Valérie, and others (2008). Enhanced seamount location database for the western and central Pacific Ocean: screening and cross-checking of 20 existing datasets. *Deep Sea Research Part I: Oceanographic Research Papers*, vol. 55, No. 8, pp. 1035–1047.

Almada, Gustavo Vaz de Mello Baez, and Angelo Fraga Bernardino (2017). Conservation of deep-sea ecosystems within offshore oil fields on the Brazilian margin, SW Atlantic. *Biological Conservation*, vol. 206, pp. 92–101.

Anderson, Owen F., and others (2016a). Field validation of habitat suitability models for vulnerable marine ecosystems in the South Pacific Ocean: implications for the use of broad-scale models in fisheries management. *Ocean & Coastal Management*, vol. 120, pp. 110–126.

Anderson, Owen F., and others (2016b). Habitat suitability models for predicting the occurrence of vulnerable marine ecosystems in the seas around New Zealand. *Deep Sea Research Part I: Oceanographic Research Papers*, vol. 115, pp. 265–292.

Baco, Amy R., and others (2017). Defying dissolution: discovery of deep-sea scleractinian coral reefs in the North Pacific. *Scientific Reports*, vol. 7, No.1, art. 5436.

Baco, Amy R., and others (2019). Amid fields of rubble, scars, and lost gear, signs of recovery observed on seamounts on 30- to 40-year time scales. *Science Advances*, vol. 5, No. 8, eaaw4513.

Beeston, Mark A., and others (2018). Hydrological features above a Southern Ocean seamount inhibit larval dispersal and promote speciation: evidence from the bathyal mytilid *Dacrydium alleni* sp. nov. (Mytilidae: Bivalvia). *Polar Biology*, vol. 41, No. 7, pp. 1493–1504.

Bensch, Alexis, and others (2008). *Worldwide Review of Bottom Fisheries in the High Seas*. vol. 522. Food and Agriculture Organization of the United Nations. Rome.

Bernardino, Angelo F., and Paulo Y.G. Sumida (2017). Deep risks from offshore development. *Science*, vol. 358, No. 6361, pp. 312–312.

Bernardino, Angelo Fraga, and others (2016). Bathymetric and regional changes in benthic macrofaunal assemblages on the deep Eastern Brazilian margin, SW Atlantic. *Deep Sea Research Part I: Oceanographic Research Papers*, vol. 111, pp. 110–120.

Bruckner, A.W. (2014). Advances in management of precious corals in the family Corallidae: are new measures adequate? *Current Opinion in Environmental Sustainability*, vol. 7, pp. 1–8.

Carmo, Vanda, and others (2013). Variability of zooplankton communities at Condor seamount and surrounding areas, Azores (NE Atlantic). *Deep Sea Research Part II: Topical Studies in Oceanography*, vol. 98, pp. 63–74.

Clark, Malcolm R. (2009). Deep-sea seamount fisheries: a review of global status and future prospects. *Latin American Journal of Aquatic Research*, vol. 37, No. 3, pp. 501–512.

Clark, Malcolm R., and others (2007). Large-scale distant-water trawl fisheries on seamounts. *Seamounts: Ecology, Fisheries, and Conservation*, vol. 12, pp. 361–399.

Clark, Malcolm R., and others (2010). The ecology of seamounts: structure, function, and human impacts. *Annual Review of Marine Science*, vol. 2, pp. 253–278.

Clark, Malcolm R., and others (2012). Science priorities for seamounts: research links to conservation and management. *PloS One*, vol. 7, No. 1, e29232.

Clark, Malcolm R., and others (2015). The impacts of deep-sea fisheries on benthic communities: a review. *ICES Journal of Marine Science*, vol. 73, No. suppl. 1, pp. i51–i69.

Clark, Malcolm R., and others (2016). *Biological Sampling in the Deep Sea*. Oxford: Wiley-Blackwell.

Clark, Malcolm R., and others (2017). Biodiversity of the Kermadec Islands and offshore waters of the Kermadec Ridge: report of a coastal, marine mammal and deep-sea survey (TAN1612). *New Zealand Aquatic Environment and Biodiversity Report*, No. 179, pp. 95.

Clark, Malcolm R., and others (2019). Little evidence of benthic community resilience to bottom trawling on seamounts after 15 years. *Frontiers in Marine Science*, vol. 6, p. 63.

Da Silva, Helder Marques, and Mário Rui Pinho (2007). Small-scale fishing on seamounts. In *Seamounts: Ecology Fisheries and Conservation, Fisheries and Aquatic Resource Series, Blackwell Scientific*, T.J. Pitcher and others, eds., pp. 335–360. Fish and Aquatic Resources Series. Oxford: Blackwell Science.

Davies, Andrew J., and John M. Guinotte (2011). Global habitat suitability for framework-forming cold-water corals. *PloS One*, vol. 6, No. 4, e18483.

Davies, Jaime S., and others (2015). Benthic assemblages of the Anton Dohrn Seamount (NE Atlantic): defining deep-sea biotopes to support habitat mapping and management efforts with a focus on vulnerable marine ecosystems. *PloS One*, vol. 10, No. 5, e0124815.

Denda, A., and Bernd Christiansen (2014). Zooplankton distribution patterns at two seamounts in the subtropical and tropical NE Atlantic. *Marine Ecology*, vol. 35, No. 2, pp. 159–179.

Denda, A., and others (2017). Microzooplankton and meroplanktonic larvae at two seamounts in the subtropical and tropical NE Atlantic. *Journal of the Marine Biological Association of the United Kingdom*, vol. 97, No. 1, pp. 1–27.

Djurhuus, A., and others (2017). The spatial distribution of particulate organic carbon and microorganisms on seamounts of the South West Indian Ridge. *Deep Sea Research Part II: Topical Studies in Oceanography*, vol. 136, pp. 73–84.

Dong, Dong, and others (2017). Three squat lobsters (Crustacea: Decapoda: Anomura) from tropical West Pacific seamounts, with description of a new species of *Uroptychus* Henderson, 1888. *Zootaxa*, vol. 4311, No. 3, pp. 389–398.

Du Preez, Cherisse, and others (2016). The structure and distribution of benthic communities on a shallow seamount (Cobb Seamount, Northeast Pacific Ocean). *PloS One*, vol. 11, No. 10, e0165513.

Dueñas, Luisa F., and others (2016). The Antarctic Circumpolar Current as a diversification trigger for deep-sea octocorals. *BMC Evolutionary Biology*, vol. 16, No. 1, art. 2.

Food and Agriculture Organization of the United Nations (FAO) (2009). *International Guidelines for the Management of Deep-Sea Fisheries in the High-Seas*. 42. Rome.

________ (2018). *Global Review of Orange Roughy* (Hoplostethus Atlanticus), *Their Fisheries, Biology and Management*. Geoffrey Tingley and Matthew Dunn, eds. FAO Fisheries and Technical Paper 622. Rome.

Guinotte, John M., and others (2006). Will human-induced changes in seawater chemistry alter the distribution of deep-sea scleractinian corals? *Frontiers in Ecology and the Environment*, vol. 4, No. 3, pp. 141–146.

Harris, Peter, and others (2014). Geomorphology of the oceans. *Marine Geology*, vol. 352, pp. 4–24. https://doi.org/10.1016/j.margeo.2014.01.011.

Hebbeln, Dierk, and others (2019). The fate of cold-water corals in a changing world: a geological perspective. *Frontiers in Marine Science*, vol. 6, art. 119.

Hein, James R., and others (2013). Deep-ocean mineral deposits as a source of critical metals for high- and green-technology applications: comparison with land-based resources. *Ore Geology Reviews*, vol. 51, pp. 1–14.

Henry, L.-A., and others (2016). Seamount egg-laying grounds of the deep-water skate *Bathyraja richardsoni*. *Journal of Fish Biology*, vol. 89, No. 2, pp. 1473–1481.

Henry, Lea-Anne, and others (2014). Environmental variability and biodiversity of megabenthos on the Hebrides Terrace Seamount (Northeast Atlantic). *Scientific Reports*, vol. 4, art. 5589.

Holland, L.P., and others (2019). *Genetic connectivity of deep-sea corals in the New Zealand region*. New Zealand Aquatic Environment & Biodiversity Report No. 245. Wellington.

Howell, Kerry L., and others (2010). Mounting evidence: near-slope seamounts are faunally indistinct from an adjacent bank. *Marine Ecology*, vol. 31, pp. 52–62.

Ison, S., and others (2021). Tourist preferences for seamount conservation in the Galapagos Marine Reserve. *Frontiers in Marine Science*.

Jones, E.M., and others (2018). Oceanographic setting and short-timescale environmental variability at an Arctic seamount sponge ground. *Deep Sea Research I*, vol. 138, pp. 98–113.

Kennedy, Brian R.C., and others (2019). The unknown and the unexplored: insights into the Pacific Deep-Sea following NOAA CAPSTONE expeditions. *Frontiers in Marine Science*, vol. 6, art. 21.

Kvile, Kristina Ø., and others (2014). A global assessment of seamount ecosystems knowledge using an ecosystem evaluation framework. *Biological Conservation*, vol. 173, pp. 108–120.

Laptikhovsky, V., and others (2017). Cephalopods of the Southwest Indian Ocean Ridge: a hotspot of biological diversity and absence of endemism. *Deep Sea Research Part II: Topical Studies in Oceanography*, vol. 136, pp. 98–107.

Lavelle, J. William, and Christian Mohn (2010). Motion, commotion, and biophysical connections at deep ocean seamounts. *Oceanography*, vol. 23, No. 1, pp. 90–103.

Lemos, A.T., and others (2018). Annual phytoplankton blooming using satellite-derived chlorophyll-a data around the Vitória-Trindade Chain, Southeastern Brazil. *Deep Sea Research Part I: Oceanographic Research Papers*, vol. 136, pp. 62–71.

Letessier, Tom B., and others (2017). Seamount influences on mid-water shrimps (Decapoda) and gnathophausiids (Lophogastridea) of the South-West Indian Ridge. *Deep Sea Research Part II: Topical Studies in Oceanography*, vol. 136, pp. 85–97.

Levin, Lisa A., and others (2016). Defining "serious harm" to the marine environment in the context of deep-seabed mining. *Marine Policy*, vol. 74, pp. 245–259.

Maldonado, Manuel, and others (2015). Aggregated clumps of lithistid sponges: a singular, reef-like bathyal habitat with relevant paleontological connections. *PloS One*, vol. 10, No. 5, e0125378.

Matos, Lélia, and others (2017). Coral mound development at the Campeche cold-water coral province, southern Gulf of Mexico: implications of Antarctic Intermediate Water increased influence during interglacials. *Marine Geology*, vol. 392, pp. 53–65.

Meirelles, Pedro M., and others (2015). Baseline assessment of mesophotic reefs of the Vitória-Trindade seamount chain based on water quality, microbial diversity, benthic cover and fish biomass data. *PloS One*, vol. 10, No. 6, e0130084.

Meredith, Michael P., and others (2015). Circulation, retention, and mixing of waters within the Weddell-Scotia Confluence, Southern Ocean: the role of stratified Taylor columns. *Journal of Geophysical Research: Oceans*, vol. 120, No. 1, pp. 547–562.

Miller, Karen J., and Rasanthi M. Gunasekera (2017). A comparison of genetic connectivity in two deep sea corals to examine whether seamounts are isolated islands or stepping stones for dispersal. *Scientific Reports*, vol. 7, art. 46103.

Miller, Kathryn A., and others (2018). An overview of seabed mining including the current state of development, environmental impacts, and knowledge gaps. *Frontiers in Marine Science*, vol. 4, art. 418.

Montserrat, Francesc, and others (2019). Deep-sea mining on the Rio Grande Rise (Southwestern Atlantic): a review on environmental baseline, ecosystem services and potential impacts. *Deep Sea Research Part I: Oceanographic Research Papers*, vol. 145, pp. 31–58.

Morato, Telmo, and others (2008). Evidence of a seamount effect on aggregating visitors. *Marine Ecology Progress Series*, vol. 357, pp. 23–32.

Morato, Telmo, and others (2010). Can we protect seamounts for research? A call for conservation. *Oceanography*, vol. 23, No. 1, pp. 190–199.

Murton, B.J., and others (2017). Detailed description of FeMn crusts at Tropic Seamount. *Proceedings of the American Geophysical Union*, Fall Meeting 2017, abstract #OS34A-05. Washington, D.C.: American Geophysical Union.

Nanson, R., and others (2018). An eco-narrative of Gifford Marine Park: Temperate East marine region. *Report to the National Environmental Science Programme*, Marine Biodiversity Hub. Geoscience Australia.

Narayanaswamy, Bhavani E., and others (2013). First observations of megafaunal communities inhabiting George Bligh Bank, northeast Atlantic. *Deep Sea Research Part II: Topical Studies in Oceanography*, vol. 92, pp. 79–86.

Nishida, K., and others (2016). Prey use by three deep-sea fishes in the Emperor Seamount waters, North Pacific Ocean, as revealed by stomach contents and stable isotope analyses. *Environmental Biology of Fishes*, vol. 99, No. 4, pp. 335–349.

O'Hara, Timothy D., and others (2010). Environmental predictors and turnover of biota along a seamount chain. *Marine Ecology*, vol. 31, pp. 84–94.

Pante, Eric, and others (2015). An inter-ocean comparison of coral endemism on seamounts: the case of *Chrysogorgia*. *Journal of Biogeography*, vol. 42, No. 10, pp. 1907–1918.

Pereira-Filho, Guilherme H., and others (2012). Extensive rhodolith beds cover the summits of southwestern Atlantic Ocean seamounts. *Journal of Coastal Research*, vol. 28, No. 1, pp. 261–269.

Perez, Jose Angel Alvarez, and others (2018). Benthopelagic megafauna assemblages of the Rio Grande Rise (SW Atlantic). *Deep Sea Research Part I: Oceanographic Research Papers*, vol. 134, pp. 1–11.

Pitcher, Tony J., and others (2007). *Seamounts: Ecology, Fisheries & Conservation*. Oxford: Blackwell.

Pitcher, Tony J., and others (2010). Seamount fisheries: do they have a future? *Oceanography*, vol. 23, No. 1, pp. 134–144.

Preciado, Izaskun, and others (2017). Food web functioning of the benthopelagic community in a deep-sea seamount based on diet and stable isotope analyses. *Deep Sea Research Part II: Topical Studies in Oceanography*, vol. 137, pp. 56–68.

Ramiro-Sánchez, Berta, and others (2019). Characterization and mapping of a deep-sea sponge ground on the Tropic Seamount (northeast tropical Atlantic): implications for spatial management in the high seas. *Frontiers in Marine Science*, vol. 6, art. 278.

Ramos, Manuela, and others (2016). Patterns in megabenthic assemblages on a seamount summit (Ormonde Peak, Gorringe Bank, Northeast Atlantic). *Marine Ecology*, vol. 37, No. 5, pp. 1057–1072.

Read, Jane, and Raymond Pollard (2017). An introduction to the physical oceanography of six seamounts in the southwest Indian Ocean. *Deep Sea Research Part II: Topical Studies in Oceanography*, vol. 136, pp. 44–58.

Rogers, A. (2016). Pelagic ecology of the South West Indian Ocean Ridge seamounts: introduction and overview. *Deep Sea Research Part II: Topical Studies in Oceanography*, vol. 136, pp. 1–4.

________(2018). The biology of seamounts: 25 years on. *Advances in Marine Biology*, vol. 79, pp. 137–223.

________(2015). Environmental change in the deep ocean. *Annual Review of Environment and Resources*, vol. 40, pp. 1–38.

Rowden, A.A., and others (2008). *New Zealand's "SEAMOUNT" database: recent updates and its potential use for ecological risk assessment*. Aquatic Environment and Biodiversity Report No. 27.

Rowden, A.A., and others (2010a). A test of the seamount oasis hypothesis: seamounts support higher epibenthic megafaunal biomass than adjacent slopes. *Marine Ecology*, vol. 31, pp. 95–106.

Rowden, A.A., and others (2010b). Paradigms in seamount ecology: fact, fiction and future. *Marine Ecology*, vol. 31, pp. 226–241.

Rowden, A.A., and others (2017). High-resolution habitat suitability models for the conservation and management of vulnerable marine ecosystems on the Louisville Seamount Chain, South Pacific Ocean. *Frontiers in Marine Science*, vol. 4, art. 335.

Schlacher, Thomas A., and others (2014). Seamount benthos in a cobalt-rich crust region of the central Pacific: conservation challenges for future seabed mining. *Diversity and Distributions*, vol. 20, No. 5, pp. 491–502.

Secretariat of the Convention on Biological Diversity (2009). *Azores Scientific Criteria and Guidance for Identifying Ecologically or Biologically Significant Marine Areas and Designing Representative Networks of Marine Protected Areas in Open Ocean Waters and Deep Sea Habitats*. Montreal, Canada.

Sonnekus, Martinus J., and others (2017). Phytoplankton and nutrient dynamics of six South West Indian Ocean seamounts. *Deep Sea Research Part II: Topical Studies in Oceanography*, vol. 136, pp. 59–72.

Staudigel, Hubert and others (2010). Seamount sciences: quo vadis? *Oceanography*, vol. 23, No. 1, pp. 212–213.

Stefanoudis, P., and others (2019). Depth-dependent structuring of reef fish assemblages from the shallows to the rariphotic zone. *Frontiers in Marine Science*, vol. 6, art. 307.

Stocks, Karen I. (2010). BOX 10-SeamountsOnline: A Desktop Window Into the Lives of Seamounts. *Oceanography*, vol. 23, No. 1, p. 145.

Stocks, Karen I., and others (2012). CenSeam, an international program on seamounts within the census of marine life: achievements and lessons learned. *PloS One*, vol. 7, No. 2, e32031.

Sweetman, Andrew K., and others (2017). Major impacts of climate change on deep-sea benthic ecosystems. *Elementa: Science of the Anthropocene*, vol. 5, pp. 1–23.

Taylor, M.L., and others (2016). Plastic microfibre ingestion by deep-sea organisms. *Scientific Reports*, vol. 6, art. 33997.

Tittensor, Derek P., and others (2010). Seamounts as refugia from ocean acidification for cold-water stony corals. *Marine Ecology*, vol. 31, pp. 212–225.

Tracey, Dianne M., and others (2011). Habitat-forming cold-water corals show affinity for seamounts in the New Zealand region. *Marine Ecology Progress Series*, vol. 430, pp. 1–22.

Turnewitsch, Robert, and others (2016). Tidal influence on particulate organic carbon export fluxes around a tall seamount. *Progress in Oceanography*, vol. 149, pp. 189–213.

United Nations (2017a). Chapter 51: Biological communities on seamounts and other submarine features potentially threatened by disturbance. In *The First Global Integrated Marine Assessment: World Ocean Assessment I*. Cambridge: Cambridge University Press.

________(2017b). *The First Global Integrated Marine Assessment: World Ocean Assessment I*. Cambridge: Cambridge University Press.

Victorero, Lissette, and others (2018). Species replacement dominates megabenthos beta diversity in a remote seamount setting. *Scientific Reports*, vol. 8, No. 1, art. 4152.

Vieira, Rui P., and others (2015). Lost fishing gear and litter at Gorringe Bank (NE Atlantic). *Journal of Sea Research*, vol. 100, pp. 91–98.

Wang, Dexiang, and others (2016). Three new species of glass sponges Pheronematidae (Porifera: Hexactinellida) from the deep-sea of the northwestern Pacific Ocean. *Zootaxa*, vol. 4171, No. 3, pp. 562–574.

Watson, Reg, and others (2007). Catches from world seamount fisheries. In *Seamounts: Ecology, Fisheries & Conservation*, Tony J. Pitcher and others, eds., pp. 400–412. Oxford: Blackwell Publishing.

Wessel, Paul, and others (2010). The global seamount census. *Oceanography*, vol. 23, No. 1, pp. 24–33.

Woodall, Lucy C., and others (2014). The deep sea is a major sink for microplastic debris. *Royal Society Open Science*, vol. 1, No. 4, 140317.

Woodall, Lucy C., and others (2015). Deep-sea litter: a comparison of seamounts, banks and a ridge in the Atlantic and Indian Oceans reveals both environmental and anthropogenic factors impact accumulation and composition. *Frontiers in Marine Science*, vol. 2, art. 3.

Woodall, Lucy C., and others (2018). A multidisciplinary approach for generating globally consistent data on mesophotic, deep-pelagic, and bathyal biological communities. *Oceanography*, vol. 31, No. 3, pp. 76–89.

Yesson, Chris, and others (2011). The global distribution of seamounts based on 30 arc seconds bathymetry data. *Deep Sea Research Part I: Oceanographic Research Papers*, vol. 58, No. 4, pp. 442–453.

Yesson, Chris, and others (2012). Global habitat suitability of cold-water octocorals. *Journal of Biogeography*, vol. 39, No. 7, pp. 1278–1292.

Yesson, Chris, and others (2017). The global distribution of deep-water Antipatharia habitat. *Deep Sea Research Part II: Topical Studies in Oceanography*, vol. 145, pp. 79–86.

Zeng, Cong, and others (2017). Population genetic structure and connectivity of deep-sea stony corals (Order Scleractinia) in the New Zealand region: Implications for the conservation and management of vulnerable marine ecosystems. *Evolutionary Applications*, vol. 10, No. 10, pp. 1040–1054.

Chapter 7M
Abyssal plains

Contributors: Jeroen Ingels (convener), Diva Amon, Angelo F. Bernardino, Punyasloke Bhadury, Holly Bik, Malcolm R. Clark, Thomas Dahlgren, Daniel O.B. Jones, Craig McClain, Clifton Nunnally, Paul Snelgrove, Joshua T. Tuhumwire (subchapter lead member) and Moriaki Yasuhara.

Keynote points

- The abyss lies between 3 and 6 km water depth and covers more of the Earth's surface than all other habitats combined.
- The present chapter is the first in the *World Ocean Assessment* that is dedicated to the abyss, covering biodiversity, regional differences, biogeography, and changes and impacts as a result of natural stressors and anthropogenic activity.
- Abyssal biodiversity is not well understood, and many gaps exist in the current understanding of abyssal evolution and biogeography, as well as the distributions, connectivity and responses of organisms to changing conditions.
- Fragmentary knowledge of abyssal taxonomy is largely the result of difficulties in sampling the vast and remote area and, hence, limited research effort, which hinders the advancement of scientific knowledge.
- Most abyssal environments support the processes that drive deep-sea and global ecosystem functioning and are linked closely to surface production and pelagic processes.
- Climate change and anthropogenic impacts affect the abyss, despite its remoteness.

1. Introduction

1.1. Situation as recorded in the first *World Ocean Assessment*

The first *World Ocean Assessment* (United Nations, 2017d) contained a brief description of abyssal environments, in chapter 1 (United Nations, 2017a), chapter 36F (United Nations, 2017c) and the chapters on biodiversity in different ocean regions. The dependence of abyssal habitats on the flux of food from above, the possible impacts of climate change and the likely impact of deep seabed mining were noted. There is continued uncertainty about abyssal biodiversity and its potential connections to pelagic and surface water organisms and future changes therein. The first Assessment did not contain the comprehensive description of abyssal biodiversity provided in the present chapter.

1.2. General

The abyssal zone (3–6 km water depth) (Gage and Tyler, 1991) encompasses the largest area on Earth (about 58 per cent of the planet's surface). It mainly comprises vast areas of sea floor plains covered in generally fine sediments, punctuated by sporadic hard substrate at topographic highs in the form of knolls, seamounts, mid-ocean ridges and island arcs, as well as lows in the form of valleys and deeper trenches. The total absence of sunlight penetration and in situ primary production, apart from some chemosynthesis (see chap. 7P), characterize an ecosystem based on a variable rain of material from shallower euphotic zones. Although it is food-limited, with low abundances compared with most deep-sea habitats (Gage and Tyler, 1991), the abyss supports high levels of alpha and beta diversity of meiofauna, macrofauna and megafauna (Rex and Etter, 2010). The quantity and quality of food particles sinking from the ocean surface strongly modulate ecosystem structure and function (Smith and others, 2008; McClain and others, 2012a), but feedback mechanisms through nutrient cycling back into the water column are poorly understood (Thurber and others, 2014). Abyssal regions differ from each other in physical variables, surface water characteristics and biogeographical distinctions, which are reflected in their organisms, communities and biodiversity.

Abyssal biodiversity varies in space (Glover and others, 2002; Woolley and others, 2016; Simon-Lledó and others, 2019a) and time (Ruhl and others, 2008). Despite poorly known biodiversity patterns on regional to global scales, some regions, such as the abyssal Southern Ocean (Brandt and others, 2006; Griffiths, 2010) and the equatorial Pacific (Glover and others, 2002; Amon and others, 2016a), house major biodiversity reservoirs. For the few taxa studied, connectivity appears high (Baco and others, 2016; Taboada and others, 2018), whereas studies of deep-sea functional diversity have just begun (e.g., Chapman and others, 2019), including of the abyssal sea floor (e.g., Christodoulou and others, 2019; O'Hara and others, 2019). Biodiversity knowledge varies by region and, in recent years, interest in seabed mining (see chap. 18) has helped to generate new information for regions such as the Clarion-Clipperton Fracture Zone in the central Pacific (e.g., Dahlgren and others, 2016; Glover and others, 2016a; Amon and others, 2017a, 2017b; Marsh and others, 2018; Wiklund and others, 2019), with evidence of biodiverse, yet vulnerable life (Vanreusel and others, 2016).

Climate change will likely affect the abyss (Yasuhara and Danovaro, 2016; Sweetman and others, 2017). Projections suggest increased abyssal ocean temperatures and acidification, and decreased oxygen concentrations and downward flux of organic matter. Other oceanographic processes will likely respond, increasing stratification and reducing water mass exchange. Given the narrow environmental niches of abyssal biota, such changes could produce geographic shifts and increase the vulnerability of abyssal organisms to other anthropogenic impacts (Levin and others, 2020). The current understanding of anthropogenic impacts on abyssal ecosystems remains poor but highlights a vulnerability that will very likely increase in the future.

2. Shifting baselines and documenting status and change in abyssal biodiversity

The challenges of sampling in remote locations at depths of more than 3,000 m contribute to abyssal undersampling (Glover and others, 2018). Biodiversity records reflect that deficiency (figures I and II). Sampling effort has also focused more on the sea floor than on the highly variable and vast pelagic realm.

2.1. Benthic abyssal biodiversity and benthic-pelagic coupling

Biogenic habitat comprises much of the fine-scale habitat structure on sediments. The patchy food resource also contributes fine-scale structure to the sea floor (McClain and Schlacher, 2015). Characteristically low current speeds result in minimal sediment erosion (Smith and others, 2008) but affect sediment composition (McCave, 2017). Abyssal waters are cold (< 5°C) and relatively constant in temperature (Sweetman and others, 2017), and they are characterized by extremely high hydrostatic pressure.

The transfer of organic matter to the abyss occurs mainly through sinking particulate organic carbon, which is largely produced in surface waters through photosynthetic primary production and zooplankton, the latter generating secondary production and by-products (Cavan and others, 2015). In addition, carcasses of marine vertebrates can sink to the abyssal plain within a few days, temporarily increasing local food

(Amon and others, 2016b). Surface export of organic material can reach abyssal depths within a few days but rates fluctuate (Smith and others, 2008). Particle export dynamics, such as summer export from upper layers, can strongly influence abyssal biogeochemical processes (Bouef and others, 2019). However, remineralization throughout the water column results in very low quantities of organic matter reaching the abyssal sea floor (about 0.5–5.0 per cent of surface production) (Lutz and others, 2007; Smith and others, 2008; Smith and others, 2009). The arrival of food influences abyssal communities and their diversity, abundance, density and composition, whereas important microbial groups affect processes such as carbon and nitrogen cycling, and vertical organic matter transport shapes the composition and biogeography of deep ocean prokaryotic (and eukaryotic) communities (Mestre and others, 2018). The low energy availability results in generally low abyssal abundances, biomasses and biological rates (metabolism, growth and reproduction) (Smith and others, 2008; Wei and others, 2010).

The total biomass of all benthic size classes generally declines with increasing water depth, except for bacteria and archaea, which dominate the biomass of the abyssal plain and deeper (Wei and others, 2010). Modelling estimates suggest global prokaryotic biomass on the sea floor of approximately 35 megatons of carbon (Wei and others, 2010). Thus, the activities of microbial communities strongly influence the type and abundance of nutrients released back into the pelagic realm. The microbes also experience top-down forcing from viral populations (Suttle, 2005) and grazing by animals of various sizes (e.g., Howell and others, 2003; Ingels and others, 2010).

Figure I.A
Number of Ocean Biodiversity Information System (OBIS) records plotted against ocean depth

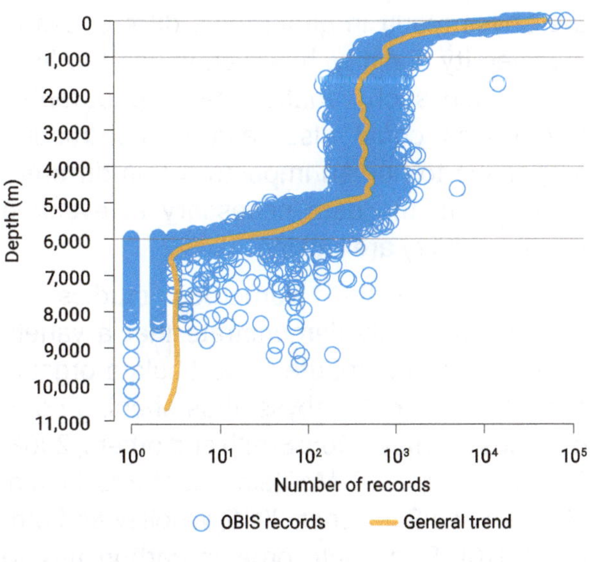

Figure I.B
Proportion of all OBIS records occurring in different depth zones, plotted against the proportion of the global ocean that occurs at those depths

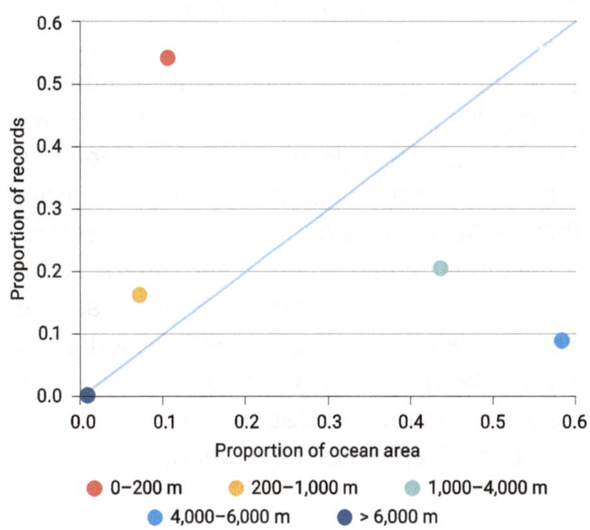

Source: Webb and others, 2010.
Note: The 1:1 line identifies those areas of the ocean with proportionately more (points above the line) or fewer (points below the line) records than expected given their area. The depiction provides a conservative view of under- and overrepresentation based on the volume of each habitat.

Figure II.A
World map of Ocean Biodiversity Information System abyssal records

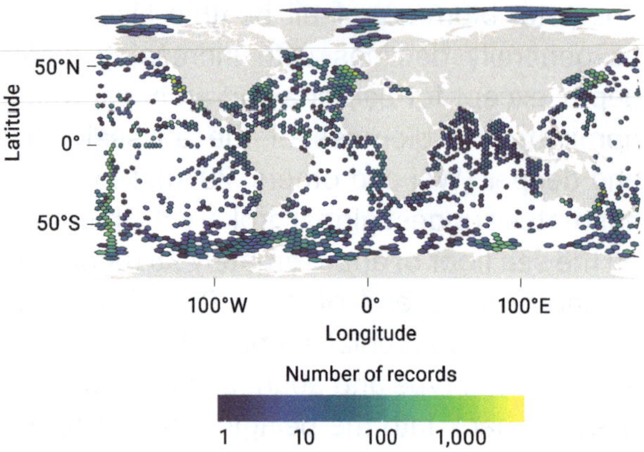

Figure II.B
World map of Ocean Biodiversity Information System species presence between 3,000 and 6,000 m depth

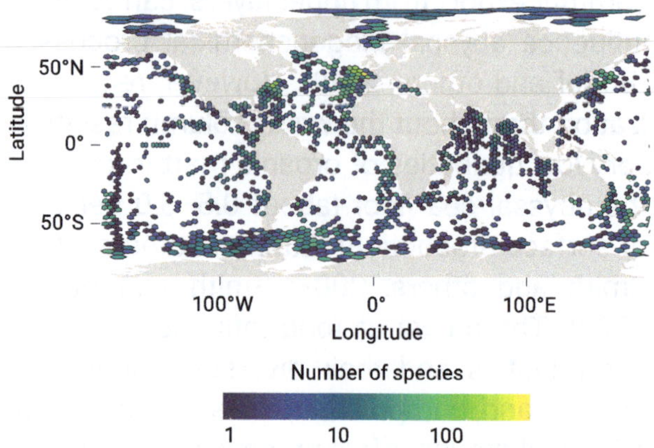

Source: Ocean Biodiversity Information System (OBIS), 16 May 2019; Intergovernmental Oceanographic Commission of the United Nations Educational, Scientific and Cultural Organization. Retrieved from https://obis.org).
Notes: A: Data available rarely exceed 1,000 records per 75,000 m^2 area; gaps exist, especially in the Pacific Ocean, the South Atlantic Ocean and the South Indian Ocean. B: The North-East Atlantic Ocean has more samples as compared with all other oceans. Note the correlation between records and species presence.

2.2. Abyssopelagic zone

Much less is known about the pelagic fauna that primarily occupy depths between 3 and 6 km and that live more than 200 m above the sea floor. The Ocean Biodiversity Information System shows minimal sampling of those ecosystems, which results in major knowledge gaps spanning over a billion km^3 of habitat – potentially the largest reservoir of unknown diversity on Earth (Robison, 2009). The abyssopelagic zone facilitates the largest carbon sink on the planet, a critical ecosystem service of the global ocean (Atwood and others, 2020). Daily vertical migration between deep-sea pelagic layers can move dissolved nutrients that contribute to primary production in the photic zone (Houghton and Dabiri, 2019), along with long-term deep ocean circulation.

2.3. Key region-specific differences or contrasts

Broad-scale variation in physical and chemical environments (e.g., organic flux, oxygen) in the abyss result in geographic differences in biodiversity. Salinity, however, varies too little to produce such variation. Such geographic differences could also lead to contrasting responses to human impacts within different regions, but the data necessary to evaluate that possibility are lacking.

Carbon availability. Numerous studies on carbon availability demonstrate that a variety of processes contribute to particulate organic carbon levels in the abyss, thus shaping communities (Carney, 2005; Smith and others, 2008; Rex and Etter, 2010; McClain and others, 2012a; McClain and Schlacher, 2015; Woolley and others, 2016). Particulate organic carbon flux to the deep varies in time and space (Lampitt and Antia, 1997; Lutz and others, 2007; figure III). Such factors as depth, distance from productive coastal waters and/or upwelling regions can produce considerable local effects, generally limiting particulate organic carbon flux to the deep sea floor. For example, upwelling in the equatorial Pacific results in high levels of particulate organic carbon flux (2–6 g/m²/year) compared with the extremely low particulate organic carbon (< 1 g/m²/year) in adjacent regions to the south (Watling and others, 2013). Intense areas of coastal upwelling, combined with narrow continental shelves, place abyssal habitats in the North-East Pacific and the South-East Atlantic closer to productive coastal waters, resulting in higher particulate organic carbon input (Lutz and others, 2007; Lampitt and Antia, 1997). Moderately high particulate organic carbon fluxes also occur in the North Atlantic (6.6 gm²/year) because of spring bloom pulses (Lampitt and Antia, 1997).

Figure III.A
Particulate organic carbon flux to the bottom at depths between 3,500 and 6,500 m

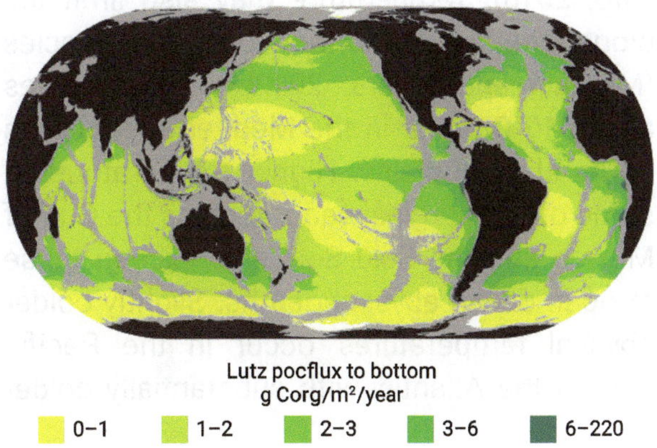

Source: Data from Lutz and others, 2007; adapted from Watling and others, 2013.

Figure III.B
Depth-particulate organic carbon plot illustrating differences in particulate organic carbon flux and flux variability between bathyal and abyssal regions (main) and variability between abyssal regions (temp-particulate organic carbon flux) (insert)

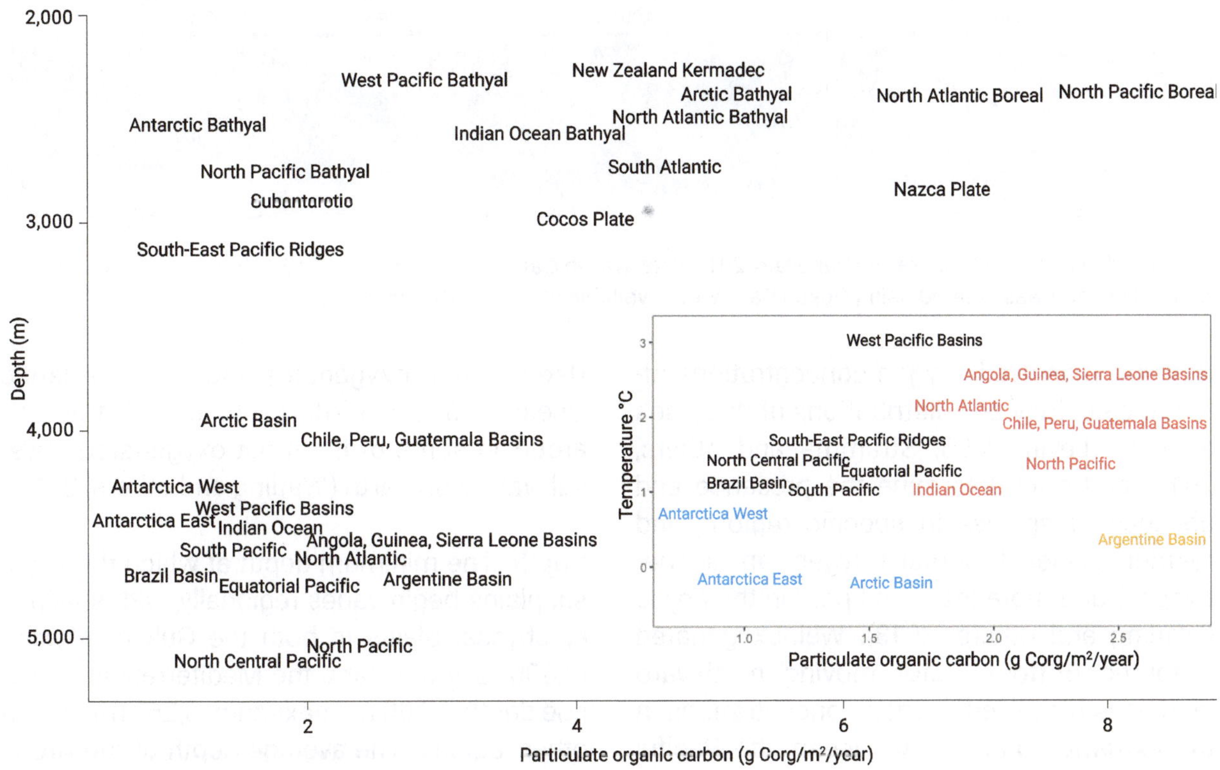

Source: Based on data from Watling and others, 2013.

Temperature. Temperature often shows a statistically significant relationship with diversity in the abyss (Cronin and Raymo, 1997; Hunt and others, 2005; Yasuhara and Danovaro, 2016). Temperature may also limit the biogeographic distribution of some species (McClain and others, 2012b). Temperatures above 10°C occur in the Mediterranean, even on its abyssal plains. Higher temperatures in some other marginal seas, such as the Gulf of Mexico and the Sulu Sea, also exceed those at open ocean abyssal depths. Slightly colder abyssal temperatures occur in the Pacific than in the Atlantic, with substantially colder polar sea abyssal temperatures (see figure IV; Yasuhara and Danovaro, 2016). Gebbie and Huybers (2019) recently reported a significant difference between Pacific (continuing to cool as a result of the Little Ice Age) and Atlantic (beginning to warm because of recent climate change) circulation. The changes may alter carbonate compensation depths (water depth at which carbonate supply and dissolution are equal) within the different basins. Regional differences result from influences of depth and bottom-water formation, downwelling and other water mass exchanges.

Figure IV
Global temperature distributions at different water depths (°C)

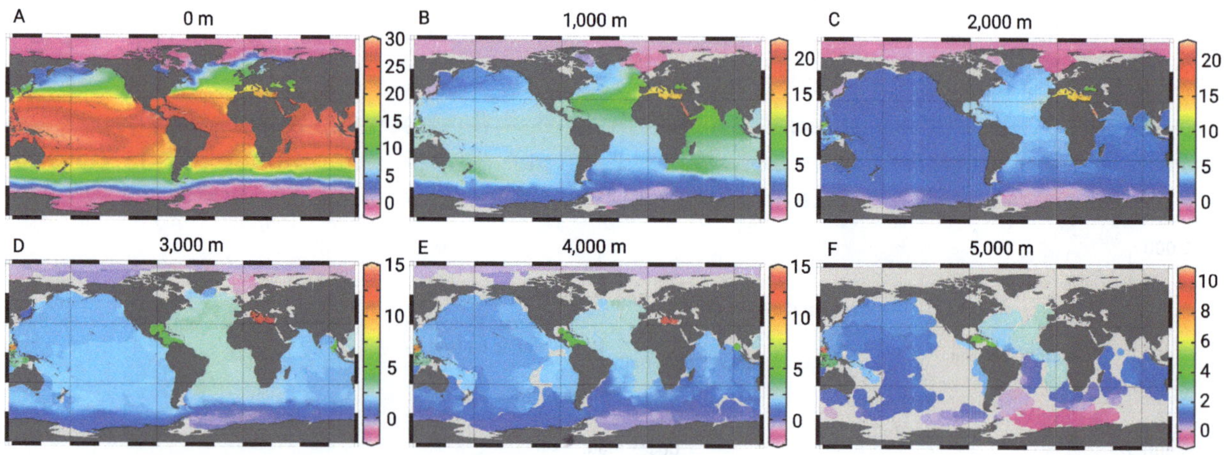

Source: Adapted from Yasuhara and Danovaro, 2016; data available at www.nodc.noaa.gov/OC5/WOA09/pr_woa09.html.
Notes: The map was created using Ocean Data View, available at http://odv.awi.de.

Oxygen. Dissolved oxygen concentrations affect the ecology and distributions of deep-sea animals (Levin, 2003; Stramma and others, 2010) and may determine the presence and absence of species in specific regions and restrict species dispersal. Oxygen can vary between 1 and more than 6 ml per l in the abyss (Watling and others, 2013). Well-oxygenated Antarctic Bottom Water moving northward increases dissolved oxygen concentrations in the extreme south Indian Ocean, the Pacific Ocean and the Atlantic Ocean (3–4 ml/l). Likewise, the southward-flowing North Atlantic Deep Water oxygenates the North Atlantic Ocean (5.5–6.5 ml/l), resulting, with the Antarctic, in some of the most oxygenated abyssal waters on Earth (Watling and others, 2013).

Depth. The minimum depth at which the abyssal plains begin varies regionally, with shallower abyssal plains in both the Gulf of Mexico (3,000–3,900 m) and the Mediterranean (average depth, 1,500 m; maximum, 5,267 m) than in other regions. The average depth of the Arctic Ocean and the Chile, Peru and Guatemala basins approach 4,000 m, in contrast to average depths closer to 5,000 m in the northern and central Pacific. Greater depth, ceteris paribus, reduces particulate organic carbon flux. Depth, as a proxy for pressure, may also limit biogeographic distributions (Somero, 1992; Carney, 2005). Regional abyssal depth differences may thus cause taxonomic compositional shifts and influence biodiversity. Nonetheless, despite broad biogeographic differences among regions, little evidence points to depth as a strong correlate of diversity within abyssal plains.

Topography. Topographic features can impede exchange of individuals between deep-sea populations and influence biogeographic classification (McClain and Hardy, 2010). The Pacific and Atlantic share only 15–20 per cent of species (Vinogradova, 1997). The Strait of Gibraltar limits colonization of the relatively species-poor Mediterranean by Atlantic fauna (Sardà and others, 2004). Mid-oceanic ridges may also limit dispersal on the abyssal plains. Half of the known species of deep-sea bivalves are restricted to either the eastern or western Atlantic (McClain and others, 2011), likely because of the Mid-Atlantic Ridge.

Researchers recently recognized that abyssal hills rising less than 1,000 m off the sea floor create topographic, depth and sediment differences that support different taxonomic assemblages and higher biomass levels (Yesson and others, 2011; Durden and others, 2015) than in flatter abyssal sediments.

Sediment and substrate. Sediment types can vary dramatically in composition within different abyssal regions. Most diatom oozes occur at abyssal depths but radiolarian oozes occur, inter alia, in the Southern Ocean, the equatorial Pacific and the Peru basin. Sponge spicules form a major component of sediments in the Australian-Antarctic basin. Clay dominates large seabed regions off South America and in the Indian Ocean, and it dominates the South Australian basin (Dutkiewicz and others, 2015). Sediment diversity affects biodiversity, but linkages between sediment type and biodiversity patterns remain underexplored. In abyssal plain sediments, polymetallic nodules can also affect biodiversity. Assemblages on nodules differ fundamentally from both near-bottom seawater and sediment communities (Shulse and others, 2017; Simon-Lledó and others, 2019a). Increased nodule presence promotes increased megafaunal and xenophyophore abundance (Simon-Lledó and others, 2019b). Thus, increased habitat complexity generated by polymetallic nodules increases diversity in all levels of abyssal biota.

Riverine influences. Riverine input can influence the abyss through: (a) input of terrestrial carbon; (b) creation of a dispersal barrier, thus affecting biogeography; and (c) disturbance that alters deep-sea sediments. Significant discharges are shown in table 1.

Table 1
Riverine influences

Recipient	River	Megatons carbon/year
Indian Ocean	Ganges and others	30.0
South-East Atlantic	Congo and others	30.0
South-West Atlantic	Amazon	37.6
North-West Pacific	Yangtze, Yellow and Mekong	16.2
Arctic	Siberian rivers	12.8
Gulf of Mexico	Mississippi	3.6
South-West Pacific	Indonesian rivers (with high annual rainfall)	90.0

Sediments from large rivers may also deliver substantial loads of anthropogenic contaminants, with unknown effects on abyssal biodiversity (Davies and Moore, 1970). Organic matter influx from large rivers to the continental margins, slopes and canyons is easily channelled through various processes to the abyss, where it may disturb and drive sea floor biomass and community diversity.

Ice cover. Polar ice cover influences primary production and thus particulate organic carbon flux to the abyss. Permanent ice cover reduces or prevents surface production, thus limiting biodiversity and biomass in the Arctic Ocean, where known species richness of polychaetes may be lower than in other, similar-sized basins (Bodil and others, 2011). Summer ice absence can bolster surface production and increase biodiversity and biomass (Wlodarska-Kowalczuk and Pearson, 2004).

Geological age. Geological changes likely affected the distribution of abyssal biodiversity by altering connectivity among ocean regions, including those shown in table 2.

Table 2
Connectivity among ocean regions

Connection	Opening	Closing	Source
	Millions of years ago (approximately)		
Mediterranean/Atlantic Ocean and Indian Ocean (Tethys Seaway)		19–14	Harzhauser and others, 2007
Drake Passage	30		Lawver and Gahagan, 2003; Livermore and others, 2007; Scher and Martin, 2006
Central American Seaway		3	Schmidt and others, 2007; O'Dea and others, 2016; Schmidt and others, 2016
Bering Strait (Arctic/Pacific)	4.8–7.4		Marincovich and Gladenkov, 2001; Hu and others, 2012
Fram Strait (Arctic/Atlantic)	10–20		Engen and others, 2008; Ehlers and Jokat, 2013

Source: Yasuhara and others, 2019a.

2.4. Abyssal biogeography

In contrast to well-recognized boundaries among benthic assemblages on continental margins, uncertainty remains as to whether such abyssal boundaries exist (Carney, 2005). Researchers have attempted to establish biogeographic realms below 3,000 m. Some early attempts based on temperature, topography or faunal similarities suggested Atlantic, Indo-Pacific, Antarctic and Arctic divisions; others linked the Arctic and Atlantic, or questioned such linkages and split the Indian Ocean and Pacific Ocean or proposed more subregions (Menzies and others, 1973; Vinogradova, 1979, 1997; Carney, 1994).

A scheme sponsored by the United Nations Educational, Scientific and Cultural Organization, the Global Open Oceans and Deep Seabed (GOODS) biogeographic classification, used environmental parameters, including temperature, salinity, dissolved oxygen, carbon flux, primary production, bathymetry and plate boundary layers, to delineate biogeographic provinces, resulting in 14 abyssal provinces (Briones and others, 2009).[1]

A more recent proposal revised the 14 abyssal provinces (figure V) by giving greater weight to hydrographic patterns, particulate organic carbon flux, dissolved oxygen and the effects of cold Antarctic waters and warmer North Atlantic waters (Watling and others, 2013).

Figure V
Proposed biogeographical regions

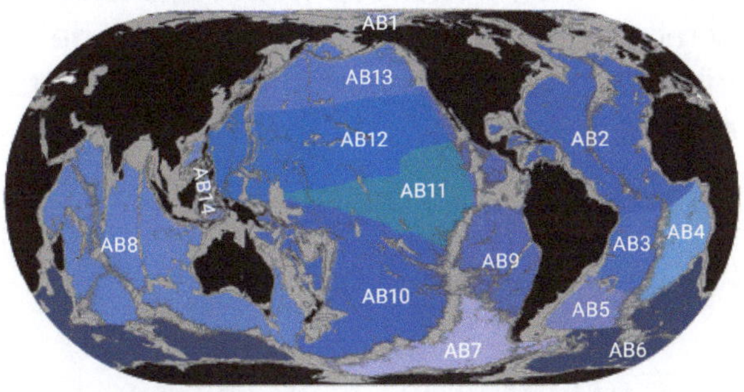

AB1: Arctic basin
AB2: North Atlantic
AB3: Brazil basin
AB4: Angola, Guinea, Sierra Leone basins
AB5: Argentine basin
AB6: Antarctica East
AB7: Antarctica West
AB8: Indian
AB9: Chile, Peru, Guatemala basins
AB10: South Pacific
AB11: Equatorial Pacific
AB12: North Central Pacific
AB13: North Pacific
AB14: West Pacific basins

Source: Based on Watling and others, 2013.

2.5. Documented change in abyssal biodiversity

2.5.1. Evidence from palaeoecological studies

Fossil records from deep-sea sediment cores provide the only time series data longer than a few decades (Yasuhara and others, 2017, 2019b), and those palaeoecological records clearly point to long-term impacts of climatic change on abyssal biodiversity. Abyssal diversity correlates positively with bottom-water temperature over 1,000–10,000 years (Cronin and Raymo, 1997; Cronin and others, 1999; Yasuhara and Cronin, 2008; Yasuhara and Danovaro, 2016). Dynamic deepwater circulation and associated temperature changes have occurred even at multi-decadal–centennial timescales (Yashayaev and others, 2007; Hoffmann and others, 2018; Thornalley and others, 2018; Yasuhara and others, 2019b). Researchers infer great stability in abyssal biotic and environmental conditions compared with those at bathyal or shallower depths. Over larger timescales, fossil data show that present-day deep-sea fauna were established during the Miocene epoch, approximately 13 million years ago (Thomas and others, 2000; Thomas, 2007). Latitudinal diversity gradients in the deep sea that were established during the late Eocene epoch, approximately 37 million years ago, persist today (Thomas and Gooday, 1996).

2.5.2. Evidence from long-term observatories

Few long-term research programmes have obtained sufficient data to draw conclusions regarding long-term natural versus anthropogenic changes. Those that do indicate a strong connection between surface production and abyssal sea floor communities, often with a high degree of dynamism. The studies suggest that one-time or short-term investigations in the abyss cannot adequately assess biological community changes mechanistically, in particular in the context of deep-sea stewardship.

The monitoring studies of Station M off central California since 1989 strongly correlated surface ocean processes and particulate organic carbon supply to the abyss, where fluctuations affect community structure and processes. Short-term variations in Station M abyssal communities (Kuhnz and others, 2014) link to inter-annual variation in climate (El Niño/La Niña) (Ruhl and others, 2014), but long-term consequences are poorly understood. Sporadic, intense food pulses to the abyss could provide food surplus following many years of undersupply.

1 See Intergovernmental Oceanographic Commission, IOC Technical Series, No. 84 (IOC/2009/TS/84 and Corr.).

Porcupine Abyssal Plain Sustained Observatory. Sustained observations at a depth of 4,850 m in the North-East Atlantic have produced high-resolution surface-to-sea floor data since 1989. Dramatic community and abundance shifts occur in response to organic matter influx changes (e.g., Billett and others, 2001), resulting from the tight correlation between surface productivity and export fluxes (Frigstad and others, 2015). The shifts (1989–2005) dramatically alter carbon storage. Most abyssal biota respond to food influx, environmental change and competitive interactions (Gooday and others, 2010; Kalogeropoulou and others, 2010; Lampitt and others, 2010; Soto and others, 2010). Biogeochemical results show that the partial pressure of CO_2 (pCO_2) has decreased with increasing anthropogenic CO_2 emissions (Hartman and others, 2015).

Long-term ecological research observatory Hausgarten. Data from the observatory (Fram Strait, Arctic, 250–5,500 m depth, since 1999), point to seasonal forcing of communities related to regional sea ice and hydrodynamic conditions (Soltwedel and others, 2005, 2016). Fifteen years of pelagic and benthic data indicate rapid responses of the entire ecosystem to water column changes. However, uncertainty remains as to whether trends should be attributed to anthropogenic changes or natural multi-year variability.

3. Major natural and anthropogenic pressures

3.1. Natural pressures

Natural disturbances in the form of near-bottom currents, sediment resuspension or settling food particles can dramatically alter benthic communities (Hessler and Jumars, 1974; Snelgrove and Smith, 2002). In the Atlantic, the mass movement of sediment downslope can affect the transport of organic matter to adjacent abyssal basins (Levin and Gooday, 2003). Similar processes occur during dense shelf-water cascading through canyons and slopes to abyssal depths, triggered by increased salinity and winter cooling (Carney, 2005; Company and others, 2008). Such disturbances may increase organic matter transport to abyssal depths (Canals and others, 2006; Ulses and others, 2008; Palanques and others, 2011).

Similarly, heterogeneous sea floor topography can modify the composition and abundance of species, as well as carbon remineralization rates. Such abyssal hills likely play a significant role in Pacific deep-sea communities and organic matter cycling, given their large number and the limited continental sediment supply (Smith and Demopoulos, 2003).

3.2. Anthropogenic pressures

3.2.1. Climate change

Climate change will affect abyssal physical (e.g., salinity, temperature), biogeochemical (nutrients, CO_2, oxygen (O_2), sedimentology) and biological processes and functions (Mora and others, 2013; Sweetman and others, 2017). Abyssal temperatures could increase by 1° over the next 80 years, whereas abyssal sea floor habitats beneath regions of deepwater formation may experience reductions in water column oxygen concentrations by as much as 0.03 ml per l by 2100. Such changes could affect food supply and sediment transport (Cheung and Levin, 2019; Food and Agriculture Organization of the United Nations (FAO), 2019). Climate-induced changes in ocean circulation and hydrodynamics may affect abyssal connectivity by altering distributions of the pelagic larvae of abyssal organisms (acknowledging that larvae of some abyssal taxa do not reach the upper ocean). Questions persist about how such changes impact deep ocean communities but decadal studies in the northern Pacific demonstrate significant linkages (Ruhl and others, 2008). Assessments of climate change impacts, as well as synergistic or cumulative impacts with other anthropogenic activities, must therefore consider abyssal ecosystem responses (Smith and others, 2008; Levin and Le Bris, 2015; Sweetman and others, 2017).

The food-limited nature of abyssal ecosystems suggests high sensitivity of all biota, from microbes to megafauna, to changes in phytoplankton community structure and productivity and the quantity and quality of export flux (Ruhl and Smith, 2004; Ruhl and others, 2008; Billett and others, 2010; Smith and others, 2013). Climate warming will likely increase ocean stratification, reduce primary production, increase acidity and shift dominant phytoplankton community structure, driving biotic changes over major regions of the abyss, such as the equatorial Pacific (Smith and others, 2008; Levin and others, 2020). Predictions of significant decline in organic matter flux to the deep sea floor in most oceans (Sweetman and others, 2017) contrast with predictions of increased production of water column and sea floor biomass in polar seas (Jones and others, 2014). Threats to abyssopelagic environments also include the deepening of oxygen minimum zones.

3.2.2. Plastics and other forms of pollution

Pollution has long affected abyssal depths (Chiba and others, 2018). High levels of plastic debris have been found, along with benthic organisms contaminated with organic pollutants, even at ocean depths of over 10,000 m (see chaps. 11 and 12). Few studies have documented interactions of abyssal life with debris and other pollutants but the research topic is rapidly gaining interest. Other examples of abyssal pollution include the dumping of nuclear waste prior to 1983, as described in chapter 24, section 3, of the first Assessment (United Nations, 2017b).

3.2.3. Mining

In the past few decades, interest in mineral reserves at abyssal depths has grown considerably. The future extraction of sea floor minerals, in the form of polymetallic nodules, cobalt-rich crusts, and polymetallic sulphides, pose a significant potential threat to abyssopelagic and benthic communities, directly and indirectly (Christiansen and others, 2020). Chapter 18 of the present Assessment discusses the environmental, social and economic aspects of seabed mining.

3.2.4. Anthropogenic pressures on abyssopelagic biodiversity

Although currently rare, bioprospecting and oil extraction activities on abyssal plains pose additional threats to the health of abyssopelagic and benthic habitats. Commercial fishing and fish farming on the high seas could threaten abyssopelagic diversity if poorly managed nationally and internationally. Poor management of both activities can reduce prey populations, affect food downflux and undermine biodiversity, including targeted and non-targeted resources. Although currently rare, bioprospecting and oil extraction activities on abyssal plains pose additional threats to the abyssopelagic environment.

4. Consequences of the changes on human communities, economies and well-being

Despite its apparent remoteness and inhospitality, the deep ocean plays a crucial role in human social and economic well-being through its ecosystem functions and services on a regional to global scale (Van den Hove and Moreau, 2007; Armstrong and others, 2012; Thurber and others, 2014; tables 3 and 4).

Table 3
Susceptibility of the abyssal sea floor and abyssopelagic zone to climate change-affected environmental drivers and pressures

	Abyssal sea floor impacts	Abyssopelagic impacts
Changes in temperature, acidity, salinity and oxygen patterns	Medium to high	Low
Changes in sea level	Low (through terrestrial influence)	Low (through terrestrial influence)
Changes in severity of storms and intensity of extreme events	Low	Low
Changes in ultraviolet radiation	Low, indirect through benthopelagic coupling	Low, indirect through benthopelagic coupling
Changes in the physical and chemical aspects of the ocean	Low	Low
Food input	Medium to high	Medium to high

4.1. Impacts on abyssal ecosystem services

Compared with other deep-sea habitats, abyssal plains provide ecosystem services that are limited in scope but important in magnitude and reach. Few abyssal services, such as mineral resources, could directly benefit humans, whereas most abyssal environments support the processes that drive deep-sea and global ecosystem and Earth climate system functioning on such vast scales that they influence the entire Earth system.

The "biological pump" provides the most important supporting and regulating ecosystem service of the abyssopelagic zone by accelerating the transfer of carbon, nutrients and other compounds from surface waters to the deep sea. Changes in fauna, trophic links or community composition, or physical alterations in water masses (e.g., stratification, warming, deoxygenation, acidification), can disrupt associated biological processes, with abyssal impacts through benthic-pelagic coupling. Stress imposed by low oxygen, acidification or elevated temperature can reduce species and ecosystem resilience through shifts in organism tolerance (Pörtner and Farrell, 2008; Pörtner, 2010), thus retarding recovery from disturbance caused by human activities, such as seabed mining. Climate change effects could exacerbate anthropogenic impacts and compromise deep-sea ecosystem structure and function and, ultimately, their benefits for human welfare (Mora and others, 2013).

Table 4
Threats and pressures on abyssal ecosystem services and their importance in the abyss

	Abyssal plain threats	Abyssopelagic zone threats
Provisioning services		
Fisheries	Currently none	Currently none
Oil and gas	Currently some; also indirect impact through dispersal from shelf and bathyal activity	Currently none, but indirect impact through dispersal from shelf and bathyal activity
Methane reserves/potential for gas hydrate extraction	Gulf of Mexico, potentially other areas	Not applicable
Hydrogen generation and subseabed storage for future carbon capture and disposal	Presently unknown	Not applicable
Mining (metal-rich sediments, polymetallic nodules, rare earth metals, massive sulphides)	Moderate to high in future (potential)	Moderate to high in future (potential) through mining waste and processing water discharge
Waste disposal	High (widespread)	Moderate to high (present)
Bioprospecting	Present, potentially high	High potential, unknown
Military activities and use	Unknown	Unknown
Other energy provision	Currently none	Currently none
Supporting services		
Habitat	Low to moderate and high in the future	Low to moderate and high in the future
Nutrient cycling	Moderate	Moderate
Water circulation and exchange	Moderate	Moderate
Chemosynthetic primary production	Moderate	Moderate
Resilience	High	High
Regulating services		
Gas and climate regulation	Moderate	Moderate
Waste absorption and detoxification	Moderate	Moderate
Biological regulation	Moderate	Moderate
Nutrient cycling	Moderate	Moderate
Cultural services		
Scientific knowledge	Moderate	Moderate
Educational value	Moderate	Moderate
Economic benefits	Potentially high	Potentially high
Aesthetic, inspirational, ethical, indigenous	High	High
Climatic record in deep-sea sediments	Moderate	Not applicable

5. Outlook

Many unknowns remain regarding abyssal ecosystems, but related research has increased significantly in the past decade, with more anticipated, in particular given the increasing interest in deep-sea mineral extraction. The United Nations Decade of Ocean Science for Sustainable Development (2021–2030) also includes plans for more deep-sea research.

The emergence of potential deep seabed mining to exploit polymetallic nodules poses a risk to abyssal ecosystems. However, data collected during current exploration activities may increase deep-sea knowledge in several regions over the next 10 years. Researchers frequently lament the substantial lack of taxonomic biodiversity data for most abyssal fauna. Work to collect such data is under way, but it will require much more time and resources (Glover and others, 2018).

Studies demonstrate the sensitivity of the abyss to climate change. Despite the difficulties of predicting precise climate change effects over the next 10–20 years, rising temperatures, declining oxygen concentrations, shallowing of the aragonite saturation horizon and changes in benthopelagic coupling can be expected (Rogers, 2015; Sweetman and others, 2017). Considering the slow growth rates of organisms and the fact that they are well adapted to the abyssal conditions of cold, high pressure, stability and food poverty, the impacts of predicted changes on abyssal communities will likely be more severe than those at shallower depths. Predictions of significant decreases in the flux of organic material to the deep sea floor in most oceans may be especially problematic for abyssal areas. Future research will enhance abyssal biodiversity knowledge and increase our understanding of how climate change and anthropogenic activities will affect abyssal ecosystems.

Globally, the protection of abyssal environments may increase. The Convention on Biological Diversity classification of ecologically or biologically significant areas (Secretariat of the Convention on Biological Diversity, 2008) includes those environments, and further efforts are under way through the regional environmental management plans of the International Seabed Authority in connection with seabed mining, as well as legislative developments to manage biodiversity beyond national jurisdiction.

6. Key remaining knowledge gaps

Despite recent advances in the knowledge of abyssal ecosystems, many gaps exist in understanding abyssal biodiversity, evolution, biogeography and the distributions, connectivity and responses to changing conditions and anthropogenic impacts.

The current poor state of taxonomic, natural history and biodiversity knowledge of the fauna on abyssal plains limits environmental impact monitoring and exposes the need for baseline studies that provide species lists and numbers. Given that more than 95 per cent of species in planned mining areas are undescribed, current monitoring protocols are inadequate. Despite ongoing efforts to create the necessary faunal catalogues and taxonomic knowledge (Dahlgren and others, 2016; Glover and others, 2016b; Wiklund and others, 2017), future efficient monitoring requires sustained resources.

Very few studies have examined abyssal hard-bottom habitats and, although some megafauna information exists, there is almost no information available on associated microbes, protists, meiofauna or macrofauna.

Vast areas of the abyssal sea floor remain completely unsampled. Records in international databases (e.g., the Ocean Biodiversity Information System), suggest particularly severe undersampling of the southern Pacific Ocean, as well as the deep Indian Ocean and Bay of Bengal.

Knowledge about species geographic ranges, connectivity patterns or the resilience of assemblages to climate stressors or direct human disturbance in the abyss is limited. The effective management of human activities to sustain deep-sea biodiversity hinges upon such information. In addition, poor characterization of abyssal contributions to ecosystem goods and services limits the availability of appropriate tools to value human benefits adequately (Jobstvogt and others, 2014a, 2014b; Thurber and others, 2014).

The lack of documentation on and relating to the management of human impacts on such a vast, dynamic space, almost all of which is located beyond national jurisdictions, may represent the single most important knowledge gap.

References

Amon, Diva J., and others (2016a). Insights into the abundance and diversity of abyssal megafauna in a polymetallic-nodule region in the eastern Clarion-Clipperton Zone. *Scientific Reports*, vol. 6, art. 30492.

Amon, Diva J., and others (2016b). Observations of organic falls in the abyssal Clarion-Clipperton Zone, tropical eastern Pacific Ocean. *Marine* Biodiversity. https://doi.org/10.1007/s12526-016-0572-4.

Amon, Diva J., and others (2017a). Megafauna of the UKSRL exploration contract area and eastern Clarion-Clipperton Zone in the Pacific Ocean: Annelida, Arthropoda, Bryozoa, Chordata, Ctenophora, Mollusca. Jeffrey C. Drazen and others, eds. *Biodiversity Data Journal*, vol. 5, e14598. https://doi.org/10.3897/BDJ.5.e14598.

Amon, Diva J., and others (2017b). Megafauna of the UKSRL exploration contract area and eastern Clarion Clipperton Zone in the Pacific Ocean: Echinodermata. *Biodiversity Data Journal*, vol. 5, e11794. https://doi.org/10.3897/BDJ.5.e11794.

Armstrong, Claire W., and others (2012). Services from the deep: Steps towards valuation of deep sea goods and services. *Ecosystem Services*, vol. 2, pp. 2–13. https://doi.org/10.1016/j.ecoser.2012.07.001.

Atwood T.B., and others (2020) Global Patterns in Marine Sediment Carbon Stocks. *Frontiers in Marine Science*, vol. 7, art. 165. https://doi.org/10.3389/fmars.2020.00165.

Baco, Amy R., and others (2016). A synthesis of genetic connectivity in deep-sea fauna and implications for marine reserve design. *Molecular Ecology*, vol. 25, No. 14, pp. 3276–3298.

Billett, D.S.M., and others (2001). Long-term change in the megabenthos of the Porcupine Abyssal Plain (NE Atlantic). *Progress in Oceanography*, vol. 50, Nos. 1–4, pp. 325–348.

Billett, D.S.M., and others (2010). Long-term change in the abyssal NE Atlantic: The 'Amperima Event' revisited. *Deep Sea Research Part II: Topical Studies in Oceanography*, vol. 57, No. 15, pp. 1406–1417.

Bodil, Bluhm A., and others (2011). Diversity of the arctic deep-sea benthos. *Marine Biodiversity*, vol. 41, No. 1, pp. 87–107. https://doi.org/10.1007/s12526-010-0078-4.

Bouef, Dominique, and others (2019). Biological composition and microbial dynamics of sinking particulate organic matter at abyssal depths in the oligotrophic open ocean. *Proceedings of the National Academy of Sciences*, vol. 116, No. 24, pp. 11824–11832.

Brandt, Angelika, and others (2006). The biodiversity of the deep Southern Ocean benthos. *Philosophical Transactions of the Royal Society B: Biological Sciences*, vol. 362, No. 1477, pp. 39–66.

Brlones, Elva Escobar, and others (2009). Global open oceans and deep seabed (GOODS) biogeographic classification. IOC Technical Series No. 84, UNESCO-IOC.

Canals, Miquel, and others (2006). Flushing submarine canyons. *Nature*, vol. 444, No. 7117, p. 354.

Carney, Robert S. (1994). Consideration of the oasis analogy for chemosynthetic communities at Gulf of Mexico hydrocarbon vents. *Geo-Marine Letters*, vol. 14, Nos. 2–3, pp. 149–159.

________ (2005). Zonation of deep biota on continental margins. In *Oceanography and Marine Biology*, pp. 221–288. CRC Press.

Cavan, E.L., and others (2015). Attenuation of particulate organic carbon flux in the Scotia Sea, Southern Ocean, is controlled by zooplankton fecal pellets. *Geophysical Research Letters*, vol. 42, No. 3, pp. 821–830.

Chapman, Abbie S.A., and others (2019). sFDvent: A global trait database for deep-sea hydrothermal-vent fauna. *Global Ecology and Biogeography*, vol. 28, No. 11, pp. 1538–1551. https://doi.org/10.1111/geb.12975.

Cheung, William, and Lisa Levin (2019). Ecosystem considerations. In *Deep-Ocean Climate Change Impacts on Habitat, Fish and Fisheries*. FAO Fisheries and Aquaculture Technical Paper 638. Rome.

Chiba, S., and others (2018). Human footprint in the abyss: 30 year records of deep-sea plastic debris. *Marine Policy*, vol. 96, pp. 204–212.

Christiansen, B., and others (2020). Potential effects of deep seabed mining on pelagic and benthopelagic biota. *Marine Policy*, vol. 114, 103442.

Christodoulou, Magdalini, and others (2019). Dark Ophiuroid Biodiversity in a Prospective Abyssal Mine Field. *Current Biology*, vol. 29, No. 22, pp. 3909–3912.e3. https://doi.org/10.1016/j.cub.2019.09.012.

Company, Joan B., and others (2008). Climate influence on deep sea populations. *PLoS One*, vol. 3, No. 1, pp. 1–8. https://doi.org/10.1371/journal.pone.0001431.

Cronin, Thomas M., and Maureen E. Raymo (1997). Orbital forcing of deep-sea benthic species diversity. *Nature*, vol. 385, No. 6617, p. 624.

Cronin, Thomas M., and others (1999). Deep-sea ostracode species diversity: response to late Quaternary climate change. *Marine Micropaleontology*, vol. 37, Nos. 3–4, pp. 231–249.

Dahlgren, Thomas G., and others (2016). Abyssal fauna of the UK-1 polymetallic nodule exploration area, Clarion-Clipperton Zone, central Pacific Ocean: Cnidaria. *Biodiversity Data Journal*, No. 4.

Davies, David K., and W. Richard Moore (1970). Dispersal of Mississippi sediment in the Gulf of Mexico. *Journal of Sedimentary Research*, vol. 40, No. 1.

Durden, Jennifer M., and others (2015). Abyssal hills – hidden source of increased habitat heterogeneity, benthic megafaunal biomass and diversity in the deep sea. *Progress in Oceanography*, vol. 137, pp. 209–218.

Dutkiewicz, Adriana, and others (2015). Census of seafloor sediments in the world's ocean. *Geology*, vol. 43, No. 9, pp. 795–798.

Ehlers, Birte-Marie, and Wilfried Jokat (2013). Paleo-bathymetry of the northern North Atlantic and consequences for the opening of the Fram Strait. *Marine Geophysical Research*, vol. 34, No. 1, pp. 25–43.

Engen, Øyvind, and others (2008). Opening of the Fram Strait gateway: A review of plate tectonic constraints. *Tectonophysics*, vol. 450, Nos. 1–4, pp. 51–69.

Food and Agriculture Organization of the United Nations (FAO) (2019). *Deep-Ocean Climate Change Impacts on Habitat, Fish and Fisheries*. Fisheries and Aquaculture Technical Paper 638. Rome.

Frigstad, H., and others (2015). Links between surface productivity and deep ocean particle flux at the Porcupine Abyssal Plain sustained observatory. *Biogeosciences*, vol. 12, No. 19, pp. 5885–5897.

Gage, John D., and Paul A. Tyler (1991). *Deep-Sea Biology: A Natural History of Organisms at the Deep-Sea Floor.* Cambridge University Press.

Gebbie, G., and P. Huybers (2019). The Little Ice Age and 20th-century deep Pacific cooling. *Science*, vol. 363, No. 6422, pp. 70–74.

Glover, Adrian G., and others (2002). Polychaete species diversity in the central Pacific abyss: local and regional patterns, and relationships with productivity. *Marine Ecology Progress Series*, vol. 240, pp. 157–170.

Glover, Adrian G., and others (2016a). Abyssal fauna of the UK-1 polymetallic nodule exploration claim, Clarion-Clipperton Zone, central Pacific Ocean: Echinodermata. *Biodiversity Data Journal*, No. 4.

Glover, Adrian G., and others (2016b). An end-to-end DNA taxonomy methodology for benthic biodiversity survey in the Clarion-Clipperton Zone, central Pacific abyss. *Journal of Marine Science and Engineering*, vol. 4, No. 1, art. 2.

Glover, Adrian G., and others (2018). Point of View: Managing a sustainable deep-sea 'blue economy' requires knowledge of what actually lives there. *ELife*, vol. 7, e41319.

Gooday, Andrew J., and others (2010). Decadal-scale changes in shallow-infaunal foraminiferal assemblages at the Porcupine Abyssal Plain, NE Atlantic. *Deep Sea Research Part II: Topical Studies in Oceanography*, vol. 57, No. 15, pp. 1362–1382.

Griffiths, Huw J. (2010). Antarctic marine biodiversity – what do we know about the distribution of life in the Southern Ocean? *PloS One*, vol. 5, No. 8, e11683.

Hartman, S.E., and others (2015). Biogeochemical variations at the Porcupine Abyssal Plain sustained Observatory in the northeast Atlantic Ocean, from weekly to inter-annual timescales. *Biogeosciences*, vol. 12, No. 3, pp. 845–853.

Harzhauser, Mathias, and others (2007). Biogeographic responses to geodynamics: a key study all around the Oligo–Miocene Tethyan Seaway. *Zoologischer Anzeiger*, vol. 246, No. 4, pp. 241–256.

Hessler, Robert R., and Peter A. Jumars (1974). Abyssal community analysis from replicate cores in the central North Pacific. In *Deep Sea Research and Oceanographic Abstracts*, vol. 21, pp. 185–209. Elsevier.

Hoffmann, Sharon S., and others (2018). Evidence for stable Holocene basin-scale overturning circulation despite variable currents along the deep western boundary of the North Atlantic Ocean. *Geophysical Research Letters*, vol. 45, No. 24, pp. 13,427–13,436.

Houghton, Isabel A., and John O. Dabiri (2019). Alleviation of hypoxia by biologically generated mixing in a stratified water column. *Limnology and Oceanography*, vol. 64, No. 5, pp. 2161–2171. https://doi.org/10.1002/lno.11176.

Howell, Kerry L., and others (2003). Feeding ecology of deep-sea seastars (Echinodermata: Asteroidea): a fatty-acid biomarker approach. *Marine Ecology Progress Series*, vol. 255, pp. 193–206.

Hu, Aixue, and others (2012). The Pacific-Atlantic seesaw and the Bering Strait. *Geophysical Research Letters*, vol. 39, No. 3.

Hunt, Gene, and others (2005). Species-energy relationship in the deep sea: a test using the Quaternary fossil record. *Ecology Letters*, vol. 8, No. 7, pp. 739–747.

Ingels, Jeroen, and others (2010). Preferred use of bacteria over phytoplankton by deep-sea nematodes in polar regions. *Marine Ecology Progress Series*, vol. 406, pp. 121–133.

Jobstvogt, Niels, and others (2014a). How can we identify and communicate the ecological value of deep-sea ecosystem services? *PloS One*, vol. 9, No. 7, e100646.

Jobstvogt, Niels, and others (2014b). Twenty thousand sterling under the sea: estimating the value of protecting deep-sea biodiversity. *Ecological Economics*, vol. 97, pp. 10–19.

Jones, Daniel O.B., and others (2014). Global reductions in seafloor biomass in response to climate change. *Global Change Biology*, vol. 20, No. 6, pp. 1861–1872.

Kalogeropoulou, V., and others (2010). Temporal changes (1989–1999) in deep-sea metazoan meiofaunal assemblages on the Porcupine Abyssal Plain, NE Atlantic. *Deep Sea Research Part II: Topical Studies in Oceanography*, vol. 57, No. 15, pp. 1383–1395.

Kuhnz, Linda A., and others (2014). Rapid changes and long-term cycles in the benthic megafaunal community observed over 24 years in the abyssal northeast Pacific. *Progress in Oceanography*, vol. 124, pp. 1–11.

Lampitt, R.S., and A.N. Antia (1997). Particle flux in deep seas: regional characteristics and temporal variability. *Deep Sea Research Part I: Oceanographic Research Papers*, vol. 44, No. 8, pp. 1377–1403.

Lampitt, R.S., and others (2010). The sustained observatory over the Porcupine Abyssal Plain (PAP): Insights from time series observations and process studies (preface). *Deep Sea Research Part II: Topical Studies in Oceanography*, vol. 57, No. 15, pp. 1267–1271.

Lawver, Lawrence A., and Lisa M. Gahagan (2003). Evolution of Cenozoic seaways in the circum-Antarctic region. *Palaeogeography, Palaeoclimatology, Palaeoecology*, vol. 198, Nos. 1–2, pp. 11–37.

Levin, Lisa A. (2003). Oxygen minimum zone benthos: Adaptation and community response to hypoxia. *Oceanography and Marine Biology: An Annual Review*, vol. 41, pp. 1–45.

Levin, Lisa A., and Andrew J. Gooday (2003). The deep Atlantic Ocean. In *Ecosystems of the World*, pp. 111–178. Elsevier.

Levin, Lisa A., and Nadine Le Bris (2015). The deep ocean under climate change. *Science*, vol. 350, No. 6262, pp. 766–768.

Levin, Lisa A., and others (2020). Climate change considerations are fundamental to management of deep-sea resource extraction. *Global Change Biology*. https://doi.org/10.1111/gcb.15223.

Livermore, Roy, and others (2007). Drake Passage and Cenozoic climate: An open and shut case? *Geochemistry, Geophysics, Geosystems*, vol. 8, No. 1.

Lutz, Michael J., and others (2007). Seasonal rhythms of net primary production and particulate organic carbon flux to depth describe the efficiency of biological pump in the global ocean. *Journal of Geophysical Research: Oceans*, vol. 112, No. C10.

Marincovich, Louie, Jr., and Andrey Y. Gladenkov (2001). New evidence for the age of Bering Strait. *Quaternary Science Reviews*, vol. 20, Nos. 1–3, pp. 329–335.

Marsh, Leigh, and others (2018). Geomorphological evidence of large vertebrates interacting with the seafloor at abyssal depths in a region designated for deep-sea mining. *Royal Society Open Science*, vol. 5, No. 8, 180286.

McCave, Ian Nicholas (2017). Formation of sediment waves by turbidity currents and geostrophic flows: a discussion. *Marine Geology*, vol. 390, pp. 89–93.

McClain, Craig R., and others (2011). Dispersal, environmental niches and oceanic-scale turnover in deep-sea bivalves. *Proceedings of the Royal Society B: Biological Sciences*, vol. 279, No. 1735, pp. 1993–2002.

McClain, Craig R., and others (2012a). Energetics of life on the deep seafloor. *Proceedings of the National Academy of Sciences*, vol. 109, No. 38, pp. 15366–15371.

McClain, Craig R., and others (2012b). Increased energy promotes size-based niche availability in marine mollusks. *Evolution: International Journal of Organic Evolution*, vol. 66, No. 7, pp. 2204–2215.

McClain, Craig R., and Sarah Mincks Hardy (2010). The dynamics of biogeographic ranges in the deep sea. *Proceedings of the Royal Society B: Biological Sciences*, vol. 277, No. 1700, pp. 3533–3546.

McClain, Craig R., and Thomas A. Schlacher (2015). On some hypotheses of diversity of animal life at great depths on the sea floor. *Marine Ecology*, vol. 36, No. 4, pp. 849–872.

Menzies, Robert James, and others (1973). Abyssal environment and ecology of the world oceans.

Mestre, Mireia, and others (2018). Sinking particles promote vertical connectivity in the ocean microbiome. *Proceedings of the National Academy of Sciences*, vol. 115, No. 29, pp. E6799–E6807.

Mora, Camilo, and others (2013). Biotic and human vulnerability to projected changes in ocean biogeochemistry over the 21st century. *PLoS Biology*, vol. 11, No. 10, e1001682.

O'Dea, Aaron, and others (2016). Formation of the Isthmus of Panama. *Science Advances*, vol. 2, No. 8, e1600883.

O'Hara, Timothy D., and others (2019). Contrasting processes drive ophiuroid phylodiversity across shallow and deep seafloors. *Nature*, vol. 565, No. 7741, p. 636.

Palanques, Albert, and others (2011). Effects of storm events on the shelf-to-basin sediment transport in the southwestern end of the Gulf of Lions (Northwestern Mediterranean).

Pörtner, Hans-O., and Anthony P. Farrell (2008). Physiology and climate change. *Science*, vol. 322, No. 5902, pp. 690–692.

Pörtner, H-O. (2010). Oxygen- and capacity-limitation of thermal tolerance: a matrix for integrating climate-related stressor effects in marine ecosystems. *Journal of Experimental Biology*, vol. 213, No. 6, pp. 881–893.

Rex, Michael A., and Ron J. Etter (2010). *Deep-Sea Biodiversity: Pattern and Scale*. Harvard University Press.

Robison, Bruce H. (2009). Conservation of deep pelagic biodiversity. *Conservation Biology*, vol. 23, No. 4, pp. 847–858.

Rogers, Alex David (2015). Environmental change in the deep ocean. *Annual Review of Environment and Resources*, vol. 40, pp. 1–38.

Ruhl, Henry A., and Kenneth L. Smith (2004). Shifts in deep-sea community structure linked to climate and food supply. *Science*, vol. 305, No. 5683, pp. 513–515.

Ruhl, Henry A., and others (2014). Links between deep-sea respiration and community dynamics. *Ecology*, vol. 95, No. 6, pp. 1651–1662.

Ruhl, Henry A., and others (2008). Connections between climate, food limitation, and carbon cycling in abyssal sediment communities. *Proceedings of the National Academy of Sciences*, vol. 105, No. 44, pp. 17006–17011.

Sardà, Francisco, and others (2004). An introduction to Mediterranean deep-sea biology. *Scientia Marina*, vol. 68, No. S3, pp. 7–38.

Scher, Howie D., and Ellen E. Martin (2006). Timing and climatic consequences of the opening of Drake Passage. *Science*, vol. 312, No. 5772, pp. 428–430.

Schmidt, Daniela N., and others (2007). The closure history of the Central American seaway: evidence from isotopes and fossils to models and molecules. *Deep Time Perspectives on Climate Change Marrying the Signal from Computer Models and Biological Proxies: London, Geological Society of London*, pp. 427–442.

Schmidt, Daniela N., and others (2016). Morphological response of planktic foraminifers to habitat modifications associated with the emergence of the Isthmus of Panama. *Marine Micropaleontology*, vol. 128, pp. 28–38.

Shulse, Christine N., and others (2017). Polymetallic nodules, sediments, and deep waters in the equatorial North Pacific exhibit highly diverse and distinct bacterial, archaeal, and microeukaryotic communities. *Microbiology Open*, vol. 6, No. 2, e00428.

Secretariat of the Convention on Biological Diversity (2008). *Conference of the Parties to the Convention on Biological Diversity, Decision IX/20. Marine and Coastal Biodiversity. UNEP/CBD/COP/DEC/IX/20. Annex I. Scientific Criteria for Identifying Ecologically or Biologically Significant Marine Areas in Need of Protection in Open-Ocean Waters and Deep-Sea Habitats.*

Simon-Lledó, Erik, and others (2019a). Ecology of a polymetallic nodule occurrence gradient: Implications for deep-sea mining. *Limnology and Oceanography*, vol. 64, No. 5, pp. 1883–1894. https://doi.org/10.1002/lno.11157.

Simon-Lledó, Erik, and others (2019b). Megafaunal variation in the abyssal landscape of the Clarion Clipperton Zone. *Progress in Oceanography*, vol. 170, pp. 119–133.

Smith, Craig R., and others (2008). Abyssal food limitation, ecosystem structure and climate change. *Trends in Ecology & Evolution*, vol. 23, No. 9, pp. 518–528.

Smith, Craig R., and Amanda W.J. Demopoulos (2003). The deep Pacific ocean floor. In *Ecosystems of the World*, pp. 179–218.

Smith, Kenneth L., Jr., and others (2009). Climate, carbon cycling, and deep-ocean ecosystems. *Proceedings of the National Academy of Sciences*, vol. 106, No. 46, pp. 19211–19218.

Smith, Kenneth L., Jr., and others (2013). Deep ocean communities impacted by changing climate over 24 y in the abyssal northeast Pacific Ocean. *Proceedings of the National Academy of Sciences*, vol. 110, No. 49, pp. 19838–19841.

Snelgrove, Paul V.R., and C.R. Smith (2002). A riot of species in an environmental calm: the paradox of the species-rich deep-sea floor. *Oceanography and Marine Biology: An Annual Review*, vol. 40, pp. 311–42.

Soltwedel, Thomas, and others (2005). HAUSGARTEN: multidisciplinary investigations at a deep-sea, long-term observatory in the Arctic Ocean. *Oceanography*, vol. 18, No. 3.

Soltwedel, Thomas, and others (2016). Natural variability or anthropogenically-induced variation? Insights from 15 years of multidisciplinary observations at the arctic marine LTER site HAUSGARTEN. *Ecological Indicators*, vol. 65, pp. 89–102.

Somero, G.N. (1992). Biochemical ecology of deep-sea animals. *Experientia*, vol. 48, No. 6, pp. 537–543.

Soto, Eulogio H., and others (2010). Temporal variability in polychaete assemblages of the abyssal NE Atlantic Ocean. *Deep Sea Research Part II: Topical Studies in Oceanography*, vol. 57, No. 15, pp. 1396–1405.

Stramma, Lothar, and others (2010). Ocean oxygen minima expansions and their biological impacts. *Deep Sea Research Part I: Oceanographic Research Papers*, vol. 57, No. 4, pp. 587–595.

Suttle, Curtis A. (2005). Viruses in the sea. *Nature*, vol. 437, No. 7057, p. 356.

Sweetman, Andrew K., and others (2017). Major impacts of climate change on deep-sea benthic ecosystems. *Elementa: Science of the Anthropocene*, vol. 5, art. 4.

Taboada, Sergi, and others (2018). Implications of population connectivity studies for the design of marine protected areas in the deep sea: an example of a demosponge from the Clarion-Clipperton Zone. *Molecular Ecology*, vol. 27, No. 23, pp. 4657–4679.

Thomas, Ellen (2007). Cenozoic mass extinctions in the deep sea: what perturbs the largest habitat on Earth? In *Large Ecosystem Perturbations: Causes and Consequences*, Simonetta Monechi, and others, eds. Geological Society of America, vol. 424. https://doi.org/10.1130/2007.2424(01).

Thomas, Ellen, and Andrew J. Gooday (1996). Cenozoic deep-sea benthic foraminifers: tracers for changes in oceanic productivity? *Geology*, vol. 24, No. 4, pp. 355–358.

Thomas, Ellen, and others (2000). Deep-sea environments on a warm earth: latest Paleocene-early Eocene. In *Warm Climates in Earth History*, eds. Brian T. Huber and others, pp. 132–160. Cambridge University Press.

Thornalley, David J.R., and others (2018). Anomalously weak Labrador Sea convection and Atlantic overturning during the past 150 years. *Nature*, vol. 556, No. 7700, p. 227.

Thurber, Andrew R., and others (2014). Ecosystem function and services provided by the deep sea. *Biogeosciences*, vol. 11, No. 14, pp. 3941–3963.

Ulses, C., and others (2008). Impact of storms and dense water cascading on shelf-slope exchanges in the Gulf of Lion (NW Mediterranean). *Journal of Geophysical Research: Oceans*, vol. 113, No. C2. https://doi.org/10.1029/2006JC003795.

United Nations (2017a). Chapter 1: Introduction – Planet, oceans and life. In *The First Global Integrated Marine Assessment: World Ocean Assessment I*. Cambridge: Cambridge University Press.

________ (2017b). Chapter 24: Solid waste disposal. In *The First Global Integrated Marine Assessment: World Ocean Assessment I*. Cambridge: Cambridge University Press.

________ (2017c). Chapter 36F: Open ocean deep sea. In *The First Global Integrated Marine Assessment: World Ocean Assessment I*. Cambridge: Cambridge University Press.

________ (2017d). *The First Global Integrated Marine Assessment: World Ocean Assessment I*. Cambridge: Cambridge University Press.

Van den Hove, Sybille, and Vincent Moreau (2007). *Deep-Sea Biodiversity and Ecosystems: A Scoping Report on Their Socio-Economy, Management and Governanace*. UNEP Regional Seas Reports and Studies No. 184. UNEP/Earthprint.

Vanreusel, Ann, and others (2016). Threatened by mining, polymetallic nodules are required to preserve abyssal epifauna. *Scientific Reports*, vol. 6, art. 26808.

Vinogradova, Nina G. (1979). The geographical distribution of the abyssal and hadal (ultra-abyssal) fauna in relation to the vertical zonation of the ocean. *Sarsia*, vol. 64, Nos. 1–2, pp. 41–50.

________ (1997). Zoogeography of the abyssal and hadal zones. In *Advances in Marine Biology*, vol. 32, pp. 325–387. Elsevier.

Watling, Les, and others (2013). A proposed biogeography of the deep ocean floor. *Progress in Oceanography*, vol. 111, pp. 91–112.

Webb, Thomas J., and others (2010). Biodiversity's big wet secret: the global distribution of marine biological records reveals chronic under-exploration of the deep pelagic ocean. *PLoS One*, vol. 5, No. 8, e10223.

Wei, Chih-Lin, and others (2010). Global patterns and predictions of seafloor biomass using random forests. *PloS One*, vol. 5, No. 12, e15323.

Wiklund, Helena, and others (2017). Abyssal fauna of the UK-1 polymetallic nodule exploration area, Clarion-Clipperton Zone, central Pacific Ocean: Mollusca. *ZooKeys*, vol. 707, pp. 1–46. https://doi.org/10.3897/zookeys.707.13042.

Wiklund, Helena, and others (2019). Abyssal fauna of polymetallic nodule exploration areas, eastern Clarion-Clipperton Zone, central Pacific Ocean: Annelida: Capitellidae, Opheliidae, Scalibregmatidae, and Travisiidae. *ZooKeys*, vol. 883, pp. 1–82. https://doi.org/10.3897/zookeys.883.36193.

Wlodarska-Kowalczuk, Maria, and Thomas H. Pearson (2004). Soft-bottom macrobenthic faunal associations and factors affecting species distributions in an Arctic glacial fjord (Kongsfjord, Spitsbergen). *Polar Biology*, vol. 27, No. 3, pp. 155–167.

Woolley, Skipton N.C., and others (2016). Deep-sea diversity patterns are shaped by energy availability. *Nature*, vol. 533, No. 7603, p. 393.

Yashayaev, Igor, and others (2007). Spreading of the Labrador Sea Water to the Irminger and Iceland basins. *Geophysical Research Letters*, vol. 34, No. 10.

Yasuhara, Moriaki, and others (2017). Combining marine macroecology and palaeoecology in understanding biodiversity: microfossils as a model. *Biological Reviews*, vol. 92, No. 1, pp. 199–215.

Yasuhara, Moriaki, and others (2018). Marine biodiversity in space and time: what tiny fossils tell. *Mètode Science Studies Journal - Annual Review*. https://doi.org/10.7203/metode.9.11404.

Yasuhara, Moriaki, and others (2019a). Quaternary deep-sea ostracods from the north-western Pacific Ocean: global biogeography and Drake-Passage, Tethyan, Central American and Arctic pathways. *Journal of Systematic Palaeontology*, vol. 17, No. 2, pp. 91–110. https://doi.org/10.1080/14772019.2017.1393019.

Yasuhara, Moriaki, and others (2019b). North Atlantic intermediate water variability over the past 20,000 years. *Geology*, vol. 47, No. 7, pp. 659–63. https://doi.org/10.1130/G46161.1.

Yasuhara, Moriaki, and Thomas M. Cronin (2008). Climatic Influences on deep-sea ostracode (Crustacea) diversity for the last three million years. *Ecology*, vol. 89, No. sp11, pp. S53–S65.

Yasuhara, Moriaki, and Roberto Danovaro (2016). Temperature impacts on deep-sea biodiversity. *Biological Reviews*, vol. 91, No. 2, pp. 275–287.

Yesson, Chris, and others (2011). The global distribution of seamounts based on 30 arc seconds bathymetry data. *Deep Sea Research Part I: Oceanographic Research Papers*, vol. 58, No. 4, pp. 442–453.

Chapter 7N
Open ocean

Contributors: Peter Croot (convener), Osman Keh Kamara, Joseph Montoya, Tracy T. Sutton and Michael Vecchione.

Keynote points

- Global warming is already affecting the open ocean, and marine heatwaves are likely to increase in frequency and strength in the future.
- Climate change-induced changes in the open ocean biological pump will alter the ocean's ability to take up anthropogenic carbon.
- Deoxygenation of the open ocean is already leading to habitat compression for some pelagic species, with subsequent impacts on their vertical and horizontal distribution in the ocean.
- Increasing fluxes of plastic litter from the land are affecting open ocean ecosystems.
- A critical knowledge gap exists for deep pelagic (e.g., mesopelagic and bathypelagic) environments, as they are poorly sampled and understood.

1. Introduction

1.1. Topic scope

Chapter 36F of the first *World Ocean Assessment* (United Nations, 2017a) included an assessment of both the open ocean (pelagic zone) and deep-sea (benthic zone) ecosystems seaward of the continental shelf (200 m depth). In the present, updated Assessment, benthic ecosystems are treated separately, and the present chapter focuses solely on the pelagic realm throughout the full water column.

Previously, it was reported in the first Assessment (United Nations, 2017b) that the open ocean provided essential marine ecosystem goods and services, despite its relative inaccessibility. Furthermore, there was great potential for mineral, energy and living resources from the pelagic areas, although they were poorly studied spatially and temporally, which made conservation issues complicated, as so little was known about biodiversity and ecosystem function.

1.2. Pelagic realm

The main physical drivers structuring pelagic ecosystems are depth and pressure, light, temperature, nutrient inputs (e.g., nitrogen, iron), dissolved oxygen and currents. The surface zone of the open ocean (epipelagic, down to 200 m) is defined by sufficient sunlight penetration to support primary production. Below that zone is the mesopelagic or "twilight" zone, generally extending from the bottom of the epipelagic zone to about 1,000 m – the maximum depth of sunlight penetration and the bottom of the permanent thermocline. The mesopelagic zone is important for active vertical migration and the microbial degradation of organic matter sinking from the surface – two key elements of the biological pump (Robinson and others, 2010). The daily vertical migration of deep-sea organisms from the mesopelagic zone to the epipelagic zone and back is increasingly appreciated as a driver of carbon flux owing to the increased estimation of mesopelagic nekton biomass (Irigoien and others, 2014).

The planet's largest ecosystem comprises the bathypelagic domain, a dark and cold (0–5°C) zone found between 1,000 and 4,000 m, comprising almost 75 per cent of ocean volume (Costello and others, 2010). The deep pelagic ocean is underobserved and undersampled owing to a combination of factors: (a) limited access to open ocean sampling platforms for obtaining deepwater samples; (b) the large ocean volume involved; and (c) the widely dispersed populations. Data on mesopelagic ecosystems are improving, but still very little is known regarding organisms from the deeper zones, including the bathypelagic, abyssopelagic (4,0–6,000 m) and hadalpelagic (> 6,000 m) zones. Initial results indicate that the overall diversity of species may be less than in other ecosystems, although new microbial studies are revealing great diversity in the deep ocean. Surveys have shown that lateral connectivity also occurs between the deep zones of the open ocean and is not just limited to the mesopelagic and surface (Sutton, 2013).

Crustaceans (e.g., copepods, amphipods, ostracods) are important contributors to the abundance and species numbers of zooplankton in the deep ocean. Gelatinous animals, such as salps, jellyfishes and colonial siphonophores, are also very important. Larger organisms present include many species of fishes, sharks, crustaceans (e.g., shrimps, krill) and cephalopods (e.g., squids). Biomass estimates based on acoustic measurements indicate that mesopelagic fishes may have been seriously underestimated in the past. They likely represent approximately 10,000–15,000 million tons and are responsible for respiring up to 10 per cent of primary productivity (Irigoien and others, 2014). Deep pelagic fish biomass is likely the overwhelming majority of fish biomass on the planet (Sutton, 2013). Such species are also important prey for mammals (toothed whales and seals), tunas, seabirds and deep demersal fishes.

1.3. Pressures on the pelagic realm

The open ocean is affected by multiple environmental stressors, most notably ocean warming, acidification and deoxygenation. Those stressors are likely to promote shifts in the latitudinal and vertical distribution of open ocean and deep-sea fish stocks (Brander, 2010), whereas deoxygenation may result in habitat compression for aerobic organisms (Stramma and others, 2012) while expanding the volume of water supporting anaerobic processes. Changes in bottom-up or top-down forcing will likely result in complex and indirect effects on open ocean ecosystem services, most notably the biological carbon pump, although the overall impact is unclear, as so little is known about microbial diversity, function and processes in the deep ocean.

1.4. Knowledge gaps identified in the first *World Ocean Assessment*

It was noted in the first Assessment that key information about pelagic ecosystems was lacking, as data were available only from a few geographic areas and on a fraction of overall biodiversity. The first Assessment highlighted the fact that information on ecosystem structure and processes were insufficient to assess the potential performance of conservation and management measures that had been developed for shelf and coastal marine ecosystems.

1.5. Scientific advances since the first *World Ocean Assessment*

The ocean, including the open ocean, was the subject of a recent special report by the Intergovernmental Panel on Climate Change (Intergovernmental Panel on Climate Change (IPCC), 2019). A key advance has been the acknowledgement that the multi-stressor nature of climate change on the open ocean will require new tools for analysing the impacts of each stress on the ecosystem and the synergistic interactions among stressors, as the response may be very non-linear (Boyd and others, 2015).

Diel vertical migrations have been inferred from space using a satellite-mounted light detection and ranging instrument (Behrenfeld and others, 2019), providing new insights into that biogeochemically important process. Knowledge of the global distribution of trace elements and their isotopes in all ocean basins has been significantly advanced by the international GEOTRACES programme (Schlitzer and others, 2018). The understanding of the physics and biogeochemistry of the open ocean has expanded both spatially and temporally through the increased use of Argo floats (Roemmich and others, 2019) and ocean gliders (Rudnick, 2016). Those data, along with repeat hydrography surveys through the Global Ocean Ship-based Hydrographic Investigations Program (GO-SHIP) (Sloyan and others, 2019), have helped to inform new insights into the function of the biological pump, with a further pathway discerned, namely, the particle injection pump (Boyd and others, 2019), which acts together with the traditional biological gravitation pump.

Over the past decade, rapid advances in omics have been quickly applied to open ocean studies, which has allowed for near real-time, at-sea sequencing of the microbial community (Bennke and others, 2016); the application of environmental DNA (eDNA) to detect white sharks in the open ocean (Truelove and others, 2019); or combining eDNA with autonomous underwater vehicles to examine biodiversity (Yamahara and others, 2019). The increasing use of smart tags and sensors attached to organisms (Harcourt and others, 2019), passive acoustic sensors (Delory and others, 2014) and new visualization tools for marine particles (Lombard and others, 2019) are providing new data on the biodiversity and function of open ocean ecosystems. In particular, the understanding of the mesopelagic zone has evolved to establish global biogeographic and biogeochemical provinces (Reygondeau and others, 2018).

2. Environmental changes in the open ocean since 2010

2.1. Changes in the overall status, including physical or biological state

2.1.1. Ocean warming, marine heatwaves and wind patterns

The evidence is now clear that the ocean has been warming over recent decades (Cheng and others, 2019) and, while the surface ocean has absorbed most of the extra heat, the warming signal is also observable in the intermediate and deep ocean (Cheng and others, 2017). Surface warming has been predicted to result in greater stratification of near-surface waters, although recent work indicates that, while sea surface temperatures are increasing at middle latitudes, stratification is not increasing, nor are mixed layer depths shoaling (Somavilla and others, 2017). In fact, winter mixed-layer depths have been increasing owing to changes in Ekman pumping (Somavilla and others, 2017).

As global temperatures have increased in recent years, marine heatwaves (Hobday and others, 2016) have become longer and more frequent (Oliver and others, 2018). Modelling studies suggest that marine heatwaves are also very likely to increase in the future under global warming (Frölicher and others, 2018). Marine heatwaves in the tropical Pacific Ocean and Indian Ocean are driven by the El Niño Southern Oscillation and related teleconnections (Holbrook and others, 2019), while, in higher latitudes, they are associated with shifts in warm ocean currents, mesoscale eddy activity and atmosphere-ocean dynamics (Rodrigues and others, 2019). The thermohaline circulation has been weakened in recent years owing to global warming, and its consequences for temperature and climate patterns in countries bordering the Atlantic and the equatorial zone around the globe and in the ecosystem services are increasingly profound (Rahmstorf and others, 2015).

Wind patterns have also changed over the open ocean over the past three decades, with small increases in the average wind speed and wave height. Larger increases were seen for extreme conditions (wind speed or wave height exceeding 90th percentiles) (Young and Ribal, 2019). Satellite observations from 1995 to 2018 showed the strongest increases in wind speed and wave height in the Southern Ocean.

2.1.2. Ocean acidification

Most recent estimates of the uptake of anthropogenic carbon by the open ocean (1994–2007) indicate that the uptake is increasing but with important regional deviations. For example, uptake was slower than expected in the North Atlantic, but faster in the South Atlantic (Gruber and others, 2019). Open ocean time series studies, gathered through the inclusion of autonomous marine carbon sensors (e.g., pH and the partial pressure of CO_2 in seawater (pCO_2)), going back almost 20 years at some sites, show clearly observable trends in pH (decreasing) and pCO_2 (increasing) (Sutton and others, 2019).

2.1.3. Ocean deoxygenation

Oxygen loss from the open ocean is expected to increase in a warming world through a complex set of biogeochemical and physical processes (Levin, 2018). The capability to measure dissolved oxygen at nanomolar levels has indicated that anoxic (zero oxygen) regions may have been underestimated previously in the open ocean (Tiano and others, 2014). The impact of climate change on respiration is poorly understood at present, in particular for microbes (Robinson, 2019), for which complex feedback may result in the redistribution of bacterial and archaeal species in the ocean (Beman and Carolan, 2013), as organisms align to specific niches within the different redox zones (Bertagnolli and Stewart, 2018). Despite being hypoxia-tolerant, some open ocean zooplankton are already living close to their physiological limits, and continued deoxygenation may cause unanticipated changes to ecosystem structure and function in the mesopelagic zone (Wishner and others, 2018).

2.1.4. Human impacts: remaining wilderness areas and the rise of plastic pollution

The ocean has been strongly affected by human activities (Jones and others, 2018), and most of the remaining wilderness areas identified lie outside of exclusive economic zones (i.e., areas beyond national jurisdiction). Despite being far from land-based sources, the abundance and extent of plastic litter in the open ocean is increasing (Van Sebille and others, 2015). Open ocean gyres are acting as zones of accumulation for plastic pollution, the so-called garbage patches (Lebreton and others, 2018). The flux of microplastic litter to the open ocean is predicted to increase greatly in the future (Lebreton and Andrady, 2019).

2.2. Factors associated with the changes

2.2.1. Ocean warming and changes to the ocean carbon pump

Warming of the open ocean reduced marine fisheries production by up to an estimated 4.1 per cent between 1930 and 2010, based on hindcast models (Free and others, 2019). Ocean warming is suggested to have more of an impact than ocean acidification on global circumpolar fish stocks (Watson and others, 2018). Overall, the open ocean is likely to have a greater vulnerability to thermal stress compared with land (Pinsky and others, 2019), which may result in a higher sensitivity to warming and faster rates of colonization, resulting in faster species turnover. Warming has already promoted a poleward shift in the distribution of some species (Pinsky and others, 2020), including commercial species, such as tuna (Monllor-Hurtado and others, 2017). While some seabird distributions appear to be insensitive to shifting ocean temperatures (Keogan and others, 2018), a marine heatwave in the North-East Pacific caused extreme mortality of common murres (Piatt and others, 2020). In that context, marine heatwaves are likely to greatly affect biodiversity in the open ocean (Smale and others, 2019).

Large uncertainties remain as to whether a warming ocean will alter primary productivity (Behrenfeld and others, 2016), although modelling suggests small decreases with increasing temperatures in the tropical ocean (Kwiatkowski and others, 2017). However, the strong temperature dependence of metabolic rates in the upper ocean is likely to affect the biological carbon pump, in particular microbial species (Cavan and others, 2019), and may act as a positive feedback to climate by reducing the net sequestration of carbon by the ocean (Boscolo-Galazzo and others, 2018).

Global warming may also be affecting the timing (phenology) of phytoplankton blooms in the open ocean (Barton and others, 2016), although changes in insolation are the main driver for phytoplankton (Boyce and others, 2017) and may ultimately limit poleward migration of species (Sundby and others, 2016). Warming may be reducing the time lag between phytoplankton, protozoan production (Aberle and others, 2012) and zooplankton abundance, with implications for higher trophic levels (Sundby and others, 2016) and for the biological carbon pump and its resulting ecosystem services (Barange and others, 2017).

While changes in pH and carbonate concentration are likely to be less dramatic in the open ocean than in coastal waters (Duarte and others, 2013), biodiversity may be negatively affected in regions where the uptake of anthropogenic CO_2 is greatest (e.g., North Atlantic) (Gehlen and others, 2014). Other biogeochemical processes will be affected by decreasing pH (Gehlen and others, 2011) and, already, evidence exists for decreases in nitrification rates in the open ocean (Beman and others, 2011), which could alter the microbial community and nitrogen cycling in the future.

2.2.2. Deoxygenation and habitat compression

Ocean deoxygenation is resulting in the expansion, both vertically and horizontally, of oxygen minimum zones (Levin, 2018), which can lead to the habitat compression of some pelagic organisms (Stramma and others, 2012) through metabolic constraints (Deutsch and others, 2015). Habitat compression may also lead to increases in the catchability of some billfishes in the eastern Pacific, with potential for overexploitation if not managed carefully (Pohlot and Ehrhardt, 2017).

2.2.3. Direct human impacts

The cumulative impact of human activities on the open ocean is changing both temporally and spatially (Halpern and others, 2015), resulting in only a small proportion of marine wilderness areas of the open ocean remaining (Jones and others, 2018). Aside from the impacts of anthropogenic carbon emissions, other human activities also directly affect the open ocean.

Fishing activities. Marine foods are now sourced farther away from where they are consumed (Watson and others, 2015), thus expanding the global footprint of fishing activities on the open ocean (Kroodsma and others, 2018). Ultimately, however, marine productivity limits the amount of fishes that are available (Chassot and others, 2010) and, currently, global marine capture fisheries appear to have reached a plateau (see chap. 15).

Human-induced contaminant plumes in the open ocean. Industrial activities in the open ocean are affecting biodiversity through the continuous emission of pollutants (Tournadre, 2014), as well as transient events. Oil spills, such as the Deepwater Horizon disaster in 2010 in the Gulf of Mexico, showed that the impact of such an event is felt across all trophic levels for many years (McClain and others., 2019). Modelling efforts suggest that the impact on the ecosystem could last for decades

(Ainsworth and others, 2018). Deep-sea mining and the marine disposal of mining wastes are also likely to affect the open ocean (Vare and others, 2018). The effects of dumping mine tailings in deep water (Ramirez-Llodra and others, 2015) are still poorly understood with regard to the impact on (meso)pelagic organisms.

2.3. Impacts of the changes on and interactions with other components of the marine system

2.3.1. Changes in ecosystem services

Currently, there is a lack of information with regard to the ecosystem services of the mesopelagic and bathypelagic communities (Martinetto and others, 2020; St. John and others, 2016). Similarly, there are few examples of studies examining the extent of benthic-pelagic coupling in the ocean (Trueman and others, 2014).

2.3.2. Indirect climate change impacts on higher trophic levels

Changes in circulation owing to climate warming in the North Atlantic have resulted in a northward shift of the copepod (*Calanus finmarchicus*), the main food source for the endangered northern right whale. The changes have led to the whales altering their seasonal foraging pattern to follow the copepods. Unfortunately, the redistribution of whales into regions where protections from ship strikes or entanglement with fishing gear are not yet in place results in a stall in the recovery of the species (Record and others, 2019).

3. Consequences of the changes for human communities, economies and well-being

3.1. Currently observed consequences

The open ocean is a special case when it comes to assessing the consequences of ecosystem change to humans, as there are no permanent human communities living in or on the open ocean at present. However, there are many coastal communities that depend on resources extracted from the open ocean, and they will be affected by the effects of climate change on ecosystems there. Some impacts are already starting to be observed, such as changes in species distributions resulting from poleward shifts in taxa (Barton and others, 2016) and habitat compression owing to deoxygenation (Stramma and others, 2012).

Recent work suggests that the adoption of the Paris Agreement[1] has benefits for fisheries (Sumaila and others, 2019) and that improved fisheries management could offset some of the impacts of climate change to fisheries (Gaines and others, 2018).

3.2. Implications for achieving the Sustainable Development Goals[2]

3.2.1. Reducing marine pollution (Sustainable Development Goal indicator 14.1.1)

The avalanche of plastic materials entering the oceans requires action by States to eliminate or reduce plastic use where possible. Recent initiatives in the European Union and elsewhere, to reduce plastic usage, increase

---

1 See FCCC/CP/2015/10/Add.1, decision 1/CP.21, annex.

2 See General Assembly resolutions 70/1 and 71/313, annex.

recycling, to ultimately limit plastic waste entering the environment, in response to public pressure, are welcome but more needs to be done by all States if a significant reduction is to be achieved by 2025.

3.2.2. Assessing biogeochemical changes in the ocean (Sustainable Development Goal indicators 14.2.1 and 14.3.1)

Modelling work has indicated the timescales required to assess and detect climate change-driven trends in the open ocean (Henson and others, 2016), referred to as a time of trend emergence. Continuous time series data are required for between 14 (pH) and 32 (primary productivity) years to distinguish climate change from natural variability. Similarly, pH and pCO_2 data collected at open ocean time series sites suggest 8 to 15 years to achieve time of trend emergence (Sutton and others, 2019). Developing new and extending existing time series sites in the open ocean will require investments in both technology and human capacity (Miloslavich and others, 2019).

3.2.3. Marine reserves (Sustainable Development Goal indicators 14.5.1 and 14.c.1)

Increasingly large areas of the open ocean are being designated as marine reserves or marine protected areas, most notably in the Pacific (e.g., the Rapa Nui Marine Protected Area in Chile and the Papahānaumokuākea Marine National Monument in the United States), and managing those pelagic protected areas will be challenging (Norse, 2005). All of the marine protected areas will contribute to Goal 14, although more progress is still to be made (Lubchenco and Grorud-Colvert, 2015), as few existing marine protected areas are in areas beyond national jurisdiction. The outcome of the intergovernmental conference on an international legally binding instrument under the United Nations Convention on the Law of the Sea on the conservation and sustainable use of marine biological diversity of areas beyond national jurisdiction will be a key step in establishing the legal framework for marine protected areas in the high seas (see also chap. 27).

3.2.4. Fisheries (Sustainable Development Goal indicator 14.4.1)

Achieving sustainable development of fisheries in the open ocean is complicated even with the existence of several legal frameworks in each region (see Sustainable Development Goal indicators 14.5.1 and 14.c.1, mentioned previously; indicator 14.6 (subsidies) also applies here), most probably owing to insufficient knowledge of the overall structure and function of open ocean ecosystems. In particular, potential development of an industry based on mesopelagic fishes will require the refinement of stock assessment methods and the inclusion of new technologies and modelling approaches (Hidalgo and Browman, 2019). Climate change will also affect open ocean ecosystems and their related fisheries (Barange and others, 2018). Illegal, unreported and unregulated fishing remains one of the greatest threats to sustainable fisheries and a global challenge.

3.2.5. Research resources for marine technology (Sustainable Development Goal indicator 14.a.1)

More resources are needed for marine research and technology and for building capacity, both technical and human, for collecting, interpreting and disseminating knowledge on the open ocean. That includes developing ocean basin-wide cooperation through the Global Ocean Observing System and related organizations. Activities within the United Nations Decade of Ocean Science for Sustainable Development (2021–2030)[3] should greatly benefit the achievement of the indicator.

3 See General Assembly resolution 72/73, para. 292.

4. Key region-specific changes and consequences

4.1. Arctic Ocean

The Arctic continues to warm rapidly, with the subsequent loss of multi-year sea ice affecting ecosystems in the Arctic open ocean, combined with global warming and ocean acidification. That could lead to major changes in primary productivity, biodiversity and ecosystem function. In addition, the poleward shift of many North Atlantic species (see below) is impinging increasingly into the polar Arctic.

4.2. North Atlantic Ocean, Baltic Sea, Black Sea, Mediterranean Sea and North Sea

Ocean warming is resulting in the poleward shift of many species, with potential changes in ecosystem function. Regional seas are likely to be further affected by pollution from microplastics unless the sources of those materials to the open ocean are significantly reduced.

4.3. Tropical North Atlantic and Caribbean

Sargassum blooms are likely to further affect the Caribbean in years to come (Putman and others, 2018), and the impact there is likely to affect tourism negatively, although in other regions, it might provide new opportunities (Milledge and Harvey, 2016). Large *Sargassum* blooms provide extensive physical structure at the surface while shading the waters below, possibly affecting phytoplankton productivity but also aggregating fishes in the shadows below the *Sargassum* rafts. The gyre structure of North Atlantic currents is concentrating macroplastics and microplastics into a "garbage patch" (Poulain, 2019). Intensification of hurricanes may be affecting meso- to large-scale mixing processes.

4.4. South Atlantic Ocean

Marine heatwaves are becoming more common in the South Atlantic, with serious implications for climate patterns and hence for the distribution and abundance of fish stocks. Another adverse phenomenon involves the increasing frequency of South Atlantic tropical storms under climate change.

4.5. Indian Ocean, Arabian Sea, Bay of Bengal, Gulf of Aden and Persian Gulf

As the Indian Ocean warms in response to climate change, more marine heatwaves are likely in the future, resulting in changes to ecosystem function and dynamics, with subsequent impacts on the communities in the region that rely on fishing.

4.6. North Pacific Ocean

The North Pacific gyre is likely to become more heavily affected by microplastic pollution, with the potential impacts on the marine ecosystems not yet fully known or understood. Ocean warming, acidification and deoxygenation are also likely to cause changes in marine ecosystems and productivity. Increased marine heatwaves are affecting ecosystems of the north-eastern Pacific, inter alia, leading to pyrosome blooms and the starvation of fish-eating birds (Piatt and others, 2020).

4.7. South Pacific Ocean

Marine heatwaves are likely to become more frequent and pronounced in the future. The oxygen minimum zone in the eastern tropical South Pacific is expected to expand both horizontally and vertically, affecting the distribution of pelagic fish stocks. For ocean-dependent Pacific islands, climate change is likely to have a sizeable impact on livelihoods, health and culture.

4.8. Southern Ocean

The Southern Ocean is continuing to warm, and it accounts for the largest proportion of the global increase in heat in the ocean (IPCC, 2019). A likely consequence of that is the further contraction of the sea ice habitat for krill southward and a shift towards salp-dominated states, affecting higher trophic organisms (seals, whales and penguins) for which krill are a key prey species. The Southern Ocean will continue to remove CO_2 from the atmosphere, resulting in a lower pH, with potential for further decreasing calcification rates.

5. Outlook

5.1. The open ocean in the near future

Over the next decades, the open ocean will become warmer, deoxygenated and more acidic owing to the impacts of climate change (IPCC, 2019). The different timescales of those multiple stressors and the interactions among them will lead to changes in ecosystem function and structure on a variety of temporal and spatial scales. The time frame of those impacts will vary regionally and will take longer to manifest at depth owing to the volume and slow circulation of the deep ocean.

5.2. Ecosystem consequences of continued changes in the open ocean

Global change is affecting ecosystems services in the open ocean (blue economy), but it is difficult to predict future changes owing to the overlapping effects of multiple stressors (Boyd and others, 2018). Habitat compression owing to expanding oxygen minimum zones and the continued poleward migration of some key taxa are expected to cause ongoing change in open ocean ecosystems. Changes in productivity and the biological pump will affect the sequestration of carbon into the deep sea.

5.3. Socioeconomic consequences of ongoing changes in the open ocean

Ongoing changes in the open ocean will have a wide-ranging socioeconomic impact over time, as related activities will need to adapt (e.g., migration of fish species owing to warming and deoxygenation, increases in *Sargassum* seaweed) and mitigate (e.g., microplastic pollution, maritime emissions) in response. The development of large-scale marine reserves and protected areas in the open ocean as actions to meet the Sustainable Development Goals will also require new international agreements regarding the establishment and monitoring of those areas. Increased demand for global observations of the open ocean (Levin and others, 2019; Miloslavich and others, 2019) will require greater investment in both equipment and human capacity to interpret data and provide informed assessments for advising and implementing policy.

6. Key remaining knowledge gaps

Still, very little is known about open ocean ecosystems and the impact of physical drivers on the biodiversity found in them. Critically, the mesopelagic and deeper zones of the ocean are severely underexplored and poorly understood, including exchanges between the deeper and upper ocean, chiefly the daily vertical migration of organisms. There is also a need to ensure that basic information (e.g., traditional taxonomy) is collected about species that live in such environments, as omic data are only one of many strands that inform biodiversity (Boero, 2010).

7. Key remaining capacity-building gaps

Further development of deepwater Argo and associated sampling platforms (e.g., underwater video profilers, eDNA samplers, biogeochemical sensors) for the mesopelagic (Martin and others, 2020) and deeper waters is critical to improving the understanding of that vast area of the ocean. Recent activities, such as the ocean observing conference, OceanObs'19, have helped to identify capacity gaps on the ocean basin scale, and efforts should therefore be targeted towards States working together to achieve adequate coverage for observational oceanography on such scales. Education and training of the next generation of researchers in all aspects of marine research is key to developing the human capacity to maximize the utilization of new technologies for that purpose (Levin and others, 2019).

References

Aberle, Nicole, and others (2012). Warming induces shifts in microzooplankton phenology and reduces time-lags between phytoplankton and protozoan production. *Marine Biology*, vol. 159, No. 11, pp. 2441–2453.

Ainsworth, Cameron H., and others (2018). Impacts of the Deepwater Horizon oil spill evaluated using an end-to-end ecosystem model. *PloS One*, vol. 13, No. 1, e0190840.

Barange, Manuel, and others (2017). The cost of reducing the North Atlantic Ocean biological carbon pump. *Frontiers in Marine Science*, vol. 3, art. 290.

Barange, Manuel, and others, eds. (2018). Impacts of climate change on fisheries and aquaculture: synthesis of current knowledge, adaptation and mitigation options. FAO Fisheries and Aquaculture Technical Paper 627. Rome: FAO.

Barton, A.D., and others (2016). Anthropogenic climate change drives shift and shuffle in North Atlantic phytoplankton communities. *Proceedings of the National Academy of Sciences*, vol. 113, pp. 2964–2969.

Behrenfeld, Michael J., and others (2016). Revaluating ocean warming impacts on global phytoplankton. *Nature Climate Change*, vol. 6, No. 3, p. 323.

Behrenfeld, Michael J., and others (2019). Global satellite-observed daily vertical migrations of ocean animals. *Nature*, vol. 576, No. 7786, pp. 257–61. https://doi.org/10.1038/s41586-019-1796-9.

Beman, J. Michael, and others (2011). Global declines in oceanic nitrification rates as a consequence of ocean acidification. *Proceedings of the National Academy of Sciences*, vol. 108, No. 1, pp. 208–213.

Beman, J. Michael, and Molly T. Carolan (2013). Deoxygenation alters bacterial diversity and community composition in the ocean's largest oxygen minimum zone. *Nature Communications*, vol. 4, art. 2705.

Bennke, Christin M., and others (2016). Modification of a high-throughput automatic microbial cell enumeration system for shipboard analyses. *Applied and Environmental Microbiology*, vol. 82, No. 11, pp. 3289–3296.

Bertagnolli, Anthony D., and Frank J. Stewart (2018). Microbial niches in marine oxygen minimum zones. *Nature Reviews. Microbiology*, vol. 16, No. 12, pp. 723–729.

Boero, Ferdinando (2010). The Study of Species in the Era of Biodiversity: A Tale of Stupidity. *Diversity*, vol. 2. https://doi.org/10.3390/d2010115.

Boscolo-Galazzo, Flavia, and others (2018). Temperature dependency of metabolic rates in the upper ocean: A positive feedback to global climate change? *Global and Planetary Change*, vol. 170, pp. 201–212.

Boyce, Daniel G., and others (2017). Environmental structuring of marine plankton phenology. *Nature Ecology & Evolution*, vol. 1, No. 10, p. 1484.

Boyd, Philip W., and others (2015). Biological ramifications of climate-change-mediated oceanic multi-stressors. *Nature Climate Change*, vol. 5, No. 1, p. 71.

Boyd, Philip W., and others (2018). Experimental strategies to assess the biological ramifications of multiple drivers of global ocean change—a review. *Global Change Biology*, vol. 24, No. 6, pp. 2239–2261.

Boyd, Philip W., and others (2019). Multi-faceted particle pumps drive carbon sequestration in the ocean. *Nature*, vol. 568, No. 7752, pp. 327–335.

Brander, Keith (2010). Impacts of climate change on fisheries. *Journal of Marine Systems*, vol. 79, Nos. 3–4, pp. 389–402.

Cavan, Emma Louise, and others (2019). The sensitivity of subsurface microbes to ocean warming accentuates future declines in particulate carbon export. *Frontiers in Ecology and Evolution*, vol. 6, pp. 1–10.

Chassot, Emmanuel, and others (2010). Global marine primary production constrains fisheries catches. *Ecology Letters*, vol. 13, No. 4, pp. 495–505.

Cheng, Lijing, and others (2017). Improved estimates of ocean heat content from 1960 to 2015. *Science Advances*, vol. 3, No. 3, e1601545.

Cheng, Lijing, and others (2019). How fast are the oceans warming? *Science*, vol. 363, No. 6423, pp. 128–129.

Costello, Mark John, and others (2010). Surface area and the seabed area, volume, depth, slope, and topographic variation for the world's seas, oceans, and countries. *Environmental Science & Technology*, vol. 44, No. 23, pp. 8821–8828.

Delory, Eric, and others (2014). Developing a new generation of passive acoustics sensors for ocean observing systems. In *2014 IEEE Sensor Systems for a Changing Ocean (SSCO)*, pp. 1–6. IEEE.

Deutsch, Curtis, and others (2015). Climate change tightens a metabolic constraint on marine habitats. *Science*, vol. 348, No. 6239, pp. 1132–1135.

Duarte, Carlos M., and others (2013). Is ocean acidification an open-ocean syndrome? Understanding anthropogenic impacts on seawater pH. *Estuaries and Coasts*, vol. 36, No. 2, pp. 221–236.

Free, Christopher M., and others (2019). Impacts of historical warming on marine fisheries production. *Science*, vol. 363, No. 6430, pp. 979–983.

Frölicher, Thomas L., and others (2018). Marine heatwaves under global warming. *Nature*, vol. 560, No. 7718, p. 360.

Gaines, Steven D., and others (2018). Improved fisheries management could offset many negative effects of climate change. *Science Advances*, vol. 4, No. 8, eaao1378.

Gehlen, Marion, and others (2011). Biogeochemical consequences of ocean acidification and feedbacks to the Earth system. *Ocean Acidification*, vol. 1, pp. 230–248.

Gehlen, Marion, and others (2014). Projected pH reductions by 2100 might put deep North Atlantic biodiversity at risk.

Gruber, Nicolas, and others (2019). The oceanic sink for anthropogenic CO_2 from 1994 to 2007. *Science*, vol. 363, No. 6432, pp. 1193–1199.

Halpern, Benjamin S., and others (2015). Spatial and temporal changes in cumulative human impacts on the world's ocean. *Nature Communications*, vol. 6, art. 7615.

Harcourt, Rob, and others (2019). Animal-borne telemetry: an integral component of the ocean observing toolkit. *Frontiers in Marine Science*.

Henson, Stephanie A., and others (2016). Observing climate change trends in ocean biogeochemistry: when and where. *Global Change Biology*, vol. 22, No. 4, pp. 1561–1571.

Hidalgo, Manuel, and Howard I. Browman (2019). *Developing the Knowledge Base Needed to Sustainably Manage Mesopelagic Resources*. Oxford University Press.

Hobday, Alistair J., and others (2016). A hierarchical approach to defining marine heatwaves. *Progress in Oceanography*, vol. 141, pp. 227–238.

Holbrook, Neil J., and others (2019). A global assessment of marine heatwaves and their drivers. *Nature Communications*, vol. 10, No. 1, art. 2624.

Intergovernmental Panel on Climate Change (IPCC) (2019). *IPCC Special Report on the Ocean and Cryosphere in a Changing Climate,* H-O. Pörtner and others, eds.

Irigoien, Xabier, and others (2014). Large mesopelagic fishes biomass and trophic efficiency in the open ocean. *Nature Communications*, vol. 5, art. 3271.

Jones, Kendall R., and others (2018). The location and protection status of Earth's diminishing marine wilderness. *Current Biology*, vol. 28, No.15, pp. 2506–2512.

Keogan, Katharine and others (2018). Global phenological insensitivity to shifting ocean temperatures among seabirds. *Nature Climate Change*, vol. 8, No. 4, pp. 313–18. https://doi.org/10.1038/s41558-018-0115-z.

Kroodsma, David A., and others (2018). Tracking the global footprint of fisheries. *Science*, vol. 359, No. 6378, pp. 904–908.

Kwiatkowski, Lester, and others (2017). Emergent constraints on projections of declining primary production in the tropical oceans. *Nature Climate Change*, vol. 7, No. 5, p. 355.

Lebreton, Laurent, and others (2018). Evidence that the Great Pacific Garbage Patch is rapidly accumulating plastic. *Scientific Reports*, vol. 8, No. 1, art. 4666.

Lebreton, Laurent, and Anthony Andrady (2019). Future scenarios of global plastic waste generation and disposal. *Palgrave Communications*, vol. 5, No. 1, art. 6.

Levin, Lisa A. (2018). Manifestation, drivers, and emergence of open ocean deoxygenation. *Annual Review of Marine Science*, vol. 10, pp. 229–260.

Levin, Lisa A., and others (2019). Global observing needs in the deep ocean. *Frontiers in Marine Science*, vol. 6, art. 241.

Lombard, Fabien, and others (2019). Globally consistent quantitative observations of planktonic ecosystems. *Frontiers in Marine Science*, vol. 6, art. 196.

Lubchenco, Jane, and Kirsten Grorud-Colvert (2015). Making waves: the science and politics of ocean protection. *Science*, vol. 350, No. 6259, pp. 382–383.

Martin, Adrian, and others (2020). The oceans' twilight zone must be studied now, before it is too late. *Nature*, vol. 580, pp. 26–28.

Martinetto, Paulina, and others (2020). Linking the scientific knowledge on marine frontal systems with ecosystem services. *Ambio*, vol. 49, No. 2, pp. 541–556. https://doi.org/10.1007/s13280-019-01222-w.

McClain, Craig R., and others (2019). Persistent and substantial impacts of the Deepwater Horizon oil spill on deep-sea megafauna. *Royal Society Open Science*, vol. 6, No. 8, 191164.

Milledge, John J., and Patricia J. Harvey (2016). Golden tides: problem or golden opportunity? The valorisation of *Sargassum* from beach inundations. *Journal of Marine Science and Engineering*, vol. 4, No. 3, art. 60.

Miloslavich, Patricia, and others (2019). Challenges for global ocean observation: the need for increased human capacity. *Journal of Operational Oceanography*, vol. 12, No. sup2, pp. S137–S156.

Monllor-Hurtado, Alberto, and others (2017). Shift in tuna catches due to ocean warming. *PloS One*, vol. 12, No. 6, e0178196.

Norse, Elliott (2005). Pelagic protected areas: the greatest parks challenge of the 21st century. *Parks*, vol. 15, pp. 32–39.

Oliver, Eric C.J., and others (2018). Longer and more frequent marine heatwaves over the past century. *Nature Communications*, vol. 9, No. 1, art. 1324.

Piatt, John F., and others (2020). Extreme mortality and reproductive failure of common murres resulting from the northeast Pacific marine heatwave of 2014–2016. *PLoS One*, vol. 15, No. 1, pp. 1–32. https://doi.org/10.1371/journal.pone.0226087.

Pinsky, Malin L., and others (2019). Greater vulnerability to warming of marine versus terrestrial ectotherms. *Nature*, vol. 569, No. 7754, p. 108.

Pinsky, Malin L., and others (2020). Climate-driven shifts in marine species ranges: scaling from organisms to communities. *Annual Review of Marine Science*, vol. 12, No. 1, pp. 153–79. https://doi.org/10.1146/annurev-marine-010419-010916.

Pohlot, Bruce G., and Nelson Ehrhardt (2017). An analysis of sailfish daily activity in the Eastern Pacific Ocean using satellite tagging and recreational fisheries data. *ICES Journal of Marine Science*, vol. 75, No. 2, pp. 871–879.

Poulain, Marie, and others (2019). Small microplastics as a main contributor to plastic mass balance in the North Atlantic subtropical gyre. *Environmental Science & Technology*, vol. 53, pp. 1157–1164.

Putman, Nathan F., and others (2018). Simulating transport pathways of pelagic *Sargassum* from the Equatorial Atlantic into the Caribbean Sea. *Progress in Oceanography*, vol. 165, pp. 205–214.

Rahmstorf, S., and others (2015). Exceptional twentieth-century slowdown in Atlantic Ocean overturning circulation. *Nature Climate Change*, vol. 5, No. 5, pp. 475–480.

Ramirez-Llodra, Eva, and others (2015). Submarine and deep-sea mine tailing placements: a review of current practices, environmental issues, natural analogs and knowledge gaps in Norway and internationally. *Marine Pollution Bulletin*, vol. 97, Nos. 1–2, pp. 13–35.

Record, Nicholas, and others (2019). Rapid climate-driven circulation changes threaten conservation of endangered North Atlantic right whales. *Oceanography*, vol. 32, No. 2.

Reygondeau, Gabriel, and others (2018). Global biogeochemical provinces of the mesopelagic zone. *Journal of Biogeography*, vol. 45, No. 2, pp. 500–514.

Robinson, Carol (2019). Microbial respiration, the engine of ocean deoxygenation. *Frontiers in Marine Science*, vol. 5, art. 533.

Robinson, Carol, and others (2010). Mesopelagic zone ecology and biogeochemistry – a synthesis. *Deep Sea Research Part II: Topical Studies in Oceanography*, vol. 57, No. 16, pp. 1504–1518.

Rodrigues, Regina, and others (2019). Common cause for severe droughts in South America and marine heatwaves in the South Atlantic. *Nature Geoscience*, vol. 12. https://doi.org/10.1038/s41561-019-0393-8.

Roemmich, Dean, and others (2019). On the future of Argo: a global, full-depth, multi-disciplinary array. *Frontiers in Marine Science*, vol. 6.

Rudnick, Daniel L. (2016). Ocean research enabled by underwater gliders. *Annual Review of Marine Science*, vol. 8, pp. 519–541.

Schlitzer, Reiner and others (2018). The GEOTRACES Intermediate Data Product 2017. *Chemical Geology*, vol. 493, pp. 210–223. https://doi.org/10.1016/j.chemgeo.2018.05.040.

Sloyan, B.M., and others (2019). The Global Ocean Ship-Based Hydrographic Investigations Program (GO-SHIP): a platform for integrated multidisciplinary ocean science. *Frontiers in Marine Science*, vol. 6, art. 445. https://doi.org/10.3389/fmars.2019.00445.

Smale, Dan A., and others (2019). Marine heatwaves threaten global biodiversity and the provision of ecosystem services. *Nature Climate Change*, vol. 9, No. 4, p. 306.

Somavilla, R., and others (2017). The warmer the ocean surface, the shallower the mixed layer. How much of this is true? *Journal of Geophysical Research: Oceans*, vol. 122, No. 9, pp. 7698–7716.

St. John, Michael A., and others (2016). A dark hole in our understanding of marine ecosystems and their services: perspectives from the mesopelagic community. *Frontiers in Marine Science*, vol. 3, art. 31.

Stramma, Lothar, and others (2012). Expansion of oxygen minimum zones may reduce available habitat for tropical pelagic fishes. *Nature Climate Change*, vol. 2, No. 1, p. 33.

Sumaila, U. Rashid, and others (2019). Benefits of the Paris Agreement to ocean life, economies, and people. *Science Advances*, vol. 5, No. 2, eaau3855.

Sundby, Svein, and others (2016). The North Atlantic spring-bloom system – where the changing climate meets the winter dark. *Frontiers in Marine Science*, vol. 3, art. 28.

Sutton, Adrienne J., and others (2019). Autonomous seawater pCO_2 and pH time series from 40 surface buoys and the emergence of anthropogenic trends. *Earth System Science* Data, p. 421.

Sutton, T.T. (2013). Vertical ecology of the pelagic ocean: classical patterns and new perspectives. *Journal of Fish Biology*, vol. 83, No. 6, pp. 1508–1527.

Tiano, Laura, and others (2014). Oxygen distribution and aerobic respiration in the north and south eastern tropical Pacific oxygen minimum zones. *Deep Sea Research Part I: Oceanographic Research Papers*, vol. 94, pp. 173–183.

Tournadre, Jean (2014). Anthropogenic pressure on the open ocean: The growth of ship traffic revealed by altimeter data analysis. *Geophysical Research Letters*, vol. 41, No. 22, pp. 7924–7932.

Truelove, Nathan K., and others (2019). A rapid environmental DNA method for detecting white sharks in the open ocean. *Methods in Ecology and Evolution*, vol. 10, No. 8, pp. 1128–1135. https://doi.org/10.1111/2041-210X.13201.

Trueman, Clive N., and others (2014). Trophic interactions of fish communities at midwater depths enhance long-term carbon storage and benthic production on continental slopes. *Proceedings of the Royal Society B: Biological Sciences*, vol. 281.

United Nations (2017a). Chapter 36F: Open ocean deep sea. In *The First Global Integrated Marine Assessment: World Ocean Assessment I*. Cambridge: Cambridge University Press.

________ (2017b). *The First Global Integrated Marine Assessment: World Ocean Assessment I*. Cambridge: Cambridge University Press.

Van Sebille, Erik, and others (2015). A global inventory of small floating plastic debris. *Environmental Research Letters*, vol. 10, No. 12, 124006.

Vare, Lindsay L., and others (2018). Scientific considerations for the assessment and management of mine tailings disposal in the deep sea. *Frontiers in Marine Science*, vol. 5, art. 17.

Watson, Reg A., and others (2015). Marine foods sourced from farther as their use of global ocean primary production increases. *Nature Communications*, vol. 6, art. 7365.

Watson, Sue-Ann, and others (2018). Ocean warming has a greater effect than acidification on the early life history development and swimming performance of a large circumglobal pelagic fish. *Global Change Biology*, vol. 24, No. 9, pp. 4368–4385.

Wishner, Karen F., and others (2018). Ocean deoxygenation and zooplankton: very small oxygen differences matter. *Science Advances*, vol. 4, No. 12, eaau5180.

Yamahara, Kevan M., and others (2019). In situ autonomous acquisition and preservation of marine environmental DNA Using an autonomous underwater vehicle. *Frontiers in Marine Science*, vol. 6, art. 373. https://doi.org/10.3389/fmars.2019.00373.

Young, Ian R., and Agustinus Ribal (2019). Multiplatform evaluation of global trends in wind speed and wave height. *Science*, vol. 364, No. 6440, pp. 548–552.

Chapter 70
Ridges, plateaux and trenches

Contributors: Ana Colaço (convener), Angelika Brandt, Ana Hilario, Nuno Lourenço, Bhavani E. Narayanaswamy, Imants George Priede, Joshua T. Tuhumwire (subchapter lead member), Michael Vecchione and Hiromi Watanabe.

Keynote points

- Most recent studies of ridge biology address chemosynthetic environments (see chap. 7P).
- Climate change models show that bathyal environments will suffer a reduction in pH, which will affect benthic communities.
- Ridges, rises, plateaux and banks are under human pressure arising from existing and potential exploitation of resources, while evidence of pollution in trenches is accumulating.
- The vulnerability of those ecosystems to human pressures has triggered both increased societal awareness and new regulations.

1. Introduction and summary of the first *World Ocean Assessment*

Chapter 51 of the first *World Ocean Assessment* (United Nations, 2017a) was devoted to deep-sea features that were considered to be potentially threatened by human disturbance. Those features, including seamounts, ridges and plateaux, submarine canyons and hadal trenches, are all topographically and hydrographically complex. Of those features, in the present Assessment, seamounts (chap. 7L) and canyons (chap. 7J) are considered elsewhere, along with hydrothermal vents and other chemosynthetic ecosystems (chap. 7P), which were also covered separately in the first Assessment (United Nations, 2017b). For each of the features considered, chapter 51 provided detailed descriptions, including geology and physical oceanographic characteristics, extent (i.e., numbers of each feature, as well as percentage of oceanic area) and ecological characteristics, such as biodiversity and biogeography. It also documented anthropogenic impacts on the features, highlighting fishing (including the removal of species and biomass, as well as the physical effects of fishing activities on structure-forming benthic communities), climate change (including acidification and deoxygenation, as well as rising temperatures), pollution, dumping and mining. The present chapter builds on that background by focusing on changes and new knowledge regarding ridges, plateaux and hadal trenches gained since 2010.

1.1. Ridges

Mid-ocean ridges subdivide the major ocean basins (figure I), but fracture zones at intervals permit the movement of deep water and abyssal organisms (see chap. 7M) between basins on the two sides of a ridge. Chapter 7P deals with active hydrothermal vents, which are often associated with mid-ocean spreading ridges (Beaulieu and others, 2013), whereas chapter 7L addresses seamounts associated with ridges. Biogeographically, ridge faunas generally appear to be related to the faunas of adjacent basins or continental slopes (Alt and others, 2019; Watling and others, 2013). Ridges can host structure-forming benthic invertebrates, such as cold water corals (chap. 7E) and sponges, some of which are classified as indicator species of vulnerable marine ecosystems, as defined by the Food and Agriculture Organization of the United Nations (Food and Agriculture Organization of the United Nations (FAO), 2009). The island slopes and summits of seamounts (chap. 7L) associated with ocean ridges are important areas for fishing activities. Most ridges are in areas outside exclusive economic zones (Harris and Whiteway, 2009).

1.2. Hadal trenches

Hadal environments are relatively unknown because of the challenge of accessing extreme depths. However, because of their locations on the subduction edges of tectonic plates, the trenches are all close to land masses or islands and entirely or partly within exclusive economic zones. For example, the Mariana Trench, near territory claimed by the United States, lies almost entirely with the exclusive economic zone of the United States, while the Kermadec-Tonga Trench traverses the exclusive economic zones of New Zealand and Tonga (Flanders Marine Institute, 2018). Thus, while trenches all potentially occur within the limits of national jurisdiction, some are shared by more than one State, and the technical capacity of the different States to monitor and manage threats to hadal environments varies widely. The hadal zone, with depths below 6,000 m, occupies 45 per cent of the total ocean depth range, but only 0.404 per cent of total ocean area. Within that area, 95 distinct basins or trenches have maximum depths of 7,000 m or more and are dominated by 15 recognized trenches (Priede, 2017; figure I). For biological purposes, the United Nations Educational, Scientific and Cultural Organization redefined the upper boundary of the hadal zone as a depth of 6,500 m to reflect the boundary between the globally widespread abyssal fauna and the upper limit of the specialized hadal fauna (United Nations Educational, Scientific and Cultural Organization (UNESCO), 2009). Except for higher pressure, the environment in hadal trenches is typical of much of the deep sea, with slow currents of 1–8 cm per second carrying water at approximately 2°C with sufficient oxygen (3.43 ml/l) to support aerobic life and a slow rain of particulate organic matter from the surface providing food (Jamieson and others, 2010).

Figure I
World map with hadal trenches and mid-ocean ridge system

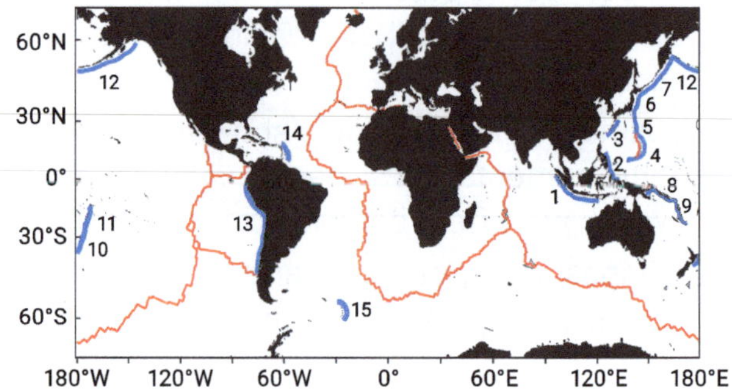

Source: Adapted from Bird, 2003.
Notes: Blue segments represent hadal trenches (depths greater than 6,000 m below sea level): 1–Java, 2–Philippines, 3– Ryukyu, 4–Marianas, 5–Izu-Bonin, 6–Japan, 7–Kuril-Kamchatka, 8–New Britain, 9–South Solomon, 10–Kermadec, 11–Tonga, 12–Aleutian, 13–Ecuador to Chile, 14–Puerto Rico, 15–South Sandwich/Islas Sándwich del Sur. Red segments represent the world's ocean ridge system.

1.3. Plateaux, rises and banks

Plateaux, rises and banks are large and relatively flat topographic features, identified as continental fragments or microcontinents and often separated from major continents by deepwater channels. Currently, 184 plateaux have been mapped (Harris and others, 2014), covering approximately 5 per cent of the world's oceans. Although they are found in all the world's oceans, plateaux are most prevalent in the Indian Ocean (e.g., Kerguelen and Mascarene plateaux) and the South Pacific (Challenger and Campbell plateaux), which reflects their recently recognized origin from the tectonic break-up of the Gondwana supercontinent (Mortimer and others, 2017). Faunal diversity and composition can closely resemble those of nearby continental shelves, slopes and banks (Narayanaswamy and others, 2013). However, topographic and oceanographic complexity, combined with food availability, can have a major influence on community composition and diversity (Compton and others, 2013; Knox and others, 2012). Compton and others (2013), investigating amphipod communities around New Zealand, reported higher abundance and diversity on the more complex Chatham Rise compared with the western end of the Challenger plateau, with its comparatively lower food supply. However, Leduc and others (2012) rejected food availability as a primary driver of nematode community composition.

Deepwater fishing targets the slopes of plateaux, rises and banks (e.g., Johnson and others, 2019). Emerging activities, such as mining, also pose threats to those environments (e.g., Leduc and others, 2015).

2. Description of the environmental changes between 2010 and 2020

2.1. New knowledge acquired since 2010 and how it can be used to evaluate changes

Changes owing to human pressures have been observed and, consequently, some measures of protection for bottom habitats (ridges, trenches and plateaux) are being applied (see table below).[1] A huge advance in knowledge resulted from the Census of Marine Life that ended in 2010 but for which publication has continued well into the current decade. Many of those publications (2010–2014) were reviewed for the first Assessment. Other advances are summarized below.

2.1.1. Ridges: biodiversity and ecosystem function

Substantial progress in the study of ridges has been reported over the past decade. Mid-ocean ridges increase environmental heterogeneity and influence biological communities (Alt and others, 2019). On the Indian Ocean ridges, the first detailed investigation of megafaunal assemblages (Sautya and others, 2017) showed that abundances were higher at the upper bathyal zone but lower at deeper zones in rift valley walls and floors. Recent seabed mapping in the southern Indian Ocean has improved the resolution of large-scale features and has revealed an unknown diversity and complexity of seabed morphology that will likely be reflected in the biodiversity of benthic communities (Picard and others, 2018).

Further discoveries of deep ocean seabed complexity will be made as mapping continues, in particular through global initiatives, such as the GEBCO-Nippon Foundation Seabed 2030 Project.

Large portions of the mid-ocean ridge system fit the criteria of the Food and Agriculture Organization of the United Nations for defining vulnerable marine ecosystems (Morato and others, 2018), while others are considered priority habitats in need of protection by regional conventions, such as the Convention for the Protection of the Marine Environment of the North-East Atlantic.[2] Recent ridge studies have shown the importance of vulnerable marine ecosystem indicator species. Both cold water coral communities and sponge grounds are important for global biogeochemical cycles and the ocean's benthic-pelagic coupling loop, which is responsible for nearly 30 per cent of the transfer of organic matter produced at the ocean surface and the sea floor (Cathalot and others, 2015).

The diverse benthic communities along the northern Mid-Atlantic Ridge provide a complex

1 See www.mpatlas.org/map/high-seas.
2 United Nations, *Treaty Series*, vol. 2354, No. 42279.

three-dimensional structural habitat that provides refuge, feeding opportunities, and spawning and nursery areas for a wide range of associated sessile and mobile species, including commercially important fish and crustacean species (Beazley and others, 2013; Pham and others, 2015; Gomes-Pereira and others, 2017). For example, deepwater sharks were found to lay eggs among cold water corals (Henry and others, 2013). The presence of large black coral colonies with high longevity (several millenniums) in the Mid-Atlantic Ridge is also indicative of well-preserved environments.

Large fracture zones not only allow for the communication of water masses between basins separated by the Mid-Atlantic Ridge but also can act as a conduit for larval dispersal. Along the Vema Fracture Zone, macrofauna abundances were generally higher on the eastern than the western side (Brandt and others, 2018). Much new knowledge of ridges in the past 10 years has been acquired in relation to interest in finding polymetallic sulphides and cobalt-rich manganese crusts enriched in valuable metals for mining. However, because most of the data relate to hydrothermal vents, they are not dealt with in the present subchapter (see chap. 7P). Geological, geochemical and geophysical studies conducted for submission to the Commission on the Limits of the Continental Shelf provide another major source of new information about ridges. Although the data are not directly collected for environmental understanding, they may address that need in the future. For example, such data may be used to model the distribution of suitable habitat for fauna, which may provide valuable input for management (Lecours, 2017). Some work on habitat suitability models already show that, together with the margins, mid-ocean ridges contain important and suitable habitats for seven suborders of Octocorallia (Yesson and others, 2012) and scleractinian corals (Davies and Guinotte, 2011), with the northern Mid-Atlantic Ridge being particularly important to those taxa.

2.1.2. Deep-sea fishing on mid-ocean ridges

Deep-sea trawling has direct impacts on deep-sea benthic communities where the gear touches the sea floor, whereas bottom longlines have a much smaller impact but still affect some of the oldest continuously living organisms on the northern Mid-Atlantic Ridge (Pham and others, 2014).

2.1.3. Climate

Climate change projections for the deep sea indicate substantial effects in bathyal habitats (200–3,000-m depths), including ridges, and their communities (Levin and others, 2019a). Recent model projections by Sweetman and others (2017) indicate that bathyal depths worldwide will experience significant reductions in pH (0.29 to 0.37 pH units) in all oceans by 2100, and oxygen concentrations will decline by as much as 3.7 per cent in the bathyal North-East Pacific Ocean and the Southern Ocean. The flux of particulate organic matter (marine snow) to the sea floor will decline significantly in most oceans, most notably in the bathyal Indian Ocean, with a predicted decrease of 40–55 per cent by the end of the century. Models also predict a decreased calcium carbonate saturation rate throughout the world's oceans (Zheng and Cao, 2014). Marine calcifying organisms inhabiting cold waters and deep areas may be particularly sensitive to projected changes in carbonate chemistry (Levin and others, 2019a). Where ridges occur at bathyal depths, ridge fauna will likely experience all of the aforementioned climate-related effects (Levin and others, 2019a).

2.1.4. Hadal trenches

The past decade has seen a great increase in sampling and research at hadal depths, driven by renewed interest and new technologies (Jamieson, 2015; Jamieson and others, 2018). New low-cost lander vehicles can be deployed from small ships, without full ocean-depth winches (Jamieson and others, 2019). A major effort was the international KuramBio II expedition to the Kuril-Kamchatka Trench (Brandt and others, 2016, 2018). A parallel development has been research and expeditions sponsored by private individuals, such as the descent to the bottom of the Mariana Trench by the Deepsea Challenger in 2015 or the Five Deeps Expedition (Five Deeps Expedition, 2019; Stewart and Jamieson, 2019).

Such activity has provided new insight into the environment and life in hadal trenches. The findings include evidence that no fish can survive at depths greater than 8,400 m (Yancey and others, 2014), which confines trench-endemic fishes to the slopes around the edges of the deepest trenches. At depths greater than 6,800 m, the only fishes present are snailfishes of the family Liparidae. New species in several taxa have been discovered in the Mariana Trench (Gerringer and others, 2017) and the Atacama Trench (also known as the Peru-Chile Trench) (Priede, 2017), and more await description.

Generally, biodiversity decreases with increasing depth in the trenches (Jamieson, 2015). For invertebrates, no fixed maximum depth exists. Nematodes, polychaetes, molluscs, crustaceans and echinoderms all occur at the bottom of the deepest trenches. A funnelling effect concentrates organic matter along the trench axis (Ichino and others, 2015; Luo and others, 2017), potentially resulting in the highest abundances and biomass at the maximum depth. Leduc and others (2016) reported six times more infaunal nematodes at the bottom of the Tonga Trench (10,800 m depth) than at the trench edge. Jamieson and others (2009) found the largest numbers of mobile scavenging lysianassoid amphipods at the greatest depths. In the Kuril-Kamchatka Trench, bivalves and holothurians dominated hadal depths (Brandt and others, 2018).

In contrast to the endemicity of hadal snailfishes, the same species of invertebrates tend to occur in different trenches (Ritchie and others, 2017), although new evidence indicates genetic differentiation between species in some trenches (Zhang and others, 2019) and discoveries of new species (Eustace and others, 2016).

Hadal trenches can act as faunal barriers between different parts of the deep sea. In the Kuril-Kamchatka Trench, the hadal fauna differs from the abyssal fauna of the North-West Pacific and its marginal seas (Brandt and others, 2016). The Trench isolates species of the marginal seas from the North-West Pacific Ocean. It also hampers faunal dispersal for some species of desmosomatid, nannoniscid and ischnomesid isopods (Bober and others, 2018; Jennings and others, 2020). However, for some species, there is no evidence of a strict biogeographic barrier between the Sea of Okhotsk and the open North-West Pacific. Jamieson and others (2011) recognized a transition zone between abyssal and hadal fauna in the Kermadec Trench, and there is evidence of community structure within the trenches (Jamieson and others, 2013; Fujii and others, 2013; Gallo and others, 2015; Lacey and others, 2016; Leduc and Rowden, 2018).

Trenches are in seismically active zones, and the giant great east Japan earthquake (also known as the Tohoku-Oki earthquake) in 2011 resulted in the almost instantaneous slumping of 0.2 cubic km of sediment, containing over 1 teragram of organic carbon, into the Japan Trench (Kioka and others, 2019; Oguri and others, 2013). The input of carbon changed the composition and distribution of meiofauna on the landward slope of the trench (Kitahashi and others, 2014) and altered the sediment and subsea floor hydrogeologic structure (Kawagucci and others, 2012). In addition, radioactive isotopes from the Fukushima Daiichi nuclear power plant disaster were transferred to a depth of 4,800 m about a month after the earthquake (Honda and others, 2013) and deposited on the sea floor below 7,000 m in depth within four months (Oguri and others, 2013).

Proximity to land and human habitation increases hadal trench vulnerability to anthropogenic impacts, a situation magnified by the funnelling effect that concentrates sedimentation into the trench axis. Pollen from terrestrial trees occurs in significant quantities in South-West Pacific trenches, providing a potential food source for hadal organisms (Leduc and Rowden, 2018). Jamieson and others (2017) found extraordinarily high levels of polychlorinated biphenyls and polybrominated diphenyl ethers in amphipods at the most extreme depths in the Mariana Trench and the Kermadec Trench. Concentrations were much higher than those observed in areas of high industrialization, indicating long-term bioaccumulation at those depths. Chiba and others (2018) also observed plastic debris down to the bottom of the Mariana Trench, and microplastic particles have been reported in the hind guts of amphipods from six Pacific Ocean trenches at depths from 7,000 to 10,890 m (Jamieson and others, 2019).

2.1.5. Plateaux, rises and banks

Prior to 2010, a dedicated Global Census of Marine Life on Seamounts project (CenSeam) studied seamounts, but it targeted very few banks and plateaux. In the North-East Atlantic, a national strategic environmental assessment of the region investigated George Bligh Bank, Hatton Ridge and Rockall Bank. The George Bligh Bank community composition resembled that observed on hard substrata elsewhere in the North-East Atlantic (Narayanaswamy and others, 2013). Recent studies of the Kerguelen plateau indicate potentially contrasting benthic faunal distribution shifts (poleward shift, latitudinal reduction and local extinction) of, for example, the echinoids *Abatus cordatus*, *Brisaster antarcticus*, *Ctenocidaris nutrix* and *Sterechinus diadema* in response to environmental change (Guillaumot and others, 2018). However, interpreting and predicting future response to climate change requires careful consideration. Predictions regarding changes in water temperature and salinity typically span large spatial scales that may not reflect local scales, such as the Kerguelen plateau, where differences in the location of fronts and heat flux (Vivier and others, 2015) may lead to changes in species distributions in that location in the future (Guillaumot and others, 2018).

Investigating the smaller macrofauna along the Chatham Rise in the South-West Pacific, Leduc and others (2015) linked community structure most strongly to the density of phosphorite nodules, sometimes forming a nodule-specific community. The nodules are likely to be mined in the future (Leduc and others, 2015). In the absence of nodules, an alternate community will develop and the nodule-specific community will be lost where mining has taken place (Bluhm, 2001).

3. Description of economic and social changes between 2010 and 2020

The vulnerability of deep-sea ecosystems to anthropogenic pressures and related impacts has increased during the past decade because of the increasing economic value of oceanic resources. Such pressures have triggered both an increased societal awareness and new regulations (figure II). The regulations tackle such issues as illegal, unreported and unregulated fishing, exploration for deep-sea mining, bioprospecting and exploitation of genetic resources, the definition of marine protected areas, and the distribution and protection of vulnerable marine ecosystems or other ecologically and biologically significant areas.

Figure II.A
Close-up of the Atlantic area. Areas identified as ecologically and biologically significant and areas with protection measures at the bottom in the Atlantic. (United Nations Environment Programme World Conservation Monitoring Centre and International Union for Conservation of Nature and Natural Resources, 2019)

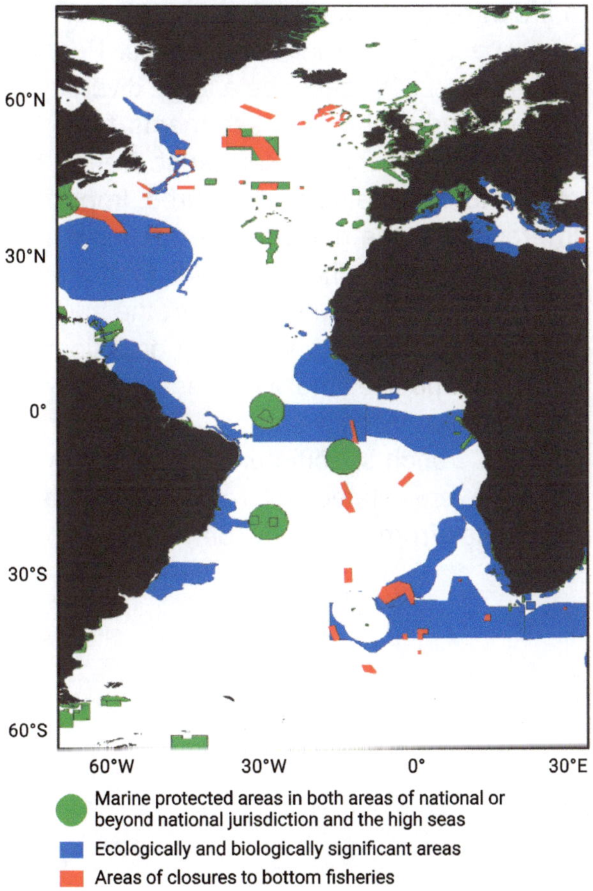

Figure II.B
Boundaries of the Convention on the Conservation of Antarctic Marine Living Resources within the Antarctic Treaty System and regional fisheries management organizations

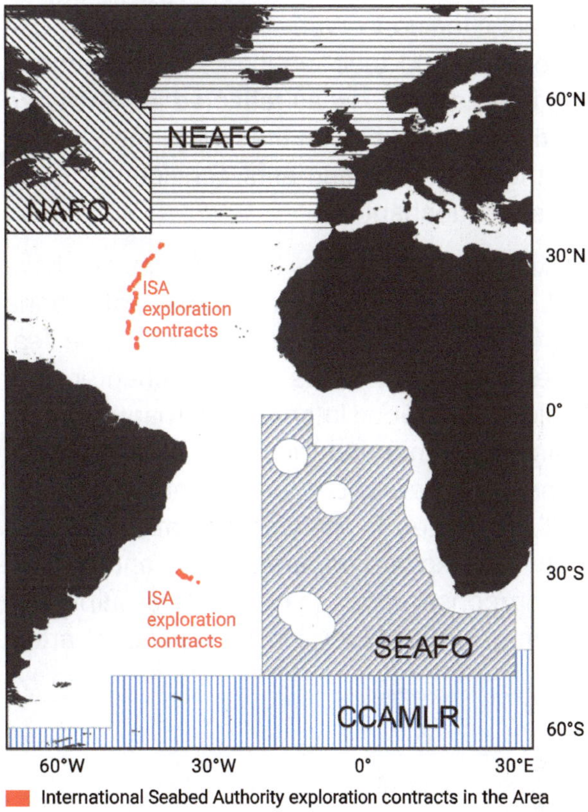

Abbreviations: CCAMLR, Convention on the Conservation of Antarctic Marine Living Resources; ISA, International Seabed Authority; NAFO, Northwest Atlantic Fisheries Organization; NEAFC, North-East Atlantic Fisheries Commission; SEAFO, South-East Atlantic Fisheries Organization.

The new regulations are contributing to Sustainable Development Goal 14: (a) to conserve and sustainably use the oceans, seas and marine resources for sustainable development[3] and related targets, such as those aimed at preventing and significantly reducing marine pollution of all kinds, as it relates to hadal trenches; (b) to sustainably manage and protect marine and coastal ecosystems to avoid significant adverse impacts; (c) to minimize and address the impacts of ocean acidification, including through enhanced scientific cooperation at all levels; (d) to effectively regulate harvesting and end overfishing, illegal, unreported and unregulated fishing and destructive fishing practices and implement

3 See General Assembly resolution 70/1.

science-based management plans; (e) to conserve at least 10 per cent of coastal and marine areas, consistent with national and international law and based on the best available scientific information; (f) also to increase scientific knowledge, develop research capacity and transfer marine technology; and (g) to enhance the conservation and sustainable use of oceans and their resources by implementing international law as reflected in the United Nations Convention on the Law of the Sea[4] for all the other environments considered in the present subchapter.

In the past 10 years, those initiatives have led, inter alia, to the establishment of marine protected areas that specifically include features discussed in the present chapter (see table below). Good international marine spatial planning (within countries' exclusive economic zones and in areas beyond national jurisdiction will be needed to manage potential conflicts between exploration/exploitation and preservation/conservation (e.g., massive sulphides, ecologically and biologically significant areas and marine protected areas in the Mid-Atlantic Ridge; see figure II).

Hadal trenches contain no resources currently considered for direct human exploitation. The biomass of the fish populations present is too low and remote from the surface to sustain any fisheries exploitation, and the sediment-draped sides lack any known mineral resources. Bioprospecting could target microbes adapted to life at high pressure (piezophiles) that may have special industrial applications. Peoples and others (2019) describe a large diversity of bacteria and archaea in sediments from the Mariana Trench and the Kermadec Trench, but few could be isolated and cultured. Important differences in the microbial communities between the two trenches could link to differences in organic matter supply from the surface; the Kermadec Trench, with high organic matter input, supported more taxa associated with organic matter degradation. However, taxa were not trench-specific and those that were isolated were related to previously identified piezophiles from other environments.

Notes for table, "4. Key region-specific changes and consequences", p. 505:
Abbreviations: EBSA, ecologically and biologically significant areas; MPA, marine protected areas; OSPAR, Convention for the Protection of the Marine Environment of the North-East Atlantic.

[a] See www.mpatlas.org/map/high-seas.

[4] United Nations, *Treaty Series*, vol. 1833, No. 31363.

4. Key region-specific changes and consequences

Region	Observed climate changes	Human pressures	Areas identified as ecologically and biologically significant and areas with protection measures at the bottom (ridges, trenches and plateaux)[a]
Arctic Ocean	Reduced oxygenation, acidification and warming of deep waters; change in export flux of organic carbon (Sweetman and others, 2017)	**Ridges and rises, plateaux, banks:** exploration for mineral resources (chap. 18); expansion of fisheries (chap. 15)	
North Atlantic Ocean, Baltic Sea, Black Sea, Mediterranean Sea and North Sea	Reduced oxygenation, acidification and warming of deep waters; change in export flux of organic carbon (Sweetman and others, 2017); potential effects on larval dispersal in the water column, affecting population connectivity (Levin and others, 2019a); warming, deoxygenation of intermediate and deep Mediterranean waters (Stendardo and others, 2015), reduction of ventilation of Mediterranean abyssal waters, with consequences on vent and seep taxa and ecosystem functions	**Ridges and rises, plateaux, banks:** exploration for mineral resources (chap. 18); offshore hydrocarbons (chap. 19); expansion of fisheries (chap. 15) **Trenches:** pollution	**EBSA:** North-West Atlantic Hydrothermal Vent Fields; Atlantic Equatorial Fracture Zone and high productivity system **MPA:** Azores Marine Park **OSPAR:** Charlie-Gibbs North High Seas MPA; Charlie-Gibbs South High Seas MPA; Mid-Atlantic Ridge North of the Azores High Seas MPA **Bottom fishing closure:** North-East Atlantic Fisheries Commission bottom fishing closures
South Atlantic Ocean and wider Caribbean	Not enough knowledge to observe change	**Ridges and rises, plateaux, banks:** exploration for mineral resources (chap. 18); offshore hydrocarbons (chap. 19); expansion of fisheries (chap. 15) **Trenches:** pollution	**EBSA:** Subtropical Convergence Zone **Bottom fishing closure:** South-East Atlantic Fisheries Organization bottom fishing closure
Indian Ocean, Arabian Sea, Bay of Bengal, Red Sea, Gulf of Aden and Persian Gulf	Not enough knowledge to observe change	**Ridges and rises, plateaux, banks:** exploration for mineral resources (chap. 18); offshore hydrocarbons (chap. 19); expansion of fisheries (chap. 15) **Trenches:** pollution	**EBSA:** East Broken Ridge Guyot
North Pacific Ocean	Warming trend in the North-East Pacific	**Ridges and rises, plateaux, banks:** exploration for mineral resources (chap. 18); expansion of fisheries (chap. 15) **Trenches:** pollution	**EBSA:** Kyushu Palau Ridge; West Kuril Trench, Japan Trench, Izu-Ogasawara Trench and North of Mariana Trench; Ryukyu Trench area
South Pacific Ocean	Not enough knowledge to observe change	**Ridges and rises, plateaux, banks:** exploration for mineral resources (chap. 18); expansion of fisheries (chap. 15) **Trenches:** pollution	**EBSA:** Salas y Gómez and Nazca Ridges, Kermadec-Tonga-Louisville Junction
Southern Ocean	Warming, change in the circulation and particulate organic carbon fluxes; (East Scotia Ridge), species distribution shifts	**Ridges and rises, plateaux, banks:** expansion of fisheries (chap. 15) **Trenches:** pollution	Commission for the Conservation of Antarctic Marine Living Resources: conservation measure 22-06 (2019) Bottom fishing in the Convention Area

5. Outlook

Many unknowns remain regarding deep-sea environments and ecosystems, but significant research has occurred in the past decade, with more anticipated in the next decade. The United Nations proclaimed the Decade of Ocean Science for Sustainable Development (2021–2030)[5] to support efforts to reverse the cycle of decline in ocean health and to gather ocean stakeholders worldwide behind a common framework that will ensure that ocean science can fully support countries in creating improved conditions for sustainable development of the ocean. The Intergovernmental Oceanographic Commission will coordinate the effort. The European Commission has created five European Research and Innovation Missions, including one dedicated to the oceans and seas.

Recently, the high media profile of some private initiatives has increased interest in ridge and trench exploration, potentially boosting interest in those ecosystems for a wide audience.

Ridges and plateaux support numerous vulnerable marine ecosystems (with benthic communities dominated by corals and sponges and hydrothermal vents) which receive protection under regulations of the Food and Agriculture Organization of the United Nations. However, the emergence of potential deep-sea mining on ridges and rises to exploit polymetallic sulphides and cobalt-rich manganese crusts poses new threats to those ecosystems. Presently, the International Seabed Authority is developing regulations for exploitation, including impact assessments and protections. Final recommendations are expected by 2020 (chap. 18).

Under the Convention on Biological Diversity,[6] work has been undertaken to promote international agreement on the development of representative networks of marine protected areas and other effective area-based conservation measures, with a 2020 target of 10 per cent of the total marine area (Secretariat of the Convention on Biological Diversity, 2010). A programme has also been initiated to identify ecologically and biologically significant areas (ibid., 2009).

Marine genetic resources in areas beyond national jurisdiction, the high seas and the Area (the seabed, ocean floor and subsoil thereof beyond the limits of national jurisdiction) are unregulated, which is a particularly significant issue given: (a) the unknown economic potential in such fields as pharmaceuticals, bioremediation, cosmetics, nutraceutical or biomedical innovation; and (b) the uneven capacity worldwide to use marine genetic resources. Since access to marine genetic resources mostly begins with marine scientific research, which, as a conditional freedom of the high seas, is subject to the relevant provisions of Part XIII of the United Nations Convention on the Law of the Sea (e.g., sharing information and knowledge), there exists a partial basis for the development of new legal regimes for the management of marine genetic resources in areas beyond national jurisdiction (Broggiato and others, 2014).

An international legally binding instrument on the conservation and sustainable use of marine biological diversity of areas beyond national jurisdiction is currently being negotiated under the United Nations Convention on the Law of Sea, focusing on a package of issues composed of area-based management tools, including marine protected areas, environmental

5 See General Assembly resolution 72/73.
6 United Nations, *Treaty Series*, vol. 1760, No. 30619.

impact assessments, capacity-building and the transfer of marine technology, and marine genetic resources, including questions on the sharing of benefits (Rabone and others, 2019).[7] However, even with all of that and planned reductions in single-use plastic and marine disposal, the deep sea ultimately becomes a sink.

Dumping and pollution are of special concern in the hadal trenches because of the tendency for the concentration of such material along the trench axis, affecting the organisms living there (Jamieson and others, 2017).

6. Key remaining knowledge gaps

In the past decade, most research on ridges has been associated with studies on hydrothermal vents and individual seamounts (chaps. 7P and 7L), which cover a small portion of the global ridge system. For ridges, plateaux and trenches, major gaps will remain in the scientific understanding of biodiversity patterns and spatial scales, species composition and abundance.

As with most of the deep ocean, knowledge on the pelagic realm is particularly limited, including basic aspects of biodiversity, such as species composition and abundance, as well as spatial and temporal variations, but certain aspects of the benthic ecosystems are also still poorly understood. Ecological and environmental data, such as life history patterns, substrate topography and mesoscale ocean dynamics, are needed to inform particle flux, food web and habitat suitability models that can address ecosystem responses to disturbance. The lack of knowledge means that the deep sea remains the "black box" in global model simulations.

Furthermore, science has barely begun to understand how human impacts will affect deep-sea ecosystem functions and, consequently, the services those ecosystems provide to society (Thurber and others, 2014). Such knowledge is critical to effective ocean management. Recently, the deep-sea biology community, through several initiatives (e.g., Deep-Ocean Stewardship Initiative, Deep Ocean Observing Strategy), has identified the four key questions that need to be answered in order to achieve sustainable management of the deep sea (Deep-Ocean Stewardship Initiative (DOSI), 2019), namely: (a) what is the diversity of life in the deep ocean?; (b) how are populations and habitats connected?; (c) what is the role of living organisms in ecosystem function and service provision?; and (d) how do species, communities and ecosystems respond to disturbance? Although they are presented as general questions on the deep sea, the questions are well suited to be posed in the specific context of ridges, plateaux and trenches and should be considered a priority in future research.

7. Key remaining capacity-building gaps

Access to the deep ocean is constrained to a few developed countries, even though a vast portion of it is within the exclusive economic zones of developing nations and the high seas.

The availability of technology, such as deep-sea vehicles, is the most limiting factor for exploration. That gap is probably the most difficult to overcome for financial and technical

7 See General Assembly resolution 72/249.

reasons, but collaborative and interdisciplinary research networks have been suggested as an effective way to optimize time at sea (Levin and others, 2019b). A multidisciplinary approach is also necessary to develop new ways to create easy-access modelling to forecast changes and vulnerability for a better environmental assessment.

Another major gap is in the field of expertise, especially in developing countries. Training is needed for a new generation of scientists, including on best practices, taxonomic abilities, the ecosystem approach and how to explore, manage and conserve the deep sea using the latest tools.

International collaborations through existing programmes, such as those offered by the International Oceanographic Data and Information Exchange of the Intergovernmental Oceanographic Commission of the United Nations Educational, Scientific and Cultural Organization or the Learn Education and Training Programme of the World Meteorological Organization, or newly formed initiatives dedicated to deep-sea research, can contribute to facilitating access to technology and training materials, including dedicated courses, participation in research cruises, training internships on field research, instrumentation development and data analysis.

References

Alt, Claudia H.S., and others (2019). Bathyal benthic megafauna from the Mid-Atlantic Ridge in the region of the Charlie-Gibbs fracture zone based on remotely operated vehicle observations. *Deep Sea Research Part I: Oceanographic Research Papers*, vol. 145, pp. 1–12.

Beaulieu, Stace E., and others (2013). An Authoritative Global Database for Active Submarine Hydrothermal Vent Fields. *Geochemistry, Geophysics, Geosystems*, vol. 14, No. 11, pp. 4892–4905.

Beazley, Lindsay I., and others (2013). Deep-Sea Sponge Grounds Enhance Diversity and Abundance of Epibenthic Megafauna in the Northwest Atlantic. *ICES Journal of Marine Science*, vol. 70, No. 7, pp. 1471–1490.

Bird, Peter (2003). An updated digital model of plate boundaries. *Geochemistry, Geophysics, Geosystems*, vol. 4, No. 3, pp. 1–52.

Bluhm, H. (2001). Re-establishment of an abyssal megabenthic community after experimental physical disturbance of the seafloor. *Deep Sea Research Part II: Topical Studies in Oceanography*, vol. 48, pp. 3841–3868.

Bober, Simon, and others (2018). Does the Mid-Atlantic Ridge affect the distribution of abyssal benthic crustaceans across the Atlantic Ocean? *Deep Sea Research Part II: Topical Studies in Oceanography*, vol. 148, pp. 91–104.

Brandt, A., and others (2018). First insights into macrofaunal composition from the SokhoBio expedition (Sea of Okhotsk, Bussol Strait and northern slope of the Kuril-Kamchatka Trench). *Deep Sea Research Part II: Topical Studies in Oceanography*, vol. 154, pp. 106–120.

Brandt, A., and others (2016). *RV Sonne SO-250 Cruise Report / Fahrtbericht: Tomakomai-Yokohama (Japan) 16.08.-26.09.2016 – SO-250 KuramBio II*. Kuril Kamchatka Biodiversity Studies.

Broggiato, A., and others (2014) Fair and equitable sharing of benefits from the utilization of marine genetic resources in areas beyond national jurisdiction: bridging the gaps between science and policy. *Marine Policy*, vol. 49, pp. 176–185.

Cathalot, Cécile, and others (2015). Cold-water coral reefs and adjacent sponge grounds: hotspots of benthic respiration and organic carbon cycling in the deep sea. *Frontiers in Marine Science*, vol. 2, art. 37.

Chiba, Sanae, and others (2018). Human footprint in the abyss: 30 year records of deep-sea plastic debris. *Marine Policy*, vol. 96, pp. 204–212.

Compton, Tanya J., and others (2013). Biophysical Patterns in Benthic Assemblage Composition across Contrasting Continental Margins off New Zealand. *Journal of Biogeography*, vol. 40, No. 1, pp. 75–89.

Davies, Andrew J., and John M. Guinotte (2011). Global habitat suitability for framework-forming cold-water corals. *PloS One*, vol. 6, No. 4.

Deep-Ocean Stewardship Initiative (DOSI) (2019). Deep-sea research in the Decade of Ocean Science: Mapping the role of the deep ocean in human society. www.dosi-project.org/wp-content/uploads/2019/07/DOSI_Decade_Position_Final-1.pdf.

Eustace, Ryan M., and others (2016). Morphological and ontogenetic stratification of abyssal and hadal *Eurythenes gryllus* sensu lato (Amphipoda: Lysianassoidea) from the Peru–Chile Trench. *Deep Sea Research Part I: Oceanographic Research Papers*, vol. 109, pp. 91–98.

Five Deeps Expedition (2019). Accessed 12 September 2019. https://fivedeeps.com.

Food and Agriculture Organization of the United Nations (FAO) (2009). *International Guidelines for the Management of Deep-Sea Fisheries in the High Seas*. Rome.

Fujii, Toyonobu, and others (2013). Deep-sea amphipod community structure across abyssal to hadal depths in the Peru-Chile and Kermadec trenches. *Marine Ecology Progress Series*, vol. 492, pp. 125–138.

Gallo, Natalya D., and others (2015). Submersible- and lander-observed community patterns in the Mariana and New Britain trenches: influence of productivity and depth on epibenthic and scavenging communities. *Deep Sea Research Part I: Oceanographic Research Papers*, vol. 99, pp. 119–133.

Gerringer, Mackenzie E., and others (2017). *Pseudoliparis swirei* sp. nov.: a newly-discovered hadal snailfish (Scorpaeniformes: Liparidae) from the Mariana Trench. *Zootaxa*, vol. 4358, No. 1, pp. 161–177.

Gomes-Pereira, José Nuno, and others (2017). Cold-water corals and large hydrozoans provide essential fish habitat for *Lappanella fasciata* and *Benthocometes robustus*. *Deep Sea Research Part II: Topical Studies in Oceanography*, vol. 145, pp. 33–48.

Guillaumot, Charlène, and others (2018). Benthic species of the Kerguelen Plateau show contrasting distribution shifts in response to environmental changes. *Ecology and Evolution*, vol. 8, No. 12, pp. 6210–6225.

Harris, Peter T., and others (2014). *Geomorphology of the Oceans. Marine Geology*, vol. 352, pp. 4–24.

Harris, Peter T., and Tanya Whiteway (2009). High seas marine protected areas: benthic environmental conservation priorities from a GIS analysis of global ocean biophysical data. *Ocean & Coastal Management*, vol. 52, No. 1, pp. 22–38.

Henry, Lea-Anne, and others (2013). Cold-water coral reef habitats benefit recreationally valuable sharks. *Biological Conservation*, vol. 161, pp. 67–70.

Honda, M.C., and others (2013). Concentration and vertical flux of Fukushima-derived radiocesium in sinking particles from two sites in the Northwestern Pacific Ocean. *Biogeosciences*, vol. 10, No. 6, pp. 3525–3534.

Ichino, Matteo C., and others (2015). The distribution of benthic biomass in hadal trenches: a modelling approach to investigate the effect of vertical and lateral organic matter transport to the seafloor. *Deep Sea Research Part I: Oceanographic Research Papers*, vol. 100, pp. 21–33.

Jamieson, A.J. (2015). *The Hadal Zone: Life in the Deepest Oceans*. Cambridge University Press.

Jamieson, Alan J., and others (2009). HADEEP: Free-falling landers to the deepest places on Earth. *Marine Technology Society Journal*, vol. 43, No. 5, pp. 151–160.

Jamieson, Alan J., and others (2010). Hadal trenches: the ecology of the deepest places on Earth. *Trends in Ecology & Evolution*, vol. 25, No. 3, pp. 190–197.

Jamieson, Alan J., and others (2011). Bait-attending fauna of the Kermadec Trench, SW Pacific Ocean: evidence for an ecotone across the abyssal-hadal transition zone. *Deep Sea Research Part I: Oceanographic Research Papers*, vol. 58, No. 1, pp. 49–62.

Jamieson, Alan J., and others (2019). Microplastics and synthetic particles ingested by deep-sea amphipods in six of the deepest marine ecosystems on Earth. *Royal Society Open Science*, vol. 6, No. 2, 180667.

Jamieson, A.J., and others (2013). The supergiant amphipod *Alicella Gigantea* (Crustacea: Alicellidae) from hadal depths in the Kermadec Trench, SW Pacific Ocean. *Deep Sea Research Part II: Topical Studies in Oceanography*, vol. 92, pp. 107–113.

Jamieson, A.J., and others (2018). Exploring the hadal zone: recent advances in hadal science and technology. *Deep Sea Research Part II: Topical Studies in Oceanography*, vol. 155, pp. 1–3.

Jamieson, A.J., and others (2017). Bioaccumulation of persistent organic pollutants in the deepest ocean fauna. *Nature Ecology & Evolution*, vol. 1, No. 3, pp. 1–4.

Jennings, R., and others (2020). Integrative species delimitation of desmosomatid and nannoniscid isopods from the Kuril-Kamchatka trench, with description of a hadal species. *Progress in Oceanography*, vol. 182, art. 102236.

Johnson, David Edwards, and others (2019). Rockall and Hatton: resolving a super wicked marine governance problem in the high seas of the northeast Atlantic Ocean. *Frontiers in Marine Science*, vol. 6, art. 69.

Kawagucci, Shinsuke, and others (2012). Disturbance of deep-sea environments induced by the M9.0 Tohoku Earthquake. *Scientific Reports*, vol. 2, art. 270.

Kioka, A., and others (2019). Megathrust earthquake drives drastic organic carbon supply to the hadal trench. *Scientific Reports*, vol. 9, No. 1, pp. 1–10.

Kitahashi, Tomo, and others (2014). Effect of the 2011 Tohoku Earthquake on deep-sea meiofaunal assemblages inhabiting the landward slope of the Japan Trench. *Marine Geology*, vol. 358, pp. 128–137.

Knox, Matthew A., and others (2012). Mitochondrial DNA (COI) analyses reveal that amphipod diversity is associated with environmental heterogeneity in deep-sea habitats. *Molecular Ecology*, vol. 21, No. 19, pp. 4885–4897.

Lacey, Nichola C., and others (2016). Community structure and diversity of scavenging amphipods from bathyal to hadal depths in three South Pacific Trenches. *Deep Sea Research Part I: Oceanographic Research Papers*, vol. 111, pp. 121–137.

Lecours, Vincent (2017). On the use of maps and models in conservation and resource management (warning: results may vary). *Frontiers in Marine Science*, vol. 4, art. 288.

Leduc, Daniel, and Ashley A. Rowden (2018). Not to be sneezed at: does pollen from forests of exotic pine affect deep oceanic trench ecosystems? *Ecosystems*, vol. 21, No.2, pp. 237–247.

Leduc, Daniel, and others (2012). Nematode beta diversity on the continental slope of New Zealand: spatial patterns and environmental drivers. *Marine Ecology Progress Series*, vol. 454, pp. 37–52.

Leduc, Daniel, and others (2015). Distribution of macro-infaunal communities in phosphorite nodule deposits on Chatham Rise, Southwest Pacific: implications for management of seabed mining. *Deep Sea Research Part I: Oceanographic Research Papers*, vol. 99, pp. 105–118.

Leduc, Daniel, and others (2016). Comparison between infaunal communities of the deep floor and edge of the Tonga Trench: possible effects of differences in organic matter supply. *Deep Sea Research Part I: Oceanographic Research Papers*, vol. 116, pp. 264–275.

Levin, Lisa A., and others (2019a). Climate change: overview and drivers. Chapter 2 in: *Deep Ocean Climate Change Impacts on Habitat, Fish and Fisheries*, L. Levin and others, eds. FAO Fisheries and Aquaculture Technical Paper. Rome, FAO, pp. 56–79.

Levin, Lisa A., and others (2019b). Global Observing Needs in the Deep Ocean. *Frontiers in Marine Science*, vol. 6, art. 241.

Luo, Min, and others (2017). Provenances, distribution, and accumulation of organic matter in the southern Mariana Trench rim and slope: implication for carbon cycle and burial in hadal trenches. *Marine Geology*, vol. 386, pp. 98–106.

Morato, Telmo, and others (2018). A Multi Criteria Assessment Method for Identifying Vulnerable Marine Ecosystems in the North-East Atlantic. *Frontiers in Marine Science*, vol. 5, art. 460.

Mortimer, Nick, and others (2017). Zealandia: Earth's hidden continent. *GSA Today*, vol. 27, No. 3, pp. 27–35.

Narayanaswamy, Bhavani E., and others (2013). First observations of megafaunal communities inhabiting George Bligh Bank, northeast Atlantic. *Deep Sea Research Part II: Topical Studies in Oceanography*, vol. 92, pp. 79–86.

Oguri, Kazumasa, and others (2013). Hadal disturbance in the Japan Trench induced by the 2011 Tohoku-Oki Earthquake. *Scientific Reports*, vol. 3, art. 1915.

Peoples, Logan Maxwell, and others (2019). Microbial community diversity within sediments from two geographically separated hadal trenches. *Frontiers in Microbiology*, vol. 10, art. 347.

Pham, Christopher K., and others (2015). The importance of deep-sea vulnerable marine ecosystems for demersal fish in the Azores. *Deep Sea Research Part I: Oceanographic Research Papers*, vol. 96, pp. 80–88.

Pham, Christopher K., and others (2014). Deep-water longline fishing has reduced impact on Vulnerable Marine Ecosystems. *Scientific Reports*, art. 4837.

Picard, Kim, and others (2018). Malaysia Airlines flight MH370 search data reveal geomorphology and seafloor processes in the remote southeast Indian Ocean. *Marine Geology*, vol. 395, pp. 301–319.

Priede, Imants G. (2017). *Deep-Sea Fishes: Biology, Diversity, Ecology and Fisheries*. Cambridge University Press.

Rabone, Muriel, and others (2019). Access to marine genetic resources (MGR): raising awareness of best-practice through a new agreement for biodiversity beyond national jurisdiction (BBNJ). *Frontiers in Marine Science*, vol. 6, art. 520.

Ritchie, H., and others (2017). Population genetic structure of two congeneric deep-sea amphipod species from geographically isolated hadal trenches in the Pacific Ocean. *Deep Sea Research Part I: Oceanographic Research Papers*, vol. 119, pp. 50–57.

Sautya, Sabyasachi, and others (2017). First quantitative exploration of benthic megafaunal assemblages on the mid-oceanic ridge system of the Carlsberg Ridge, Indian Ocean. *Journal of the Marine Biological Association of the United Kingdom*, vol. 97, No. 2, pp. 409–417.

Secretariat of the Convention on Biological Diversity (2009). *Azores Scientific Criteria and Guidance for Identifying Ecologically or Biologically Significant Marine Areas and Designing Representative Networks of Marine Protected Areas in Open Ocean Waters and Deep Sea Habitats*. Montreal: CBD.

________ (2010). Decision adopted by the Conference of the Parties to the Convention on Biological Diversity at its Tenth Meeting X/2. The Strategic Plan for Biodiversity 2011–2020 and the Aichi Biodiversity Targets.

Stendardo, Ilaria, and others (2015). Interannual to decadal oxygen variability in the mid-depth water masses of the eastern North Atlantic. *Deep Sea Research Part I: Oceanographic Research Papers*, vol. 95, pp. 85–98.

Stewart, Heather A., and Alan J. Jamieson (2019). The five deeps: the location and depth of the deepest place in each of the world's oceans. *Earth-Science Reviews*, vol. 197, 102896.

Sweetman, Andrew K., and others (2017). Major impacts of climate change on deep-sea benthic ecosystems. *Elementa: Science of the Anthropocene*, vol. 5, pp. 1–23.

Thurber, Andrew R., and others (2014). Ecosystem function and services provided by the deep sea. *Biogeosciences*, vol. 11, No. 14, pp. 3941–3963.

United Nations (2017a). Chapter 51: Biological communities on seamounts and other submarine features potentially threatened by disturbance. In *The First Global Integrated Marine Assessment: World Ocean Assessment I*. Cambridge: Cambridge University Press.

________ (2017b). *The First Global Integrated Marine Assessment: World Ocean Assessment I*. Cambridge: Cambridge University Press.

United Nations Educational, Scientific and Cultural Organization (UNESCO) (2009). *Global Open Oceans and Deep Seabed (GOODS) Bioregional Classification*. UNESCO-IOC, Technical Series 84. Paris: UNESCO.

United Nations Environment Programme World Conservation Monitoring Centre and International Union for Conservation of Nature and Natural Resources (2019). Protected Planet: The World Database on Protected Areas. Cambridge, United Kingdom: UNEP-WCMC and IUCN. www.protectedplanet.net.

Vivier, Frédéric, and others (2015). Variability of the Antarctic Circumpolar Current transport through the Fawn Trough, Kerguelen plateau. *Deep Sea Research Part II: Topical Studies in Oceanography*, vol. 114, pp. 12–26.

Watling, Les, and others (2013). A proposed biogeography of the deep ocean floor. *Progress in Oceanography*, vol. 111, pp. 91–112.

Yancey, Paul H., and others (2014). Marine fish may be biochemically constrained from inhabiting the deepest ocean depths. *Proceedings of the National Academy of Sciences*, vol. 111, No. 12, pp. 4461–4465.

Yesson, Chris, and others (2012). Global habitat suitability of cold-water octocorals. *Journal of Biogeography*, vol. 39, No. 7, pp. 1278–1292.

Zhang, Weipeng, and others (2019). Gut microbial divergence between two populations of the hadal amphipod *Hirondellea gigas*. *Applied and Environmental Microbiology*, vol. 85, No. 1, e02032–18.

Zheng, Mei-Di, and Long Cao (2014). Simulation of global ocean acidification and chemical habitats of shallow-and cold-water coral reefs. *Advances in Climate Change Research*, vol. 5, No. 4, pp. 189–196.

Chapter 7P
Hydrothermal vents and cold seeps

Contributors: Nadine Le Bris (convener), Hilconida Calumpong (lead member), Sanae Chiba (subchapter lead member), Ana Colaço, Elva Escobar, Anna Metaxas, Paraskevi Nomikou, Julia Sigwart, Verena Tunnicliffe and Hiromi Watanabe.

Keynote points

- Hydrothermal vents and cold seeps have uniquely complex habitats and communities, diverse endemic species, high biomass and productivity supported by chemosynthesis.
- Such ecosystems are sources of biotechnological and biomedical innovation.
- They play a significant role in global ocean processes, sequestering CO_2 and methane, and contributing to surface ocean productivity through iron export.
- In the past five years, explorations using new tools to detect water column signals located thousands of new vent fields and cold seeps.
- Resource exploration (polymetallic sulphides and methane hydrates) and the need to map and protect vulnerable habitats and species support recent investigations.
- The International Seabed Authority has issued seven polymetallic sulphide exploration contracts since 2011, encompassing vent fields from the mid-ocean ridges of the Indian Ocean and the Atlantic Ocean.
- Vulnerable marine ecosystems and marine protected areas in exclusive economic zones and in areas beyond national jurisdiction protect some vents and seeps.
- The conservation status of vents or seeps identified eight areas protected under national law within exclusive economic zones.
- Knowledge gaps include spatial and temporal patterns, impacts from direct disturbance, changes to deepwater circulation, deoxygenation, warming and acidification.
- Ocean warming triggering gas hydrate dissociation is a major stressing factor to cold seepage activity and ecosystems.
- Capacity-building is a priority, in particular in island States.

1. Introduction

1.1. Scope and summary of the baseline state from the first *World Ocean Assessment*

Hydrothermal vents occur wherever a heat source drives seawater circulation through the subsea floor. Cold seeps refer to hydrocarbon-rich fluids emanating from buried organic matter, fossil fuel reservoirs or methane hydrates. Both environments encompass a wide range of fluid compositions and habitat types (Cordes and others, 2009; Watanabe and others, 2010; Levin and Sibuet, 2012; Le Bris and others, 2019).

The present chapter focuses on marine life and sea floor habitats that are influenced by fluid emissions, including shallow seeps and vents, which are important to local biodiversity and biogeography and to the flux of greenhouse gases to the atmosphere, with long-distance effects on both the seabed and the water column. In chapter 45 of the first *World Ocean Assessment* (United Nations, 2017a), the accelerating exploration of hydrothermal vents was noted, in particular in relation to mineral resource exploration, and information was provided on the development of the conservation status of vents or seeps within exclusive economic zones and beyond national jurisdictions.

1.2. How the topic is affected by and affects other marine components

High local biomass of chemosynthetic microbial primary producers and associated fauna are sustained by fluxes of methane, hydrogen sulphide, hydrogen, or reduced iron and manganese. Many specialist taxa host bacterial symbionts (Dubilier and others, 2008) and act as ecosystem foundation species (Govenar, 2010). Global communities are similar at the family level, forming 11 biogeographic regions by endemic species (Rogers and others, 2012; Moalic and others, 2012). Peripheral habitats benefit from chemosynthetic resources (Levin and others, 2016). Hydrothermal plumes export metals and organic material and contribute to regional and global iron budgets (Resing and others, 2015; German and others, 2016; Tagliabue and Resing, 2016).

1.3. How the topic is relevant to human communities and well-being

1.3.1. Fishing grounds

Sessile organisms, including habitat-forming species (sponges, soft and hard corals), benefit from hard substrata formed at vents and seeps, contributing to essential habitats for groundfishes (United States Pacific Fishery Management Council, 2019). Chemosynthetic primary production can contribute to commercial fisheries stocks productivity, for example, exploited crabs off the coast of British Columbia, Canada, that assimilate carbon from chemosynthetic sources (Seabrook and others, 2019). Off the coast of California, United States, seep population densities of species targeted by fisheries increase (Grupe and others, 2015).

1.3.2. Greenhouse gas flux regulation

Vents are natural sources of CO_2 and methane issuing from magma degassing, mantle serpentinization and diagenetic organic matter degradation in buried sediments. Chemoautotrophy and methanotrophy contribute to trapping the emissions at the sea floor (Orcutt and others, 2011; Wankel and others, 2011; Römer and others, 2014a; Ruppel and Kessler, 2017). Anaerobic methane oxidation by archaea is a key sequestration pathway (Boetius and Wenzhöffer, 2013). Hydrothermal iron can locally fertilize the surface waters (Guieu and others, 2018; Ardyna and others, 2019) and globally supports the oceanic phytoplankton-driven CO_2 sink.

1.3.3. Ecological models for adaptation and resilience

Hydrothermal vents and cold seeps provide models for the study of animal stress responses in high CO_2/low pH conditions, extreme temperature, hypoxia and exposure to sulphides, toxic metals and metalloids and are helpful in furthering an understanding of biochemical, physiological and behavioural adaptations (e.g., Hall-Spencer and others, 2008; Tunnicliffe and others, 2009; Childress and Girguis, 2011; Di Carlo and others, 2017; Rossi and Tunnicliffe, 2017). Colonization patterns provide insights into larval dispersal capacities, species dependencies and resilience to disturbances (Gollner and others, 2017; Mullineaux and others, 2018). Functional traits approaches address the contributions of common and rare species (Chapman and others, 2018).

1.3.4. Biotechnological and biomedical innovation

Biotechnical discoveries reflecting unique microbe-animal interaction and extreme habitat conditions include: antibiotic molecules in hydrothermal worms (Tasiemski and others, 2014; Papot and others, 2017); metal resistance genes in vent microbes implicating enzymatic detoxication pathways in polluted environments (Vetriani and others, 2005; Colaço and others, 2006); and chemoautotrophic carbon fixation pathways relevant to CO_2 emissions sequestration (Scott and others, 2018; Rubin-Blum and others, 2019).

1.3.5. Public engagement with the ocean

The discoveries and imagery of such iconic ecosystems reach large, global audiences, supplemented by telepresence cruises, books, films, theatre, games and toys, inspiring citizen science projects. The ecosystems showcase the diverse adaptations of deep ocean taxa, the roles of microbes and the origin of life on Earth.

1.4. Advances in knowledge and capacity

1.4.1. Exploration and mapping

Systematic mapping with autonomous underwater vehicles tracking anomalies in the water column (e.g., temperature, redox potential or methane, gas bubbles or particles) improved the capacity to locate seeps or vents (Baker and others, 2016; James and others, 2016; Andreassen and others, 2017; Baumberger and others, 2018).

The exploration of vent and seep systems has extended in subduction zones of the North-West Pacific (e.g., Baker and others, 2017) and on Arctic ridges (Marques and others, 2020), in the Southern Ocean (Linse and others, 2019) and on the Central, Western and Eastern Indian Ridges (Copley and others, 2016; Zhou and others, 2018; Gerdes and others, 2019; figure I). Identified methane seepage areas have increased since the first Assessment (United Nations, 2017b) along the coasts of the United States (Quattrini and others, 2015; Baumberger and others, 2018), in the South China Sea (Feng and others, 2018), in Brazil (South-West Atlantic) (Ketzer and others, 2019), in the Caribbean (Digby and others, 2016) and in India (Bay of Bengal) (Mazumdar and others, 2019). Improved seabed mapping resolution produced detailed baselines of faunal assemblage distributions that can be used for assessing responses to human activities (Thornton and others, 2016; Gerdes and others, 2019).

Figure I
Active vent sites, seep areas and currently designated marine protected areas

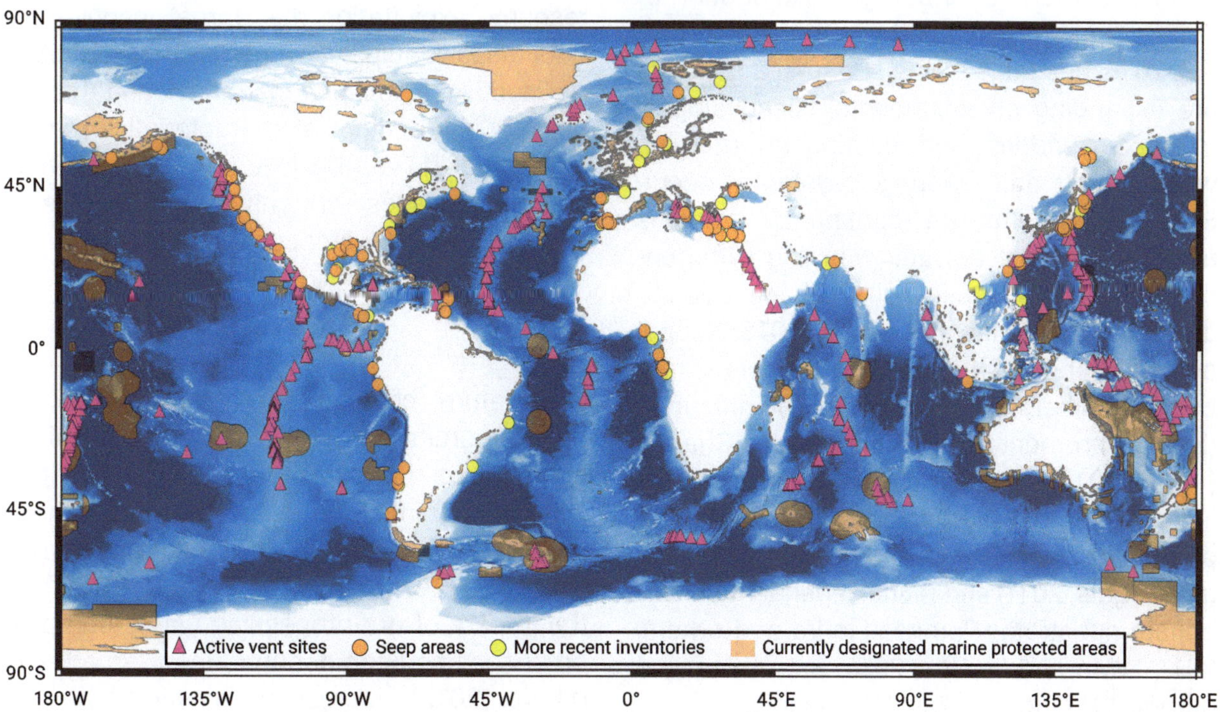

Sources: Active vent sites are from the InterRidge Vent Database (Beaulieu and Szafrański, 2020). Seep areas are from the Biogeography of Deep-Water Chemosynthetic Ecosystems database (ChEssBase, 2019). More recent inventories are from Olu and others (2010), Quattrini and others (2015), Baumberger and others (2018), Feng and others (2018), Etiope and others (2019) and MacDonald and others (2020). Marine protected areas are from the United Nations Environment Programme World Conservation Monitoring Centre and International Union for Conservation of Nature and Natural Resources (2019). Bathymetry is from www.gebco.net/data_and_products/gridded_bathymetry_data.
Note: Map created using Q-GIS version 2.18.20 (QGIS Development Team, 2018).

1.4.2. Variability in space and time

Repeated surveys show that communities can be stable over decades on slow spreading ridges (Cuvelier and others, 2011) and some back-arc basins (Du Preez and Fisher, 2018). That challenges the paradigm, based on fast-spreading ridges, that vent communities are dynamic and resilient to disturbance. Variable geothermal and geochemical energy sources fuel both ecosystems, including hybrid systems on sedimented margins (Goffredi and others, 2017). Genetic and hydrodynamic models reveal population connectivity patterns that are critical in the context of managing seabed resource development (Mullineaux and others, 2018; Suzuki and others, 2018).

2. Environmental changes since the first *World Ocean Assessment*

2.1. Changes in overall status

2.1.1. Drivers and pressures

Vent and seep drivers of change include growing economic demands for energy, strategic metals and food (figure II). Fossil fuel demand is driving offshore oil and gas exploitation at depths greater than 1,500 m (Cordes and others, 2016; see table below). Deep-sea fisheries are expanding on seamounts, island slopes, mid-ocean ridges and continental margins where vent and seep ecosystems occur. Seep ecosystems on continental margins are exposed to deepwater warming with concomitant enhancement of methane hydrate dissociation (James and others, 2016; Ruppel and Kessler, 2017), expanding hypoxia (Breitburg and others, 2018) and ocean acidification (Intergovernmental Panel on Climate Change (IPCC), 2019; see table below). Vulnerabilities include chemosynthetic holobiont species with particularly high oxygen demand (Childress and Girguis, 2011) and fauna dependent on carbonate substrata (Ramirez-Llodra and others, 2011; Levin and Le Bris, 2015; Sweetman and others, 2017; figure II, see table below).

Exploration for sea floor massive sulphides (Petersen and others, 2016) and extractive technology testing (Okamoto and others, 2019) are developing (see table below). By 2018, the International Seabed Authority had issued seven polymetallic sulphide exploration contracts in areas beyond national jurisdiction in parts of the Atlantic and Indian mid-ocean ridges hosting active and inactive vents (chap. 18). Vent ecosystems have also been reported in exclusive economic zones covered by exploration licences for sea floor massive sulphides. While resource exploitation may target inactive sea floor massive sulphides (chap. 18), the definition of "inactive" is still unconstrained, and their biological and ecological characteristics remain understudied (Van Dover and others, 2019), in particular at sites with only diffuse fluid vents that may escape water column detection.

2.1.2. State of changes associated with pressures and potential impacts

Substratum changes, vent emissions and faunal recruitment were reported near drilling holes (Nakajima and others, 2015), contrasting with another impact study (Copley and others, 1999). Destruction by trawling damage has been predicted (Bowden and others, 2013) and the impact of ore-processing waste dumping on seeps has been documented (Samadi and others, 2015). Increasing plastic debris at seeps has been reported (Chiba and others, 2018) and hypoxia expansion, acidification and warming have been documented in regions hosting seeps and vents (IPCC, 2019).

Figure II
Synthesis of the drivers, pressures, state, impact and response approach applied to vents and seeps

Currently significant (globally):

- Warming, hydrate dissociation
- Deoxygenation and acidification
- Trawling
- Ocean circulation changes

Minor/growing:

- Prospection and impact studies
- Pollution (oil spills)
- Waste dumping (from mining)

Potential threats (future):

- Mineral extraction
- Bioresources harvesting
- Subsea floor CO_2 sequestration

- Development of extractive industries (mining, fishing, energy)
- CO_2 emissions and climate change
- Geothermal energy
- Climate geoengineering (potential)

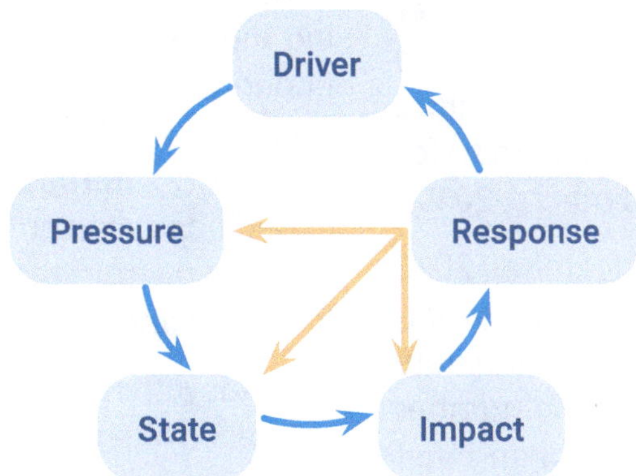

- CO_2 emission reduction
- Marine protected area designation and management
- Seabed mapping and marine spatial planning
- Knowledge advancement and monitoring
- Capacity-building and public engagement

- Habitat loss and fragmentation
- Reduction of species populations and dispersal capacity
- Abiotic stressors affecting ecosystem stability and functions (acidification, deoxygenation, warming, turbidity, hydrodynamics)

- Loss of genetic diversity
- Loss of cultural heritage
- Loss of greenhouse gas sequestration capacity
- Loss of nursery grounds for fished species
- Accumulation of toxins in commercial species
- Impact on tourism

Note: Summarizing the information developed in sections 2 and 3.

3. Economic and social consequences

3.1. Area-based management tools

Eight vent or seep areas are protected under national law within the exclusive economic zones of Canada, France, Mexico, Portugal and the United States (see table below). Beyond national jurisdictions, the OSPAR Commission established by the Convention for the Protection of the Marine Environment of the North-East Atlantic[1] categorized "oceanic ridges with hydrothermal vents/fields" as threatened and/or endangered habitats to be protected by its network of marine protected areas in Region V of the maritime area (Arctic and North Atlantic) (OSPAR Commission for the Protection of the Marine Environment of the North-East Atlantic (OSPAR), 2014, 2018; see table below).

Several vent fields have been described as ecologically and biologically significant areas under the process of the Convention on Biological Diversity (Dunn and others, 2014; Bax and others, 2016; see table below). Ecologically and biologically significant areas are not area-based management tools, but they provide information that

[1] United Nations, *Treaty Series*, vol. 2354, No. 42279.

may play a role in decision-making processes. Seeps have been recommended as ecologically and biologically significant areas, but few have specific status (see table below).

Habitats hosting chemosynthetic ecosystems are identified as specially protected areas under the 1995 Convention for the Protection of the Marine Environment and the Coastal Region of the Mediterranean[2] (United Nations Environment Programme, Mediterranean Action Plan, 2017; see table below). Another approach to biodiversity conservation is the International Union for Conservation of Nature Red List assessment, as recently executed for Indian Ocean vents (Sigwart and others, 2019).

In its resolution 71/123 of 7 December 2016 on sustainable fisheries, the General Assembly welcomed actions by States to eliminate destructive fishing practices, and the number of protected vents increased (Menini and Van Dover, 2019; figure I; see table below). Following the International Guidelines for the Management of Deep-Sea Fisheries in the High Seas (Food and Agriculture Organization of the United Nations (FAO), 2009), regional fisheries management organizations recognize hydrothermal vents as vulnerable marine ecosystems (FAO, 2016). Developing International Seabed Authority regional environmental management plans may ensure the protection of active vents as areas of particular environmental interest, but the paucity of data renders the task difficult (Dunn and others, 2018; chap. 27 of the present Assessment).

3.2. Implications for achieving the Sustainable Development Goals

Sustainable Development Goal 14[3] calls for the conservation and sustainable use of the oceans and their resources by implementing international law as reflected in the United Nations Convention on the Law of the Sea.[4] Achieving that goal requires an assessment of the cumulative pressures of climate change and human activities at vents and seeps (Levin and Le Bris, 2015), involving strong investments in capacity-building, technology transfer and development for deep-sea investigations (Goal 4, with target 4B related to knowledge and technology transfer in developing countries). Ocean literacy societies promote ocean conservation and sustainable use by supporting multi-stakeholder partnerships and public understanding of vent and seep ecosystems (Sustainable Development Goal targets 17.16 and 17.17) through telepresence research expeditions, citizen science initiatives, policy briefs (e.g., from the Deep-Ocean Stewardship Initiative) and education.

Chemosynthetic ecosystems provide "regulatory ecosystem services" for climate change impact mitigation by acting as filters for natural methane and CO_2 emissions (Thurber and others, 2014; James and others, 2016; Goal 7). The development of the renewable energy industry contributes to the projected deficit of secure supplies of certain metals and has stimulated exploration of deep-sea mineral resources. If mining for sea floor massive sulphides occurs at vents, it will imply trade-offs with Goal 14, with the introduction of pollutants in the trophic chain and essential habitat degradation, incompatible with sustainable fishing activities. Costs to biodiversity, biotechnological innovations and cultural values may arise (see chap. 18; see also section 2.1.1).

2 Ibid., vol. 1102, No. 16908.

3 See General Assembly resolution 70/1.

4 United Nations, *Treaty Series*, vol. 1833, No. 31363.

4. Key region-specific changes and consequences

Emerging threats, risks and conservation assessments and efforts at vents and seeps since 2014

Basin	Resource exploration and exploitation	Risk of increased cumulative pressures (including climate change)	Conservation assessment and efforts
Arctic	Mineral resource exploration on Arctic ridges Gas exploitation extension Fisheries expansion	Accelerated exploitation with sea ice retreat combined with warming (Sweetman and others, 2017) through methane hydrate destabilization (James and others, 2016)	Iceland: Eyjafjörður hydrothermal vents 1 and 2 nominated components to OSPAR MPA network (OSPAR, 2018)
North Atlantic	Three International Seabed Authority exploration contracts signed for northern Mid-Atlantic Ridge in area hosting active vent fields (Russian Federation, 2012; France, 2014; Poland, 2018)	North Atlantic abyssal/intermediate water warming and acidification (Gehlen and others, 2014) Water column larval dispersal effects on connectivity (FAO, 2018)	Portugal EEZ: Menez Gwen, Lucky Strike and Rainbow vent fields protected, Natura 2000 sites, Azores Marine Park Mid-Atlantic Ridge north Azores High Seas MPA: OSPAR water column protection; seabed and subsoil protection by Portugal (OSPAR, 2018), Spain: Gulf of Cádiz, mud volcanoes Site of Community Importance (European Union Habitat Directive on the Conservation of Natural Habitats and Wild Fauna and Flora) (2014) United Nations Environment Programme/Convention on Biological Diversity: Lost City, Broken Spur and TAG vent fields as EBSAs
Mediterranean	Natural gas extraction extension, eastern and south-western Mediterranean	Intermediate/deep water warming, deoxygenation; abyssal waters ventilation reduction (Adloff and others, 2015)	General Fisheries Commission for the Mediterranean of the Food and Agriculture Organization of the United Nations (GFCM-FAO) (2018): Nile delta cold hydrocarbon seeps vulnerable marine ecosystem fishery closure
Black Sea		Salinity change enhances methane hydrate destabilization (Riboulot and others, 2018) Anoxia expansion, seep ecosystem threats	
South Atlantic	Deep oil and gas exploration and exploitation development off the coast of Brazil (Almada and Bernardino, 2017)	Continental margin extensive oil production off the coast of Brazil, seep exploration limited (Bernardino and Sumida, 2017)	

Basin	Resource exploration and exploitation	Risk of increased cumulative pressures (including climate change)	Conservation assessment and efforts
Gulf of Mexico, Caribbean	Deep oil and gas exploration and exploitation in the Gulf of Mexico and off the coast of Guyana	Climate change and eutrophication-driven extension of dead zones with impacts on seeps at intermediate depths (Johnson and Purkey, 2009; Breitburg and others, 2018) Oil exploration in course with seep exploration limited in south-western Caribbean (Digby and others, 2016)	
Indian	Four exploration contracts issued by the International Seabed Authority in areas hosting active vent fields of the Southwest Indian Ridge (China, 2011) and Central Indian Ridge (Republic of Korea, 2014; Germany, 2015; India, 2016)	Sensitivity of seep ecosystems to regional oxygen decrease on Pakistan margins (Fischer and others, 2012)	IUCN Red List of Threatened Species: Scaly-foot gastropod listed as endangered, draft assessments completed for endemic species to regional vents (Sigwart and others, 2019)
North Pacific	Japan Oil, Gas and Metals National Corporation (JOGMEC) tests SMS extraction in Okinawa Trough (Okamoto and others, 2019) South China Sea: production test of seep gas hydrate extraction (Li and others, 2018)	Trawling pressures increase combined with ocean warming trend in the North-East Pacific Increasing risk of methane hydrate dissociation (Ruppel and Kessler, 2017; Hautala and others, 2014)	Canada: EEZ Pacific vents offshore Pacific area of interest; Canadian Pacific cold seeps are EBSAs (Fisheries and Oceans Canada (DFO), 2018) Mexico: EEZ MPA system Guaymas Basin and Eastern Pacific Rise Sanctuary (decreed in 2009, management plan published in 2014) EBSA: Guaymas Basin Hydrothermal Vents Sanctuary (in 2016); Deep Pacific Mexican Biosphere Reserve (in 2018) United States Pacific Fishery Management Council (2019) designated seeps as essential fish habitat EBSA: Cold seeps in south-western Taiwan Basin
South Pacific	Vent SMS exploration licences in EEZs of the South-West Pacific	Landmine tailings vent/seeps impact (Samadi and others, 2015)	New Zealand EEZ: 88 per cent active hydrothermal vents are MPAs France – New Caledonia EEZ (2014): Shallow vents and unexplored ridge systems in the MPA Parc Marin de la Mer de Corail Mining regulation and environmental policies in several island nations
Southern Ocean		East Scotia Ridge vents and Antarctic seeps influenced by warming, circulation and carbon flux changes (Römer and others, 2014b)	Vent and seep species identified in the Commission for the Conservation of Antarctic Marine Living Resources vulnerable marine ecosystems taxa classification guide (2009)

5. Outlook

Vent and seep biota and habitat conditions remain poorly documented, including vent areas where mineral exploration is under way (e.g., Indian Ocean). The biogeography of seeps is not described (Olu and others, 2010). The effects of the extraction of sea floor massive sulphides on active vent ecosystems and their peripheries are unclear. Expected impacts include sediment plumes, toxic compound release, habitat loss and metapopulation connectivity disruption (Dunn and others, 2018). Models of larval dispersal have begun to be developed and the latest outcomes indicate limited interregional connectivity in areas such as the western Pacific (Mitarai and others, 2016). The great longevity of foundation species of seep ecosystems (up to 200 years) further suggests slow disturbance recovery (Fisher and others, 2016). Both trawling (see section 2.1.2) and fossil fuel exploitation could also have long-term impacts (Amon and others, 2017). The lack of seep baseline data still limits the ability to predict the resilience capacity of the communities (Cordes and others, 2016).

In addition to acidification, oxygen depletion, methane flux and hydrodynamic condition changes resulting from climate disturbance affect large regions hosting vent and seep ecosystems and likely interact with key biological processes, although specific impacts have not yet been reported. A direct impact from warming to seeps is expected to be methane hydrate dissociation (Ruppel and Kessler, 2017). Changes in the upper water column may also affect vent and seep propagule dispersal (Yahagi and others, 2017; Mullineaux and others, 2018).

6. Key remaining knowledge gaps

Vent species discovery rates remain high, suggesting species richness undersampling (Thaler and Amon, 2019). Chemosynthetic fauna databases (ChEssBase, 2019; Chapman and others, 2019) will enable global scale analyses and the use of new techniques, such as eDNA high-throughput sequencing, and metacommunity models will help to resolve connectivity patterns (Chen and others, 2015; Breusing and others, 2016; Mullineaux and others, 2018).

Long-term ecosystem research is essential to elucidating processes linking vent and seep communities to climate change stressors (IPCC, 2019). In situ measures of physiological responses and changes in ecosystem function are required to assess climate change vulnerability. Tolerance thresholds to climate stressors are largely unknown, in particular for species that thrive at the hypoxic areas of the periphery (Fischer and others, 2012).

Carbon storage in vents and seeps lacks quantitative estimates, including ranges and controlling factors for vent and seep productivity (Marlow and others, 2014; Le Bris and others, 2019) and the role of viruses as mediators of prokaryote ecology (Corinaldesi and others, 2012; Ortmann and Suttle, 2005). Vent effluents in ocean biogeochemical cycles require integrated assessments to evaluate impacts on surface water productivity from iron injection (Guieu and others, 2018; Ardyna and others, 2019).

7. Key remaining capacity-building gaps

The capacity gap is larger in less developed nations, in particular in small island developing States that host resources derived from vents and seeps. Training new deep-sea scientists in environment-related disciplines is a challenge recognized by Intergovernmental Oceanographic Commission programmes (Intergovernmental Oceanographic Commission of the United Nations Educational, Scientific and Cultural Organization (UNESCO-IOC), 2016). Ecological assessments require faunal knowledge, while molecular tools need field validation of results. Oceanographic data and species inventories underpin ecosystem function models, helping to predict vulnerability and recovery. Knowledge transfer includes taxonomic capacity and the development of low-cost technologies for deep-sea research and monitoring (Levin and others, 2019) and mapping, while the exploration and development of monitoring strategies involves the training of scientists, young students and, in particular, women.

References

Adloff, Fanny, and others (2015). Mediterranean Sea response to climate change in an ensemble of twenty first century scenarios. *Climate Dynamics*, vol. 45, pp. 2775–2802.

Almada, Gustavo Vaz de Mello Baez, and Angelo Fraga Bernardino (2017). Conservation of deep-sea ecosystems within offshore oil fields on the Brazilian margin, SW Atlantic. *Biological Conservation*, vol. 206, pp. 92–101. https://doi.org/10.1016/j.biocon.2016.12.026.

Amon, Diva J., and others (2017). Characterization of methane-seep communities in a deep-sea area designated for oil and natural gas exploitation off Trinidad and Tobago. *Frontiers in Marine Science*, vol. 4, art. 342.

Andreassen, Karin, and others (2017). Massive blow-out craters formed by hydrate-controlled methane expulsion from the Arctic seafloor. *Science*, vol. 356, No. 6341, pp. 948–953.

Ardyna, Mathieu, and others (2019). Hydrothermal vents trigger massive phytoplankton blooms in the Southern Ocean. *Nature Communications*, vol. 10, No. 1, pp. 1–8.

Baker, Edward T., and others (2016). How many vent fields? New estimates of vent field populations on ocean ridges from precise mapping of hydrothermal discharge locations. *Earth and Planetary Science Letters*, vol. 449, pp. 186–196.

Baker, Edward T., and others (2017). The effect of arc proximity on hydrothermal activity along spreading centers: new evidence from the Mariana Back Arc (12.7°N–18.3°N). *Geochemistry, Geophysics, Geosystems*, vol. 18, No. 11, pp. 4211–4228.

Beaulieu, Stace E., and Kamil M. Szafrański (2020). InterRidge Global Database of Active Submarine Hydrothermal Vent Fields Version 3.4. PANGAEA. https://doi.org/10.1594/PANGAEA.917894.

Baumberger, Tamara, and others (2018). Mantle-derived helium and multiple methane sources in gas bubbles of cold seeps along the Cascadia Continental Margin. *Geochemistry, Geophysics, Geosystems*, vol. 19, No. 11, pp. 4476–4486.

Bax, Nicholas J., and others (2016). Results of efforts by the Convention on Biological Diversity to describe ecologically or biologically significant marine areas. *Conservation Biology*, vol. 30, No. 3, pp. 571–581.

Bernardino, Angelo F., and Paulo Y.G. Sumida (2017). Deep risks from offshore development. *Science*, vol. 358, No. 6361, pp. 312–312.

Boetius, Antje, and Frank Wenzhöfer (2013). Seafloor oxygen consumption fuelled by methane from cold seeps. *Nature Geoscience*, vol. 6, No. 9, pp. 725–734.

Bowden, David A., and others (2013). Cold seep epifaunal communities on the Hikurangi Margin, New Zealand: composition, succession, and vulnerability to human activities. *PLoS One*, vol. 8, No. 10, e76869.

Breitburg, Denise, and others (2018). Declining oxygen in the global ocean and coastal waters. *Science*, vol. 359, No. 6371, eaam7240.

Breusing, Corinna, and others (2016). Biophysical and population genetic models predict the presence of "phantom" stepping stones connecting Mid-Atlantic Ridge vent ecosystems. *Current Biology*, vol. 26, No. 17, pp. 2257–2267.

Chapman, Abbie S.A., and others (2019). sFDvent: A global trait database for deep-sea hydrothermal-vent fauna. *Global Ecology and Biogeography*, vol. 28, No. 11, pp. 1538–1551.

Chapman, Abbie S.A., and others (2018). Both rare and common species make unique contributions to functional diversity in an ecosystem unaffected by human activities. *Diversity and Distributions*, vol. 24, No. 5, pp. 568–578.

Chen, Chong, and others (2015). Low connectivity between "scaly-foot gastropod" (Mollusca: Peltospiridae) populations at hydrothermal vents on the Southwest Indian Ridge and the Central Indian Ridge. *Organisms Diversity & Evolution*, vol. 15, No. 4, pp. 663–670.

ChEssBase (2019). http://ipt.vliz.be/eurobis/resource?r=chessbase.

Chiba, Sanae, and others (2018). Human footprint in the abyss: 30 year records of deep-sea plastic debris. *Marine Policy*, vol. 96, pp. 204–212.

Childress, J.J., and Peter R. Girguis (2011). The metabolic demands of endosymbiotic chemoautotrophic metabolism on host physiological capacities. *Journal of Experimental Biology*, vol. 214, No. 2, pp. 312–325.

Colaço, Ana, and others (2006). Bioaccumulation of Hg, Cu, and Zn in the Azores triple junction hydrothermal vent fields food web. *Chemosphere*, vol. 65, No. 11, pp. 2260–2267.

Copley, Jonathan T.P., and others (1999). Subannual temporal variation in faunal distributions at the TAG hydrothermal mound (26° N, Mid-Atlantic Ridge). *Marine Ecology*, vol. 20, Nos. 3–4, pp. 291–306.

Copley, J.T., and others (2016). Ecology and biogeography of megafauna and macrofauna at the first known deep-sea hydrothermal vents on the ultraslow-spreading Southwest Indian Ridge. *Scientific Reports*, vol. 6, art. 39158.

Cordes, Erik E., and others (2016). Environmental impacts of the deep-water oil and gas industry: a review to guide management strategies. *Frontiers in Environmental Science*, vol. 4, art. 58.

Cordes, Erik E., and others (2009). Macro-ecology of Gulf of Mexico cold seeps. *Annual Review of Marine Science*, vol. 1, pp. 143–168.

Corinaldesi, Cinzia, and others (2012). Viral infections stimulate the metabolism and shape prokaryotic assemblages in submarine mud volcanoes. *The ISME Journal*, vol. 6, No. 6, pp. 1250–1259. https://doi.org/10.1038/ismej.2011.185.

Cuvelier, Daphne, and others (2011). Community dynamics over 14 years at the Eiffel Tower hydrothermal edifice on the Mid-Atlantic Ridge. *Limnology and Oceanography*, vol. 56, No. 5, pp. 1624–1640.

Di Carlo, Marta, and others (2017). Trace elements and arsenic speciation in tissues of tube dwelling polychaetes from hydrothermal vent ecosystems (East Pacific Rise): an ecological role as antipredatory strategy? *Marine Environmental Research*, vol. 132, pp. 1–13.

Digby, Adrian, and others (2016). Cold seeps associated with structured benthic communities: more accurate identification and evaluation using a new multibeam survey methodology in the offshore Southern Colombian Caribbean. *International Journal of Geosciences*, vol. 7, No. 5, pp. 761–774.

Dubilier, Nicole, and others (2008). Symbiotic diversity in marine animals: the art of harnessing chemosynthesis. *Nature Reviews Microbiology*, vol. 6, No. 10, p. 725.

Du Preez, Cherisse, and Charles R. Fisher (2018). Long-term stability of back-arc basin hydrothermal vents. *Frontiers in Marine Science*, vol. 5, art. 54.

Dunn, Daniel C., and others (2014). The Convention on Biological Diversity's ecologically or biologically significant areas: origins, development, and current status. *Marine Policy*, vol. 49, pp. 137–145.

Dunn, Daniel C., and others (2018). A strategy for the conservation of biodiversity on mid-ocean ridges from deep-sea mining. *Science Advances*, vol. 4, No. 7.

Etiope, Giuseppe, and others (2019). Gridded maps of geological methane emissions and their isotopic signature. *Earth System Science Data*, vol. 11, pp. 1–22.

Fisheries and Oceans Canada (DFO) (2018). *Assessment of Canadian Pacific Cold Seeps against Criteria for Determining Ecologically and Biologically Significant Areas. DFO Canadian Science Advisory Secretariat. Science Response 2018/002.*

Food and Agriculture Organization of the United Nations (FAO) (2009). *International Guidelines for the Management of Deep-Sea Fisheries in the High-Seas*. Rome.

________ (2016). *Vulnerable Marine Ecosystems: Processes and Practices in the High Seas*, Anthony Thompson and others, eds. Fisheries and Aquaculture Technical Paper 595. Rome.

________ (2018). *Deep-Ocean Climate Change Impacts on Habitat, Fish and Fisheries*, Lisa Levin, Maria Baker and Anthony Thompson, eds. FAO Fisheries and Aquaculture Technical Paper 638. Rome.

Feng, Dong, and others (2018). Cold seep systems in the South China Sea: An overview. *Journal of Asian Earth Sciences*, vol. 168, pp. 3–16.

Fischer, David, and others (2012). Interaction between hydrocarbon seepage, chemosynthetic communities, and bottom water redox at cold seeps of the Makran accretionary prism: insights from habitat-specific pore water sampling and modeling. *Biogeosciences*, vol. 9, No. 6, pp. 2013–2031.

Fisher, Charles R., and others (2016). How Did the Deepwater Horizon Oil Spill Impact Deep-Sea Ecosystems?, *Oceanography*, vol. 29, No. 3, pp. 182–195.

Gehlen, Marion, and others (2014). Projected pH reductions by 2100 might put deep North Atlantic biodiversity at risk. *Biogeosciences*, vol. 11, pp. 6955–6967.

General Fisheries Commission for the Mediterranean of the Food and Agriculture Organization of the United Nations (GFCM-FAO) (2018). *Forty-Second Session of the Commission. Final Report English (before editing). FAO Headquarters, Rome, Italy, 22–26 October 2018.*

Gerdes, Klaas, and others (2019). Detailed mapping of hydrothermal vent fauna: a 3D reconstruction approach based on video imagery. *Frontiers in Marine Science*, vol. 6, art. 96.

German, Christopher R., and others (2016). Hydrothermal impacts on trace element and isotope ocean biogeochemistry. *Philosophical Transactions of the Royal Society A: Mathematical, Physical and Engineering Sciences*, vol. 374, No. 2081, 20160035.

Goffredi, Shana K., and others (2017). Hydrothermal vent fields discovered in the southern Gulf of California clarify role of habitat in augmenting regional diversity. *Proceedings of the Royal Society B: Biological Sciences*, vol. 284, No. 1859, 20170817.

Gollner, Sabine, and others (2017). Resilience of benthic deep-sea fauna to mining activities. *Marine Environmental Research*, vol. 129, pp. 76–101.

Govenar, Breea (2010). Shaping vent and seep communities: habitat provision and modification by foundation species. In *The Vent and Seep Biota*, pp. 403–432. Springer.

Grupe, Benjamin M., and others (2015). Methane seep ecosystem functions and services from a recently discovered southern California seep. *Marine Ecology*, vol. 36, pp. 91–108.

Guieu, Cécile, and others (2018). Iron from a submarine source impacts the productive layer of the Western Tropical South Pacific (WTSP). *Scientific Reports*, vol. 8, No. 1, art. 9075.

Hall-Spencer, Jason M., and others (2008). Volcanic carbon dioxide vents show ecosystem effects of ocean acidification. *Nature*, vol. 454, No. 7200, pp. 96–99.

Hautala, Susan L., and others (2014). Dissociation of Cascadia margin gas hydrates in response to contemporary ocean warming. *Geophysical Research Letters*, vol. 41, No. 23, pp. 8486–8494.

Intergovernmental Oceanographic Commission of the United Nations Educational, Scientific and Cultural Organization (UNESCO-IOC) (2016). *UNESCO-IOC Capacity Building Strategy 2015-2021. Paris. IOC/INF-1332.*

Intergovernmental Panel on Climate Change (IPCC) (2019). Summary for Policymakers. In *IPCC Special Report on the Ocean and Cryosphere in a Changing Climate*, H-O. Pörtner and others, eds.

James, Rachael H., and others (2016). Effects of climate change on methane emissions from seafloor sediments in the Arctic Ocean: a review. *Limnology and Oceanography*, vol. 61, No. S1, pp. S283–S299.

Johnson, Gregory C., and Sarah G. Purkey (2009). Deep Caribbean Sea warming. *Deep Sea Research Part I: Oceanographic Research Papers*, vol. 56, No. 5, pp. 827–834.

Ketzer, Marcelo, and others (2019). Gas seeps at the edge of the gas hydrate stability zone on Brazil's continental margin. *Geosciences*, vol. 9, No. 5, art. 193.

Le Bris, Nadine, and others (2019). Hydrothermal energy transfer and organic carbon production at the deep seafloor. *Frontiers in Marine Science*, vol. 5.

Levin, Lisa A., and others (2016). Hydrothermal vents and methane seeps: rethinking the sphere of influence. *Frontiers in Marine Science*, vol. 3, art. 72.

Levin, Lisa A., and others (2019). Global observing needs in the deep ocean. *Frontiers in Marine Science*, vol. 6, art. 241.

Levin, Lisa A., and Myriam Sibuet (2012). Understanding continental margin biodiversity: a new imperative. *Annual Review of Marine Science*, vol. 4, pp. 79–112.

Levin, Lisa A., and Nadine Le Bris (2015). The deep ocean under climate change. *Science*, vol. 350, No. 6262, pp. 766–768.

Li, Jin-fa, and others (2018). The first offshore natural gas hydrate production test in South China Sea. *China Geology*, vol. 1, No. 1, pp. 5–16.

Linse, Katrin, and others (2019). Fauna of the Kemp Caldera and its upper bathyal hydrothermal vents (South Sandwich Arc, Antarctica). *Royal Society Open Science*, vol. 6, No. 11, 191501.

MacDonald, Ian R., and others (2020). The Asphalt Ecosystem of the Southern Gulf of Mexico: Abyssal Habitats Across Space and Time. In: *Scenarios and Responses to Future Deep Oil Spills*, pp. 132–146. Springer.

Marques, Ana F., and others (2020). The Seven Sisters Hydrothermal System: first record of shallow hybrid mineralization hosted in mafic volcaniclasts on the Arctic mid-ocean ridge. *Minerals*, vol. 10, No. 5, art. 439. https://doi.org/10.3390/min10050439.

Marlow, Jeffrey J., and others (2014). Carbonate-hosted methanotrophy represents an unrecognized methane sink in the deep sea. *Nature Communications*, vol. 5, art. 5094.

Mazumdar, A., and others (2019). The first record of active methane (cold) seep ecosystem associated with shallow methane hydrate from the Indian EEZ. *Journal of Earth System Science*, vol. 128, No. 1, art. 18.

Menini, Elisabetta, and Cindy Lee Van Dover (2019). An atlas of protected hydrothermal vents. *Marine Policy*, vol. 108, art. 103654.

Mitarai, Satoshi, and others (2016). Quantifying dispersal from hydrothermal vent fields in the western Pacific Ocean. *Proceedings of the National Academy of Sciences*, vol. 113, No. 11, pp. 2976–2981.

Moalic, Yann, and others (2012). Biogeography revisited with network theory: retracing the history of hydrothermal vent communities. *Systematic Biology*, vol. 61, No. 1, p. 127.

Mullineaux, Lauren S., and others (2018). Exploring the ecology of deep-sea hydrothermal vents in a metacommunity framework. *Frontiers in Marine Science*, vol. 5, art. 49.

Nakajima, Ryota, and others (2015). Post-drilling changes in seabed landscape and megabenthos in a deep-sea hydrothermal system, the Iheya North field, Okinawa Trough. *PLoS One*, vol. 10, No. 4, e0123095.

Okamoto, Nobuyuki, and others (2019). World's first lifting test for seafloor massive sulphides in the Okinawa Trough in the EEZ of Japan. In *The 29th International Ocean and Polar Engineering Conference*. International Society of Offshore and Polar Engineers.

Olu, Karine, and others (2010). Biogeography and potential exchanges among the Atlantic equatorial belt cold-seep faunas. *PloS One*, vol. 5, No. 8, e11967.

Orcutt, Beth N., and others (2011). Microbial ecology of the dark ocean above, at, and below the seafloor. *Microbiology and Molecular Biology Reviews*, vol. 75, No. 2, pp. 361–422.

Ortmann, Alice C., and Curtis A. Suttle (2005). High abundances of viruses in a deep-sea hydrothermal vent system indicates viral mediated microbial mortality. *Deep Sea Research Part I: Oceanographic Research Papers*, vol. 52, No. 8, pp. 1515–27. https://doi.org/10.1016/j.dsr.2005.04.002.

OSPAR Commission for the Protection of the Marine Environment of the North-East Atlantic (OSPAR) (2014). *Recommendation 2014/11 on Furthering the Protection and Conservation of Hydrothermal Vents/Fields Occurring on Oceanic Ridges in Region V of the OSPAR Maritime Area*. OSPAR 14/21/1, Annex 16.

________ (2018). *Status Report on the OSPAR Network of Marine Protected Areas.*

Papot, Claire, and others (2017). Antagonistic evolution of an antibiotic and its molecular chaperone: how to maintain a vital ectosymbiosis in a highly fluctuating habitat. *Scientific Reports*, vol. 7, No. 1, art. 1454.

Petersen, Sven, and others (2016). News from the seabed – geological characteristics and resource potential of deep-sea mineral resources. *Marine Policy*, vol. 70, pp. 175–187.

QGIS Development Team (2018). QGIS Geographic Information System. Open Source Geospatial Foundation Project. http://qgis.osgeo.org.

Quattrini, Andrea M., and others (2015). Exploration of the canyon-incised continental margin of the northeastern United States reveals dynamic habitats and diverse communities. *PLoS One*, vol. 10, No. 10, e0139904.

Ramirez-Llodra, Eva, and others (2011). Man and the last great wilderness: human impact on the deep sea. *PLoS One*, vol. 6, No. 8, e22588.

Resing, Joseph A., and others (2015). Basin-scale transport of hydrothermal dissolved metals across the South Pacific Ocean. *Nature*, vol. 523, No. 7559, pp. 200–203.

Riboulot, Vincent, and others (2018). Freshwater lake to salt-water sea causing widespread hydrate dissociation in the Black Sea. *Nature Communications*, vol. 9, No. 1, art. 117.

Rogers, Alex D., and others (2012). The discovery of new deep-sea hydrothermal vent communities in the Southern Ocean and implications for biogeography. *PLoS Biology*, vol. 10, No. 1, e1001234.

Römer, Miriam, and others (2014a). First evidence of widespread active methane seepage in the Southern Ocean, off the sub-Antarctic island of South Georgia. *Earth and Planetary Science Letters*, vol. 403, pp. 166–177.

Römer, Miriam, and others (2014b). Methane fluxes and carbonate deposits at a cold seep area of the Central Nile Deep Sea Fan, Eastern Mediterranean Sea. *Marine Geology*, vol. 347, pp. 27–42.

Rossi, Giulia S., and Verena Tunnicliffe (2017). Trade-offs in a high CO_2 habitat on a subsea volcano: condition and reproductive features of a bathymodioline mussel. *Marine Ecology Progress Series*, vol. 574, pp. 49–64.

Rubin-Blum, and others (2019). Genetic Evidence for Two Carbon Fixation Pathways (the Calvin-Benson-Bassham Cycle and the Reverse Tricarboxylic Acid Cycle) in Symbiotic and Free-Living Bacteria. *MSphere*, vol. 4, No. 1, e00394–18.

Ruppel, Carolyn D., and John D. Kessler (2017). The interaction of climate change and methane hydrates. *Reviews of Geophysics*, vol. 55, No. 1, pp. 126–168.

Samadi, Sarah, and others (2015). Patchiness of deep-sea communities in Papua New Guinea and potential susceptibility to anthropogenic disturbances illustrated by seep organisms. *Marine Ecology*, vol. 36, pp. 109–132.

Scott, Kathleen M., and others (2018). Diversity in CO_2-concentrating mechanisms among chemolithoautotrophs from the genera *Hydrogenovibrio*, *Thiomicrorhabdus*, and *Thiomicrospira*, ubiquitous in sulfidic habitats worldwide. *Applied and Environmental Microbiology*, vol. 85, No. 3.

Seabrook, Sarah, and others (2019). Heterogeneity of methane seep biomes in the Northeast Pacific. *Deep Sea Research Part II: Topical Studies in Oceanography*, vol. 150, pp. 195–209.

Sigwart, Julia D., and others (2019). Red Listing can protect deep-sea biodiversity. *Nature Ecology & Evolution*, vol. 3, No. 8, p. 1134.

Suzuki, Kenta, and others (2018). Mapping the resilience of chemosynthetic communities in hydrothermal vent fields. *Scientific Reports*, vol. 8, No. 1, art. 9364.

Sweetman Andrew K., and others (2017). Major impacts of climate change on deep-sea benthic ecosystems. *Elementa Science of the Anthropocene*, vol. 5, No. 4, art. 203.

Tagliabue, Alessandro, and Joseph Resing (2016). Impact of hydrothermalism on the ocean iron cycle. *Philosophical Transactions of the Royal Society A: Mathematical, Physical and Engineering Sciences*, vol. 374, 20150291.

Tasiemski, Aurélie, and others (2014). Characterization and function of the first antibiotic isolated from a vent organism: the extremophile metazoan *Alvinella pompejana*. *PLoS One*, vol. 9, No. 4, e95737.

Thaler, Andrew D., and Diva Amon (2019). 262 Voyages Beneath the Sea: a global assessment of macro-and megafaunal biodiversity and research effort at deep-sea hydrothermal vents. *PeerJ*, vol. 7, e7397.

Thornton, Blair, and others (2016). Biometric assessment of deep-sea vent megabenthic communities using multi-resolution 3D image reconstructions. *Deep Sea Research Part I: Oceanographic Research Papers*, vol. 116, pp. 200–219.

Thurber, Andrew R., and others (2014). Ecosystem function and services provided by the deep sea. *Biogeosciences*, vol. 11, No. 14, pp. 3941–3963.

Tunnicliffe, Verena, and others (2009). Survival of mussels in extremely acidic waters on a submarine volcano. *Nature Geoscience*, vol. 2, No. 5, p. 344.

United Nations (2017a). Chapter 45: Hydrothermal vents and cold seeps. In *The First Global Integrated Marine Assessment: World Ocean Assessment I*. Cambridge: Cambridge University Press.

________(2017b). *The First Global Integrated Marine Assessment: World Ocean Assessment I*. Cambridge: Cambridge University Press.

United Nations Environment Programme, Mediterranean Action Plan (2017). *Draft Guidelines for Inventoring and Monitoring of Dark Habitats*. UNEP(DEPI)/MED WG. 431/Inf.12. Nairobi: UNEP.

United Nations Environment Programme World Conservation Monitoring Centre and International Union for Conservation of Nature and Natural Resources (2019). Protected Planet: The World Database on Protected Areas, Cambridge, United Kingdom. www.protectedplanet.net.

United States Pacific Fishery Management Council (2019). *Pacific Coast Groundfish Fishery Management Plan For The California, Oregon, And Washington Groundfish Fishery. Appendix B Part 2. Groundfish Essential Fish Habitat And Life History Descriptions, Habitat Use Database Description, and Habitat Suitability Probability Information.* www.pcouncil.org/wp-content/uploads/2019/06/Appendix-B2-FINAL-Am28.pdf.

Van Dover, Cindy Lee (2019). Inactive sulfide ecosystems in the deep sea: a review. *Frontiers in Marine Science*, vol. 6, art. 461.

Vetriani, Costantino, and others (2005). Mercury adaptation among bacteria from a deep-sea hydrothermal vent. *Applied and Environmental Microbiology*, vol. 71, No. 1, pp. 220–226.

Wankel, Scott D., and others (2011). Influence of subsurface biosphere on geochemical fluxes from diffuse hydrothermal fluids. *Nature Geoscience*, vol. 4, No. 7, p. 461.

Watanabe, Hiromi, and others (2010). Japan: vents and seeps in close proximity. In *The Vent and Seep Biota*, pp. 379–401. Springer.

Yahagi, Takuya, and others (2017). Do larvae from deep-sea hydrothermal vents disperse in surface waters? *Ecology*, vol. 98, No. 6, pp. 1524–1534.

Zhou, Yadong, and others (2018). Characterization of vent fauna at three hydrothermal vent fields on the Southwest Indian Ridge: implications for biogeography and interannual dynamics on ultraslow-spreading ridges. *Deep Sea Research Part I: Oceanographic Research Papers*, vol. 137, pp. 1–12.

Chapter 7Q
Sargasso Sea

Contributors: Howard S.J. Roe (convener), David Freestone, Laurence Kell, Brian E. Luckhurst and Tammy Warren.

Keynote points

- The Sargasso Sea is an area of the high seas that is internationally recognized as a fundamentally important part of the global ocean because of its role in climate regulation and its unique ecosystems.
- The Bermuda Atlantic Time-series Study continues to collect observations enabling inferences on the impact of climate change in the ocean and increased understanding of ocean processes. Continuation of that fundamental long-term research is essential.
- Mass blooms and strandings of *Sargassum* since 2011 are attributable to a previously rare form of *Sargassum natans*. They are causing major socioeconomic problems for the region and may also adversely affect unique oceanic *Sargassum* communities.
- The importance of the Sargasso Sea as a spawning area for both the European eel (*Anguilla anguilla*) and the American eel (*Anguilla rostrata*) has been emphasized by satellite tracking of adults and widespread larval surveys. An increased understanding of the ecology of commercial tuna and tuna-like species and awareness of the use of the area by endangered and threatened species is increasing the need for ecosystem-based fisheries management.
- Most changes and threats, including climate change, overfishing of eels, plastic pollution and mass blooms of *Sargassum*, are externally driven. Those increasing threats will adversely affect the contribution of the Sargasso Sea to Sustainable Development Goal 14 to conserve and sustainably use the oceans, seas and marine resources for sustainable development and, therefore, other Goals.[1]
- The increasing activity in the Sargasso Sea demonstrates the importance of addressing the cumulative impacts of human activities on the high seas.

1. Introduction

The present chapter builds on developments and knowledge of the previous baseline state, as reported in chapter 50 of the first *World Ocean Assessment* (United Nations, 2017). Continuing research around the ocean time series hosted by the Bermuda Institute of Ocean Sciences underpins the understanding of fundamental ocean processes, including the importance of microbes and the effects of climate change. Progress in understanding the broader ecology of the Sargasso Sea is described, in particular: *Sargassum* weed, its distribution, associated fauna and the real and potential impact of recent blooms on coastal communities; ongoing research into the life cycle of the European eel (*Anguilla Anguilla*) and the American eel (*Anguilla rostrata*); increased awareness of the biology of some commercial fish species and ongoing developments in ecosystem modelling; and increased threats from plastic pollution. Ongoing and recent international developments pertaining to the Sargasso Sea are outlined in the light of the ongoing intergovernmental conference on an international legally binding instrument under the United Nations Convention on the Law of the Sea[2] on the conservation and sustainable use of marine biological diversity of areas beyond national jurisdiction.

1 See General Assembly resolution 70/1.
2 United Nations, *Treaty Series*, vol. 1833, No. 31363.

The 2016 baseline case contained a description of the background oceanography of the Sargasso Sea, the unique surface ecosystem and communities based upon floating aggregations of two species of *Sargassum* and their role as feeding and nursery areas for fishes, juvenile turtles and seabirds. Many animals migrate through the Sargasso Sea and many migrate to it to breed. It is the only known spawning area for the European eel (*Anguilla Anguilla*) and the American eel (*Anguilla rostrata*). Many of the species inhabiting the Sargasso Sea are endangered or threatened and are listed as such in the International Union for Conservation of Nature Red List of Threatened Species, the Convention on International Trade in Endangered Species of Wild Fauna and Flora,[3] and the Protocol concerning Specially Protected Areas and Wildlife to the Convention for the Protection and Development of the Marine Environment of the Wider Caribbean Region[4] (Laffoley and others, 2011). Threats, economic values and conservation responses were summarized.

There have been changes and developments to the baseline case, but the Sargasso Sea remains a fundamentally important part of the global ocean owing to an interdependent mix of physical oceanography, its ecosystems and its role in global-scale ocean and Earth-system processes. It contributes significantly to local as well as global economies both directly from fisheries for highly migratory species (including the European eel and the American eel), coral reefs, whale watching and "turtle tourism", and indirectly from its role in climate regulation, conservation of genetic diversity and nutrient cycling (Laffoley and others, 2011; Pendleton and others, 2015). On the other hand, it is also threatened by climate change, pollution, increased fishing activities and increased shipping.

2. Change of state

2.1. Ocean time series

The continuing importance of long-term ocean time series in understanding variability in the ocean and ocean processes has been reinforced both locally by the results from Hydrostation S and the Bermuda Atlantic Time-series Study, and globally through numerous reviews (Neuer and others, 2017; O'Brien and others, 2017). The Bermuda Atlantic Time-series Study programme is one of the few ocean time series with long enough records to enable anthropogenic change to be distinguished from natural variability (Henson and others, 2016). The breadth of research utilizing those data is summarized on the Bermuda Atlantic Time-series Study website.[5]

2.2. *Sargassum*

The baseline case contained a description of the role of two species, namely, *Sargassum natans* and *Sargassum fluitans*, primarily for their role in hosting specialized communities of animals and acting as nursery and feeding areas. Advances in the knowledge of those communities have implications for future conservation measures. The communities vary in both time and space. Considerable variability over a 40-year period, and also between samples taken a year apart, was found by Huffard and others (2014). The reasons are unknown, but increasing ocean acidity may be the cause of the reduction in calcareous epibionts, such as bryozoans. Variability at the molecular level

3 United Nations, *Treaty Series*, vol. 993, No. 14537.
4 Ibid., vol. 2180, No.25974.
5 See http://bats.bios.edu.

within species occurs in the widely dispersed slender Sargassum shrimp (*Latreutes fucorum*), and it is suggested that conservation measures for such species should cover large areas or have networks of protected areas (Sehein and others, 2014).

Since the mass stranding of thousands of tons of *Sargassum* on beaches in the Caribbean and the Gulf of Mexico and on the coasts of West Africa and South America in 2011, there have been considerable efforts to identify the blooms, their causes and their movements using satellite tracking, modelling, direct sampling at sea and a combination of different techniques (Schell and others, 2015; Franks and others, 2016; Djakouré and others, 2017; Brooks and others, 2018; Putnam and others, 2018). The blooms were identified as a previously rare form of *Sargassum* (*S. natans* VIII) by Schell and others (2015). The identification was subsequently confirmed by genetic studies (Amaral-Zettler and others, 2017). *S. natans* VIII was described from the Caribbean by Parr (1939), but it was then largely forgotten. It differs morphologically from both *S. fluitans* and *S. natans* and hosts reduced communities of animals, which, in turn, make it less attractive to fishes, turtles and seabirds, which feed on or beneath the *Sargassum* mats (Martin, 2016). Consequently, changes in *Sargassum* type or distribution could affect species diversity and abundance. The distributions of the different species and forms of *Sargassum* differ both spatially and temporally, and *S. natans* VIII is believed to be limited by temperature, since it is most abundant in warm water to the south of the Sargasso Sea and the Caribbean. It is rare further north, but it has been found off Bermuda since 2016 (Clover, 2017). So far, the blooms have not affected the Sargasso Sea directly, but they have the potential to do so through reduced *Sargassum* communities and because they are preventing the successful nesting of turtles on the affected beaches around the Caribbean.

The blooms originate in the North Equatorial Recirculation Region south of the Sargasso Sea and, from there, are carried into the Caribbean (Johnson and others, 2013; Franks and others, 2016; Djakouré and others, 2017; Putnam and others, 2018) Such blooms have been an annual event since they were first observed in 2011. High levels of dead *Sargassum* that has sunk from the surface have also been reported on the sea floor in the Vema Fracture Zone beneath the North Equatorial Recirculation Region, potentially providing a food source to deep-sea benthic ecosystems (Baker and others, 2018). The causes of the blooms are the subject of ongoing research and may include modifications induced by climate change, such as increased temperature and changes in ocean currents, enhanced nutrient levels originating from the Congo, Orinoco and especially Amazon Rivers, equatorial upwelling and dust from the Sahara (Djakouré and others, 2017). The question is whether the regime shift in the tropical and subtropical Atlantic is primarily caused by human activity. Various monitoring satellites, which feed information to the Sargassum Watch System, for example, inform communities on the location of blooms and warn them of potential beaching events (Hu and others, 2016).

2.3. Fishes

The importance of the Sargasso Sea to the European eel and the American eel has been reinforced. The larvae of both species were known to occur in the south-west of the Sargasso Sea in the vicinity of the seasonal subtropical convergence (Munk and others, 2010; Miller and others, 2015). Satellite tagging tracked migrating European eels from European rivers as far as the Azores (Righton and others, 2016). Similar tagging of American eels showed migration from Canada to the Sargasso Sea (Béguer-Pons and others, 2015). More recently, it has been shown that European eels spawn across a 2,000 km swathe of the southern Sargasso

Sea in an area bounded by temperature fronts (Miller and others, 2019). The wide spawning area may reflect different starting times of migrations, different swimming abilities or larval drift in ocean currents. Recruitment levels to fisheries for both species have collapsed, and the reduction is matched by declines in numbers of eel larvae in the Sargasso Sea (Hanel and others, 2014). Climate change, rising sea temperatures, changes in ocean currents and the North Atlantic Oscillation all potentially adversely affect the marine life cycle of eels (Miller and others, 2016).

New information on food webs and spawning sites of tuna and tuna-like species managed by the International Commission for the Conservation of Atlantic Tunas has reinforced the importance of the Sargasso Sea in providing habitat, foraging and spawning grounds, and migratory corridors for those species (Luckhurst, 2015a; Luckhurst and Arocha, 2016; Anonymous, 2016). The Northwest Atlantic Fisheries Organization has acted to protect seamounts in the northern Sargasso Sea by closing the area to bottom trawling until 2020 (Northwest Atlantic Fisheries Organization (NAFO), 2015).

Despite decreases in catches worldwide and in the relative abundance of the main commercial pelagic species, the capacity of the global fishing fleet has continued to increase (Rousseau and others, 2019). Spatial estimates of fishing effort are not available from the Commission for the Sargasso Sea. In their absence, the percentage of the catches of the main Commission species (stock boundaries defined by the Commission) has been plotted as an indicator of the level of fishing in the Sargasso Sea over time.

The figure below contains an updated analysis of Luckhurst (2015b), which provided catch analyses from 1992 to 2011 for the principal Commission species, with the latest catches reported by the Commission.[6] The latest catches reported are up to and including 2017; more recent data are not available owing to delays in reporting to and processing by the Commission. Up to a maximum of 12 per cent of North Atlantic albacore and 10 per cent of West Atlantic bluefin catches are taken in the region. Catches of tropical tunas (bigeye, yellowfin and skipjack) and billfishes (swordfishes, sailfishes, blue marlin and white marlin) are smaller but still significant. The proportion of the catches taken from the Sargasso Sea show considerable variability over time, potentially owing to changes in targeting by the longline fleets, but effort data to evaluate that are not available from the Commission. The longline fleet is shown because of the ecological importance of by-catch species, such as billfishes, sharks, seabirds and sea turtles. To move towards ecosystem-based fisheries management in the Sargasso Sea, it will be important to understand the spatial overlap between fishing effort and the behaviour of non-target species, including the migration routes, aggregating behaviour and habitat use of all species that use it (Kell and Luckhurst, 2018; Boerder and others, 2019).

2.4. Plastic pollution

Plastic pollution in the Sargasso Sea was first observed in 1972 (Carpenter and Smith, 1972). Concentrations of microplastics in the surface of the Sea were found to be orders of magnitude greater than previously recorded, with the greatest concentration in the subtropical convergence zone (Law and others, 2010). The accumulation of surface particles in the zone was forecast by models used to guide sampling by the "7th Continent" expedition in 2014. It was estimated that the North Atlantic subtropical gyre, that is, the Sargasso Sea, had about 56,000 tons of floating plastic in 2014 (Eriksen and others, 2014). Presumably, more

6 See www.iccat.int/en/accesingdb.html.

is present today. There have been extensive reviews of plastic pollution and its effects in the global ocean and in ocean gyres (Joint Group of Experts on the Scientific Aspects of Marine Environmental Protection (GESAMP), 2014, 2016; Law, 2017; Eriksen and others, 2016). More recently, laboratory experiments have found adverse effects of plastic leachates on the photosynthetic bacterium *Prochlorococcus* (Tetu and others, 2019). *Prochlorococcus* produces up to 20 per cent of atmospheric oxygen. If the results are confirmed in situ, plastic pollution poses a threat to global oxygen production by marine bacteria. The adverse impacts described in the various reviews will apply to the Sargasso Sea, and the concentrating effects of the ocean gyre and the subtropical convergence trap plastic within mats of *Sargassum*, which makes the Sargasso Sea particularly vulnerable.

Time series showing the percentage of the total and longline catches that come from the Sargasso Sea region

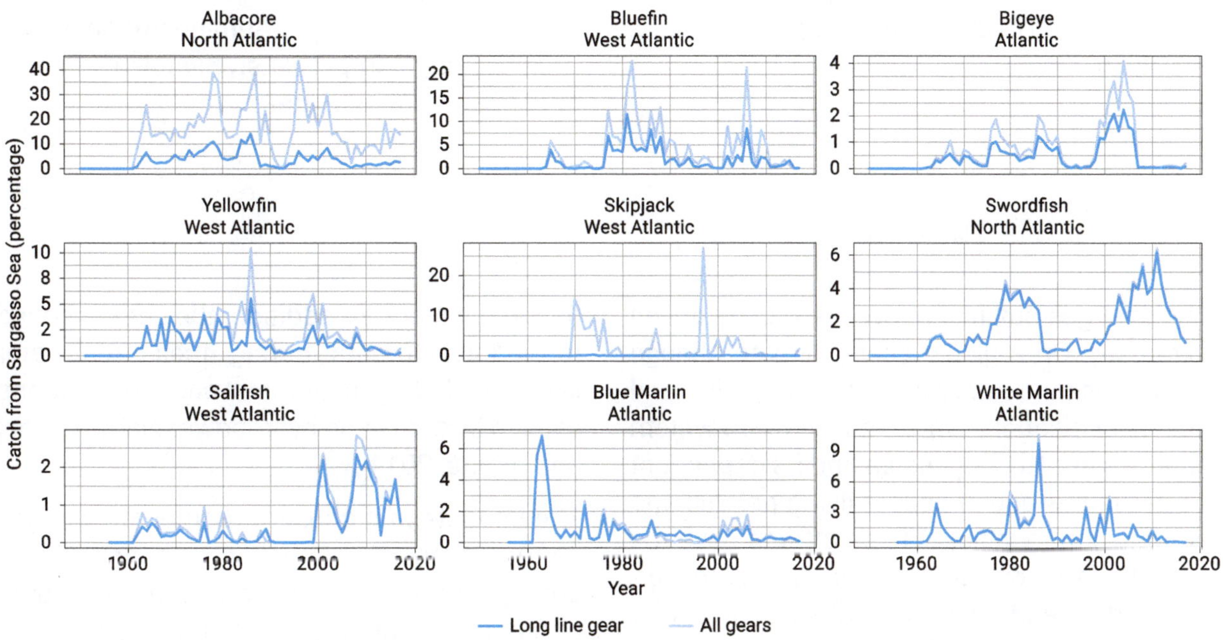

Source: Data are from the International Commission for the Conservation of Atlantic Tunas catch-at-size database (CATDIS).

3. Institutional arrangements

One of the major challenges facing the Sargasso Sea is a legal one. The Sargasso Sea falls within the high seas – the 50 per cent of the planet that is outside national jurisdiction (Freestone, 2015). To address that challenge, five Governments came together in 2014 to sign the Hamilton Declaration on Collaboration for the Conservation of the Sargasso Sea and to establish the Sargasso Sea Commission to act as steward for the extraordinary area (Freestone and Morrison, 2014). Five more Governments have since joined and others may follow (Sargasso Sea Commission, 2018).

The Sargasso Sea Commission is based on a new paradigm for the conservation of areas beyond national jurisdiction, convening stakeholders from multiple countries and organizations to address issues that fall outside national agendas. Parties to the Convention on Biological Diversity[7] have agreed that the Sargasso Sea be included on a list of ecologically or biologically significant areas (Secretariat of the Convention on Biological Diversity, 2012). Using that as a basis, in 2015, the Northwest Atlantic Fisheries Organization agreed conservation measures by declaring a moratorium on bottom trawling on Sargasso Sea seamounts in the Organization's area, together with gear restrictions on midwater trawling (NAFO, 2015; Diz, 2016).

The Commission is working to protect the Sargasso Sea alongside a number of Governments and partners. In collaboration with the secretariat of and the range States parties to the Convention on the Conservation of Migratory Species of Wild Animals,[8] the Commission is working to protect the migratory range of the European eel through the Sargasso Sea. It is also exploring ways to regulate impacts of vessel activities and to work with the International Commission for the Conservation of Atlantic Tunas to use the Sargasso Sea as a pilot project on the ecosystem approach to fisheries management (Kell and Luckhurst, 2018), and it is working with the United States National Aeronautics and Space Administration, which is developing comprehensive satellite imagery of the Sargasso Sea area.

4. Consequences of changes

The changes outlined above are mostly driven externally. On a global scale, climate change affects ocean temperature, ocean acidity and ocean circulation, which causes ecosystem changes in both *Sargassum* and its dependent communities and in deeper living pelagic and benthic communities. Those effects have the potential to adversely affect spawning, larval feeding and migrations of eels and other fishes. Concurrent with the overall warming of the global ocean is an increase in frequency of global marine heatwaves, which adversely affect biodiversity and threaten to disrupt ecosystem services in certain areas of the ocean (Smale and others, 2019). The southern Sargasso Sea has been identified as an area that has been significantly affected. Most of the pollution, including plastic, comes from land and is concentrated by ocean currents in the Sargasso Sea. Eel populations are affected by overfishing in exclusive economic zones and national waters. In addition, they are exposed to various threats during their freshwater stage, including pollutants and obstructions caused by dams and hydropower plants (Hanel and others, 2019).

The impacts of global environmental changes on the oceans, future predictions for fisheries and governance issues are summarized in a report by the Nippon Foundation-Nereus Programme (2015), and ocean issues related to the Sustainable Development Goals are summarized in a second report by the Nippon Foundation-Nereus Programme (2017). Monitoring changes in ocean temperature and chemistry and understanding the impact of those changes on ecosystems are directly related to Goal 13 on climate action and to Goal 14. The ongoing time series stations off Bermuda are central to that global monitoring (Neuer and others, 2017). Mass strandings

7 United Nations, *Treaty Series*, vol. 1760, No. 30619.
8 Ibid., vol. 1651, No. 28395.

of *Sargassum* on beaches cause widespread socioeconomic problems to local communities, adversely affecting tourism, fishing and health, and killing biota, including turtles and fish. The costs of cleaning up beaches run into millions of dollars, and affected countries are developing management plans and technologies to minimize impacts and seek potential uses for the *Sargassum* (Milledge and Harvey, 2016; Wabnitz and others, 2019). Because of the widespread impacts to both humans and to local and ocean ecology, the blooms directly affect all Goals. The American eel and the European eel support valuable fisheries in many countries on both sides of the Atlantic, as well as lucrative aquaculture operations in Asia, but the populations of both species have crashed in recent years (Hanel and others, 2019; Atlantic States Marine Fisheries Commission, 2018). The causes are many and varied, and the ecological and socioeconomic changes will also affect all Goals.

Changes in trophic webs brought on by a warming ocean and increased acidity may significantly affect populations of top predators, such as highly migratory tunas and swordfishes (Fernandes and others, 2013). Changes caused by the weakening of the Atlantic Ocean's overturning circulation may result in shifts in species distributions (Caesar and others, 2018). Ocean warming, ocean acidification and deoxygenation, combined with other stresses, could change the primary productivity, growth and distribution of fish populations (Barange and others, 2018). That, in turn, will result in changes in the potential yield of exploited marine species and the associated economic and social benefits that they provide (Gattuso and others, 2015). Those impacts will, in turn, affect all Goals. Finally, the impacts and potential impacts of the rising amounts of plastic in the oceans are well documented (e.g., Beaumont and others, 2019) and will affect all Goals.

5. Outlook

The outlook for the Sargasso Sea, in both the short and long term, depends upon international decisions, priorities and cooperation. The importance of the Sargasso Sea is recognized internationally, and because it is in the high seas, its protection falls within the competence of a number of organizations. The remoteness and size of the Sea mean that, in open ocean terms, it remains relatively pristine despite the concentrating effects of its rotating currents. However, its integrity is threatened both by the changes outlined above and by others, including the increased fishing activity over the past three years by some 28 countries that has been estimated using automatic identification system data (Sargasso Sea Commission, 2019), and by increased shipping activities through the region. The development of deep-sea mining in areas adjacent to the Sargasso Sea poses new threats (Dunn and others, 2018). The Sargasso Sea demonstrates the challenges faced by existing sectoral bodies to govern a high seas ecosystem in a holistic manner.

References

Amaral-Zettler, Linda A., and others (2017). Comparative mitochondrial and chloroplast genomics of a genetically distinct form of *Sargassum* contributing to recent "Golden Tides" in the Western Atlantic. *Ecology and Evolution*, vol. 7, No. 2, pp. 516–525.

Anonymous (2016). An assessment of the ecological importance of the Sargasso Sea to tuna and tuna-like species and ecologically associated species. *ICCAT Collective Volume of Scientific Papers*, vol. 72, No. 28, pp. 2007–2015.

Atlantic States Marine Fisheries Commission (2018). American Eel. www.asmfc.org/species/american-eel.

Baker, Philip, and others (2018). Potential contribution of surface-dwelling *Sargassum* algae to deep-sea ecosystems in the southern North Atlantic. *Deep Sea Research Part II: Topical Studies in Oceanography*, vol. 148, pp. 21–34.

Barange, Manuel, and others (2018). Impacts of climate change on fisheries and aquaculture. *Synthesis of Current Knowledge, Adaptation and Mitigation Options. Rome: Food and Agriculture Organization of the United Nations.*

Beaumont, Nicola J., and others (2019). Global ecological, social and economic impacts of marine plastic. *Marine Pollution Bulletin*, vol. 142, pp. 189–195.

Béguer-Pon, Mélanie, and others (2015). Direct observations of American eels migrating across the continental shelf to the Sargasso Sea. *Nature Communications*, vol. 6, art. 8705.

Bermuda Atlantic Time-series Study (BATS) (n.d.). Accessed 8 November 2019. http://bats.bios.edu.

Boerder, Kristina, and others (2019). Not all who wander are lost: improving spatial protection for large pelagic fishes. *Marine Policy*, vol. 105, pp. 80–90.

Brooks, Maureen T., and others (2018). Factors controlling the seasonal distribution of pelagic *Sargassum*. *Marine Ecology Progress Series*, vol. 599, pp. 1–18.

Caesar, Levke, and others (2018). Observed fingerprint of a weakening Atlantic Ocean overturning circulation. *Nature*, vol. 556, No. 7700, p. 191.

Carpenter, Edward J., and K.L. Smith (1972). Plastics on the Sargasso Sea surface. *Science*, vol. 175, No. 4027, pp. 1240–1241.

Clover, Charles (2017). *Sargassum* is weird stuff – and it gets weirder. Blue Marine Foundation. 16 May 2017. www.bluemarinefoundation.com/2017/05/16/sargassum-is-weird-stuff-and-it-gets-weirder.

Diz, Daniela (2016). The Sargasso Sea. *International Journal of Marine and Coastal Law*, vol. 31, No. 2, pp. 359–370.

Djakouré, Sandrine, and others (2017). On the potential causes of the recent Pelagic *Sargassum* blooms events in the tropical North Atlantic Ocean. *Biogeosciences*.

Dunn, Daniel C., and others (2018). A strategy for the conservation of biodiversity on mid-ocean ridges from deep-sea mining. *Science Advances*, vol. 4, No. 7, eaar4313.

Eriksen, Marcus, and others (2014). Plastic pollution in the world's oceans: more than 5 trillion plastic pieces weighing over 250,000 tons afloat at sea. *PloS One*, vol. 9, No.12, e111913.

Eriksen, Marcus, and others (2016). Nature of plastic marine pollution in the subtropical gyres. In *Hazardous Chemicals Associated with Plastics in the Marine Environment*, pp. 135–162. Springer.

Fernandes, Jose A., and others (2013). Modelling the effects of climate change on the distribution and production of marine fishes: accounting for trophic interactions in a dynamic bioclimate envelope model. *Global Change Biology*, vol. 19, No. 8, pp. 2596–2607.

Franks, James S., and others (2016). Pelagic *Sargassum* in the tropical North Atlantic. *Gulf and Caribbean Research*, vol. 27, No. 1, pp. SC6–SC11.

Freestone, David (2015). Governance of Areas Beyond National Jurisdiction: An Unfinished Agenda of the 1982 Convention. *UNCLOS At 30*.

Freestone, David, and Kate Killerlain Morrison (2014). The signing of the Hamilton Declaration on Collaboration for the Conservation of the Sargasso Sea: a new paradigm for high seas conservation? *International Journal of Marine and Coastal Law*, vol. 29, pp. 345–362.

Gattuso, J.-P., and others (2015). Contrasting futures for ocean and society from different anthropogenic CO_2 emissions scenarios. *Science*, vol. 349, No. 6243, aac4722.

Hanel, Reinhold, and others (2014). Low larval abundance in the Sargasso Sea: new evidence about reduced recruitment of the Atlantic eels. *Naturwissenschaften*, vol. 101, No. 12, pp. 1041–1054.

Hanel, Reinhold, and others (2019). *Research for PECH Committee – Environmental, Social and Economic Sustainability of European Eel Management*. Brussels: European Parliament, Policy Department for Structural and Cohesion Policies.

Henson, Stephanie A., and others (2016). Observing climate change trends in ocean biogeochemistry: when and where. *Global Change Biology*, vol. 22, No. 4, pp. 1561–1571.

Hu, Chuanmin, and others (2016). *Sargassum* watch warns of incoming seaweed. *Eos*, vol. 97, pp. 10–15.

Huffard, C.L., and others (2014). Pelagic *Sargassum* community change over a 40-year period: temporal and spatial variability. *Marine Biology*, vol. 161, No. 12, pp. 2735–2751.

International Commission for the Conservation of Atlantic Tunas (ICCAT). Access to ICCAT Statistical Databases. www.iccat.int/en/accesingdb.htm.

Johnson, Donald R., and others (2013). The *Sargassum* Invasion of the Eastern Caribbean and Dynamics of the Equatorial North Atlantic; pp. 102–103 in Proceedings of the 65th Gulf and Caribbean Fisheries Institute Conference, 5–9 November 2012. Gulf and Caribbean Fisheries Institute, Santa Marta, Colombia.

Joint Group of Experts on the Scientific Aspects of Marine Environmental Protection (GESAMP) (2014). *Sources, Fate and Effects of Microplastics in the Marine Environment: A Global Assessment*. GESAMP Reports and Studies 90.

________ (2016). *Sources, Fate and Effects of Microplastics in the Marine Environment: A Global Assessment*. GESAMP Reports and Studies 93.

Kell, L., and B.E. Luckhurst (2018). Extending the indicator-based Ecosystem Report Card to the whole ecosystem; a preliminary example based on the Sargasso Sea. *ICCAT Collective Volume of Scientific Papers*, vol. 75, No. 67, pp. 258–275.

Laffoley, D.d'A., and others (2011). The Protection and Management of the Sargasso Sea. Sargasso Sea Alliance.

Law, Kara Lavender, and others (2010). Plastic accumulation in the North Atlantic subtropical gyre. *Science*, vol. 329, No. 5996, pp. 1185–1188.

Law, K.L. (2017). Plastics in the Marine Environment. *Annual Review of Marine Science*, vol. 9, No. 1, pp. 205–229. https://doi.org/10.1146/annurev-marine-010816-060409.

Luckhurst, B.E., and Freddy Arocha (2016). Evidence of spawning in the southern Sargasso Sea of fish species managed by ICCAT – albacore tuna, swordfish and white marlin. *Collection Volume of Scientific Papers ICCAT*, vol. 72, No. 8, pp. 1949–1969.

Luckhurst, Brian E. (2015a). A preliminary food web of the pelagic environment of the Sargasso Sea with a focus on the fish species of interest to ICCAT. *Collected Volume of Scientific Papers, International Commission for the Conservation of Atlantic Tuna*, vol. 71, pp. 2913–2932.

Luckhurst, Brian E. (2015b) Analysis of ICCAT reported catches of tuna and swordfish in the Sargasso Sea (1992–2011). *Collected Volume of Scientific Papers, International Commission for the Conservation of Atlantic Tuna*, vol. 71, pp. 2900–2912.

Martin, Lindsay Margaret (2016). Pelagic *Sargassum* and Its Associated Mobile Fauna in the Caribbean, Gulf of Mexico, and Sargasso Sea. PhD Thesis. Texas A & M University.

Milledge, John J., and Patricia J. Harvey (2016). Golden tides: problem or golden opportunity? The valorisation of *Sargassum* from beach inundations. *Journal of Marine Science and Engineering*, vol. 4, No. 3, art. 60.

Miller, Michael J., and others (2015). A century of research on the larval distributions of the Atlantic eels: a re-examination of the data. *Biological Reviews*, vol. 90, No. 4, pp. 1035–1064.

Miller, Michael J., and others (2016). Did a "perfect storm" of oceanic changes and continental anthropogenic impacts cause northern hemisphere anguillid recruitment reductions? *ICES Journal of Marine Science*, vol. 73, No. 1, pp. 43–56.

Miller, Michael J., and others (2019). Spawning by the European eel across 2000 km of the Sargasso Sea. *Biology Letters*, vol. 15, No. 4, 20180835.

Munk, Peter, and others (2010). Oceanic fronts in the Sargasso Sea control the early life and drift of Atlantic eels. *Proceedings of the Royal Society B: Biological Sciences*, vol. 277, No. 1700, pp. 3593–3599.

Neuer, Susanne, and others (2017). Monitoring Ocean Change in the 21st Century. *Eos*, vol. 98. https://eos.org/features/monitoring-ocean-change-in-the-21st-century.

Nippon Foundation-Nereus Programme (2015). *Predicting Future Oceans : Climate Change, Oceans & Fisheries.*

________ (2017). *Oceans and the Sustainable Development Goals: Co-Benefits, Climate Change & Social Equity.*

Northwest Atlantic Fisheries Organization (NAFO) (2015). Report of the Fisheries Commission and its Subsidiary Body (STACTIC), 37th Annual Meeting of NAFO, 21–25 September 2015, Halifax, Canada. NAFO/FC Doc. 15/23.

O'Brien, T.D., and others (2017). *What Are Marine Ecological Time Series Telling Us about the Ocean? A Status Report. IOC-UNESCO.* IOC Technical Series 129.

Parr, Albert Eide (1939). Quantitative observations on the pelagic *Sargassum* vegetation of the western North Atlantic with preliminary discussion of morphology and relationships. *Bulletin of the Bingham Oceanographic Collection*, vol. 7, pp. 1–94.

Pendleton, L., and others (2015). *Assessing the economic contribution of marine and coastal ecosystem services in the Sargasso Sea*. Nicholas Institute for Environmental Policy Solutions, Report NI R 14-05. Durham, North Carolina, United States: Duke University.

Putnam, Nathan F., and others (2018). Simulating transport pathways of pelagic *Sargassum* from the Equatorial Atlantic into the Caribbean Sea. *Progress in Oceanography*, vol. 165, pp. 205–214.

Righton, David, and others (2016). Empirical observations of the spawning migration of European eels: the long and dangerous road to the Sargasso Sea. *Science Advances*, vol. 2, No. 10, e1501694.

Rousseau, Yannick, and others (2019). Evolution of global marine fishing fleets and the response of fished resources. *Proceedings of the National Academy of Sciences*, vol. 116, No. 25, pp. 12238–12243.

Sargasso Sea Commission (2018). www.sargassoseacommission.org/about-the-commission/hamilton-declaration.

Sargasso Sea Commission (2019). www.sargassoseacommission.org/storage/Strengthening_Stewardship_of_the_Sargasso_Sea.pdf.

Schell, Jeffrey M., and others (2015). Recent *Sargassum* inundation events in the Caribbean: shipboard observations reveal dominance of a previously rare form. *Oceanography*, vol. 28, No. 3, pp. 8–11.

Secretariat of the Convention on Biological Diversity (2012).UNEP/CBD/CoP/11/35.

Sehein, Taylor, and others (2014). Connectivity in the slender Sargassum shrimp (*Latreutes fucorum*): implications for a Sargasso Sea protected area. *Journal of Plankton Research*, vol. 36, No. 6, pp. 1408–1412.

Smale, Dan A., and others (2019). Marine heatwaves threaten global biodiversity and the provision of ecosystem services. *Nature Climate Change*, vol. 9, No. 4, p. 306.

Tetu, Sasha G., and others (2019). Plastic leachates impair growth and oxygen production in *Prochlorococcus*, the ocean's most abundant photosynthetic bacteria. *Communications Biology*, vol. 2, No. 1, art. 184.

United Nations (2017). Chapter 50: Sargasso Sea. In *The First Global Integrated Marine Assessment: World Ocean Assessment I*. Cambridge: Cambridge University Press.

Wabnitz, Colette, and others (2019). The *Sargassum* Mass-Bloom of 2018. *Nereus Program - The Nippon Foundation* (blog). https://nereusprogram.org/works/the-sargassum-mass-bloom-of-2018.